THE SCIENCES OF THE SOUL

THE SCIENCES OF THE SOUL

The Early Modern Origins of Psychology

FERNANDO VIDAL
Translated by SASKIA BROWN

THE UNIVERSITY OF CHICAGO PRESS
CHICAGO AND LONDON

The University of Chicago Press, Chicago 60637
The University of Chicago Press, Ltd., London
© 2011 by The University of Chicago
Published 2011
Paperback edition 2020

29 28 27 26 25 24 23 22 21 20   1 2 3 4 5

ISBN-13: 978-0-226-85586-8 (cloth)
ISBN-10: 0-226-85586-4 (cloth)
ISBN-13: 978-0-226-71036-5 (paper)
ISBN-13: 978-0-226-85588-2 (e-book)
DOI: https://doi.org/10.7208/chicago/9780226855882.001.0001

Originally published as *Les sciences de l'âme XVIe–XVIIIe siècle*
© 2006. Éditions Champion, Paris.

*Ouvrage publié avec le soutien du Centre national du livre—ministère français
chargé de la culture* / This work is published with support from the National
Center of the Book—French Ministry of Culture.
*Cet ouvrage a bénéficié du soutien des Programmes d'aide à la publication de
l'Institut français* / This work is supported by the Publication Aid Programs from
the Institut Français.

Library of Congress Cataloging-in-Publication Data

Vidal, Fernando.
    [Sciences de l'âme. English]
    The sciences of the soul : the early modern origins of psychology / Fernando
Vidal ; translated by Saskia Brown.
        p. cm.
    Translation of: Les sciences de l'âme: XVIe–XVIIIe siècle.
    Includes bibliographical references and index.
    ISBN-13: 978-0-226-85586-8 (alk. paper)
    ISBN-10: 0-226-85586-4 (alk. paper)
        1. Psychology—History. 2. Psychology—History—18th century. 3. Soul.
    I. Brown, Saskia. II. Title
    BF101.V5313 2011
    150.9—dc22

                                                                2011014851

CONTENTS

FIGURES AND TABLES

Figures

Tables

The Sciences of the Soul may be read as an answer to a question I have often been asked—basically, Did psychology exist in the eighteenth century? The question is understandable insofar as psychology is still widely held to have been "founded as a scientific discipline" in the last third of the nineteenth century. The reply is that psychology did exist as such in the eighteenth century, and under that name. My interlocutors have frequently reacted with curiosity or suspicion: either this "psychology" was illegitimately usurping its name, or else it designates a "prehistory" of the discipline, a sort of "history of psychological ideas," perhaps a "prescientific," "metaphysical" approach to the soul. These are some of the misconceptions I wish to address. Between the first attested uses of the term "psychology," in the last third of the sixteenth century, and the end of the eighteenth century, psychology existed essentially as a "physics of the soul," and it belonged as much to natural philosophy as to Christian anthropology. It remained so throughout this period, despite the break caused by the collapse of Aristotelian frameworks and the transformations of the discipline caused by—while also contributing to—the demise of the Aristotelian notion of the soul.

Yet the reaction mentioned above is by no means unjustified. At least since the histories of psychology written in the wake of psychology's alleged foundation as a scientific discipline, the very same version of events has been rehearsed time and again. This may explain why, for the period prior to that supposed moment, the vast majority of histories of psychology, historical profiles of the discipline, encyclopedias, dictionaries, and general works on early modern and Enlightenment thought fail to address psychology, whether in name or in substance. The same goes for histories of science, in which psychology prior to the late nineteenth century does not seem to merit inclusion. Such is the state of affairs this book seeks to redress.

I would also like to suggest that what is at stake in the transition from early modern to Enlightenment psychology is not simply historiographical issues but fundamental questions concerning the advent of "modern identity" and the relations between bodily and psychological identity. We shall address these at the end of our inquiry. But our starting point is elsewhere.

Since words and things are closely linked, and since terminology plays a constitutive role in the development of disciplines and fields of knowledge, we shall explore the history of the term "psychology" from its invention in the sixteenth century to its redefinition at the beginning of the eighteenth century. This history is of course not purely lexical but allows us to track the transformation of psychology's object, from the soul as the "form" of all living organisms to the soul as an exclusively human mind (*mens*). The soul as form could account for the structure and functions of living beings; as mind it became an *explanandum*, a substance whose operations and contents had to be described through an analysis of its "commerce," or interaction, with the body. Although the new psychology borrowed from natural philosophy as reshaped during the Scientific Revolution, its autonomy derived largely from its claim to provide methodological and epistemic principles for all the other "sciences of man."

It would be premature to attempt an overview of psychology itself in the eighteenth century. This is partly for conceptual reasons and partly because different regional and national situations would have to be taken into account, and the necessary historiography is still lacking. Since my focus is on the emergence of psychology as a field claiming to be new and autonomous, I will not concentrate on theories and systems but examine the creation of frameworks within which these developed, and mechanisms that legitimated them as "psychological." I use the term "discipline" advisedly, although it is not usually associated with psychology before the last third of the nineteenth century. Psychology, I argue, existed as a discipline before the Enlightenment. The broad cultural field within which it was recast in the eighteenth century was marked by certain core themes, but also by the development of specific forms and contents, and diverse ways of linking theory and practice.

I consequently identify intellectual practices and conceptual configurations that were decisive for defining psychology's borders and subject matter: the invention of a psychological tradition through the writing of its history, the inclusion of psychology within the "science of man" and the "history of humankind," the theorization of psychological methodology, and the elaboration of psychology's place within the organization of knowledge. These intellectual mechanisms went hand in hand with the appropriation by psychology and anthropology of logic, metaphysics, and morals—an appropriation thanks to which Enlightenment psychology came to regard itself as "the most useful of all the sciences."

ACKNOWLEDGMENTS

I am grateful to the Athena program of the Swiss National Science Foundation for providing the conditions to explore questions which were new to me at the start of this project. Some of the preliminary research was carried out when I was a visiting scholar in the Department of the History of Science at Harvard University, which I thank for its hospitality. The Max Planck Institute for the History of Science, Berlin, has been an ideal place to pursue this research. I thank in particular the institute's librarians for their kind and efficient assistance, and Zoe Schlepfer for help on this English-language edition. Some of the ideas discussed here were addressed in lectures, seminars, and classes given at the École des Hautes Études en Sciences Sociales, the Universidade do Estado do Rio de Janeiro, and the Pontifícia Universidade Católica do Rio de Janeiro. I would like to thank my hosts at these institutions, namely, Jacqueline Carroy and François Azouvi, Ana Jacó, and Maria Apparecida Mamede. Many friends and colleagues accompanied this work in very different ways. I can mention but a few of them here: Bronislaw Baczko, Vincent Barras, Antonio Battro, Jacques Berchtold, Claude Blanckaert, Marie-Noëlle Bourguet, Caroline Bynum, Mary Campbell, Andrea Carlino, Lorraine Daston, Claire Gantet, Mechthild Fend, Tomás Fernández, Abigail Lustig, Francisco Ortega, Katharine Park, Françoise Parot, Cristina Pitassi, Michel Porret, the late Roy Porter, Sonu Shamdasani, Patrick Singy, Jean Starobinski, Friedrich Steinle, Claudia Swan, and Anke te Heesen.

ON CITATIONS AND THE TRANSLATION

Articles in encyclopedias are referred to by the article's headword in small capital letters, the classificatory term(s) or "field indicator(s)" in italics and in parenthesis (spelled out in full where there is an abbreviation in the original), the volume number, the page range for the article, and the page number(s) cited. Additionally, for the Paris and Yverdon *Encyclopédies*, the name of the author of the article will be given. I refer to these as "Diderot and d'Alembert's *Encyclopédie*," the "Paris *Encyclopédie*," or simply "the *Encyclopédie*," abbreviated to EP; and the "Yverdon *Encyclopédie*" or "the

Swiss *Encyclopédie,"* abbreviated to EY. When no authors are mentioned, it is because they are unknown; names in square brackets derive from John Lough, "The problem of the unsigned articles in the *Encyclopédie*," *Studies on Voltaire and the eighteenth century*, 32, 1965, 327–390. In the Yverdon *Encyclopédie*, articles marked with an "R" are articles taken from the Paris *Encyclopédie* and rewritten; articles marked with an "N" are new (and therefore have no equivalent headword in the Paris work).

Original-language quotations and references reproduce the original punctuation, spelling, and accents; I have not added *sic* when these differ from contemporary usage. Publication place-names are translated, but the original forms of printers' and publishers' names are retained. Titles in the notes appear first in full, thereafter in abbreviated form. All references appear in the bibliography, with the exception of strictly biographical reference material and some works that are mentioned but not quoted; whenever possible and appropriate I have referred to English translations. I have reviewed the entire translation and introduced a few changes with respect to the French original. Needless to say, the frequent use of "man" and "science of man" belongs to the vocabulary of the period dealt with in this book.

NOTE TO THE PAPERBACK EDITION

Important work relevant to early modern psychology and anthropology has been carried out since the first publication of this book. Special mention should be made of Davide Cellamare's *Psychology in the Age of Confessionalisation: A Case Study of the Interaction between Psychology and Theology c. 1517–c. 1640* (PhD thesis, Radboud University Nijmegen, 2015) and related publications by the same author. Early modern psychology was indeed a major locus of transformation of the Aristotelian tradition and the pursuit of centuries-old questions about the soul across confessional divides. Cellamare nevertheless shows that the context where the term "psychology" initially prospered was more intimately related to Lutheranism and the dynamics of confessionalization than suggested in *The Sciences of the Soul*. A discussion and update of the historiography can be found in my entry "Psychology" for the *Encyclopedia of Early Modern Philosophy and the Sciences*, edited by Dana Jalobeanu and Charles T. Wolfe (Springer, forthcoming 2020).

CHAPTER ONE

The "Century of Psychology"

In 1774, an article in a Swiss encyclopedia wondered, "What science or art deserving of our attention does not have psychology as its foundation, its source and its guide?" Indeed,

> It is to ourselves that we relate all things, it is the influence of things upon ourselves that leads us to applaud or condemn them; it is therefore the relation of things to ourselves that makes them of interest to us; and without knowledge of the nature, faculties, qualities, state, relations and destination of the human soul, we can pass judgment on nothing, decide nothing, determine nothing, choose nothing, reject nothing, prefer nothing and do nothing with certainty and without error. Psychology is consequently the first and most useful of all the sciences, the source, the principle and the foundation of them all, as well as the guide which leads to each.[1]

This triumphant passage in praise of "the most useful of all the sciences" is representative of what psychology became in the eighteenth century. It is also indicative of a state of mind. Enlightenment psychologists were convinced that psychology was the queen of the sciences, the science which laid the foundations necessary for action and thought. While the importance of self-knowledge and of the science of the soul as the most valuable and noble of knowledges was an earlier commonplace, it was redefined in

---

1. Gabriel Mingard, PSYCHOLOGIE (*Métaphysique*), in Fortunato Bartolomeo de Felice, ed., *Encyclopédie, ou Dictionnaire universel raisonné des connoissances humaines* (Yverdon: Société Typographique, 1770–1775); *Supplément* (1775–1776), vol. 35, 511b–513a, pp. 512b–513a. See the complete text for this article from the Yverdon *Encyclopédie* in appendix II.

the eighteenth century. Henceforth, knowing oneself would involve a new empirical science, increasingly called "psychology," which corresponded to the methodological and epistemological ideals of the Enlightenment. This science was not "new" in the sense that it was created ex nihilo in the eighteenth century. It already existed within an Aristotelian universe in which it served as an introduction to the different sciences of living beings, describing the vegetative and sensitive functions, as well as, for humans, a relatively stable set of faculties (the external senses, the common sense, imagination, memory, and the intellect).

So psychology was not invented in the eighteenth century but remade.[2] Its object was transformed by the critique of Aristotelian frameworks: the soul ceased to be the principle of life responsible for generation, growth, sensation, and thought and was reduced to mind (mens). The science of the soul, often termed "psychology" from the last third of the sixteenth century, was redefined as the science of the mind. As an empirical science, it was to be based on observation and experimentation, dealing with the soul only in its relation to the body. It distanced itself from the theological and metaphysical discourses on the nature, origin, and ultimate end of an immaterial substance.

As regards the organization of knowledge, eighteenth-century psychology incorporated subjects from logic, metaphysics, and morals and positioned itself at the center of another uncharted field, that of anthropology, or the general science of the human being. The changes involved were by no means purely structural and lexical; on the contrary, they accompanied, sustained, and promoted psychological ways of understanding the human being and of grounding knowledge, from logic to legislation and from aesthetics to pedagogy. They were to propel humanity into enlightenment and enable human perfectibility to be realized. Such transformations also led to the creation of a new conceptual and social space, not that of psychology as a profession with its associated institutions but that of a discipline which broke with the Aristotelian *scientia de anima* and claimed a value and an autonomy of its own.

2. Gary Hatfield, "Remaking the science of mind: Psychology as a natural science," in Christopher Fox, Roy Porter, and Robert Wokler, eds., *Inventing human science: Eighteenth-century domains* (Berkeley: University of California Press, 1994). See also Wolfgang Riedel, "Erster Psychologismus: Umbau des Seelenbegriffs in der deutschen Spätaufklärung," in Jörn Garber and Heinz Thoma, eds., *Zwischen Empirisierung und Konstruktionsleistung: Anthropologie im 18. Jahrhundert* (Tübingen: Max Niemeyer, 2004).

PSYCHOLOGY AS A "DISCIPLINE"

I speak of psychology as a "discipline" partly in order to engage with a sterile but ongoing debate concerning, essentially, the date at which psychology became a "scientific discipline." Even if this question were relevant—and it seems to me badly framed—the answer would not be very important, and to prove this, I would anyway have to retain, rather than reject, the notion of discipline. I use the term, however, for a more essential reason.

Clearly, if the sine qua non of a discipline is its incarnation in a profession and in institutions, then psychology is not a discipline before the last third of the nineteenth century, when it was allegedly born as a science. Yet it can reasonably be considered to be one much earlier, according both to the traditional sense of "discipline," and to the history and sociology of science. For "discipline" is the concept which has ensured the continuity and substance of the history of knowledge in the Western world.[3] In medieval usage, *disciplina* could be synonymous with *ars* or *scientia*, sometimes with a connotation of rigor that restricted its application to subjects using demonstrative methods.[4] *Disciplina* is derived from *discere*, "to learn," which in classical Latin also designated the act of learning, of being taught or educated. It is what one learns from a master: seventeenth-century lexicons define it as "Scientia acquisita in discente" and "informatio mentis a Magistro accepta."[5] In this sense, what was taught from the Middle Ages onward as *animastica* or *scientia de anima* in the framework of "physics" or natural philosophy was certainly a discipline. But instead of leading to a profession, it was one of the preparatory courses for medicine, law, and theology.[6]

Early modern universities typically involved two propaedeutic cycles. The first included the "arts" of grammar and rhetoric, the second, the

---

3. Donald R. Kelley, "Introduction," and "The problem of knowledge and the concept of discipline," in D. R. Kelley, ed., *History and the disciplines: The reclassification of knowledge in early modern Europe* (Rochester: University of Rochester Press, 1997).

4. G. Schrimpf, "Disciplina" (s.v. "Disciplina, doctrina"), in Joachim Ritter, Karlfried Gründer, and Gottfried Gabriel, eds., *Historisches Wörterbuch der Philosophie* (Darmstadt: Wissenschaftliche Buchgesellschaft, 1971), vol. 2.

5. Rudolph Goclenius, *Lexicon philosophicum* (1613; Hildesheim: Olms, 1980), s.v. "Disciplina"; Stephanus Chauvin, *Lexicon philosophicum* (1692), ed. of 1713 (Düsseldorf: Stern-Verlag Janssen, 1967), s.v. "Disciplina."

6. Laurence W. B. Brockliss, *French higher education in the seventeenth and eighteenth century: A cultural history* (Oxford: Clarendon Press, 1987), pts. II–IV. See also Wilhelm Schmidt-Biggemann, "New structures of knowledge," and L. Brockliss, "Curricula," in Hilde de Ridder-Symoens, ed., *A history of the university in Europe*, vol. 2, *Universities in early modern Europe (1500–1800)* (Cambridge: Cambridge University Press, 1996).

"sciences" of logic, ethics, physics, and metaphysics. Physics, or the science of nature (*phusis*) prepared for medecine. It covered, among other subjects, the vegetative and the sensitive soul, while the rational soul fell within metaphysics. Psychology before the eighteenth century was therefore comparable to natural history before the end of the sixteenth century: it was taught almost exclusively from canonical works (Aristotle's *De anima* and its commentaries) and remained epistemically subordinate to other sciences, which it served in an instrumental and propaedeutic role.[7] The social expression of this subordination was that psychology was a discipline but not a profession; there was no community or even isolated individuals who devoted themselves exclusively to the *scientia de anima*. As Johann Georg Sulzer (1720–1779), a thinker known for his writings on aesthetics, explained, the different branches of scholarship (*Gelehrsamkeit*) are called arts and sciences, and although the name of "science" is usually reserved for those concerned with general truths drawn from the nature of things, all may be called "disciplines."[8]

A discipline, though, is obviously not reducible to its pedagogical function. It is also a social and intellectual structure characterized by the existence of scholars who devote themselves to it. It consists of a body of knowledge, and a set of issues, rules, methods, disagreements, and debates. It has a terminology of its own, a set of works and individuals associated with the field and recognized as authoritative, as well as periodicals, textbooks, and curricula. Last, a discipline may be linked to specific institutions such as faculties, departments, or societies.[9] It was in the course of the eighteenth century that psychology developed the consistency and scope that made Kant think it should be promoted to the rank of "separate university discipline" (see chap. 4). The process of "disciplinary" consolidation ultimately required administrative decisions, but Kant's observation, made

7. Paula Findlen, "The formation of a scientific community: Natural history in sixteenth-century Italy," in Anthony Grafton and Nancy Siraisi, eds., *Natural particulars: Nature and the disciplines in Renaissance Europe* (Cambridge, MA: MIT Press, 1999).

8. "Man kann aber jeden besondern Theil der Gelehrsamkeit eine *Disciplin* nennen." Johann Georg Sulzer, *Kurzer Begriff aller Wissenschaften und andern Theile der Gelehrsamkeit, worin nach seinem Inhalt, Nuzen und Vollkommenheit kürzlich beschrieben wird*, 2nd ed. (Leipzig: bey Johann Christian Langenheim, 1759), p. 9. On Sulzer's work, see Hans Erich Bödeker, "Konzept und Klassifikation der Wissenschaften bei Johann Georg Sulzer (1720–1779)," in Martin Fontius and Helmut Holzhey, eds., *Schweizer im Berlin des 18. Jahrhunderts* (Berlin: Akademie Verlag, 1996).

9. Rudolf Stichweh, *Zur Entstehung des modernen Systems wissenschaftlicher Disziplinen: Physik in Deutschland 1740–1890* (Frankfurt: Suhrkamp, 1984), chap. 1, esp. §§ 1–3.

in the 1770s, implies that the borders, contents, methods, and place of psychology within the sciences had already been theorized by that time.

A discipline may cover several different fields. This has long been the case for psychology, whose object, methods, goals, and key questions have always been defined in the most diverse ways. Psychology itself does not exist as a unitary and homogeneous entity. The singular may be useful for naming institutions and fashioning a professional identity, but one has only to open a standard psychology textbook to realize that the unity expressed in its title is, as Georges Canguilhem declared in the mid-1950s, nothing but "a pact of pacific coexistence between specialists."[10] Yet the American Psychological Association, the largest association of psychologists in the world, blithely explains that psychology is the "study of the mind and behavior," and that it addresses "all aspects of the human experience, from the functions of the brain to the actions of nations, from child development to care for the aged." It goes on: "In every conceivable setting from scientific research centers to mental health care services, 'the understanding of behavior' is the enterprise of psychologists."[11] The vague and general nature of such a definition proves that psychology is nothing other than what psychologists do, and that any definition such as that of the APA is a function of the interests of the body that formulates it. As a result, and for precisely the reason that psychology cannot really be defined as anything other than what the people who say they practice it actually do, it has a strong disciplinary identity which resides in the individuals, texts, and institutions which act in its name.

On the other hand, not every field of study becomes a discipline. For example, while research into ocular vision was a coherent field in 1860s Germany, the attempts to set up university posts and institutes failed, and its "research programs" ended up flourishing within opthalmology, psychology, and physiology. The borders of a field or a research program do not necessarily coincide with those of instituted disciplines, and even a large degree of consensus on methods and core issues is not enough to constitute one.[12]

In short, speaking of *a* discipline, in the singular, is a convenient abstraction that does not reflect the processes involved in the production of

10. Georges Canguilhem's lecture, given in 1956, "Qu'est-ce que la psychologie?," in *Etudes d'histoire et de philosophie des sciences* (Paris: Vrin, 1994), p. 366.

11. See the Web site of the American Psychological Association, http://www.apa.org/about.

12. This example comes from Timothy Lenoir, *Instituting science: The cultural production of scientific disciplines* (Stanford: Stanford University Press, 1997).

knowledge and the distribution of resources.[13] What Pierre Bourdieu calls a "scientific field" can be understood as a set of competing disciplinary programs rooted in particular contexts, which vie with one another for authority and legitimacy, as well as for the control of the social, political, and economic power which often accompany them.[14] It is in these terms that one can analyze the creation of institutions and professions, that is, the establishment of formal structures within which scientific activity takes place and groups authorized to carry it out are formed. The scientific and the social processes generally go hand in hand and involve defining requisite training and accreditation procedures, and establishing hierarchies and systems of reward and legitimation. The production of knowledge is therefore inseparable from social realms, be they disciplines, professions, institutions, or a "Republic of Letters."[15]

Psychology is not professionalized until the end of the nineteenth century. However, in the eighteenth century there were already individuals calling themselves "psychologists," publications and teachings classed under "psychology," as well as a forceful discourse that championed psychology and advocated it as the foundation of the knowledge system. Despite the variety of geographically dispersed projects which Enlightenment psychologists undertook, a common identity was emerging. Immanuel Kant could express the wish for empirical psychology to be taught by its own teaching body at universities in the 1770s because he considered its core issues, contents, and intellectual identity to be sufficiently developed to be granted an autonomous institutional existence.

There is additionally a third sense of "discipline" we must consider, beyond the discursive and epistemic aspects on which I will be focusing. For Michel Foucault, a discipline is not solely a branch of knowledge but also a set of social practices involving both the practitioners and their subjects or clients. A discipline fashions the experience and behavior of those who practice it while also dictating the practice of others. For example, the discipline of self-observation and attention to self and other to which late eighteenth-century pedagogues subjected themselves was aimed at fashioning the bodies and minds of their pupils. A discipline would therefore be

13. On this point see Timothy Lenoir, "The discipline of nature and the nature of disciplines," in Ellen Messer-Davidow, David R. Shumway and David J. Sylvan, eds., *Knowledges: Historical and critical studies in disciplinarity* (Charlotteville: University Press of Virginia, 1993).

14. Pierre Bourdieu, "Le champ scientifique et les conditions sociales du progrès de la raison," *Sociologie et sociétés*, 7, 1975, 91–117.

15. Jan Golinski, *Making natural knowledge: Constructivism and the history of science* (Cambridge: Cambridge University Press, 1998).

a way of imposing power relations, which themselves make possible the constitution of knowledge.

Writing turned out to be one of the major instruments of "discipline" in this sense. A "disciplined" person is one whose world is that of the administrative or scientific document which in turn constitutes a new source of knowledge and consolidates a discipline. So it was with the late eighteenth-century physician who drew up tables of patients admitted to the mental asylum, or the psychologist who examined himself and others, noting down everything with the conviction that no detail may be deemed secondary. For Foucault, these techniques, which were designed to "discipline" bodies and minds, heralded the establishment of disciplines such as pedagogy, psychiatry, or psychology, whose methods and epistemologies they partly defined.[16]

How should we study the emergence and development of such disciplines? Writing a history of psychological ideas (including systems and particular notions such as *imagination* or *attention*) brings to light some of psychology's contents. Examining spiritual or mystical discourses points to models of the soul whose links to empirical psychology can be explored.[17] Studying forms of sociability (academies, correspondence, master-pupil relations) confirms the existence of networks of individuals who identify with psychology. Analyzing curricula and university textbooks allows one to track how psychology developed within institutions, and research into administrative decisions, buildings, and staffing elucidates some of the material processes on which the institution and profession were grounded, while the reconstitution of practices and methods gives access to the personal and collective disciplines through which the science became instituted concretely. For the period covered here, all of these approaches are worth developing further.

The subject, however, may also be broached more obliquely. I will address the history of the concept of psychology from the sixteenth to the eighteenth century, the invention of a psychological tradition, the articulation of psychology with the "general science of man" and the "history of humankind," the classification of the sciences and the psychological appropriation of logic, morals, and metaphysics, and, last, certain consequences of Enlightenment empirical psychology for the construction of "modern identity." I will attempt to bring out the role played by the *representation*

---

16. Jan Goldstein, "Foucault among the sociologists: The 'disciplines' and the history of the professions," *History and theory*, 23, 1984, 170–192.

17. For similar models in seventeenth-century French spirituality, see Mino Bergamo, *L'anatomia dell'anima: Da François de Sales a Fénelon* (Bologna: Il Mulino, 1991).

of the organization of knowledge in establishing the forms, contents, and borders of the sciences of the soul, as well as the variety of paths which, in the eighteenth century, led to empirical psychology.

The themes explored here may at first sight appear marginal in comparison with histories of psychological ideas. I will, morever, be less concerned with the "epistemic cultures" of psychology, that is, the modes of production of psychological knowledge through localized practices, than with the mechanisms by which psychology was recast and emerged as a modern discipline.[18] Such mechanisms belonged to a cultural field that was broader than psychology alone and affected every system of knowledge and interpretation of the human being. This is why, in the wake of Jean Starobinski and his ideal of a "history of ideas without borders," historical semantics will be particularly important to us, as will a *Begriffsgeschichte* and the perspectives opened up by the history of scholarship, or *Wissensgeschichte*.[19] Such an oblique approach will help throw into relief certain aspects of the subject which could otherwise pass unnoticed.

A LONG PAST BUT A SHORT HISTORY?

Explicitly or implicitly, histories of philosophy have tended to identify the Enlightenment as the "century of psychology."[20] This, however, has had the paradoxical effect of making eighteenth-century psychology invisible.

One often reads, even in recent works, that psychology was born as a science in the last third of the nineteenth century, thanks to Wilhelm Wundt (1832–1920), professor of philosophy at the University of Leipzig. Wundt founded an Institute of Experimental Psychology in 1879, the first of its kind, which immediately spawned others across Europe and the Americas. But the received wisdom that he was the "founder of modern psychology," taking a purely experimental approach and rejecting philosophy, is no longer tenable.[21] It is the post-Wundtian "new psychology" which seems to have

---

18. On the notion of "epistemic culture," see Karin Knorr Cetina, *Epistemic cultures: How the sciences make knowledge* (Cambridge, MA: Harvard University Press, 1999), chap. 1.

19. See F. Vidal, "Jean Starobinski: The history of psychiatry as the cultural history of consciousness," in M. S. Micale and R. Porter, eds., *Discovering the History of Psychiatry* (New York: Oxford University Press, 1999).

20. Étienne Gilson and Thomas D. Langan, *Modern philosophy: Descartes to Kant* (New York: Random House, 1964), p. 225.

21. Arthur L. Blumenthal, "A reappraisal of Wilhelm Wundt," *American psychologist*, 30, 1975, 1081–1088; Blumenthal, "Wilhelm Wundt: Problems of interpretation," in Wolfgang G. Bringmann and Ryan D. Tweney, eds., *Wundt studies* (Toronto: C. J. Hogrefe, 1980); Kurt Danziger, "The positivist repudiation of Wundt," *Journal of the history of the behavioral sciences*,

instituted psychology as a natural science.[22] It was called "new" because it reacted against the "abstract" principles of the psychological thought of the Enlightenment and was committed to empirical fact and the experimental method. For the many North Americans who spent time at German universities in the 1880s and 1890s, attending Wundt's lectures and working in his laboratory was "a sort of apostolic call to the 'new psychology.'"[23] There was of course some recognition that prior to this there existed observations, questions, ideas, and theories of a psychological nature, but these were relegated to a *past* deemed of questionable relevance to the *history* of scientific psychology.

For Hermann Ebbinghaus (1850–1909), one of the pioneers of experimental psychophysics, psychology had a long past but a short history: "Die Psychologie hat eine lange Vergangenheit, doch eine kurze Geschichte."[24] The English translation reads, "Psychology has a long past, yet its real history is short"—and that is the first sentence of his 1908 "elementary textbook" of psychology. In 1929, the American psychologist and historian of psychology Edwin Boring (1886–1968) noted that histories of psychology treat the "new psychology" as the end point of centuries of philosophical reflection on the mind and focus on the discipline's past "at the expense of its short scientific history."[25] Boring, who saw Wundt as the first person who deserved the label "psychologist," sought rather to influence the future path of the discipline by stressing its autonomy and the model of the laboratory as an ideal of scientificity.[26] Ebbinghaus's phrase, repeated at second and third hand, still

15, 1979, 205–230; R. M. Farr, "Wilhelm Wundt (1832–1920) and the origins of psychology as an experimental and social science," *British journal of social psychology*, 22, 1983, 289–301; W. M. O'Neil, "The Wundt myths," *Australian journal of psychology*, 36, 1984, 285–289; Adrian Brock, "Something old, something new: The 'reappraisal' of Wilhelm Wundt in textbooks," *Theory and psychology*, 3, 1993, 235–242.

22. See, for example, the book title of one of Wundt's doctoral students, Edward Wheeler Scripture, *The new psychology* (New York: Charles Scribner's Sons, 1897). The French equivalent would be Théodule Ribot's study, *La psychologie allemande contemporaine: Ecole expérimentale* (Paris: Baillière, 1879). The expression took root in English in the 1880s and was probably first used by John Dewey, "The new psychology," *Andover review*, 2, 1884, 278–289.

23. James Mark Baldwin, "Autobiography," in Carl Murchison, ed., *A history of psychology in autobiography*, vol. 1 (Worcester, MA: Clark University Press, 1930), p. 2.

24. Hermann Ebbinghaus, *Abriß der Psychologie* (1908; Leipzig: Veit, 1910), p. 9; translated as *Psychology: An elementary text-book*, trans. and ed. Max Mayer (Boston: D. C. Heath, 1908), p. 3.

25. Edwin G. Boring, *A history of experimental psychology* (1929; New York: Appleton-Century-Crofts, 1957), p. ix.

26. Ibid., p. 316. See John M. O'Donnell, "The crisis of experimentalism in the 1920s: E. G. Boring and his uses of history," *American psychologist*, 34, 1979, 289–295; John C. Cerullo, "E. G. Boring: Reflections on a discipline builder," *American journal of psychology*, 101, 1988, 561–575;

seems to possess some incantatory power, as though merely pronouncing it were enough to convey an atemporal idea of "science," justify the periodization and contents of historical narratives, and legitimate the choice of "antecedents," "precursors," and "foundations."[27] It is invoked in histories that assimilate the "prescientific" or "philosophical" past of psychology to a vast body of "psychological ideas"—ranging from Plato or Aristotle up to the "modern" period inaugurated by Descartes, and extending into the sensualist, empiricist and associationist philosophies of Locke, Condillac, and the Scottish philosophers of the Enlightenment. When philosophical speculation on the soul is abandoned, so the story goes, the empirical approach can be instituted, from which in the nineteenth century can emerge a psychology whose *history* may at last be written.

Determining what belongs to the history of a particular science is a general problem for the history of science.[28] In the case of psychology, the narrative of its "past" generally bears on ideas, contexts, individuals, and events deemed to have been relevant for its future; it is the discipline as it stands at present which determines the elements of the narrative. This is particularly so for works, especially textbooks, which seek to narrate the "history of psychology" (with both nouns in the singular), while their contents reproduce, with very minor changes, their own previous versions. Such an approach is especially prejudicial to the eighteenth century, owing to certain issues that make it distinctive in the discipline's "prehistory," particularly relating to the history of historiography. For example, whereas in the late nineteenth century historical narratives about psychology served to legitimate the chronicler's notion of what psychology ought to be, earlier ones—the first of which appeared in the eighteenth century—were instrumental in actually fashioning the discipline.

At all events, reducing Enlightenment psychology to the psychological ideas of the century is in all respects misleading and a sort of historiographical illusion.[29] This is, in the first place, quite simply because a science of the soul called "psychology" was already in existence *before* the eighteenth

---

Luigi Antonello Armando, *L'invenzione della psicologia: Saggio sull'opera storiografica di E. G. Boring* (Rome: Nuove Edizioni Romane, 1988).

27. To cite but a recent example, this is precisely how Ebbinghaus's phrase functions in Serge Nicolas's *Histoire de la psychologie* (Paris: Dunod, 2001), p. 7. But with or without the phrase, our point holds for most textbooks on the subject.

28. Roger Smith, "Does the history of psychology have a subject?," *History of the human sciences*, 1, 1988, 147–177.

29. See Gary Hatfield's treatment of this question in his *The natural and the normative: Theories of spatial perception from Kant to Helmholtz* (Cambridge, MA: MIT Press, 1990), chaps. 2 and 7.

century. While we should be attentive to specificities of time and place, and while there was not one psychology and hence no single narrative of its history, we can nevertheless make some generalizations: "Empirical psychology" designated a science which was grounded in experience and whose object was the soul united with the body. By definition it excluded the soul as an explanatory principle and tended to account for the operations of thought in terms of sensation rather than the intrinsic properties of some immaterial mind. This science was of course not "empirical" or "scientific" if these adjectives imply quantitative laboratory research, but if viewed in context and judged by its own criteria, psychology in the eighteenth century must indeed be regarded as part of the natural history of the human being.

The opening lines of the *Essay on Psychology*, published in 1754 by the Genevan Charles Bonnet (1720–1793), formulate the core principle of Enlightenment psychology: "We know the soul only through its faculties; we know these faculties only through their effects. These effects manifest themselves through the intermediary of the body."[30] In the wake of John Locke (1632–1704), the impossibility of knowing substances in themselves was a given. Consequently, empirical research could address the manifestations of the soul as observed in oneself or in others by means of the internal and the external senses. That is why Enlightenment psychology explicitly abandoned the problem of the union of body and soul and concentrated on the interaction between them; its frequent neuropsychological orientation reflected the belief that the nerve was the intermediary between the the two substances. As a result, and despite its speculative appearance, Enlightenment psychology was clearly rooted in a natural-philosophical perspective. This is shown not only by its refusal to treat the soul as a metaphysical or theological concept but also by its appeal to observation, experience, experimentation, and introspection, as well as by its attempts to apply mathematical reasoning or physical models to mental phenomena, and to use medical knowledge, anatomy, and physiology to interpret the relations between the body and the soul.

What then of the "century of psychology"? Is it part of the *past* or the *history* of the discipline? This question must be raised, since the idea that the Enlightenment is the "century of psychology" immediately encounters the paradox that general works on the Enlightenment, on the history of science at the time and the history of psychology as a whole, hardly touch on

---

30. Charles Bonnet, *Essai de psychologie* (1754), in *Œuvres d'histoire naturelle et de philosophie*, 4th ed. (Neuchâtel: Samuel Fauche, 1779–1783), vol. 8, p. 1.

what went by the name of "psychology."[31] We must examine, then, what is understood by this term, as well as the historiographical and interpretive choices determining its use.

When "psychology" designates the whole range of psychological ideas or concepts, it effectively coincides with much of Enlightenment culture. Psychological subject matter could be found in a wide range of fields, from the natural sciences, medicine, theology, moral philosophy, metaphysics, and logic to literary works, travel writing, pedagogy, and child-rearing manuals, treatises on aesthetics, legislation, and politics, as well as dictionaries and grammars. One can then pick and choose what one calls "psychological," according to one's preconceived idea of psychology. While the notion of a "century of psychology," as advanced by the major surveys of the Enlightenment, finds its confirmation here, such an approach can lead only to inconsistent results strongly colored by individual choices—which tend to exclude precisely what was actually called "psychology."

General histories such as the ones I am referring to here assimilate psychology to theories of the functioning of the mind, of the acquisition of knowledge, or of empirical modes of thought that were inspired to various degrees by John Locke's *Essay Concerning Human Understanding* (1690). In the 1930s, Carl L. Becker described the *Essay* as the "psychological gospel" of the century;[32] Ernst Cassirer explained how psychology provided the foundations for the theory of knowledge and how the psychological origin of ideas became a logical criterion;[33] Paul Hazard noted that Locke drew the eighteenth century's attention to the "most necessary and delicious of games: psychology."[34] For Isaiah Berlin, the major intellectual project of the Enlightenment was precisely the transformation of philosophy into "some

31. There is no mention of psychology in Thomas Hankins, *Science and the enlightenment* (Cambridge: Cambridge University Press, 1985); or in William Clark, Jan Golinski, and Simon Schaffer, eds., *The sciences in enlightened Europe* (Chicago: University of Chicago Press, 1999), in which Marina Frasca-Spada ("The science and conversation of human nature") nonetheless addresses many "psychological" issues. Christa Knellwolf, "The science of man," in Martin Fitzpatrick et al., eds., *The Enlightenment world* (London: Routledge, 2004), makes no reference at all to the German context and reduces to Locke and Hume the era's attempts at analyzing "the mechanisms of the mind."

32. Carl L. Becker, *The heavenly city of the eighteenth-century philosophers* (1932; New Haven: Yale University Press, 1955), p. 64.

33. Ernst Cassirer, *The philosophy of the Enlightenment*, trans. Fritz C. A . Koelln and James P. Pettegrove (Boston: Beacon Press, 1961), chap. 3. Cassirer, who is principally interested in the problem of knowledge, and interprets Enlightenment philosophy as paving the way for Kant, points out that Locke's doctrines "never gained unchallenged recognition in Germany" (p. 120).

34. Paul Hazard, *La crise de la conscience européenne* (1935; Paris: Gallimard, 1961), vol. 2, p. 26.

kind of scientific psychology."[35] Peter Gay, writing in the late 1960s, argued that psychology was the foremost Enlightenment human science.[36] Georges Gusdorf considered Locke to be the first great name in psychology, despite not being a psychologist in the modern sense of the term.[37] And Roy Porter, in the early 1990s, maintained that the objects of what we call "psychology" could be found in moral philosophy, metaphysics, and the theory of the understanding, and that, second only to Locke's critique of innate ideas, psychology emerged as the all-important human science and the key to mankind's advancement.[38]

Within this kind of framework, investigating eighteenth-century psychology amounts to selecting the highlights from empirical and sensualist philosophies, and documenting the many applications of the "analytic" method derived from Locke. The analytic method, as defined by Étienne Bonnot de Condillac (1715–1780) in his *Essay on the Origin of Human Knowledge* (1746), consists in "composing and decomposing our ideas to create new combinations and to discover, by these means, their mutual relations and the new ideas they can produce."[39] Condillac notes that, thanks to analysis, which does not aim to study the nature of the mind but only to "know its operations," we "determine the extent and limits of our knowledge and endow human understanding with new life."[40] He was echoed by the *philosophes*, who were convinced that the systematic application of analysis would enable the sciences to be reformed and the enlightened nature of the century to be secured. A typical statement of this conviction, full of confidence and enthusiasm, can be found in the exclamations of Dugald Stewart (1753–1828), a professor of moral philosophy at Edinburgh: "How

35. Isaiah Berlin, *The Age of Enlightenment* (1956; New York: New American Library, 1984), p. 19.

36. Peter Gay, *The Enlightenment: An interpretation* (New York: W.W. Norton, 1969), vol. 2, p. 167.

37. Georges Gusdorf, *La révolution galiléenne* (Paris: Payot, 1969), vol. 2, p. 252; Gusdorf, *L'avènement des sciences humaines au siècle des lumières* (Paris: Payot, 1973), p. 32.

38. Roy Porter, "Psychology," in John W. Yolton, ed., *The Blackwell companion to the Enlightenment* (Oxford: Blackwell, 1992). There is no mention of "psychology" in Michel Delon, ed., *Dictionnaire européen des Lumières* (Paris: Presses Universitaires de France, 1997); or in Alan Charles Kors, ed., *Encyclopedia of the Enlightenment* (New York: Oxford University Press, 2003); but there is a short summary from Wolff to Kant in Werner Schneiders, ed., *Lexikon der Aufklärung: Deutschland und Europa* (1995; Munich: C. H. Beck, 2001).

39. Claude Yvon, ANALYSE (*en Logique*), in Denis Diderot and Jean Le Rond d'Alembert, eds., *Encyclopédie, ou Dictionnaire raisonné des sciences, des arts et des métiers* (Paris: Briasson [. . .]), text (1751–1765), I, p. 401. See Étienne Bonnot de Condillac, *Essay on the origin of human knowledge*, trans. and ed. Hans Aarsleff (Cambridge: Cambridge University Press, 2001), chap. 7.

40. Condillac, *Essay*, p. 5. On Condillac, see Isabel F. Knight, *The geometric spirit: The abbé de Condillac and the French Enlightenment* (New Haven: Yale University Press, 1968).

many are the threads which, even in Catholic countries, have been broken by the writings of Locke! How many still remain to be broken, before the mind of man can recover the moral liberty which, at some future period, it seems destined to enjoy!"[41] In fields as disparate as political economy, aesthetics, education, or legislation, the process of enlightenment seemed to depend on knowledge of the nature and psychology of man. Metaphysics too was to undergo a psychological reformulation, understood as the implementation of Lockean principles.

Condillac, who is often seen as the Enlightenment's foremost "psychologist," declared that since the analytic method constitutes the foundation of every science, it should be termed "metaphysics." But not even such a metaphysics could be called the "first science":

> For will it be possible to analyze all our ideas adequately if we do not know what they are and how they are formed? We must discover first of all how they originate and develop. But the science concerned with this object has as yet no name, since it is so recent. I would call it "psychology," if I only knew of some good work by that name.[42]

Condillac rejected the term because of his view of the history of discussions about the origins of knowledge: "Immediately after Aristotle comes Locke, since the other philosophers who have written on this subject do not count."[43] Aristotle had established the principle of the sensory origin of knowledge without developing it, and that was how things had stood until Locke. Condillac's summary brushes aside two thousand years of "a futile science, which deals with nothing and leads nowhere. Since we progress from particular ideas to general notions, the latter cannot be the object of the first science."[44] From Aristotle to Locke, says Condillac, human knowledge was approached deductively, starting with abstract concepts, particularly the soul. Insofar as the existing *Psychologies* illustrated such an approach, they, and their titles, were to be dismissed.

---

41. Dugald Stewart, *Dissertation, exhibiting a general view of the progress of metaphysical, ethical, and political philosophy, since the revival of letters in Europe*, in *The works* (Cambridge: Hilliard and Brown, 1829), vol. 6, pp. 437–438.

42. Étienne Bonnot de Condillac, "Des progrès de l'art de raisonner" (in *Histoire moderne*, bk. XX, chap. 12, 1775; in *Cours d'études pour l'instruction du Prince de Parme*, 1768–1773), in *Œuvres philosophiques de Condillac*, ed. Georges Le Roy (Paris: Presses Universitaires de France, 1947–1951), vol. 2, p. 299.

43. Étienne Bonnot de Condillac, "Extrait raisonné du *Traité des sensations*," in *Traité des sensations* (1754; Paris: Fayard, 1984), p. 287.

44. Condillac, "Des progrès," p. 299.

Other *philosophes* took up the same argument with a different inflection. The Marquis de Condorcet (1743–1794), who was fiercely anticlerical, wondered "what philosophy monks will teach—the art of Scholastic quibbling, what the textbooks call natural theology and psychology, that is to say, theologians' daydreams on the nature of God and of the soul."[45] Destutt de Tracy (1754–1836) preferred the term "ideology" (which he invented to designate the "analysis of thought") to "psychology" because, he explained, the latter meant "science of the soul" and evoked "a vague quest for first causes."[46] At the beginning of the nineteenth century, Dominique-Joseph Garat, who taught at the *écoles normales* created during the French Revolution, regarded the word "psychology" as unfit to replace "metaphysics," with its obscure connotations. Indeed:

["Psychology"] gains almost no clarity in our language, since it is associated with almost none of its words: etymologically, it goes back to the idea of the soul rather than to that of the operations of the human mind; it would suggest that a type of knowledge which, by its very nature, should become universal and known to all, is a separate science.[47]

Garat preferred to emulate Locke and teach the "analysis of the understanding." At all events, it was necessary to get rid of metaphysics, that "benighted science of the old schools" which "plunged into obscurity even the simplest and clearest ideas."[48] This critique was leveled at the notion of the soul almost to the same extent as at syllogisms and Latin.

Writing at the end of the Enlightenment, Condorcet, Destutt de Tracy, and Garat simply applied post-Lockean commonplaces to psychology, seen as nothing but theologico-Scholastic speculations on the nature of the soul. One need only recall the frequent references of Voltaire (1694–1778) to the subject, from the *Philosophical Letters* (1734) to the *Dialogues of Euhemerus* (1777), for example: "The word 'soul' is one of those words which

---

45. Jean-Antoine-Nicolas Caritat, marquis de Condorcet, "Petits résumés sur l'histoire de l'éducation" (1774), in *Réflexions et notes sur l'éducation*, ed. Manuela Albertone (Naples: Bibliopolis, 1983), p. 121.

46. Antoine Louis Claude Destutt de Tracy, *Sur un système méthodique de Bibliographie* (1797), in *Mémoire sur la faculté de penser; De la métaphysique de Kant et autres textes* (Paris: Fayard, 1992), p. 71.

47. [Dominique-Joseph] Garat, "Analyse de l'entendement," in *Séances des Écoles Normales, recueillies par des sténographes, et revues par les professeurs*, new ed. (Paris: À l'Imprimerie du Cercle Social, 1800), vol. 1, pp. 149–150.

48. Ibid., p. 149.

everyone uses but does not understand."[49] We cannot understand it because we have no idea of it, and we have no idea of it because we cannot trace its referent back to sense impressions. If the soul exists, it cannot be known, and insofar as we can attribute thinking and feeling to the body, the soul is a superfluous concept. Yet, Voltaire observed, *raisonneurs* on the subject have never been lacking. Happily, after so many writers of "the romance of the soul," there came "a wise man who modestly recounted its history: Locke has expounded to man the nature of human reason just as a fine anatomist explains the powers of the body."[50] As the authors of the "Preliminary Discourse" of the *Encyclopédie* subsequently claimed, Locke reduced metaphysics to an "experimental physics," a natural science of the soul. The following extract from that *Discourse*, so often considered emblematic of "Enlightenment thought," condenses epistemological principles and methodological aspirations shared by most eighteenth-century psychologists, from militant materialists to convinced Christians:

> Locke undertook and successfully carried through what Newton had not dared to, or perhaps would have found impossible. It can be said that he created metaphysics, almost as Newton had created physics. He understood that the abstractions and ridiculous questions which had been debated up to that time and which seemed to constitute the substance of philosophy were the very part most necessary to proscribe. He sought the principal causes of our errors in those abstractions and in the abuse of signs, and that is where he found them. In order to know our soul, its ideas, and its affections, he did not study books, because they would only have instructed him badly; he was content with probing deeply into himself, and after having contemplated himself, so to speak, for a long time, he did nothing more in his treatise, *Essay Concerning Human Understanding*, than to present mankind with the mirror in which he had looked at himself. In a word, he reduced metaphysics to what it ought to be: the experimental physics of the soul—a very different kind of physics from that of bodies, not only in its object, but in its way of viewing that object. In the latter study we can, and often do, discover unknown phenomena. In the former, facts as ancient as the world exist equally in all men. . . . Reasonable metaphysics can only consist, as does experimental physics, in the careful assembling of all these facts, in reducing them to a

---

49. Voltaire, *Lettres philosophiques* (1734), first draft of letter 13, in *Mélanges*, ed. Jacques van den Heuvel (Paris: Gallimard, 1961), p. 41.

50. Voltaire, "On Mr. Locke," in *Philosophical letters*, trans. Ernest Dilworth (Indianapolis, Bobbs-Merrill, 1961), p. 37.

corpus of information, in explaining some by others, and in distinguishing those which ought to hold the first rank and serve as foundations.[51]

In the same vein, but half a century later, the physician and philosopher
Pierre-Jean-Georges Cabanis (1757–1808) maintained that once the study
of the mind had been separated from that of the body, it had been obscured
by "vague metaphysical hypotheses." As a result, for Cabanis, there was
no "solid basis, no fixed point to which one might attach the results of
observation and experience."[52] In a text from 1809, he wrote that what "is
still today called metaphysics bears no relation to what once went by that
name," claiming that "since Locke, Helvetius, and Condillac, metaphysics
is but the knowledge of the operations of the human mind, the formulation
of the rules which man must follow in his search after truth . . . in a word,
the *science of methods;* methods founded on the knowledge of the faculties
of man."[53]

Such statements, of which one could cite many more examples, illustrate the *philosophes'* use of Locke and the transformation of the "metaphysical" problem of the soul into the "analytic" question of the origin of
human knowledge—a transformation to which some authors went so far as
to attribute the freedom of nations: "Almost at its birth," Garat confided to
his pupils, "the analytic art of the understanding discovered human rights;
it is because this art existed that France is free, and that Europe should
be free."[54] At the opposite end of the political spectrum, Count Joseph de
Maistre (1753–1821) referred to the same events but blamed the French
for having trusted Locke and let themselves be voluntarily imprisoned—
"LOCKED in fast," as he put it.[55] He attributed to Locke the same position

51. Jean Le Rond d'Alembert, *Preliminary discourse to the Encyclopedia of Diderot,* trans.
Richard N. Schwab and Walter E. Rex (Indianapolis: Bobbs-Merrill, 1963), p. 83–84.

52. Pierre-Jean-Georges Cabanis, *Rapports du physique et du moral de l'homme* (1800), in
Œuvres philosophiques, ed. Claude Lehec and Jean Cazeneuve (Paris: Presses Universitaires de
France, 1956), vol. 1, p. 111. The English version, *On the relations between the physical and
moral aspects of man,* trans. Margaret Duggan Saidi (Baltimore: Johns Hopkins University Press,
1981), vol. 1 (here, p. 9) is rather problematic: "l'homme moral" is not "ethical man," and "le
vague des hypothèses métaphysiques" is not "the wave of metaphysical hypotheses."

53. Pierre-Jean-Georges Cabanis, "Lettre sur un passage de la *Décade philosophique* et en
général sur la perfectibilité de l'esprit humain" (1809), in Cabanis, *Œuvres philosophiques,* vol. 2,
pp. 514 and 515.

54. Garat, "Analyse de l'entendement," pp. 162–163.

55. Joseph de Maistre, *St. Petersburg dialogues* (1821), trans. Richard A. Lebrun (Montreal:
McGill–Queen's University Press, 1993), 6th dialogue. The entire dialogue is an attack on Locke.
Maistre's *Examen de la philosophie de Bacon* (1836) is also an attack on Locke, Condillac, and the
philosophes, whom he considered—and who saw themselves as—heirs to Bacon.

and function, albeit with a negative value, as did the *Encyclopédie*, Garat, Cabanis, or a Christian liberal like Madame de Staël (1766–1817) in *On Literature Considered in Its Relationship to Social Institutions* (1800, pt. 2, chap. 6). They all observed that "Scholastic" metaphysics had given way to the "analytic" study of intellectual operations.

This is the sense in which the Age of Enlightenment was the "century of psychology." The expression also refers to the reformulation in psychological terms of important areas of scholarly culture, from logic to education, from the theory of knowledge to morals, or from religion to aesthetics.[56] The term "psychology," however, was rarely used to designate the discipline which was taking shape under that name, particularly in Germany, and which was being embodied in books, periodicals, articles, teachings and a historiography.[57]

* * *

While clearly no study on eighteenth-century psychology can limit itself to the sources and individuals who actually employed the term "psychology," it is important, in order not to offer an idiosyncratic history of psychological ideas, to take into account what actually bore that name. This is why I will essentially confine myself to the categories used by the historical protagonists themselves. Such a choice has lexical, conceptual, and social aspects, since it adopts the definitions of the time and establishes its subject matter within a psychological field largely defined by the titles of contemporary works, the rubrics of periodicals, bibliographies, and biographies, as well as self-descriptions by historical protagonists.

56. These historiographical generalizations, based on synopses of the Enlightenment (some of them mentioned above) and on countless textbooks on the history of psychology, are corroborated by the bibliographical studies of Jürgen Jahnke, "Psychologie im 18. Jahrhundert: Literaturbericht 1980 bis 1989," *Das achtzehnte Jahrhundert*, 14, 1990, 253–278; and "Neuere Arbeiten zur Psychologie im 18. Jahrhundert: Historiographische Probleme, Ergebnisse und Tendenzen," *Psychologie und Geschichte*, 2, 1990, 19–24. It is of course possible to study sensualism in Enlightenment culture without dealing specifically with the history of psychology. See, for example, John C. O'Neal, *The authority of experience: Sensationist theory in the French Enlightenment* (University Park: Pennsylvania State University Press, 1996); and Anne C. Vila, *Enlightenment and pathology: Sensibility in the literature and medicine of eighteenth-century France* (Baltimore: Johns Hopkins University Press, 1998).

57. The two approaches are not incompatible, as can be seen from the historical works of Max Dessoir (1868–1947), a professor of philosophy at Berlin and a specialist in aesthetics. He wrote both a short general history of psychology, from the Presocratics to nineteenth-century psycho-physics, *Abriß einer Geschichte der Psychologie* (Heidelberg: Winter, 1911), and a fundamental study on eighteenth-century German psychology, *Geschichte der neueren deutschen Psychologie*, 2nd ed. (1902; Amsterdam: E. J. Bonset, 1964).

This way of approaching psychology in the eighteenth century sets the received historiography of the discipline at a critical distance. It also supposes a certain disintegration of the very concept of "enlightenment."[58] The great works of the 1930s, by Ernst Cassirer, Carl L. Becker, and Paul Hazard, gave the Enlightenment a unity which served to defend reason against totalitarian utopias. This unifying model came apart in the 1970s, and the positive heritage of the Enlightenment began to be contested. The dissemination of works such as *Dialectic of Enlightenment* was indisputably a factor in this.[59] The resultant fragmentation did not, however, always bring with it a critique of the Enlightenment. But it did mean that, broadly, intellectual history was replaced by cultural history,[60] with prominence increasingly given to local specificities and national or confessional differences, to the exploration of the darker, "irrational" or authoritarian, sides of the Age of Reason, to research into social and material aspects which had not hitherto been part of the picture, and to the discovery of *antiphilosophe* or other currents which played a part in the century's processes of "enlightenment."

As the Enlightenment's coherence dissolved, so the concept of enlightenment lost historiographical usefulness or authority, and indeed it frequently has, though most often implicitly, been reduced to a purely chronological category.[61] This process, however, has considerably enriched the field of eighteenth-century studies, especially in the history of science.[62] The

58. On the question of "the Enlightenment," which we can only address briefly here, see Giuseppe Ricuperati, "Le categorie di periodizzazione e il Settecento: Per una introduzione storiografica," *Studi settecenteschi*, 14, 1994, 9–106; and Ricuperati, "Illuminismo e Settecento dal dopoguerra a oggi," in G. Ricuperati, ed., *La reinvenzione dei Lumi: Percorsi storiografici del Novecento* (Florence: Leo S. Olschki, 2000). On the *Aufklärung*, see Ian Hunter's introduction in *Rival Enlightenments: Civil and metaphysical philosophy in early modern Germany* (Cambridge: Cambridge University Press, 2001). For a survey of the different national and regional situations, see Roy Porter and Mikulas Teich, eds., *The Enlightenment in national context* (Cambridge: Cambridge University Press, 1981).

59. Max Horkheimer and Theodor Adorno, *Dialektik der Aufklärung: Philosophische Fragmente*, trans. John Cumming as *Dialectic of Enlightenment* (1947; London: Verso, New Left Books, 1999).

60. Roger Chartier, *The cultural origins of the French Revolution* (1990), trans. Lydia G. Cochrane (Durham: Duke University Press, 1991).

61. See, for example, Dorinda Outram's *The Enlightenment* (New York: Cambridge University Press, 1995), in which the term is sometimes a purely chronological indicator and at others still a meaningful concept.

62. Jan V. Golinski, "Science in the Enlightenment," *History of science*, 24, 1986, 411–424; see also the editors' introduction and the chapters by Nicholas Jardine ("Inner history; or how to end Enlightenment") and Lorraine Daston ("The ethos of Enlightenment") in Clark, Golinski, and Schaffer, *The sciences in enlightened Europe*.

question What is enlightenment? has given way to the more active form, What does "to enlighten" mean?[63] The challenge is now to understand the Enlightenment in its heterogeneity, and to reformulate the question of its identity in the light of historically situated interpretations, and the diverse meanings of the verb "to enlighten." Rather than analyze a "spirit of the Enlightenment" whose signs could be identified, scholars have begun examining the practices, values, experiences, and characteristics common to individual and collective protagonists. This method has produced different notions of the Enlightenment, each being simply the substantivized form of the adjective qualifying those who considered themselves, or were considered by others, to be "enlightened" by virtue of a certain way of life, or owing to certain ideals or philosophical and anthropological principles.

However, in the case of psychology, it is impossible to "think of the Enlightenment as a web of practices without discourse."[64] As a discipline, eighteenth-century empirical psychology had little substance outside texts. Even if it involved—and, according to psychologists, really practiced—techniques of self-observation and observation of others, and even if it did aim at improving mankind and actually reforming minds and society, it was nevertheless embodied principally in the discourses which instituted it. These discourses do not correspond exactly to those on which the "century of psychology" is based, and provide a different image of the psychology of the century. In order to describe this psychology, let us begin at the beginning, with the history of its proper name.

63. It was the philosopher Moses Mendelssohn who gave the title of "Über die Frage: was heißt aufklären?" to his reply to the question "Was ist Aufklärung?" which had been formulated in the *Berlinische Monatsschrift* in 1783. For a collection of sources from the eighteenth and twentieth centuries, and studies on this question, see James Schmidt, ed., *What is enlightenment? Eighteenth-century answers and twentieth-century questions* (Berkeley: University of California Press, 1996). See also Gérard Raulet, ed., *Aufklärung: Les Lumières allemandes* (Paris: Flammarion, 1995), pt. 1.

64. Chartier, *The cultural origins*, p. 18. The author, however, does add: "at least without those varieties of discourse traditionally and spontaneously defined as 'enlightened.'"

CHAPTER TWO

"Psychology" in the Sixteenth Century: A Project in the Making?

The word *psychologia* appeared in the last third of the sixteenth century in the writings of Protestant Scholastics, and was put into circulation through philosophy textbooks used in their universities.[1] The use of the term and its dissemination were linked, from the Reformation onward, to the development of the *cursus philosophicus* as a didactic genre, and to a philosophical climate marked by the revival of Artistotelianism and the spread of Ramist doctrines of method.[2] How are we to interpret this neologism? Was it a new name for an old idea, or the sign of a novel epistemic enterprise? The question has received contradictory responses, but it is worth dwelling on, since it is relevant to the history of ideas and of psychological knowledge, as well as to psychology as a discipline. The name given to a field of knowledge is a factor in the processes through which the field acquires an identity, limits, and contents of its own. The name partially

---

1. On Protestant Scholasticism in Germany, see Peter Petersen, *Geschichte der aristotelischen Philosophie im protestantischen Deutschland* (Leipzig: Felix Meiner, 1921); Max Wundt, *Die deutsche Schulmetaphysik des 17. Jahrhunderts* (Tübingen: J. C. B. Mohr, 1939); Ulrich Gottfried Leinsle, *Reformsversuche protestantischer Metaphysik im Zeitalter des Rationalismus* (Augsburg: Maro Verlag, 1988); and Siegfried Wollgast, *Philosophie in Deutschland zwischen Reformation und Aufklärung 1550–1650* (Berlin: Akademie-Verlag, 1988). Lewis White Beck's *Early German Philosophy: Kant and his predecessors* (1969; Bristol: Thoemmes Press, 1996) is less detailed but useful. See also Richard A. Muller, *Dictionary of Latin and Greek theological terms: Drawn principally from Protestant Scholastic theology* (Grand Rapids, MI: Baker Books, 1985).

2. For overviews of these developments, see Charles B. Schmitt, "The rise of the philosophical textbook" in C. B. Schmitt et al., eds., *The Cambridge history of Renaissance philosophy* (New York: Cambridge University Press, 1990); and Martin Elsky, "Reorganising the encyclopaedia: Vives and Ramus on Aristotle and the Scholastics," in Glyn P. Norton, ed., *The Cambridge history of literary criticism*, vol. 3, *The Renaissance* (Cambridge: Cambridge University Press, 1999). For the situation in German universities, see also Neal W. Gilbert, *Renaissance conceptions of method* (New York: Columbia University Press, 1960), chap. 10.

determines the vocabularies developed, the texts considered authoritative, the links to other disciplines, and the place assigned to the field within the general organization of knowledge. Since, despite the absence of the term at the time, we discuss the "psychology" of the ancient world, the Middle Ages, and the Renaissance in a loose and unsystematic way, and since, on the contrary, we often discuss Enlightenment psychology without examining what really bore that name, we will start by outlining the lexical and semantic history of the term, in order to identify its uses and circumscribe its meaning.

The most frequent occurrence of *psychologia* is in diagrams of the organization of knowledge and in the titles of works, but the term is very rarely defined and never developed in its own right. This fact is indicative of its function. The diagrams in question are the fruit of the logical and pedagogical reforms of the French-born Petrus Ramus (1515–1572), who converted to Protestantism in 1562 and was assassinated during the Saint Bartholomew's Day Massacre.[3] After being banned from teaching and publishing for his criticism of Aristotle, he was appointed professor of rhetoric at the French Royal College (the future Collège de France). Ramus's method stipulated that each subject should be examined in its most general and familiar aspects, before working down to the particular through a series of dichotomies. This method, which was better suited to transmitting knowledge than to producing it, was represented in branching synoptic tables read from left to right, which were supposed to reflect not only how our knowledge of the world may be arranged but also how the world itself was actually organized. The method established an intrinsic link between logic and vision. Ramist tables became a typographical vogue in the sixteenth and seventeenth centuries and, as a pedagogical tool, were extremely widespread, especially in German Protestant universities. The encyclopedias of the period also owe much to them. It was initially in these contexts, and for reasons of pedagogical convenience, that the term *psychologia* first circulated.

Within the organization of knowledge at the time, the study of the soul occupied an ambiguous position: it was the first science of Aristotelian natural philosophy, yet it did not seem to constitute an autonomous discipline. *De anima* treatises invariably repeat Aristotle's assessment:

Holding as we do that, while knowledge of any kind is a thing to be honoured and prized, one kind of it may, either by reason of its greater exact-

---

3. Walter J. Ong, *Ramus, method, and the decay of dialogue: From the art of discourse to the art of reason* (Cambridge, MA: Harvard University Press, 1958).

ness or of a higher dignity and greater wonderfulness in its objects, be more honourable and precious than another, on both accounts we should naturally be led to place in the front rank the study of the soul.[4]

Late Scholasticism assimilated the study of the soul to the *scientia de anima*, that is, following the Aristotelian notion of *scientia*, to a body of certain knowledge arrived at through demonstrative reasoning and empirical experience.[5] The corpus defining this *scientia* was made up of books which, for the most part, had *de anima* in the title and were, at least in part, commentaries on Aristotle's work of the same name.[6] For example, the Spanish Jesuit Rodrigo de Arriaga (1592–1667), who was a professor of dogmatic theology at Prague and author of one of the most important philosophy courses of the first half of the seventeenth century, examined whether such books should discuss the intellective soul. His response was that the intellective soul is indeed a prime object of "this science," since the "science" which studies a "composite object" must also examine its parts. Arriaga referred to the Aristotelian classification of knowledge and went on to explain that the *scientia de anima*, though principally "speculative," is also "practical" insofar as knowledge of the actions of the soul may help guide our behavior.[7] "Such books," Arriaga maintained, should first discuss the essence of the soul, then the living being, then the faculties of the soul in general and of the intellective soul in particular, and last the rational soul separate from the body.

True to the medieval tradition, the "science of the soul" was part of natural philosophy or "physics," the study of nature (*phusis*). It included a set of physiological and psychological questions derived principally from

4. Aristotle, *On the soul (De anima)*, trans. J. A. Smith, in *The complete works of Aristotle: The revised Oxford translation*, ed. Jonathan Barnes (Princeton: Princeton University Press, 1984), vol. 1, 402a1–5, henceforth *DA*.

5. "Scientia duobus modis accipitur. Proprie pro eo habitu, quem per demonstrationem acquirimus. [. . .] Improprie accipitur pro quibusuis aliis habitibus intellectivis"; Goclenius, *Lexicon*, pp. 1009b–1010a. Stated more elaborately: "Scientia latiore significatu sumitur pro omni cognitione certa & evidenti; sive ea ex ratiocinio, sive ex sensuum experientia ducatur, dummodo ex objectis ipsis ea cognitio sit profecta [. . .]. Strictius & magis proprie scientia accipitur pro cognitione certa & evidenti rei necessariae, per propriam causam"; Chauvin, *Lexicon*, p. 588a.

6. On the *scientia de anima* during the "second Scholasticism," see Dennis Des Chene, *Life's form: Late Aristotelian conceptions of the soul* (Ithaca: Cornell University Press, 2000). See also Sven K. Knebel, "Scientia de Anima: Die Seele in der Scholastik," in Gerd Jüttemann, Michael Sonntag, and Christoph Wulf, eds., *Die Seele: Ihre Geschichte im Abendland* (Weinheim: Psychologie-Verlag, 1991).

7. Roderico de Arriaga, "Disputationes in tres libros Aristotelis De Anima," in *Cursus philosophicus* (1632; Paris: apud Jacobum Quesnel, 1639), p. 525. (Within an Aristotelian framework, *theoretikós* is translated by *speculativus*.)

Aristotle's works *De anima* and *Parva naturalia*. *De anima* was at the time a set text for university bachelor's degrees. Commentaries on these works, as well as textbooks (Arriaga's, for instance), replicated the structure and contents of the second and third books of *De anima*, which deal with the definition of the soul and its faculties, common sense, the imagination, and the intellect.[8]

Despite the diversity of Aristotelian currents and some variation in how philosophical disciplines were classified, the theories of the soul in both Catholic and Protestant circles were always linked to the study of the living body.[9] Since the science of the soul was based on research into natural living beings, it viewed the soul in terms of its operations and faculties (from the vegetative to the cognitive), and made no clear distinction between the physiological and the psychological. Arriaga was clearly working within these parameters when he situated the *scientia de anima* in the "physics" part of his course, stating that it concerns the living body, not the soul *secundum se*.[10]

When it came to the human being, however, the *scientia de anima* could not be classed solely within natural philosophy.[11] The rational, immaterial soul, whether as the image of the divine in man, or as that which brings man closer to God, had greater worth than any body. Moreover, psychology had to address the challenge posed by Pietro Pomponazzi (1462–1525) in his *Tractatus de immortalitate animae* of 1516. Pomponazzi claimed that

8. There are literally hundreds of *de anima* treatises, theses, and disputations. See Hermann Schüling, *Bibliographisches Handbuch zur Geschichte der Psychologie: Das 17. Jahrhundert* (Giessen: Universitätsbibliothek, 1964); Schüling, *Bibliographie der psychologischen Literatur des 16. Jahrhunderts* (Hildesheim: Olms, 1967); and Wilhelm Risse, *Bibliographia philosophica vetus*, pars 5, *De anima* (Hildesheim: Olms, 1998).

9. Joseph F. Freedman, "Aristotle and the content of philosophical instruction at Central European schools and universities during the Reformation era (1500–1650)," "Encyclopedic philosophical writings in Central Europe during the high and late Renaissance (ca. 1500–ca.1700)," and "Classifications of philosophy, the sciences, and the arts in sixteenth- and seventeenth-century Europe," in Freedman, *Philosophy and the arts in Central Europe, 1500–1700* (Aldershot: Ashgate/Variorum series, 1999). For the diversity of Aristotelian thought, see Charles B. Schmitt, *Aristotle and the Renaissance* (Cambridge, MA: Harvard University Press, 1983); and also Christoph Lüthy, Cees Leijenhorst, and Johannes M. M. H. Thijseen, "The tradition of Aristotelian natural philosophy: Two theses and seventeen answers," in C. Lüthy, C. Leijenhorst, and J. M. M. H. Thijseen, eds., *The dynamics of Aristotelian natural philosophy from antiquity to the seventeenth century* (Leiden: Brill, 2002).

10. Arriaga, *Cursus*, p. 524.

11. Eckhard Kessler, "The intellective soul," in Schmitt et al., *The Cambridge history of Renaissance philosophy*; Emily Michael, "Renaissance theories of body, soul, and mind," in John P. Wright and Paul Potter, eds., *Psyche and Soma: Physicians and metaphysicians on the mind-body problem from antiquity to Enlightenment* (Oxford: Clarendon Press, 2000).

since Aristotle had located the "science of the soul" within physics, the immortality of the soul could not be demonstrated rationally from Aristotelian principles. A metaphysical perspective on the issue had to be adopted.[12] The need to prove the immortality of the soul *sub ratione entis* gave credence to the idea that the intellective soul was a reality distinct from the organic soul. We will see that this position could imply a dualism, with the body possessing its own "form," and that in such cases, psychology gave a preponderant role to the study of the body.

For all these reasons, the "science of the soul" was classified under both physics and metaphysics. It could lead not only to theology but also to moral philosophy (since the soul was the ultimate source of human actions), to rhetoric (through knowledge of the passions and their mechanisms), and to medicine (due to the union of the body with the soul).

THE FUNCTION OF THE NEOLOGISM "PSYCHOLOGY"

The word "psychology" is said to have been used for the first time in a treatise, now lost, by the Dalmatian humanist Marko Marulic (1450–1524), who composed poetry in Latin and Croatian, as well as works on morals and spirituality.[13] There are, however, some doubts about this. The first attested uses of the term date from the 1570s; but the word was at the time not defined, nor was its provenance given, and it was treated as though it were sufficiently well known or else not significant enough to be examined for itself. The neologism emerged with the development of modern Latin and

12. Charles H. Lohr, "The sixteenth-century transformation of the Aristotelian division of the speculative sciences," in Donald R. Kelley and Richard H. Popkin, eds., *The shapes of knowledge from the Renaissance to the Enlightenment* (Dordrecht: Kluwer, 1991).

13. K. Krstic, "Marko Marulic—the author of the term 'psychology,'" *Acta Instituti Psychologici Universitatis Zagrabiensis*, no. 36, 1964, 7–13. Subsequently: Edwin G. Boring, "A note on the origin of the word 'psychology,'" *Journal of the history of the behavioral sciences*, 2, 1966, 167; Joseph Brozek, "*Psychologia* of Marcus Marulus (1450–1524): Evidence in printed works and estimated date of origin," *Episteme*, 7, 1973, 125–131; Solomon Diamond, "What Marulus meant by 'psychologia,'" *Storia e critica della psicologia*, 5, 1984, 407–412; Marina Massimi, "Marcus Marulus, i suoi maestri e la 'Psychologia de ratione animae humanae,'" *Storia e critica della psicologia*, 4, 1983, 27–41. More generally: F. H. Lapointe, "Origin and evolution of the term 'psychology,'" *American psychologist*, 25, 1970, 640–646; Lapointe, "Who originated the term 'psychology,'" *Journal of the history of the behavioral sciences*, 8, 1972, 328–335; Paul Mengal, "La constitution de la psychologie comme domaine de savoir aux XVIᵉ et XVIIᵉ siècles," *Revue d'histoire des sciences humaines*, 2, 2000, 5–27; Eckart Scheerer, "Psychologie," in Ritter et al., *Historisches Wörterbuch*, vol. 7; Hendrika Vande Kemp, "Origin and evolution of the term *psychology*: Addenda," *American psychologist*, 35, 1980, 774; Vande Kemp, "A note on the term 'psychology' in English titles: Predecessors of Rauch," *Journal of the history of the behavioral sciences*, 19, 1983, 185.

of a specialist philosophical vocabulary. Post-Hellenistic Greek contained
a great many compounds of *psukhē*, but "psychology" appears neither in
the works of the Fathers of the Church nor in Roman or Byzantine writers.[14] This was also the case in Latin, even though it imported from the
Greek words such as *psychogonia* (the genesis or generation of the soul) and
psychomachia (the battle for the soul).[15] Indeed, sixteenth-century Latin
saw an explosion of erudite neologisms with Greek roots, which were often
created by retranslating well-established Latin expressions:[16] just as *physiologia* became the equivalent of *de natura*, *encyclopaedia* of (among others) *orbis doctrinae*, *anthropologia* of *de homine*—while *anthroposophia*
referred to astrology and occult philosophy—so *psychologia* corresponded
to *de anima*.

Psychologia was not the only neologism relating to the soul to be coined
in the context of the theologico-naturalist debates of the sixteenth and seventeenth centuries. The Reformers, for instance, contested two theories (attributed to the Anabaptists) concerning the state of the soul after death:
that of the sleep of the soul (condemned by the Fifth Lateran Council of
1513) and that of the death of the soul, or "thnetopsychism" (from *thnetos*,
"mortality"). They countered their adversary's arguments with the traditional doctrine that the soul, which is immortal, retains its powers of perception and intellection after death. This was the position held by John
Calvin (1509–1564) in his *Psychopannychia*, which was written in 1534
against the "error of the ignorant who think that souls sleep after death
and until the Last Judgment." More specifically, as suggested by the title of
the first edition of the work, the aim was to prove that the souls of those
who had died for their faith in Christ were not slumbering but alive and by
his side: *psychopannychia* comes from the Greek term for that which lasts
throughout the night. In calling his treatise "Vigil of the Soul," Calvin was
declaring his opposition to those he called the "hypnosophists."[17]

---

14. E. A. Sophocles, *Greek lexikon of the Roman and Byzantine periods (from B.C. 146 to
A.D. 1100)* (Boston: Little, Brown and Co., 1870); G. W. H. Lampe, *A patristic Greek lexicon*
(Oxford: Clarendon Press, 1968).

15. Alexander Souter, *A glossary of later Latin to 600 A.D.* (Oxford: Clarendon Press, 1949).

16. Josef Ijsewijn, *Companion to Neo-Latin studies*, pt. 1, *History and diffusion of Neo-Latin
literature*; pt. 2 (with Dirk Sacré), *Literary, linguistic, philological and editorial questions* (1990;
Louvain: Leuven University Press, 1998). See also Tullio Gregory, "Sul lessico filosofico latino del
Seicento e del Settecento," *Lexicon philosophicum: Quaderni di terminologia filosofica e storia
delle idee*, 5, 1991, 1–20.

17. John Calvin, *Psychopannychia qua refellitur quorundam imperitorum error qui animas
post mortem usque ad ultimum iudicium dormire putant*, published in Strasbourg in 1542 under
the title of *Vivere apud Christum non dormire animis sanctis, qui in fide Christi decedunt*; see

In the case of England, Neoplatonism proved to be a fertile soil for lexical creativity. The *Psychodia platonica* of Henry More (1614–1687)—the "Platonicall song of the soul," comprising four long poems—was published in Cambridge in 1642, a century after Calvin's rebuttals. It contained *Psychozoia*, or the life of the soul; *Psychathanasia*, or the immortality of the soul; *Antipsychopannychia*, which contested the thesis of the sleep of the soul after death; and *Antimonopsychia*, against the doctrine of the unity of the intellect.[18] Each of these titles—which, following contemporary fashion, was printed in Greek letters—was inspired by a particular debate on the soul, a fact which bears out the polemical context in which the neologism *psychologia* came into being.[19]

The question remains, however, whether the emergence of the term *psychologia* was the sign of a novel project or whether it simply gave a new name to an older concept. Opinion is divided on this point. In an illuminating overview of the late Renaissance "psychological sciences," the Genevan critic Jean Starobinski has maintained that the word neither names a new discipline nor indicates a new way of defining the objects of psychological inquiry.[20] Rather, he endorses the idea that the chiefly "theological interests" of the Protestant Scholastics who used the term in the last decades of the sixteenth century disqualify them from the history of psychology.[21] But

---

also *Brieve instruction pour armer tous bons fideles contre les erreurs de la secte commune des anabaptistes* (1544).

18. Henry More, *ΨΥΧΟΔΙΑ PLATONICA: or a platonicall song of the soul, Consisting of foure severall Poems; viz ΨΥΧΟΖΙΑ ΨΥΧΑΘΑΝΑΣΙΑ ΑΝΤΙΨΥΧΟΠΑΝΝΥΧΙΑ ΑΝΤΙΜΟΝΟΨΥΧΙΑ* (1642).

19. More's principal sources are Plotinus's *Enneads*, Marsilio Ficino's *Theologia platonica: de immortalitate animorum* (1482) and the anonymous *Theologia germanica*, published by Martin Luther in 1516. *Psychozoia* is an allegory of the life of Psyche, the soul of the world, and of the return of her microcosmic progeny, the individual soul, to her. *Psychathanasia* proclaims the eternity of the soul, whose instrument, the body, exists only due to the soul's "plasticity"; this argument was based partly on the independence of the body from the operations of the intellect "joyn'd with the Eternal Idees." *Antipsychopannychia* sets out to prove that souls do not slumber until Judgment Day. One of the arguments advanced was the existence of prophetic and mystical experiences which must occur even more intensely when the soul is freed from the body. *Antimonopsychia* attacked the Averroist thesis of the unity of the intellect, which had been condemned by the church in the thirteenth century. Since, for Aristotle (*DA*, bk. 3), the intellect is separate, immortal, and eternal, it must, according to Averroes, be one and the same for all human beings—implying that there can be no personal immortality.

20. Jean Starobinski, "Panorama succinct des sciences psychologiques entre 1575 et 1625," *Gesnerus*, 37, 1980, 3–16, p. 3.

21. Ibid., p. 4: "The school was so far limited to theological interests that its doctrines need not be discussed." See also George Sydney Brett, *A history of psychology* (1912–1921; Bristol: Thoemmes Press, 1998), vol. 2, p. 150.

as soon as one dismisses such interests, the term becomes decontextual-
ized and loses an essential part of its original meaning. It is not that what
remains has nothing to do with psychology. On the contrary, as Starobinski
demonstrates in selecting the most "psychological" aspects of the *de anima*
works, even though the late Renaissance offers no "distinct science" corresponding to the "modern concept of psychology," there is an abundance
of "psychological material" to be found in philosophy, theology, medicine,
and elsewhere. Starobinski's choice of method mirrors what is implied by
the notion of "psychological sciences": while the latter cannot be identified
with *psychologia* or *scientia de anima*, they may be arrived at "by abstraction, by selecting a posteriori the episodes which correspond to the interests
of our psychology and psychiatry."[22] Such an approach is not without its
value. However, it involves a retrospective reconstruction that does not address the specific nature of the *scientia de anima*.

Georges Gusdorf's approach, in his monumental history of the human
sciences, is altogether different. He concedes that the appearance of the
term "psychology" may have some conceptual importance while not interpreting it as the expression of a novel project. For Gusdorf, the emergence
of the term was "the sign of a new awareness"; it implied "the idea of an
exploration of mental reality considered as the object of an autonomous
field of knowledge" leading to a "natural history of the human mind."[23] Before becoming widely accepted, Gusdorf claims, the new word was used "in
limited fashion for a century and a half, maybe two."[24] Gusdorf, however,
ignores the actual uses of the term, implying that from the outset "psychology" designated the approach Montaigne developed and Francis Bacon theorized, namely, an inductive science free from metaphysics and religion.[25]
Gusdorf's interpretation belongs less to a history of science than to a normative epistemology derived from the study of entities such as "man in
and for himself," "psychological reality," or "a psychology worthy of the
name."[26]

The analyses of Paul Mengal are more relevant. Mengal refuses to dis-
solve the history of psychology into the history of philosophical discourses
on the soul. Having studied the actual uses of the term "psychology," he
notes that the body of knowledge which goes by this name in the sixteenth

22. Starobinski, "Panorama," p. 4.
23. Gusdorf, *La révolution galiléenne*, vol. 2, pp. 229, 230, 231.
24. Ibid., p. 229.
25. Ibid., pp. 239–240. "While German professors were inventing the term 'psychology,'
Montaigne was inventing the thing itself"; p. 236.
26. Ibid., pp. 237, 244, 258.

century is classed under natural philosophy. In this he sees "a precious in-
dication" of a new aim, that of "situating knowledge of the soul within the
more general framework of knowledge of nature rather than within theology."[27] For Mengal, the distinction which emerged at the time between
somatologia and *psychologia* reveals a "radical change of perspective"; it
is "a clear sign that a new domain of knowledge was being constituted," in
which the soul was "no longer an exclusively theological object of eternal
salvation but had become a possible object for empirical science."[28] Yet in
the Christian-Aristotelian context of the authors whom he cites, the *scientia de anima* was far from being a matter for theology alone; on the contrary, at the time when the term *psychologia* appeared, the science of the
soul was for the most part placed within natural philosophy.

One of these authors, the Calvinist metaphysician Clemens Timpler
(1567–1624), went so far as to dispute explicitly the division of "anthropology" into "somatology" and "psychology," arguing that man is a substantial totality distinct from its causes, that is, from the soul and the body
taken separately. Timpler explained that those who advocate the soul-body
dichotomy confuse what is proper to man with the characteristics of man's
component parts (the body and the soul). They therefore treat the human
being as though it could be reduced to one or another of these substances
taken in isolation. Yet, Timpler argued, it is not the rational soul which, in
the body, brings about the operations proper to man but rather man himself
by means of the soul.[29]

In summary, the term *psychologia* may well have named certain discourses linked to new ways of thinking about the *scientia de anima* in

27. Paul Mengal, "La mythistoire de la psychologie," *Césure*, no. 2, 1992, 127–146, p. 136.

28. Ibid., pp. 136, 137. See also P. Mengal, "La constitution de la psychologie"; and Mengal,
"Pour une histoire de la psychologie," *Revue de synthèse*, 4th ser., nos. 3–4, 1988, 485–497; as
well as his "Naissances de la psychologie: La nature et l'esprit," *Revue de synthèse*, 4th ser., nos.
3–4, 1994, 355–373. These arguments are elaborated at length in Mengal's *La naissance de la
psychologie* (Paris: L' Harmattan, 2005).

29. Clemens Timpler, *Physicae seu philosophiae naturalis systema methodicum: Pars tertiam & postrema Physicae, complectens Empsychologiam; Hoc est, Doctrinam de corporibus
naturalibus animatis* (Hanau: apud Haeredes Guilielmi Antonii, 1610), p. 262. For the issue we
have just summarized, see pp. 258–262, the *problemata* "An Anthropologia recte distribuatur
in Somatologiam & Psychologiam?" and "An homo sit substantia realiter distincta a corpore
& anima, tum separatim, tum coniunctim acceptis?" These views, classed under physics, had
corresponding ones under metaphysics: Timpler considered the faculties and operations of
the rational soul to be different from those of man taken as a whole, and in the latter case the
faculties and operations were derived from the rational soul as their proximal efficient cause.
Metaphysicae systema methodicum (1604; Hanau: apud Haeredes Guilielmi Antonii, 1612),
p. 462.

sixteenth-century Protestant Germany, but it certainly was not conceptual-
ized in terms of a radical break or presented as naming a new field of empirical knowledge.[30] On the contrary, while the term was used with various
nuances right up to the end of the seventeenth century, it invariably arose
within a conceptual field fashioned by a dual intellectual heritage—Aristotelian and Galenic—which it will be helpful to recall here.

ARISTOTELIANISM AND GALENISM

It is important to distinguish the "science of the soul" as it appears in
classifications of knowledge and university textbooks from the actual distribution of psychological theories and learning. As most historical re-
constitutions of "Renaissance psychology" point out, there was no single
discipline in which psychological subjects were addressed. The famous classification by Francis Bacon (1561–1626), for example, relates each body of
knowledge—history, poetry, and philosophy—to memory, imagination, and
reason, respectively. In his *On the Dignity and Advancement of Learning*
(1623), Bacon proposed to create within philosophy a "general science of the
nature and state of man," including man's "miseries and prerogatives," his
body, his soul, and their union. The doctrine of the soul would be divided
into two parts, one for the study of the rational soul, another for that of the
irrational soul humans share with animals. The faculties (understanding,
reason, imagination, memory, appetite, and the will) were placed under the
aegis of logic and ethics. As a result, psychological topics in Bacon's system were dispersed across theology, medicine, logic, morals, physiology,
and philosophy.

The approach that dominated in all these fields was Aristotle's conception of the living creature, which was in turn informed by his metaphysics
and the fundamental notions of form and matter. Natural change, Aristotle
wrote at various points in his *Physics* and *Metaphysics*, implies the existence of a "subject" that "persists under" changes, a primary being that
remains unaltered: *ousia*, or "being," *substantia* in Latin. As for substance,
in his treatise *On Generation and Corruption*, Aristotle explains that since
an organism's matter is constantly replaced, yet its being is not affected by
such changes, matter cannot be its substance. Substance, then, is "form,"
the actuality (*entelékheia*, entelechy, *actus primus*) of that for which matter

---

30. Katharine Park and Eckhard Kessler come to a similar conclusion, although not through
a semantic history of the term "psychology"; see their "The concept of psychology," in Schmitt
et al., *The Cambridge history of Renaissance philosophy*.

has potentiality (*dunamis*).[31] Hence Aristotle defines the soul as "substance in the sense of the form of a natural body having life potentially within it" (*DA*, 412a20).[32] A living ("animate") body is by definition endowed with a soul, "ensouled" or "besouled," that is to say, it possesses the properties that distinguish it from dead bodies or from bodies that have never been alive.

Aristotle clarified his definition through an analogy: sight is to the eye what the soul is to the body. If the eye were an animal, sight would be its soul; since the eye is "merely the matter of seeing," when seeing is removed, "the eye is no longer an eye except in name—no more than the eye of a statue or of a painted figure" (*DA*, 412b20). Therefore at least certain parts of the soul cannot be separated from the body, and it is the soul and the body together which make up the animal, "as the pupil *plus* the power of sight constitutes the eye" (*DA*, 413a2–3). Some parts of the soul, if it has parts, may be separable from the body, or be its actuality only "in the sense in which the sailor is the actuality of the ship" (*DA*, 413a9), but on this Aristotle declared his ignorance. What he could conclude, however, is that the ensouled differs from the unsouled in that it displays life (*DA*, 413a21). In this statement, which appears pleonastic to us today, ensouled ("animate") translates *empsukhon*, and unsouled ("inanimate") translates *apsukhon*.

By "life," Aristotle understands self-nutrition, growth, and decay (*DA*, 412a14), driven by an internal principle rather than by external causes. Plants are living organisms since they have the capacity, shared with all living creatures, to grow and decay. Animals possess in addition sensitive faculties, and humans alone possess intellective faculties. But only one of these faculties need be operative in an entity for it to be deemed living. As for thought or intellect (*noûs*), it is perhaps a particular kind of soul, differing from the other as the eternal from the corruptible, and capable of being separated from the body; but on this, Aristotle stated that evidence was lacking (*DA*, 413b25–27).

Aristotelian *de anima* texts thus treated the soul as the form of the living organism. They mostly enumerated and classified the soul's faculties and then examined their function and physiology, which Aristotle had addressed in the *Parva naturalia* and in his books on the parts and motion of animals. These were the main topics of the *scientia de anima* right up

---

31. *Dunamis* is translated by the Latin *potentia* or *vis* and is generally rendered by "faculty."

32. This doctrine is called "hylemorphism," from *hulē*, matter, and *morphē*, form.

to the end of the seventeenth century, with considerable variations in the importance assigned to each.

If we take a closer look, the science of the soul first defined its object and enumerated the faculties—which could vary in number—that were most directly linked to the body, before addressing the ones involved in knowledge, and last discussing questions pertaining to the intellect, such as its attributes as a substance (immortality, immateriality), the nature and manner of its union with the body, and its separate existence. The Aristotelian *scientia de anima* then followed the course of the *Parva naturalia*, a collection of nine "short treatises on natural history" in which, taking the contents of *De anima* as established, Aristotle set out to study the fundamental features shared by soul and body alike. This analysis was sometimes integrated into the discussion of the faculties.

For Aristotle, the most important attributes of all or some animals, such as sensation, memory, passion, desire, appetite in general, and pleasure and pain, always belong to the soul and body together.[33] Other essential functions may be arranged into four pairs: waking and sleeping, youth and old age, inhalation and exhalation, life and death. The *Parva naturalia* does not discuss them all but deals with sensation and sensible objects, memory and remembering, sleep and waking, dreams, divination in dreams, the longevity and brevity of life, youth and old age, life and death, and breathing. Right up to the end of the seventeenth century, the dominant *de anima* discourse, considering that these were the phenomena proper to living beings, and to the soul and body in conjunction, remained psycho-physiological and, in a sense, psychosomatic. Nevertheless, true to Aristotle, it also addressed, even if to a lesser extent, questions pertaining to the metaphysical properties of the rational soul.

The *scientia de anima* structured the soul into three sets of faculties, sometimes described as three types of soul.[34] There was no shortage of debate on the unity of the soul and its relation to the body. For example, was the faculty of vision located in the eye, or did it belong to the "organic soul" as a whole but manifest itself only in the visual organs? Certain authors advocated the unity of the soul, whereas others linked organic functions to a mortal and corporeal soul distinct from the rational, immaterial and immortal soul. Despite disagreements and variations in the number and nomenclature of the functions of the soul, the basic schema remained re-

33. Aristotle, *Sense and sensibilia* (from *Parva naturalia*), trans. J. I. Beare, in *The Complete Works of Aristotle*, 436a7–9.

34. Katharine Park, "The organic soul," in Schmitt et al., *The Cambridge history of Renaissance philosophy*.

markably stable.[35] The soul, whether conceived as threefold or as a unity, carried out three sorts of functions: the vegetative, served by the nutritive, augmentative (growth) and generative (reproduction) faculties; the sensitive, served by the internal and external senses, as well as by the faculty of movement (itself divided into physical motion and the emotional and "appetitive" drive); and last, the rational or intellective functions, served by the will and the understanding.

The faculties were located in the ventricles of the brain (hence the name of "cell theory") around which the "animal spirits" circulated (see below). They were interlinked in accordance with the principle that *nihil est in intellectu quod prius non fuerit in sensu*, that nothing is in the intellect which was not previously in the senses. The data arriving from the external senses (sight, hearing, taste, touch, smell) were unified by one of the internal senses, the "common sense" (*sensus communis*), which activated memory as well as the active and passive imagination (*vis imaginativa, fantasia*) (fig. 2.1). On the basis of the sensory images thus generated, the intellect derived through abstraction universal concepts which enabled us to understand the objects perceived; it also carried out its other specific operations, such as subdividing and combining, distinguishing, inferring, deducing, and choosing. Scholastic psychology in the seventeenth century focused on the acts of the sensitive and intellective faculties in man.[36]

As for the physiological dimension of the *scientia de anima*, it drew on the Galenic tradition. Galen (c. 129–216 A.D.), a Greek doctor and philosopher from Pergamum, remained the principal authority in medical matters in the Latin West until about the end of the seventeenth century.[37] Part of his work reached Europe through Latin versions of the Greek originals and, above all, through Arabic versions and commentaries. By the end of the fifteenth century, new Latin translations were being printed in Italy; the Greek *editio princeps* was published in Venice in 1525 and was rapidly followed by Latin editions. This "new Galen" gave pride of place to anatomy. Convinced that anatomy was the cornerstone of medicine, Galen carried out dissections and experiments, including the vivisection of monkeys, pigs, sheep, and goats. His descriptions of human anatomy were challenged

35. See for, example, E. Ruth Harvey, *The inward wits: Psychological theory in the Middle Ages and the Renaissance* (London: Warburg Institute, 1975); and Simon Kemp, *Medieval psychology* (New York: Greenwood Press, 1990).

36. Sven K. Knebel, "Scotists vs. Thomists: What seventeenth-century Scholastic psychology was about," *Modern schoolman*, 74, 1997, 219–226.

37. Owsei Temkin, *Galenism: Rise and decline of a medical philosophy* (Ithaca: Cornell University Press, 1973).

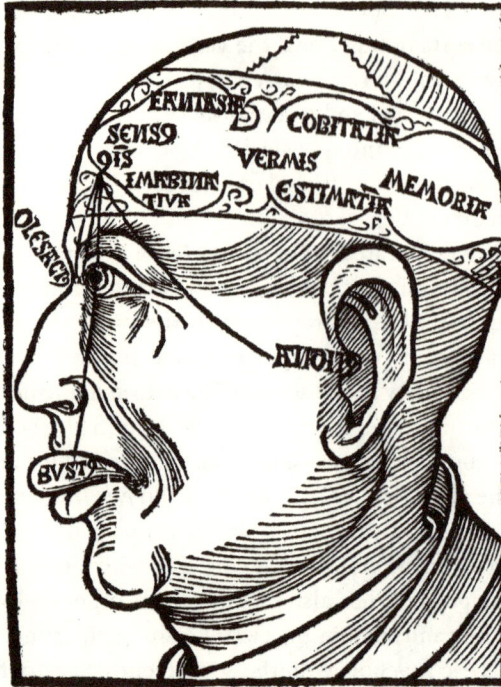

Fig. 2.1. The "cell" theory of mental functioning, in a late rendering of an image common in the Middle Ages. The tongue, ear, and nose (*gustus, auditus, olfactus*) are linked to the common sense, which is situated in the front ventricle, along with fantasy and imagination. The *vermis*, a structure supposed to regulate the flow of animal spirits, is situated between the front and median ventricles. From Hieronymus von Brunschweig, *The noble experyence of the vertuous handy warke of surgeri* (London, 1525; 1st German ed., 1497).

in the sixteenth century; as Andreas Vesalius (1514–1564) demonstrated in *De humani corporis fabrica* (1543), Galenic anatomy largely concerned animals and not humans and had to be thoroughly revised.

As demonstrated in texts on the material conditions under which the soul can exercise its functions, Galen's influence prevailed for even longer in physiology than in anatomy. He drew on older conceptions attributed to Hippocrates to define health as a balance between the four principal humors of the body: blood, yellow bile, black bile, and phlegm. The humors were composed of mixtures of the four elements (fire, air, water, earth) and hence of their fundamental qualities: hot, cold, wet, dry. Therapeutic, dietetic, and hygienic recommendations were accompanied by comments on the abundance, fluidity, or qualities of the humors. Different volumes and proportions of these fluids in the body produced "temperaments" which, in turn, determined the "habits of the soul" and corresponded to the stages of life, the seasons, and the stars. Within this all-embracing macrocosmic vision, the subject matter of psychological discourse was not the soul itself but the *empsukhon*, the living body insofar as it was ensouled. Man's essence was that he was a composite of both soul and body, two substances which functioned together and interacted in accordance with humoral cosmology.[38]

The "commerce" of body and soul was also explained in physiological terms. For Galen, the body was composed of three systems: the brain and the nerves, the heart and the arteries, and the liver and the veins. The blood, formed in the liver, was carried through the veins to the rest of the body, where it was used for nutrition and growth in the form of a "natural spirit." It was mixed with air in the lungs and passed into the heart; it then became another sort of *pneuma*, the "vital spirit," responsible for vital and locomotive functions. It subsequently underwent a last process of refining in the brain, where it became the "animal spirit" (*pneuma psychikon*) required by the sensitive and intellective functions. The qualities (temperature, density, etc.) of these spirits derived from those of the humors, particularly from blood: if, for example, the blood was too cold, the animal spirits would tend to become cold also, and the mental acts dependent on them would be weak and slow. According to this view, the animal spirits constituted the specific link between the body and the soul.

PSYCHOLOGIA AND THE SCIENTIA DE ANIMA

With these general parameters in place, let us now explore the relations between the *scientia de anima* and *psychologia*. The *scientia de anima* encompassed three areas, which were generally combined in different proportions: a naturalistic discourse on "psychosomatic" functions, the analysis of the rational soul united with the body, and the doctrine of the soul as immortal and immaterial substance separated from the body. In the sixteenth and seventeenth centuries, "psychology" could designate all of these domains, or only some of them, thus coinciding with the *scientia de anima*. In 1579, for example, the jurist Johann Thomas Freigius (1543–1583), a disciple of Ramus and rector of the Altdorf gymnasium, devoted a small part of his course on "physics" to psychology. As he explained, insofar as the soul is the animating principle of the natural body (*principium vitae in corpore*

---

38. For an exemplary study of the humoral cosmology, see Raymond Klibansky, Erwin Panofsky, and Fritz Saxl, *Saturn and melancholy: Studies in the history of natural philosophy, religion, and art* (1964; Nendeln, Lichtenstein: Thomas Nelson and Sons, 1979).

naturali), psychology deals with animate bodies. When conceived in this way and incorporated into a Ramist schema, psychology gave an overview of the three types of soul—vegetative, sensitive, and rational (*naturalis, sentiens, intelligens*)—proper to plants, animals, and human beings, respectively. Since Freigius's account of anthropology, consistent with the common understanding of the term at the time, included only human anatomy and, to a lesser extent, the bodily humors, he supplemented it with an account of the "organic" or "irrational" functions of the human soul (external and internal senses, appetitive and locomotive faculties), as well as of the "inorganic" or "rational" ones (intellect and will).[39]

Here *psychologia* was no different from the *scientia de anima*, and the use of the term was not in itself the sign of a new concept. We shall see that the first known work, published in 1590, to have *psychologia* in the title, was an anthology of texts on the traditional question of the origin (created or transmitted) of the rational soul. In 1591, a "philosophical discussion on the soul" entitled *Psychologia* took place at the (Lutheran) Academy of Strasbourg, chaired by the Aristotelian Johann Ludwig Hawenreuter (1548–1618), a doctor and professor of physics and logic. Its subject matter was drawn from Aristotle, starting with the definition of the soul as form of the living body, then briefly setting out the vegetative, sensitive, locomotive, and rational faculties.[40] In 1600, the professor Fabian Hippius (1548–1618) published a *Psychologia physica* at the University of Leipzig (which was Lutheran at the time). It presented psychology as the noblest branch of natural philosophy and defined "physical psychology" as the science of all living bodies. As a consequence, Hippius claimed, one need only deal with the soul's natural operations in the body. The first book of the *Psychologia physica* took up two-thirds of the work and was devoted to anatomy. Thereafter came books on the soul (its definition according to Aristotle, its properties in plants, animals, and humans, its faculties, unity,

39. Johann Thomas Freigius, *Quaestiones physicae: In quibus, Methodus doctrinam Physicam legitime docendi, describendique rudi Minerua descripta est, libris XXXVI* (Basel: per Sebastianum Henricpetri, 1579), liber XXVII, "De Psychologia," pp. 761–771; liber XXXV, "De Anthropologia," pp. 1147–1237; liber XXXVI, "De Anthropologia et anima hominis," pp. 1237–1290. Freigius had been professor of dialectics, ethics, and logic at the University of Freiburg before being expelled due to his sympathies for the Reform and his unconditional support of the philosophy of Ramus, of whom he also wrote a biography.

40. Johan Ludwig Hawenreuter, *ΨΥΧΟΛΟΓΙΑ: sive Philosophica de animo ΣΥΖΗΤΕΣΙΣ, ex libris tribus Aristotelis περι ψυχες, excerpta . . .* (Strasburg: Antonius Bertramus, 1591). The work is a discussion *de animo*, not *de anima*, but the terms are interchangeable here. See J. L. Hawenreuter, *Theses ex praecipuis philosophiae partibus . . .* (Strasburg: Antonius Bertramus, 1611), a series of fifty-two disputations in which the questions relating to the soul fall exclusively within "physics."

seat, origin, and union with the body), on the vital and sensitive operations of the healthy body, and on illnesses and their causes according to Galen.[41] Half a century later, again at Strasbourg, a series of disputations took place on unresolved Aristotelian issues, chaired this time by a theologian. It bore the title *Collegium psychologicum*.[42]

What is the significance of the dissemination of the term "psychology" in Lutheran universities? Although it has now been established that its first use cannot be attributed to Philipp Melanchthon (1497–1560), the humanist, reformer, and friend of Luther, some scholars argue that the term was launched to designate his doctrine of man and the soul, and that it spread initially in tandem with his work. They maintain that *psychologia* was, moreover, invented in order to mark a break with the Aristotelian *scientia de anima*. We have suggested quite the opposite, but we will now examine the issue and its significance —which is far from purely lexical— more closely, through the circulation of the term in the contexts where it appeared.

Melanchthon, who was a professor at Wittenberg from 1518 and cowrote the Augsburg Confession as well as the first systematic account of Reform theology (the *Loci communes rerum theologicarum*, 1521), was given the name of *praeceptor Germaniae* due to his academic textbooks and his contribution to the organization of Protestant schools and universities. Two of his textbooks are devoted to the soul: a *Commentarius de anima* (1540) and a *Liber de anima* (1552), both of which are notable for the space they devote to anatomy.[43] The latter was particularly influential. While it respected

41. Fabian Hippius, *ΨΥΧΟΛΟΓΙΑ Physica, sive de corpore animato, Libri quatuor, toti ex Aristotele desumti, morborum saltem doctrinis ex Medicis scriptis adiecta . . .* (Frankfurt: Typis Wolffgangi Richteri, sumptibus Ioannis Spiessij, 1600). ψυχολογια is defined in the "Epistola dedicatoria" (unpaginated), and "physical psychology" in the following section, entitled "De natura *ΤΗΣ ΨΥΧΟΛΟΓΙΑ ΘΥΣΙΚΗΣ*."

42. Johann Conrad Dannhauer, *Collegium psychologicum, in quo maxime controversae quaestiones, circa libros tres Aristotelis De Anima, proponuntur, ventilantur, explicantur* (Strasbourg: Typis Josiae Staedelii, 1660). The work contains seven disputations, with a respondent for each, on the essence and nature of the soul, and on life and death; on the vegetative soul; on vision and hearing; on the other external senses; on sleep and waking, and on the rational soul in general; on the origin of the soul (coming down in favor of traducianism); and on the intellect and the will. The Academy of Strasbourg became a university in 1621, where the controversial pastor Dannhauer (1603–1666) was a professor of oratory and theology.

43. To the best of my knowledge, there is no modern edition of the *Commentarius*. I have used Philipp Melanchthon, *Commentarius de anima* (Wittenberg, 1550), henceforth *CA* followed by the page number (numbering on recto only). The authoritative edition of the *Liber de anima* (which follows the Wittenberg edition of 1553) can be found in Carl Gottlieb Bretschneider, ed., *Philippi Melanthonis Opera quae supersunt omnia, Corpus reformatorum*, vol. 13 (1846; New York: Johnson Reprint Corp., 1963), cols. 5–178, henceforth *LDA* followed by the column

Galen's authority in matters of physiology, it drew extensively on Vesalius. It served as an elementary anatomy textbook (including for students of
medicine), and some editions are accompanied by loose anatomical plates.
In the sixteenth century it was reprinted some forty times and was the object of eight commentaries.

Melanchthon's *Commentarius* and *Liber* have captured the attention of
historians of medicine because of the place the textbooks assign to anatomy
compared with contemporary *de anima* treatises, and because of their independence from Aristotle's canonical work.[44] Some authors have concluded
that they define a new approach to psychology, henceforth conceived as an
anthropology or general theory of man.[45] Melanchthon himself, in a letter of
1534, characterized the *scientia de anima* as the part of physics "in which
the entire nature of man should be explained."[46] He justified this assertion
at the beginning of the *Commentarius*:

What does that part of philosophy contain which bears the title *De anima*? No part of Physics is richer in teachings or so pleasant as these
discussions on the soul. Although we are unable to penetrate sufficiently
the substance of the soul, its actions nevertheless show us the way to

number. A translation of *LDA* from the chapter on the internal senses onward (*LDA*, 120) can be
found in *A Melanchthon reader*, trans. Ralph Keen (New York: Peter Lang, 1988), pp. 239–289. I
have used two detailed works on Melanchthon's doctrine of the soul: Dino Bellucci, *Science de
la nature et Réformation: La physique au service de la Réforme dans l'enseignement de Philippe
Melanchthon* (Rome: Edizioni Vivere In, 1998), pt. 4, henceforth Bellucci; and Sachiko Kusukawa, *The transformation of natural philosophy: The case of Philipp Melanchthon* (Cambridge:
Cambridge University Press, 1995), chap. 3, henceforth Kusukawa. Johann Rump's thesis, although based only on *LDA*, still remains of fundamental importance for its detailed comparison
of Melanchthon with Aristotle and Galen: J. Rump, *Melanchthons Psychologie (seine Schrift de
anima) in ihrer Abhängigkeit von Aristoteles und Galenos* (Kiel: G. Marquardsen, 1897). See also
Wolfgang Holzapfel and Georg Eckardt, "Philipp Melanchthon's psychological thinking under
the influence of humanism, Reformation and empirical orientation," *Revista de historia de la
psicología*, 20, 1999, 5–34.

44. Jürgen Helm, "Zwischen Aristotelismus, Protestantismus und zeitgenössischer Medizin:
Philipp Melanchthons Lehrbuch *De anima* (1540/1552)," in Jürgen Leonhardt, ed., *Melanchthon
und das Lehrbuch des 16. Jahrhunderts* (Rostock: Universität Rostock Philosophische Fakultät,
1997); Hans-Theodor Koch, "Bartholomäus Schönborn (1530–1585): Melanchthons de anima als
medizinisches Lehrbuch," in Heinz Scheible, ed., *Melanchthon in seinen Schülern* (Wiesbaden:
Harrassowitz Verlag, 1997); Vivian Nutton, "The anatomy of the soul in early Renaissance medicine," in G. R. Dunstan, ed., *The human embryo: Aristotle and the Arabic and European traditions* (Exeter: University of Exeter Press, 1990); Nutton, "Wittenberg anatomy," in Ole Peter
Grell and Andrew Cunningham, eds., *Medicine and the Reformation* (London: Routledge, 1993).

45. Kessler, "The intellective soul," pp. 516–518.

46. Letter to Joachim Camerarius, 24 January 1534 (*Melanchthons Briefwechsel*, no. 1919),
cited in Kusukawa, p. 84, n. 40.

knowledge of it. For this reason, when we speak of its actions, and distinguish between its different faculties and forces, we will be describing the organs. In so doing, we shall have to explain at the same time the whole nature of the body, above all of the human body. That is why this part of Physics should have as its title not only *De anima* but also *On the nature of man in his entirety*.[47]

It should be clear from this why a 1603 annotated edition of the *Liber de anima* could be published under the title *Anthropologia*—a point I shall return to.

Melanchthon's approach and his thought have been interpreted in a variety of ways. On the one hand, the essentially theological nature of his anthropology, in which the image of God can be discerned within the individual mind, would point toward a modern philosophy of subjectivity;[48] on the other hand, he has been seen as paving the way for the emergence of empirical psychology, since his anthropology attributes the vegetative and sensitive functions to the body, and thought to the soul.[49] These interpretations are compatible because, in the framework of debates already mentioned on the relations between the soul and the body, Melanchthon adopted the dualist position according to which every person possesses an organic soul, inseparable from the body and hence mortal, and therefore essentially different from the spiritual or intellective soul. The functions of the organic soul were indissociable from the temperaments and the operations of the body. This would explain the place of anatomy and physiology in Melanchthon's psychology. But it does not prove that the term *psychologia* was coined to refer to his conception of the human being, or even that it was particularly associated with him.

In his texts on the soul, as elsewhere, Melanchthon was highly critical of Scholasticism. In the *Liber de anima*, he formulated several "warnings" concerning Scholastic terms, whose meanings he tried to clarify and simplify.[50] The approach of "the Schools" was repugnant to him, and the

---

47. *CA*, 11. I have essentially followed Bellucci's translation, p. 321.

48. Günter Frank, "Philipp Melanchthons 'Liber de anima' und die Etablierung der frühneuzeitlichen Anthropologie," in Michael Beyer and Günther Wartenberg, with the collaboration of Hans-Peter Hasse, eds., *Humanismus und Wittenberger Reformation* (Leipzig: Evangelische Verlagsanstalt, 1996), p. 325.

49. Mengal, "La constitution de la psychologie," p. 8.

50. For example, "Unde in scholas translata sunt haec nomina, potentia concupiscens, et potentia irascens, seu ut inepte dicunt, irascibilis?" (*LDA*, 135–136). Regarding *synderesis* (the intuitive knowledge of good and evil), Melanchthon equates it with "conscience" in order to exclude designations "de quibus alii multas peregrinas quaestiones movent" (*LDA*, 147). See also

natural philosophy which they termed "Aristotelian" seemed to him full of stupid discussions. That is why, he declared, his own account of the nature of man and the soul would emphasize anatomy, the parts of the body, and the temperaments.[51] In the *Commentarius*, Melanchthon qualified as "inept" the great struggle which took place "in the Schools" on how to differentiate the faculties. Distancing himself from the debate, he explained that the soul may be known only through its actions (*CA*, 20r). Within the commentary tradition, Melanchthon's *de anima* books are distinctive for the way they anchor psychological discourse in anatomy and physiology, and for the connection they make among a Galenic framework, an account of the passions, and the question of freedom. Such connections highlight the preeminently ethical objective of Melanchthon's psychological texts.

This was an objective that Melanchthon had in common with his Spanish contemporary Juan Luis Vives (1492–1540), one of the putative "fathers" of modern psychology, whose *De anima et vita* (1538) was sometimes published together with Melanchthon's *Liber de anima*.[52] This Catholic humanist, like his Lutheran counterpart, also criticized "the Schools." In his *De anima et vita*—which Melanchthon recommended—he stated that the ancient philosophers had lapsed into the greatest absurdities when discussing the soul. The moderns did not fare better: Vives depicted them as simply taking up the ancients and, "in order not to remain totally idle," inventing insoluble problems, or questions whose elucidation brought nothing of interest (*DAV*, [2], 84). For him, things which cannot be perceived directly can be examined only through their operations (*DAV*, 1, 88). Thus, while we cannot know the soul itself, we have access to its "works" through our internal and external senses (*DAV*, 39, 186–188). In short, as Vives declared alluding to the Delphic oracle,

---

his remarks on the distinctions between the speculative and practical, active and passive intellect (*LDA*, 146 and 147).

51. See the letter to Leonhard Fuchs of 30 April 1534 (*Melanchthons Briefwechsel*, no. 1430), cited in Kusukawa, p. 84, n. 39.

52. Juan Luis Vives, *De anima et vita* (Basel, 1538), henceforth *DAV* followed by two page numbers: that of the facsimile of the *editio princeps* (Valencia: Ayuntamiento de Valencia, 1992) and that of Mario Sancipriano's edition, *Ioannis Lodovici Vivis "De anima et vita"* (Padua: Gregoriana, 1974). I have consulted the Italian version which accompanies the Sancipriano edition as well as the Spanish version by Ismael Roca, *Juan Luis Vives, Valenciano, De anima et vita—El alma y la vida* (Valencia: Ajuntament de València, 1992). The citations are from Sancipriano, which is the only scholarly edition since Gregorio Mayans's publication of Vives's *Opera omnia* in 1782–1790.

We are not interested in knowing what the soul is, but we are concerned, and most considerably so, to know how it is and what its operations are. He who urged us to know ourselves did not mean this in respect of the essence of the soul but in respect of the actions needed to moderate our conduct, such that, rejecting vice, we should pursue virtue, which will guide us toward that place where, having become very wise and immortal, we shall live a happier life. (*DAV*, 39, 188)

Consistent with this moral standpoint, the longest chapter of *De anima* is devoted to the immortality of the soul—but also half the book to the passions, which lead the soul toward good or toward evil, and constituted, in Vives's eyes, an irrefutable proof of the union of the soul with the body.[53] Since the qualities of the passions in each individual depend to a large extent on temperament, morality is in fact rooted in empirical physiology and psychology.

For Melanchthon, anatomy and physiology were also linked to moral finalities. He replaced the question What is the soul? with What thing is the soul?; the reason he gave for turning first to Galen for an answer was that in his view, Aristotle had defined only a name, and not a thing (*CA*, 3r–3v). Nevertheless, Melanchthon kept to the Aristotelian organization of the faculties (vegetative, sensitive, appetitive, locomotive, and rational) while mapping it onto Galen's schema. What he found problematic was not so much the Aristotelian structure of the *potentiae animae* as the definition of the soul as the "form" of a body that potentially has life. Vives seemed to accept the definition but emphasized the distinction between the soul and the body.[54] He also stressed our ignorance of the essence of the soul, preferring to characterize it as an animating principle and defining it, within a Christian framework, as "the spirit by which the body to which it is united lives," a spirit "capable of knowing and loving God and, thereby, of being united with Him for eternal bliss" (*DAV*, 104, 358).

Melanchthon gave a Christian definition of the (rational) soul and stressed its source in the scriptures: the soul is an "intelligent spirit" which

53. Carlos G. Noreña, *Juan Luis Vives and the emotions* (Carbondale: Southern Illinois University Press, 1989).

54. In *De anima et vita* (bk. 1, chap. 12, "Quid sit anima"), Vives states that the soul is united with the body in the same way that other forms are united with their matter (*DAV*, 40, 192), and he justifies Aristotle's designation of the soul as *entelechy* (*DAV*, 45, 202). But he also characterizes the soul as "inhabiting" a body capable of life and as its "principal agent." (*DAV*, 42, 196). See Mario Sancipriano, "La pensée anthropologique de J. L. Vivès: L'entéléchie," in August Buck, ed., *Juan Luis Vives: Arbeitsgespräch Wolfenbüttel* (Hamburg: Dr. Ernst Hauswedell & Co., 1981).

has no physical basis, and does not die when separated from the body (*CA*, 10v; cf. *LDA*, 16). He briefly justified his position with reference to a topic the humanists debated fiercely. Melanchthon endorsed Cicero's interpretation (*Tusculan Disputations*, I.22) of Aristotle's definition of the soul as an *endelecheia* or "continuous and perpetual movement."[55] He thought it represented Aristotle's thought faithfully: *endelecheia* as the motion by which each being is alive, and is what it is. The soul was this motion, that is, the animating principle and indeed life itself.[56] And since, Melanchthon maintained, such motion stems from the spirits or temperaments ("vel spiritus vel crasin"; *CA*, 6r), it is necessarily related to the structure of the organs and the movement of the spirits.

Anatomy and physiology were therefore highly theological subjects because they were linked in essential ways to the study of the soul. They revealed *vestigia divina*, the design of the Creator, which explains why Galen, particularly in book 17 of *De usu partium*, depicted anatomy as the source of a "perfect theology." They were also instrumental in elucidating the conditions of salvation and sin. For Melanchthon, the human being possessed innate notions (*noticiae*) concerning numbers, the recognition of order and proportion, the understanding of syllogisms, and geometrical, physical, and moral principles (*LDA*, 143–144). He also maintained that we are created with appetites (hunger, thirst, sensory pleasures, and pain) and affects (the love of God, devotion to the family, benevolence toward all, a horror of demons, joy in the feeling of harmony with God, and faith in eternal life).

In the beginning, these appetites and affects obeyed the "law of the mind" (*mens*) and were "kindled by the Holy Spirit itself, which mixed its flames with the spirits born in the heart and the brain" (*LDA*, 164). This "sweetest of harmonies," however, was shattered by the Fall. Our appetites and affects departed from the law of God, due not so much to their nature as to their weakness and hostility to the law of God (*LDA*, 164–165). Yet man bears within himself an image of God: the mind (*mens*), which produces thoughts, represents the eternal Father, while the image formed through thought represents the Son, and the will, the Holy Spirit. The latter consists of "substantial love" and joy emanating from the Father and the Son; all

55. Besides Bellucci (pp. 322–332) and Kusukawa (pp. 90–91), see Eugenio Garin, "'ΕΝΔΕΛΕΧΕΙΑ e 'ΕΝΤΕΛΕΧΕΙΑ nelle discussioni umanistiche," *Atene e Roma*, 5, 1937, 177–187. For Melanchthon's discussion of this, see *CA*, 7v–10r (chap. "Recte ne vertit Cicero").

56. "Si quaeras quid est anima Bovis, repondet, est illa ipsa agitatio, qua Bos vivit, seu ipsa vita" (*CA*, 5v); cf. *LDA*, 12–16.

these elements are "in harmony with the will and the spirits in the heart, which are the flames and instruments of movement" (LDA, 170).[57]

In his Loci communes, Melanchthon disputed the existence of free will. Habit and experience proved to him that there is antagonism between the will and the "affects" (affectus) which it cannot control, such as love and hatred, joy and sorrow, envy and ambition, blasphemy and unbelief. The innate tendency of human beings to sin acts constantly within each of us, manifesting itself precisely in the affects. All men are sinners, and actually sin, due to their natural powers (per vires naturae). Sin is naturally unavoidable: "Ita fit, ut homo per vires naturales nihil possit nisi peccare."[58] This approach accounts for the profound moral and theological significance Melanchthon saw in anatomy and physiology. He was more optimistic in later works, where he conceded some influence of the cogitatio on the affectus. But this did not diminish the importance he assigned to the sciences of the body. The brain, for example, remained the most admirable of the organs, resembling God and constituting the seat of wisdom (LDA, 69). In Galen's physiology, semen was a product of the brain. Now, if the brain was "capable of light" and hence similar to celestial nature, then so was sperm; spilling one's seed violated God's will and incurred his wrath (LDA, 106).

In summary, Melanchthon's anatomy and science of the soul together formed an anthropology which was in harmony with Christian morals and a Lutheran theology of sin.[59] His De anima focused on the structure and functioning of the human soul in its union with the body, and on investigating human nature in its entirety. This is clear in the "general diagram" published in a tabular synopsis of the Liber de anima intended for students of medicine and philosophy (fig. 2.2).

The author of this work explains that philosophy's most important part deals with the nature of man, which is also fundamental to theology; without it, the different elements of Christian doctrine cannot be adequately discussed. How, for example, could one ever talk knowledgeably about the soul, whether embodied or separate, if one did not know its nature? It would be absurd for a mathematician to declare publicly he knows nothing about

57. See Jürgen Helm, "Die 'spiritus' in der medizinischen Tradition und in Melanchthons 'Liber de anima,'" in Günther Frank and Stefan Rhein, eds., Melanchthon und die Naturwissenschaften seiner Zeit (Sigmaringen: Jan Thorbecke Verlag, 1998); and D. P. Walker, "Medical spirits and God and the soul," in Marta Fattori and Massimo Bianchi, eds., Spiritus (Rome: Edizioni dell'Ateneo, 1984).

58. Philipp Melanchthon, Loci communes, ed. Horst Georg Pöhlmann (Gütersloh: Gütersloher Verlagshaus, 1997), 2, 36, and 122 (pp. 60 and 94).

59. As Kusukawa shows (chap. 3), Melanchthon's anatomy is not simply theological but specifically Lutheran.

DIAGRAMMA GENERA
LE IDEAM OMNIVM QVAE
feaquuntur, continens.

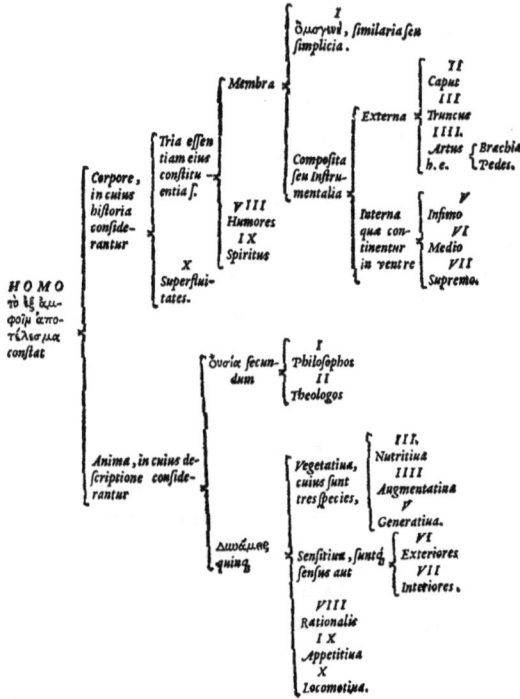

Fig. 2.2. The contents of Melanchthon's *Liber de anima*. From Johann Grün, *Liber de anima DN Philippi Melanthoni in diagrammata methodica digestus* . . . (Wittenberg, 1580). (Staatsbibliothek zu Berlin. Call number 2" Nn 1316<a> : R.)

numbers; it would be even more absurd for a theologian to say he knows nothing of the theory of the soul's faculties and actions, when the soul is the human being's form, essence, and *officina praestantissima*.[60]

The diagram above shows how the subject was treated. Man has a body and a soul. The study of the body focuses on the internal and external organs, the humors and the "spirits"; the study of the soul focuses either on the soul's substance (*ousia*) or on its faculties. As substance, the soul is to be conceived "secundum philosophos" or "secundum theologos." The first approach discusses different theories of the soul in general and the human soul in particular. The second treats the human soul as a spiritual substance

---

60. Johann Grün, *Liber de anima DN Philippi Melanthoni in diagrammata methodica digestus* . . . (Wittenberg: In Officina Typographica Simonis Gronenbergii, 1580), "Praefatio" (pages unnumbered).

and breath of life (*spiraculum vitae*); it examines its ultimate cause (God), how it differs from other spiritual substances, its origin and transmission, its properties as a rational faculty, in what sense it is the "image of God" in man, and its fate at the Last Judgment.[61] As regards its faculties (*dunamis*), the soul is divided into vegetative, sensitive, rational, appetitive, and loco-motive faculties.

The diagram illustrates what, in Melanchthon's view, a book called *De anima* should contain and shows clearly the position of anatomy and physiology within it. But nothing here authorizes us to conclude that the neologism "psychology" was created or used principally to name what the diagram summarizes.

This does not mean that the word "psychology" could not be associated with Melanchthon. In 1596 Rudolph Snellius (1546–1613), a professor of mathematics, astronomy and Hebrew at the University of Leyden (where he had studied medicine), published a commentary on Melanchthon's book "on the soul, or rather on the physiology of man." It reproduced the order of the *Liber de anima* and took the form of notes of variable length on concepts and passages from Melanchthon's text. Snellius first defined the study of the soul as a branch of psychology, the science of animate or living bodies, which itself constituted the last part of physics.[62] When psychology turns to the human being, said Snellius, it discusses man's nature and body, but since man is governed by his soul, a *de anima* work must also have two parts, dealing with each of the two substances.[63] Its title should nonetheless be *de anima*, because, Snellius explained, the soul is more fundamental and eminent (*principalior* and *praestantior*) than the body.[64] That is also why neither the theologian nor the physician can dispense with psychology. The former needs it to know post-Adamic man, whose soul is more denatured than his body and whose will, more corrupted than his intellect; and the latter, to understand the illnesses and symptoms he seeks to cure. In the last analysis, medicine is so dependent on psychology that it can be considered to be its "colony": "Medicina Psychologiae quasi colonia quaedam."[65]

---

61. The details here are taken from other diagrams: "Altera pars de essentia et potentiiis animae: Primum diagramma animae descriptionem philosophicam continens" and "Secundum diagramma descriptionem animae theologicam continens."

62. Rudolph Snellius, *In aureum Philippi Melanchthonis de anima, vel potius de hominis physiologia, libellum, commentationes utilissimae . . . Accesserunt D. Rodolphi Goclenii . . . theses quaedam ac disputationes de psychologicis selectissimae* (Frankfurt: Peter Fischer, 1596), p. 3.

63. Ibid., pp. 4–5.

64. Ibid., p. 5.

65. Ibid., p. 7.

Snellius's commentary was followed by a series of propositions and disputations "on the main psychological subjects" added by his contemporary Rudolph Goclenius (1547–1628), a professor at Marburg and author of the first known work, published in 1590, to have the word "psychology" in its title.[66] The subjects, drawn from Melanchthon's De anima, ranged without transition from the origin and propagation of souls (which was the subject of Goclenius's 1590 Psychologia, discussed below) to the problem of innate notions, and questions concerning fat, sneezing, tears, hands (including chiromancy), hunger, menstruation, the seed and temperament of women, blood, vision, the voice, echoes, the sense of smell, the internal senses, appetite—and more.

At the end of the sixteenth century, then, psychologia was not associated with any particularly new way of approaching the study of the soul. The term designated the science of the soul united with the body, a topic which justified psychology's close attention to anatomical issues. The scientia de anima could accommodate various approaches, including, of course, Melanchthon's.

An illustration of this is the volume published in 1563 by the Zurich naturalist Conrad Gesner (1516–1565), comprising his Liber de anima, the homonymous work by Melanchthon, Vives's De anima et vita, and the Quatuor libri de anima (1542) by Veit Amerbach (1503–1557).[67] Amerbach, a professor of philosophy and rhetoric, left Wittenberg after theological disagreements which prompted him to convert back to Catholicism. In most of the first book of his work (on the definition of the soul), he challenged the Ciceronian and Melanchthonian reading of entelechy as endelechy. In the passages devoted to the faculties of the soul, to the intellect, and to questions arising in the Parva naturalia, he adhered closely to Aristotelian psychology. Gesner intended his own work to be for students of philosophy and medicine; after going over the definitions of the soul, he described the vegetative and sensitive soul, also from an Aristotelian viewpoint.[68]

---

66. Disputationes ac theses praecipuis ex Libro de Anima materiis elegantissimae, a D. Rodolpho Goclenio, Philosopho celeberrimo, &c. praelectae; ibid., pp. 325–413.

67. [Conrad Gesner, ed.] Ioannis Lodovici Vives valentini de Anima & vita Libri tres: Eiusdem argumenti Viti Amerbachii de Anima Libri III: Philippi Melanthonis Liber unus: His accedit nunc primum Conradi Gesner de Anima liber . . . (Zurich: apud Jacobum Gesnerum, [1563]). See Gideon Stiening, "Psychologie," in Barbara Bauer, ed., Melanchthon und die Marburger Professoren (1527–1627) (Marburg: Universitätsbibliothek, 1999), pp. 328–333.

68. Conrad Gesner, De Anima liber, sententiosa brevitate, veluti per tabulas et aphorismos ut plurimum conscriptus, philosophiae & medicinae studiosis accommodatus (Zurich: apud Jacobum Gesnerum, 1563).

A further commentary on the *Liber de anima* was published in 1603 under the title *Anthropologia*. It was based on the teachings of Johannes Magirus (1558–1631), a physician and *professor physices* at Marburg, and was published "supplemented and enriched" by another physician, Georg Caufunger, after Magirus's death. The text reproduced Melanchthon's work up to the chapter on the will but omitted later ones on free will, the word "concupiscence," the "accidents" of the mind (*mens*) and the will, the image of God in man, and the immortality of the human soul. Magirus's "Prolegomena" mention the religious, ethical, and medical importance of knowing oneself. However, consistent with the usual sense of "anthropology" at the time, the work, after some brief considerations on man as God's creature and Aristotle's definition of the soul, was concerned chiefly with anatomy, detailing fluids and solids and mentioning at the end the sensitive faculties, the internal senses, and the "logical" faculty by which humans differ from animals. In Caufunger's definition of the soul—"εντελεχεια corporis physici vitam habentis potentia"—the Greek term is interpreted in an opposite sense to Melanchthon's; but the contradiction is not discussed, as though the definition, which is literally Aristotle's, could be taken for granted.[69]

In summary, then, sixteenth-century books of psychology sometimes partially resemble Melanchthon's and Vives's treatises, but they essentially belong to the *scientia de anima*. The term "psychology" therefore did not designate a new enterprise emblematic of the "transition" from a "metaphysical psychology" to a "descriptive and analytic" psychology.[70] If it had, then the first *Psychology* to come down to us would be truly groundbreaking.

RUDOLPH GOCLENIUS'S *PSYCHOLOGIA*

The first known title in which the term "psychology" appears is *ΨΥΧΟΛΟΓΙΑ: hoc est, de hominis perfectione, animo, et in primis ortu hujus* (Psychology: That is to say, on the perfection of man, his soul, and particularly on the

---

69. Georg Caufunger, "Dedicatio" (pages unnumbered), in Johannes Magirus, *Anthropologia, hoc est: Commentarius eruditissimus in aureum Philippi Melanchthonis libellum de Anima; Completus & locupletus Opera Georgii Caufungeri* (Frankfurt: Wolfgang Richter, 1603). See Stiening, "Psychologie," pp. 334–341.

70. These are Wilhelm Dilthey's words to describe Vives's contribution, which he did not relate to the history of the term "psychology"; "Die Funktion der Anthropologie in der Kultur des 16. und 17. Jahrhunderts" (1904), in *Gesammelte Schriften*, vol. 2, *Weltanschauung und Analyse des Menschen seit Renaissance und Reformation* (Stuttgart: B. G. Teubner / Göttingen: Vandenhoeck & Ruprecht, 1991), p. 423.

ΨΥΧΟΛΟΓΙΑ:

hoc est,

DE HOMINIS
PERFECTIONE, ANI-
MO, ET IN PRIMIS ORTU HUJUS,
commentationes ac disputationes quorundam
Theologorum & Philosophorum noſtræ ætatis,quos verſa pagina oſtendit.

Philoſophiæ ſtudioſis lectu jucundæ & utiles.

Recenſente

RODOLPHO GOCLENIO, PROFESSORE
in Academia Marpurgenſi Philoſophico.

MARPVRGI,
Ex Officina Typographica Pauli Egenolphi.
Anno M. D, LXXXX,

Fig. 2.3. The title pages of the two editions of Rudolph Goclenius's ΨΥΧΟΛΟΓΙΑ. (Staatsbibliothek zu Berlin. Call numbers Nn 1548 : S16 and 2 in: NI 4424<a> : S16.)

origin of the latter).[71] The work, first published in 1590, was a collection of
texts compiled by Rudolph Goclenius. The author, nicknamed the Plato of
Marburg and the Christian Aristotle, was the author of a vast output in logic
and metaphysics, as well as a famous philosophical lexicon. This "concilia-
tor" taught for almost fifty years at the first Protestant university, founded
in Marburg in 1527, and occupied successively the chairs in physics, logic,

71. I refer here to the second edition: Rudolph Goclenius, ΨΥΧΟΛΟΓΙΑ: hoc est, de hominis
perfectione, animo, et in primis ortu hujus, commentationes ac disputationes quorundam Theologorum & Philosophorum nostrae aetatis . . . (1590; Marburg: ex officina typographica Pauli
Egenolphi, 1597). See appendix I for a comparison of the two editions.

ΨΥΧΟΛΟΓΙΑ:

hoc est,

DE HOMINIS

PERFECTIONE, ANI-
MO, ET IN PRIMIS ORTU HU-
jus,commentationes ac difputationes quorundam Theologorum & Philofophorum noftræ
ætatis, quos proximè fequens præfa-
tionem pagina oftendit.

Philofophiæ ftudiofis lectu jucundæ & utiles.

Nunc correctæ & auctæ à
*RODOLPHO GOCLENIO, PROFESSORE IN
Academia Marpurg. Logico.*

MARPURGI,
Ex Officina Typographica Pauli Egenolphi. 1597.

mathematics, and ethics.[72] He turned out dissertations and disputations in industrial quantities, and his *Psychologia* fits into this type of large-scale academic production. The title of the work, comprising an erudite neologism with a Greek root printed in Greek characters accompanied by a Latin explanation, exemplifies what was, at the time, a well-established linguistic and typographical model (fig. 2.3).

72. To my knowledge, no monograph on Goclenius exists. For a short biography and a bibliography, see Friedrich Wilhelm Strieder et al., *Grundlage zu einer hessischen Gelehrten- und Schriftsteller-Geschichte seit der Reformation bis auf gegenwärtige Zeiten* (Cassel: Cramer, 1781–1868), vol. 4 (bibliographical supplements in vols. 5, 6, 8, 9, 13, and 15).

True to the subtitle, *de hominis animo,* Goclenius put together an anthology on *animus* (the rational, spiritual soul), rather than *anima* (the soul as animating principle).[73] The texts, by theologians and physicians, discuss the origin of the *animus,* specifically, whether it is created by God in each individual at each generation, or whether it is transmitted by parents to their offspring. The authors also address the origin of the "organic" faculties of the soul (vegetative and sensitive), which they find distinctly less problematic than the provenance of the "inorganic" powers of the rational soul, or *mens* (the intellect and free will).

The debate touched on the thorny issue of "animation," that is, of how the principle of life is infused into the body. There were several positions on this topic: a Platonic theory of the preexistence of souls, which Origen, the early church father, had adopted; a doctrine, inspired by Neoplatonism and Plotinus, which considered the human soul an emanation of the world soul or of divine substance; traducianism (sometimes called "generationism"), in a materialist version favored by Tertullian (in which, in the process of generation, the soul is transmitted through the seed, *per traducem* or *ex traduce seminis*) and in a spiritualist one (in which the infant's soul derives from that of the parents just as, in Augustine's words, a flame from another flame); and, last, creationism, in which the soul of every human being is created and infused into the body by God. Creationist animation may be immediate or delayed, depending on whether the rational soul is infused at the moment of impregnation or (as Aquinas claimed) only when the body has reached the necessary degree of development, after the vegetative and sensitive souls.[74] The latter position was revived in the sixteenth century with the circulation of the "new Galen" and particularly the previously unknown treatise on the views of Hippocrates and Plato (*De placitis Hippocratis et Platonis*).

---

73. In theory, Latin makes a distinction between *anima* and *animus.* Latin authors also used other terms for *animus,* such as *anima rationalis, mens, ratio,* and *logos.* When Chauvin wrote "animus est, quo sapimus; anima, qua vivimus" (*Lexicon,* p. 46), he was simply copying a definition attributed to the fourth-century grammarian Nonius and adopted by lexicographers such as Antonio de Nebrija in his *Vocabulario de romance en latín* (1516) and Robert Estienne in his *Thesaurus Linguae Latinae* (1543). The distinction became blurred by usage, however, and *anima* was frequently used in the sense of *animus. Spiritus,* the translation of *pneuma,* tended to replace *animus* and became widespread in the vocabulary of the Christian church. In the sixth century, Isidore of Seville declared that *anima* and *animus* were the same thing considered from two different perspectives: *anima* is that which gives life, and *animus* comprises the internal faculties (*Etymologiae* XI.1.11).

74. J. M. da Cruz Pontes, "Le problème de l'origine de l'âme de la patristique à la solution thomiste," *Revue de théologie ancienne et médiévale,* 31, 1964, 175–229.

Joannes Velcurio, a professor of natural philosophy at the University of Wittenberg, noted in his *Commentarii in Aristotelis Physicam* of 1540 that while physicians believe the embryo is initially fed and feels through the mother's soul, theologians (and Velcurio) consider the fetus to be a *corpus animatum* from the start, sustained at first by the vegetative soul and subsequently able to feel through its own sensitive faculties. According to Velcurio, physicians and theologians both agreed that God infuses the rational soul into the fetus in the sixth month.[75] In 1598, in his treatise on women's diseases, Jean Liébault (c. 1535–1596) described how the souls are formed one after the other: first the natural, or "conforming," soul, then the nutritive, sensitive and "moving," soul. As for the "intelligent and rational" soul, Liébault claimed it is "created out of nothing, not engendered," and infused in an instant into the embryo's body in the third or fourth month, after the heart and brain have been formed.[76]

While the question of the exact moment of infusion remained open, it was finally decided, at least within the Church of Rome, that the process took place all at once. In his *Quaestiones medico-legales* of 1620, a work long considered authoritative, the papal physician Paolo Zacchia (1584–1659) followed a developmental schema, insofar as he accepted that the human being lives intitially like a plant. But he differentiated between the different modes of transmission of souls: the vegetative and sensitive souls arose *ex traduce per semen a progenitoribus*, while the rational soul was created by God *ex nihilo* and *de novo* and was present from the moment of conception.[77] Zacchia's position corresponded to the doctrine of the 1513 Lateran Council: since the rational soul is the form of the body and the animating principle of the human being, the fetus's first movements must necessarily be attributed to it.

Goclenius's *Psychologia* is almost entirely taken up with these debates. The first chapter is a lengthy account of "the perfection of man." The author Hermann Vulteius, a jurist and theologian from Marburg, began with several anthropological commonplaces: the human being is composed of a spiritual and immortal soul and a material and corruptible body; the soul is

75. Vivian Nutton, "The anatomy of the soul," p. 143.

76. Jean Liébault, *Trois livres appartenans aux infirmitez et maladies des femmes* (Lyon: Jean Veyrat, 1598), bk. 3, chap. 16 ("Du temps que l'enfant formé reçoit l'ame"), pp. 666–667.

77. Paolo Zacchia, *Quaestionum medico-legalium tomus posterior: quo continentur liber nonus et decimus . . .* (Lyon: Ioan. Ant. Huguetan, & Marci-Ant. Ravaud, 1661), liber IX, "Titulus primus (De foetus humani animatione)." For a modern scholarly edition, see Zacchia, *Die Beseelung des menschlichen Fötus: Buch IX, Kapitel 1 der "Questiones medico-legales,"* ed. and trans. Beatrix Spitzer (Cologne: Böhlau, 2002).

the form of the body; it is simple and indivisible, but has three faculties, the vegetative, sensitive, and intellective; the latter faculty, and particularly its higher part, the *logos* or *ratio*, distinguishes humans from animals. Thereafter Vulteius specified the disciplines making up philosophy (*psychologia* is not mentioned), expounded upon moral issues and virtues, examined the arguments concerning the origin of the soul (favoring, like Goclenius, creationism), gave an overview of human anatomy, and ended on a legal and political discussion of the family and the republic.

The other chapters form a balanced anthology on the origin of the *animus*: in many cases the work comes down against traducianism, but its supporters are given a fair amount of space also. Creationists maintained, for instance, that making animation happen *ex traduce* contradicts the spirituality of the soul; traducianists replied that if God created each individual soul directly, he would be—implausibly—responsible for the transmission of original sin.[78] Most authors appealed to the authority of Aristotle, the church fathers, and the Bible. Some, like Melanchthon's son-in-law, the physician Caspar Peucer, who tended toward traducianism, combined medical, philosophical, and theological arguments.[79] In the end, the issue is left unresolved, and since Goclenius himself wanted to find an explanation somewhere between creation and generation, he advised "pious scholars" to investigate not how the *animus* enters the body but rather how the immortal soul could leave its mortal envelope uncorrupted.[80]

Shortly afterward, in 1606, "psychology" was used again in a similar sense by the Aristotelian physician Fortunio Liceti (1577–1657). His most famous work was on monsters, but among his many other publications (including one in support of spontaneous generation) there was a "human psychology." Liceti reviewed points of disagreement concerning the origin of the soul and argued that God creates the rational soul at the moment of conception but infuses it into the body some forty days later.[81]

In summary, no new approach to the study of the soul and no intention to define an autonomous area of knowledge emerge from Goclenius's *Psycho-*

78. The most detailed argument on this issue is that of the Hungarian theologian Petrus Monedulatus Lascovius, "Dissertatio de problemate hoc: An anima rationales, sicut corpora, per seminalem traducem propageretur: an vero quotidie a Deo creatae, corporibus nascentium infundantur?" (in Goclenius, *ΨΥΧΟΛΟΓΙΑ*). He also treated it in depth in the first book of his *De homine magno illo in rerum natura miraculo et partibus eius essentialibus* (Wittenberg, 1585).

79. Caspar Peucer, "De essentia, natura et ortu animi hominis," in Goclenius, *ΨΥΧΟΛΟΓΙΑ*.

80. Goclenius, "De Ortu animi," in *ΨΥΧΟΛΟΓΙΑ*, pp. 377–378.

81. Fortunio Liceti, *ΨΥΧΟΛΟΓΙΑ ΑΝΘΡΟΠΙΝΗ, sive de ortu animae humanae libri III: In quibus multa Arcana ac Secreta Naturae, tum de Semine, tum de Foetu, ut & assimilatione parentum & liberorum, panduntur ac revelatur . . .* (Frankfurt: Johann Saur, 1606).

logia or from the works of Liceti or others. Goclenius's edited volume is probably the type of work Christophe de Savigny (a French nobleman whom we discuss in chap. 3) had in mind when he advised those wishing to know more about the rational soul to "turn to naturalists and physicians," and to read "those who have written on the philosophers' disputes concerning the rational soul, as well as theologians' discussions of the creation of the soul."[82] The same can be said of a disputation Goclenius chaired in 1596, on why the body stops growing at a particular moment; the issue was considered to be one of the trickiest in physiology, but it was described as "psychological" because the soul is the efficient cause of physical growth.[83]

* * *

The fact that *psychologia* figured only in the title of Goclenius's work and that it was never itself glossed suggests that its function was more stylistic and didactic than conceptual. Goclenius himself did not include the term in his *Lexicon philosophicum*. Such an omission is significant in the light of his extensive presentation of the *scientia de anima* and the divisions of philosophy. The word "psychology" would doubtless have figured in the *Lexicon* if it had been the expression of a new orientation, as was the case of "ontology," coined to refer to the conception of metaphysics as a general science of the *ens reale*, distinct from the special science of immaterial things.[84] The *Lexicon*'s article on the soul and its accompanying diagram (fig. 2.4) confirm that Goclenius had no intention of reforming the *scientia de anima*.[85]

From top to bottom and left to right, the diagram first mentions authors for whom the soul is nothing at all (as enumerated in Cicero's *Tusculan Disputations*). Philosophizing about the soul, Goclenius went on, implies

---

82. Christophe de Savigny, *Tableaux accomplis de tous les arts liberaux, contenans brieuement et clerement par singuliere methode de doctrine, vne generale et sommaire partition des dicts arts, amassez et reduicts en ordre pour le soulagement et profit de la ieunesse* (Paris: par Iean & François de Gourmont freres, 1587), unpaginated.

83. Rudolph Goclenius, *Disputatio philosophica quadripartita: Prima Logica: De natura relatorum; Secunda Psychologica, de quaestione: cur accretio in adultis cesset; Tertia Ethica: An ignorantia excuset peccatum; Quarta Historica & Physica: an iris fuerit ante diluuium* (Marburg: typis Pauli Egenolphi, 1596), "Quaestio Psychologica De Accretione, quamobrum in adultis cesset."

84. Goclenius defined ontology as "philosophia de ente seu de transcendentibus" (*Lexicon*, "Abstractio," s.v. "Abstractio materiae," p. 16). See S. Kramer, "Ontologie," in Ritter et al., *Historisches Wörterbuch*, vol. 6, col. 1189; and Ulrich Gottfried Leinsle, *Das Ding und die Methode: Methodische Konstitution und Gegenstand der frühen protestantischen Metaphysik*, 2 vols. (Augsburg: Maro Verlag, 1985), vol. 1, pp. 175–195.

85. Goclenius, *Lexicon*, s.v. "Anima."

Anima Philofo-phantibus

Nihil prorfus:Sic Pherecrates, & Dicæarchus, tefte Cice-ron. 1. Tufcul.quæft.

Aliquid

Subftantia, ὑφιϛάμι-νον,ὐσία

σωματικῶς, ἔνυλος, ὑλώ-δης,

σῶμα αὐγοειδὲς ᾳ ἀιθερῶδες : fpiritus tenuis difperfus per omne corpus animantis.

Halitus igneus,eua-poratio, fanguis: Materia tali modo temperata & con-cinnata,id eft,crafi affecta.

Ἀσώματος ἄϋλος

φαιρομένη in fpiritu vitali & animali.

Separabilis à corpo-re.

Accidens ἀνυπόϛα-]ον,ἔμφασις,

ἁρμονία elemétorum, numerus har monicus.

κρᾶσις cerebri vel cordis, de qua-tuor corporibus coagmentatum crama.

συγγυμνασία αισθήσεων, id eft,exer-citium quinq; fenfuum fibi con-fonum.

Intelligétia & fenfus interiores:qui δυνάμεις funt,vel ἐνέργειαι.

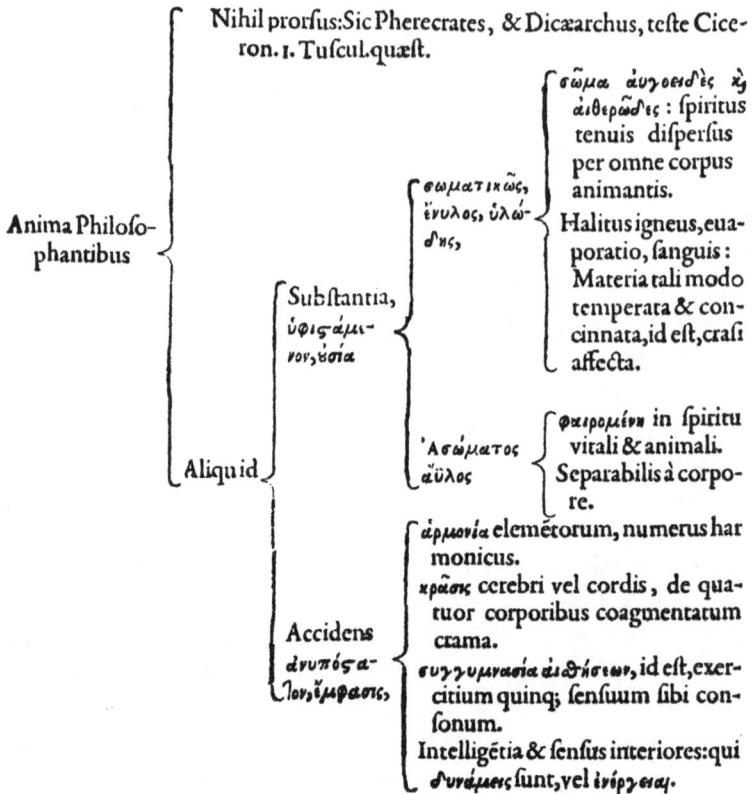

Fig. 2.4. The different ways of philosophizing on the soul. From Rudolph Goclenius, *Lexicon philosophicum* (Frankfurt, 1613), s.v. "Anima," p. 103.

examining it as a substance or in terms of its accidents (that is, things that happen to it but are not part of its essential definition). As substance, it is either incorporeal, immaterial and separable from the body, or corporeal and joined with matter; various authors from antiquity assimilate it to a subtle spirit, a breath of fire or a vital spirit in the blood. The discussion of the soul's accidents concerns the properties pertaining to its union with the body, such as the temperaments or the sensitive and intellective faculties. True to Goclenius's spirit, the article itself ends on a conciliatory compari-son of Plato and Aristotle.

It is once again clear that Goclenius had no interest in naming a new science or giving a new orientation to the theory of the soul but simply wanted to use an erudite neologism for his anthology on issues relating to the soul. His book was nonetheless a significant landmark in the history of a word which was often used subsequently in Protestant university circles, and which would later—but only in the eighteenth century—be discussed as a problematic concept.

Perhaps the first work by a single author to bear the title *Psychology*, and with the word printed in Latin characters, was *Psychologia anthropologica, sive animae humanae doctrina*, published in 1594. It was the second work of Otto Casmann (1562–1607), the first being a commentary on the logic and dialectic of Ramus and Melanchthon. Casmann was at the time a teacher at the (Calvinist) gymnasium in Steinfurt; he was Goclenius's disciple and a proponent of the Ramist method and went on to write other works on theology, logic, and natural philosophy.[86] For Casmann, anthropology, or "the doctrine of human nature," can be divided into two branches. The first is psychology, which is concerned with the nature of *anima logica*, that is, the rational soul.[87] The second is "somatotomy," or anatomy, to which the author devoted a separate treatise in 1596.[88] Casmann was at least one of the thinkers whom Timpler, his successor at Steinfurt, must have had in mind in his critique of the anatomy-psychology distinction.

At first sight, Casmann reserved the term *psychologia* for the study of the rational soul, which he characterized as substantially united with the human body.[89] He accordingly addressed the usual questions raised in a *scientia de anima*: Is the rational soul the form of the body? Is the soul a substance, or is it nothing more than its faculties? Does it have divisions and is it divisible? But Casmann's *Psychologia* had two parts. One indeed discussed the *facultas logica* and, in addition to the questions mentioned, addressed life, death, the intellect and how it works in conjunction with the body, and, last, whether language is properly "psychological."[90] The other part, which was much longer than the first, examined the *facultas aloga*, the soul's vital and animal functions.[91] It addressed generation and conception,

86. The most exhaustive study on Otto Casmann remains Dietrich Mahnke's "Beiträge zur Geistesgeschichte Niedersachsens," pt. 1, "Der Stader Rektor Casmann," *Stader Archiv*, Neue Folge, Heft 4, 1914, 142–190 (pp. 181–185 on psychology).

87. Otto Casmann, *Psychologia anthropologica; sive animae humanae doctrina . . .* (Hanau: apud Guilielmum Antonium, impensis Petri Fischeri Fr., 1594), pp. 1–2, 21–22.

88. Otto Casmann, *Secunda pars anthropologiae; hoc est de fabrica humani corporis methodice descripta* (1596). His *Somatologia physica generalis* of 1598 is a general science of bodies, their essence, and their manifest and hidden properties.

89. More precisely, Casmann described the soul and the body as "in unam hypostatin" or "in unum hyphistamenon unitae" (*hyphistamenon* is the same as *hypostasis*: subsistence, being, substance, foundation); *Psychologia*, pp. 1, 2, 8–9, 22–23.

90. He decided it was: see cap. 7, quaestio 1 ("An sit psychologicum praecipere de sermone: hoc est, an facultas loquendi insit animae, & hominis sit naturalis"), ibid., pp. 141–142.

91. Ibid., pp. 151–152.

the formation of the fetus (including controversies over the moment of animation, for which Casmann referred to Goclenius), nutrition and growth, the senses, imagination and memory, breathing and the pulse, the "appetites" and affections that prompt us to good or evil, and finally sleep and the faculty of movement.

The publication in 1662 of what to my knowledge is the first philosophical lexicon to include the entry "Psychologia" confirms the as-yet-ill-defined contours of the term's referent. Its author, the Lutheran Johannes Micraelius (1597–1658), a professor of philosophy and theology at Stettin, defined psychology as "doctrina de anima," while his article PHILOSOPHIA specified that it is "de anima separabilis."[92] Psychology, then, seems to deal with the rational soul not as separated from the body but only as *separable*; this would differentiate it from pneumatology and confirm its status as a science of the *animus* when joined to a living human being. The article ANIMA, however, expanded this field. After recalling the Aristotelian definition and stressing that the soul is one and immaterial, Micraelius briefly discussed its vegetative, sensitive, and rational functions. He opted for traducianism, declared that the soul was indivisible and that the human soul alone is immortal, and concluded: "*Vide de his omnibus in ψυχολογιαν.*"[93] In the seventeenth century, then, the term "psychology" referred to the study of the soul in all its aspects—a vast domain, with borders sufficiently elastic to extend from physiology to mysticism.[94] Despite its imprecision, the word

92. Johannes Micraelius, *Lexicon philosophicum terminorum philosophis usitatorum* (1662; Düsseldorf: Stern-Verlag Janssen, 1966), cols. 1165 and 1005.

93. Some of the theses defended around 1650 under Micraelius's tutelage show that this vocabulary merely reflected contemporary usage. In one case, the author of a thesis with "psychology" in the title first addressed how the soul should be studied (within theology according to divine Revelation or within philosophy, *ex naturae lumine*, and either metaphysically [according to its essence], or "physically" [*in ordine ad corpus*]). He went on to address its (Aristotelian) definition, its unity, its seat (*tota est in toto*), its origin (traducianist), its faculties (vegetative, sensitive, and rational), and finally the rational soul and its separation from the body. See *Psychologia per theses succintas de anima humana ejuisque potentiis et operationibus, quibus multae, eaque difficiles quaestiones circa illam doctrinam occurentes tanguntur . . .* (Stettin: Typis Georgii Gœtschii, [ca. 1650]), praes. Johannes Micraelius, auctor & respondens Caspar Voigth.

94. *Psychologia vera* is the title of the forty replies given by Jakob Böhme (1576–1624) in 1620 to his friend, the physician Balthazar Walther, concerning the soul: ΨΥΧΟΛΟΓΙΑ *vera I[acobi] B[oehmi] T[eutonici] XL: Quaestionibus explicata et rerum publicarum vero regimini: ac earum MAIESTATICO IURI applicata . . .* (Amsterdam: Apud Iohann. Ianßonium, 1632). The title is taken from Johannes Angelius Werdenhagen, the author of the Latin version in which the text became known. See Jacob Böhme, *PSYCHOLOGIA VERA, oder Vierzig Fragen von der Seelen, ihrem Urstande, Essenz, Wesen, Natur und Eigenschaft was sie von Ewigkeit in Ewigkeit sey . . .* , in J. Böhme, *Sämtliche Schriften*, ed. August Faust (Stuttgart: Fr. Frommans Verlag, 1942), vol. 3. For a clear presentation, see Alexandre Koyré, *La philosophie de Jacob Boehme* (1929; Paris: Vrin, 1971), sec. 3. Koyré notes that the work's title does not correspond to its

suggested a field of possibilities whose contours became increasingly clear: either "psychology" referred to the science of the soul *qua* animating principle of all living beings, or its sense was restricted to the rational human soul united with the body. With the collapse of Aristotelian frameworks, the second sense came to dominate. That is why, by the end of the seventeenth century, psychology and anthropology had switched places: as long as psychology's object was the animating principle of all living beings, it naturally encompassed anthropology, but as soon as psychology was redefined as a science of the human mind, it became a branch of anthropology.

---

content, and if one looks at the common uses of the term at the time, he is right about this. The same could be said of a book by Father Noël Taillepied (1540–1589)—who was known above all for his anthologies on the "antiquities and peculiarities" of various French towns—*Psichologie, ou traité de l'apparition des esprits, à sçavoir des âmes séparées, fantosmes, prodiges et accidents merveilleux qui précèdent quelquesfois la mort des grands personnages ou signifient changement de la chose publique* (Rouen, 1588). Its later editions even omitted "psychology" from the title. But this is precisely why stating that Böhme's *psychologia vera* represents a "different birth" of psychology (Mengal, "Naissances," §V) is to confuse the word with the concept. The use of the term in a mystical or hermetic context was unusual and shows that, before "psychology" named a specific science, it could refer to any and every discourse on the soul.

CHAPTER THREE

From the Science of the Living
Being to the Science of the Human Mind

By the mid-seventeenth century, psychology covered, in an Aristotelian and Galenic framework, everything relating to the soul in its union with the body. Its redefinition as the science of the *anima separabilis* (the rational soul, separable from but united with the body) had not yet come to dominate. When it did, a break occurred in the semantic history of psychology, signaling the disintegration of the Aristotelian framework within which psychology had been the generic science of living beings. As long as every science of living beings was a branch of psychology, the latter remained a generic discourse. This generic character determined the scope, function, and limits of psychology, as well as its relation to the other sciences and its articulation with other fields of knowledge. Particularly problematic was the human soul's immateriality, which complicated the relation of psychology to "physics," or the science of nature, and raised the question of psychology's status: Was it wholly part of physics, or was it also a metaphysics of the rational soul? And what would psychology's position be once the soul ceased to govern vegetative and sensitive functions and was restricted to the mind?

PSYCHOLOGY AS THE GENERIC
SCIENCE OF THE LIVING BEING

Since the soul was that which endowed the *empsukhon* with life, psychology was the generic science of the living being. We should stress that it was "generic" rather than "general," since the term "psychology" did not refer to a universalizing discourse but to a genus, in the logical and taxonomic sense of the term, of which the particular sciences of the living being were species.[1] This is the sense which the authors we have examined up to now, including Goclenius,[2] attributed to psychology, and it found a particularly elegant illustration in the *Complete Tables of the Liberal Arts* by Christophe de Savigny, a nobleman from the Ardennes.[3] De Savigny's diagrams reflected Ramist method and were in the spirit of a humanist like Guillaume Budé, who protested against the fragmentation of knowledge in faculties and professions and insisted on the "mutual connection," coherence, and affinity of the sciences and disciplines within an "encyclopedia," or structure of "circular learning."[4]

The center of the picture (fig. 3.1) shows the branches of philosophy, that "knowledge and science" of things human and divine. The sciences which investigate the nature of things have either incorporeal or corporeal objects. Metaphysics addresses the first of these, physics the second. Physics investigates bodies either quantitatively (as in arithmetic and geometry) or qualitatively. Bodies may be either simple or composite; the first are unchanging, such as the sky and the stars, the second changeable, such as the four elements. Composite bodies are either inanimate (meteors and metals) or "animate and alive." Two sciences explore the latter: psychology and medicine.

Here, psychology is the "history" or description of plants and animals. Animals are divided into creatures without reason and human beings, capable of intelligence and reason. Man is composed of a soul and a body, his earthly and corruptible part, but which is also the "abode and organ of the soul." What differentiates him from beasts is the possession of a "rational

1. Mengal has a different view: "Constructing a general theory of living creatures embracing at once plants, animals, and the corporeal part of man was probably beyond the reach of the late sixteenth century"; "Defined as the general science of living creatures, . . . *Psychologia* claimed to unite fields as vast as botany, zoology, and anthropology. It is clear that the means to realize this ambition were cruelly lacking" (Mengal, "Naissances," pp. 360–364).

2. Goclenius wrote a preface commending a work in which Rudolph Snellius (whose commentary on Melanchthon's *Liber de anima* we have mentioned) used the term in this sense. See R. Snellius, *Snellio-ramaeum philosophiae syntagma . . .* (Frankfurt: Peter Fischer, 1596), p. 27. Psychology, as the science of living bodies, was divided here into phytology and zoography; the latter was concerned first and foremost with anthropology—the science of man, his soul, and his body.

3. Savigny, *Tableaux*. The citations come from the explanations given after the diagrams "Encyclopédie" and "Physique," under the titles of, respectively, "Partition générale de tous les arts libéraux" and "Partitions de la physique."

4. Guillaume Budé, *De l'institution du prince . . .* (Paris: Imprimé à L'Arrivour . . . par maistre Nicole, 1547), p. 88; see Franco Simone, "La notion d'encyclopédie: Élément caractéristique de la Renaissance française," in Peter Sharratt, ed., *French Renaissance studies 1540–70: Humanism and the encyclopedia* (Edinburgh: Edinburgh University Press, 1976), p. 246.

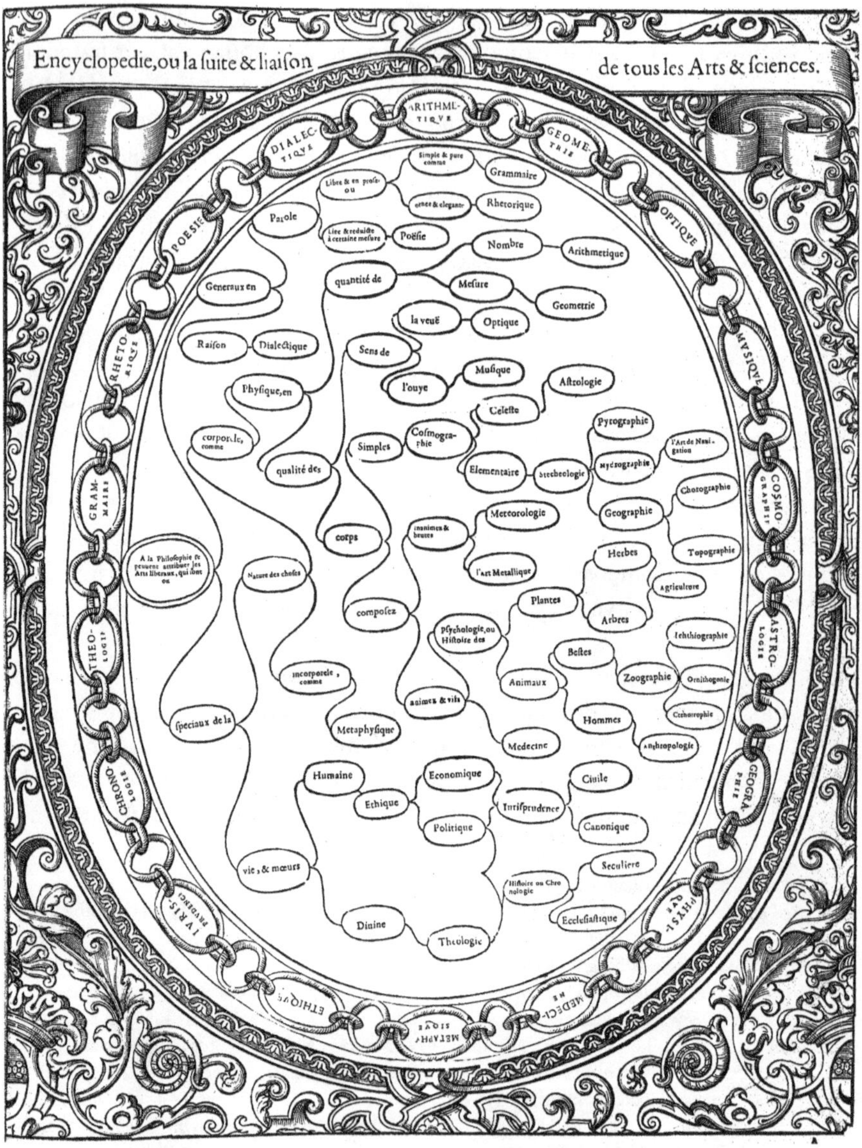

Fig. 3.1. "Encyclopédie, ou la suite & liaison de tous les Arts & sciences." From Christophe de Savigny, *Tableaux accomplis de tous les arts liberaux, contenans brieuement et clerement par singuliere methode de doctrine, vne generale et sommaire partition des dicts arts, amassez et reduicts en ordre pour le soulagement et profit de la ieunesse* (Paris, 1587). (Houghton Library, Harvard University, call number Typ 515.87.771.)

soul," a spiritual substance which "subsists and remains immortal" after its separation from the body. Although de Savigny describes it as a "rational" faculty, the human soul shares certain faculties with the souls of other living creatures: the vegetative faculty, which brings together the generative, nutritive, and growth functions; the sensitive faculty, which operates through the internal and external senses; and the appetitive and locomotive faculties.

In short, psychology examined living creatures in their corporeal functioning (largely dependent on the vegetative, sensitive, appetitive, and locomotive faculties), as well as their spiritual operations, which were the domain of the rational soul. The section dealing with animals was organized accordingly into zoography and anthropology, the latter being the science which "teaches the properties of man." As for medicine, the practical science of health, it was equally part of psychology: "The investigation into animate and living things encompasses the history of plants and animals, on which medicine also depends." The fact that psychology was a metadiscipline, a genus of which the sciences of the living being were the species, explains why anthropology was part of psychology and not vice versa.

There are echoes of this—often implicit—use of the term "psychology" right up to the end of the eighteenth century. In 1686, an English encyclopedia for gentlemen featured an illustration which looked exactly like de Savigny's.[5] A century later, in the same year as the storming of the Bastille, a "summary" of studies for the future man, dedicated to the "representatives of the Nation," divided physiography (the science of nature) into psychology (*psucologie*) and vegetology. "Psychology studies that which, being produced by generation, is endowed with life and voluntary movement"; mental phenomena were the domain of a "spiritology" that "rises above matter" and investigates sensation, the passions, the will, the intellect, and reasoning, thereby guiding man toward metaphysics. Metaphysics did not study only the mind and thought, that is, the properties and operations of the rational soul, but it also "attempt[ed] to discover if there are spiritual beings other than the one that guides man."[6]

For de Savigny, therefore, psychology was a metadiscipline lacking content of its own but encompassing several sciences and serving as an introduction to the study of living beings. While this propaedeutic function in

5. Richard Blome, *The Gentleman's recreations . . .* (1686); illustration in Richard Yeo, *Encyclopaedic visions: Scientific dictionaries and Enlightenment culture* (Cambridge: Cambridge University Press, 2001), p. 11.

6. Nicolas-Gabriel Clerc, *Abrégé des études de l'homme fait, en faveur de l'homme à former* (Paris: chez Maradan, 1789), vol. 2, pp. 275, 279–280.

no way diminished its importance, it nevertheless meant that the science of the soul was generic and not specific. Seventeenth-century courses in Scholastic philosophy state this clearly when they say that *de anima* books do not constitute a "complete treatise" but are only a part of what is to be studied concerning the living body.[7]

A second determinant factor in the bipartite organization of psychology was the division of the human being into a soul and a body which are united but separable. The *physicus* examined the soul, including the intellective soul, in its union with the body, whereas metaphysics was concerned with it in its separated state. The metaphysical discourse on the soul properly belonged to *pneumatology* (from *pneuma*, "spirit"), the science of spiritual beings as such, both uncreated (God) and created (angels, demons, and the human soul). This division, which corresponded to the Aristotelian classification of substances (*Metaphysics* XII), posed the problem of the unity of the science of the soul and its place alongside other fields.

PSYCHOLOGIA AND *EMPSYCHOLOGIA*

The distinction made within philosophy between pneumatology and a physics of the soul was a fundamental one, and a commonplace in academic circles. In the mid-seventeenth century, for example, the "gold mine of distinctions" proposed by the Leipzig professor Johann Adam Scherzer (1628–1683) to help with disputations included the following objection: "Natural philosophy (*physica*) investigates the soul. Therefore pneumatology cannot investigate the soul." The reply was that "the naturalist (*physicus*) investigates the soul as something relative. Pneumatology, by contrast, as something absolute. Hence, neither can infringe upon the other." The quality of "man" designates the "absolute" being of a male individual of the human species; the quality of father, his "relative" being. Likewise, pneumatology is concerned with the soul in its spiritual essence, whereas physics is concerned with the soul in its union with the body.[8] The potential conflict of faculties is thus avoided.

7. For example: "Libri isti de Anima non sunt integer tractatus, sed pars quaedam tractatus de corpore animatus"; Ioannes a S. Thoma, *Cursus philosophicus thomisticus* (1635), cited by Knebel, "Scientia," p. 133, n. 15.

8. Johann Adam Scherzer, *Vade mecum sive Manuale philosophicum* (1654), ed. of 1675 (Stuttgart-Bad Cannstatt: Fromann-Holzboog, 1996), pt. 4, "Aurifodina distinctionum" (littera A, distinctio X; lit. E, dist. IV). The *Vade mecum*, which went through several editions up to the beginning of the eighteenth century, was used as an Aristotelian-Scholastic school textbook. It included many Catholic sources, including Arriaga's *Cursus*, and was widely used within Protestant circles.

Many academic textbooks illustrated the distinction between a physics and a metaphysics of the soul. A perfect example of this can be found in one of the great Aristotelian textbooks of the late Renaissance, the *Corpus of Philosophy* by Scipion du Pleix (1596–1661). The last book of *Physics; or, The Science of Natural Things*, which went through some twenty editions between 1603 and 1645, is devoted to an exploration of the soul, "that most excellent, most certain and most useful part of physics."[9] Du Pleix replied in the affirmative to the question of whether the soul is an object of physics: since the major faculties and functions of living bodies depend on the soul, physics, which studies these, cannot rightly ignore it. However, insofar as physics only deals with composite things, the soul is not its "first and direct object" but belongs in it only due to the "close connection" of soul and body. Physics consequently views the soul as "form" in relation to matter, while consigning to metaphysics the soul "as substance entirely separate from matter and purely intellectual." Physics does not, however, stop at the vegetative, sensitive, and locomotive faculties but includes those which the human being alone possesses by virtue of his rational soul, namely, the understanding, the will, and memory, always considered in their union with the body.

Another treatise, *Metaphysics; or, Supernatural Science*, was published in 1610 and went through about ten editions up to 1645. The soul takes up the fifth book.[10] Books 5–11 together constitute a pneumatology (du Pleix does not use the word), with discussion of angels, demons, and God following on from discussion of the soul. The human soul, when considered as "part of a natural whole" (§ 3), is an object of physics. But it is also an object of metaphysics insofar as it can be treated "simply and absolutely in itself, distinct from matter and the contagion of natural things, being in this way a thing wholly spiritual and supernatural" (§ 4). In this case metaphysics examines the soul's properties, its infusion into, and its separation from the body, its fate after death, the knowledge it possesses when disembodied, and its reunion with the body upon resurrection.

While du Pleix did not use the term "psychology," his physical science of the soul corresponds exactly to de Savigny's sense of the term. In France, the word did not spread rapidly. In Germany, however, it was used frequently in university circles after the publication of Goclenius's

---

9. Scipion du Pleix, *La Physique, ou science des choses naturelles* (1603), ed. of 1640 (Paris: Fayard, 1990), bk. 8, "contenant la cognoissance de l'Ame."

10. S. du Pleix, *La Métaphysique, ou science surnaturelle* (1610), ed. of 1640 (Paris: Fayard, 1992), pt. 2 (= bk. 5), "qui est de l'Ame séparée," chap. 1, "Comment l'ame est de l'objet de la Physique et de la Metaphysique."

Psychologia in 1590. Its meaning at the time was sometimes closer to physics and sometimes to metaphysics. "Psychology" either investigated the rational human soul in its separation from the body, in which case it belonged to pneumatology, or, more frequently, it investigated the soul (in its entirety, or only the rational soul) in its union with the body, and so constituted, along with anatomy, one of the branches of anthropology. This distinction was sometimes expressed as the difference between *psychologia* and *empsychologia*, two words derived from the Greek for animate and inanimate, the "endowed with soul" and the "deprived of soul."

The work of the metaphysician Johann Heinrich Alsted (1588–1638), a student of Goclenius at Marburg, provides an instructive example of the lexical and conceptual situation sketched above.[11] In his *Encyclopaedia* (1630), a widely circulated monumental tribute to Ramist method, the psychological material can all be found within the theoretical part of philosophy (the other part being practical philosophy, comprising ethics, economics, and politics). Theoretical philosophy included metaphysics (which discusses the categories), pneumatology or *pneumatica* (of which psychology is a branch), physics (to which anthropology is linked), and mathematics.

Such a distribution of the disciplines, represented in a wealth of tables, proves to be incomplete when compared to the text itself.[12] For *psychologia* was linked to many different fields. It was, for example, one of the branches of *anthropologia*, or *specialis physica de hominis natura*. "General" anthropology explored the essence of the human being in relation to external or internal causes. Internal causes could be material or formal, and concerned either generation and the body (the object of somatology) or the rational soul (the domain of psychology). "Particular" anthropology again took human existence as its objet, especially the faculties; among the latter, several were linked to the understanding (*facultas intelligendi*), which was also treated under psychology.

As a branch of anthropology, psychology thus had a single object, the rational soul, but considered from two different viewpoints: as a substance

11. On Alsted see Leinsle, *Das Ding und die Methode*, vol. 1, pp. 369–393; Howard Hotson, *Johann Heinrich Alsted, 1588–1638: Between Renaissance, Reformation, and universal reform* (New York: Oxford University Press, 2000). On Alsted's enclyclopedias and particularly on the relation among psychology, didactics, and mnemonics, see Wilhelm Schmidt-Biggemann, *Topica universalis: Eine Modellgeschichte humanistischer und barocker Wissenschaft* (Hamburg: Felix Meiner, 1983), pp. 100–139.

12. Johann Heinrich Alsted, *Encyclopaedia* (1630; Stuttgart: Frommann-Holzboog, 1989–1990), vol. 2, *Philosophia theoretica*, bk. 12, *Pneumatica*, pt. 4, "Psychologia"; bk. 13, *Physica*, pt. 5, "Empsychologia et Phythologia."

DELINEATIO PSYCHOLOGIÆ.

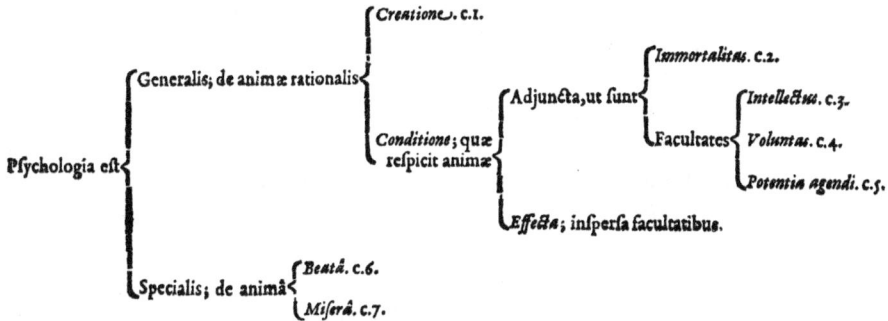

Pſychologia eſt {
- Generalis; de animæ rationalis {
  - *Creatione*. c.1.
  - *Conditione*; quæ reſpicit animæ {
    - Adjuncta, ut ſunt {
      - *Immortalitas*. c.2.
      - Facultates {
        - *Intellectus*. c.3.
        - *Voluntas*. c.4.
        - *Potentia agendi*. c.5.
    - *Effecta*; inſperſa facultatibus.
- Specialis; de animâ {
  - *Beatâ*. c.6.
  - *Miſerâ*. c.7.

Fig. 3.2. Outline of psychology. From Johann Heinrich Alsted, *Encyclopaedia* (Herborn, 1630), vol. 2, p. 667.

informing matter to make a living human being, and as an immaterial substance endowed with an intellective faculty. Alsted discusses this latter aspect of psychology in the context of pneumatology, the science of uncreated spirit (God, his existence and essence) and of created spirits (angels and the disembodied soul). Psychology was here the pneumatological doctrine of the *anima separata*, which, in Ramist fashion, was approached again from two different perspectives: the soul was either examined from the viewpoint of its immortality and its faculties (*intellectus, voluntas,* and *potentia agendi*), in which case it belonged to "general" psychology, or else was considered in its state of bliss or wretchedness (*anima beata* or *misera*), which was the theme of "special" psychology. These divisions gave rise to the diagram in figure 3.2.

Yet, as we have seen, for Alsted psychology was not only a science of the disembodied rational soul but also a general doctrine that included the soul's function as form of the living creature. The instability of his vocabulary called for conceptual distinctions. While the diagrams of the *Encyclopaedia* used *psychologia* for the two approaches, the text itself used *empsychologia*, a neologism coined by Clemens Timpler in 1610 to designate a part of natural philosophy.[13] The word was derived directly from *empsukhos,* the adjective qualifying the body when animated by the *psukhē*. We have already encountered it in Aristotle's statement that the ensouled (animate, *empsukhon*) is distinguished from the unsouled (inanimate, *apsukhon*) by

13. Ibid., p. 667.

its being alive (*DA*, 413a21). Timpler used *empsukhon*, which he found "not devoid of elegance," to refer to the physical theory of living bodies insofar as they are living.[14] *Empsychologia* therefore included zoology and phytology, the sciences of animals and plants.[15]

For Timpler, anthropology was the "physical" doctrine of man insofar as he is human, that is, endowed with a rational soul.[16] It focused principally on the intellect and the will but also on the capacity to speak, laugh, and cry. We have seen that Timpler did not develop a *psychologia*, because, while accepting that the rational soul is the specific "form" of the human being, he considered the human being to be an entity distinct from its components (the body and the soul). His metaphysics (which he separated clearly from theology and transformed into an ontology) discussed substances and substantial forms, including those of "nature" (defined as the substantial form of inanimate beings) and of the soul (the form of animate beings).[17]

Alsted, on the other hand, saw *psychologia* as the pneumatological science of the rational soul. *Empsychologia* involved for him the study of the soul as the animating principle of the body, with its vital, vegetative, and sensitive faculties. It examined the principles of the living body, as well as the vital faculties of generation, nutrition, growth, and health (*sanitas*). Yet despite his distinction between *psychologia* and *empsychologia*, Alsted ended up abandoning Timpler's neologism, integrated the "empsychological" material into psychology, and extended the latter into a universal *physica de natura animae* (see fig. 3.3).

When used in its most specific sense, *psychologia* was the science of the *anima evidens*, that is, of the soul's observable manifestations. It investigated both the "material" soul, the animating principle of vegetative

14. "Doctrina physica de corporibus animata, quatenus sunt animata"; Timpler, *Physicae seu philosophiae naturalis systema methodicum*, p. 1. He of course also created the term *apsychologia* to refer to the science of inanimate natural bodies (the sky, the stars, and the elements, and the "mixed" bodies such as winds, meteors, stones, and metallic substances). Joseph S. Freedman rightly refers to "animate physics" and "inanimate physics" in his *European academic philosophy in the late sixteenth and early seventeenth centuries: The life, significance, and philosophy of Clemens Timpler (1563/4–1624)* (Hildesheim: Olms, 1988), chap. 12.

15. Timpler, *Physicae seu philosophiae naturalis systema methodicum*, p. 61.

16. "Anthropologia est doctrina physica de homine, quatenus est homo"; "Homo est anima rationalis praeditum"; ibid., p. 253.

17. Timpler, *Metaphysicae systema methodicum*, lib. IV, cap. V, "De forma substantiali." Timpler thought the division into vegetative, sensitive, and rational soul was inadequate, since it overlooked the *anima specifica* which enables the differentiation of species within genera, "ex. gr. Leo, praeter animam vegetantem, sentientem & bestialem, habet etiam animam particularem, per quam est Leo"; ibid., p. 461.

PSYCHOLOGIA.

```
                          ┌ Generalis; de naturâ animæ in genere.
                          │   ┌ Mundi.
PSYCHOLOGIA               │   │                                              ┌ Corporibus cœleftibus.
eft Phyfica de na-  ┤     │   │            ┌ Occulta, feu pofitiva, in ┤ Elementis.
turâ animæ. Eftq́;         │   │            │                                 └ Mineralibus.
                    Specia-                                                   ┌ Vitæ.
                    lis; de                                                   │                              ┌ Princeps ┤ Nutrix.
                    animâ ┤  Partium                   ┌ Communiter │ Facultatis vita- │                     │          │ Auctrix.
                          │  mundi:     ┌ Materia-     │ ratione    ┤ lis; quæ eft    ┤                     │          └ Procreatrix.
                          │  ubi eft ┤  lis; quæ       │            │                  │          ┌ Miniftra ┤ Attractrix.
                          │  anima      │ confi-       │            └ Sanitatis.        │          │          │ Retentrix.
                          │             │ deratur ┤                                               │          │ Concoctrix.
                          │             │          │                ┌ Vegetativa; ┌ Prima: Vis germinandi.    └ Expultrix.
                          │             │          │                │ cujus facul- │                   ┌ Viriditate.
                          │             │          │ Singulariter;  │ tas eft     │ Orta: vis quæ    │ Phyllophoriâ.
                          │             │          │ ubi eft ani- ┤ │             └ cernitur in  ┤ Anthophoriâ.
                          │             │          └ ma             │                             └ Carpophoriâ.
                          │             │                           │                                      ┌ Senfus.
                          │             │                           │ Senfitiva; cu-  ┌ Senfitiva:ubi ┤ Judicium fen-
                          │             │                           │ jus facultas   │                └ fitivum.
                          │  Evidens; ┤ │                           └ eft           ┤ Appeti- ┌ Cupiditas.
                          │  eaque      │                                            │ tiva    └ Affectus.
                          │             │                                            │ Locomotiva.
                          │             │                                            └ Vis ┤ Refpirandi.
                          │             │                                                  └ Micandi.
                          │             │                                       ┌ Cognofcendi.
                          │             │                                       │ Cogitandi.
                          │             │                                       │ Docendi.
                          │             │                                       │ Difcendi.
                          │             │                                       │ Recordandi.
                          │             │          ┌ Intellectus, id │ Ratiocinandi.
                          │             │          │ eft, facultas ┤ Judicandi.
                          │             │ Immaterialis: ut eft      │ Inveniendi.
                          │             │ anima rationalis,         │ Disponendi.
                          │             └ cujus facultas eft ┤      │ Confultandi.
                          │                                         │ Contemplandi.
                          │                                         │ Approbandi.
                          │                                         └ Improbandi.
                          │                            Voluntas, id ┤ Appetendi & averfandi.
                          └                            eft, facultas └ Perfequendi & fugiendi.
```

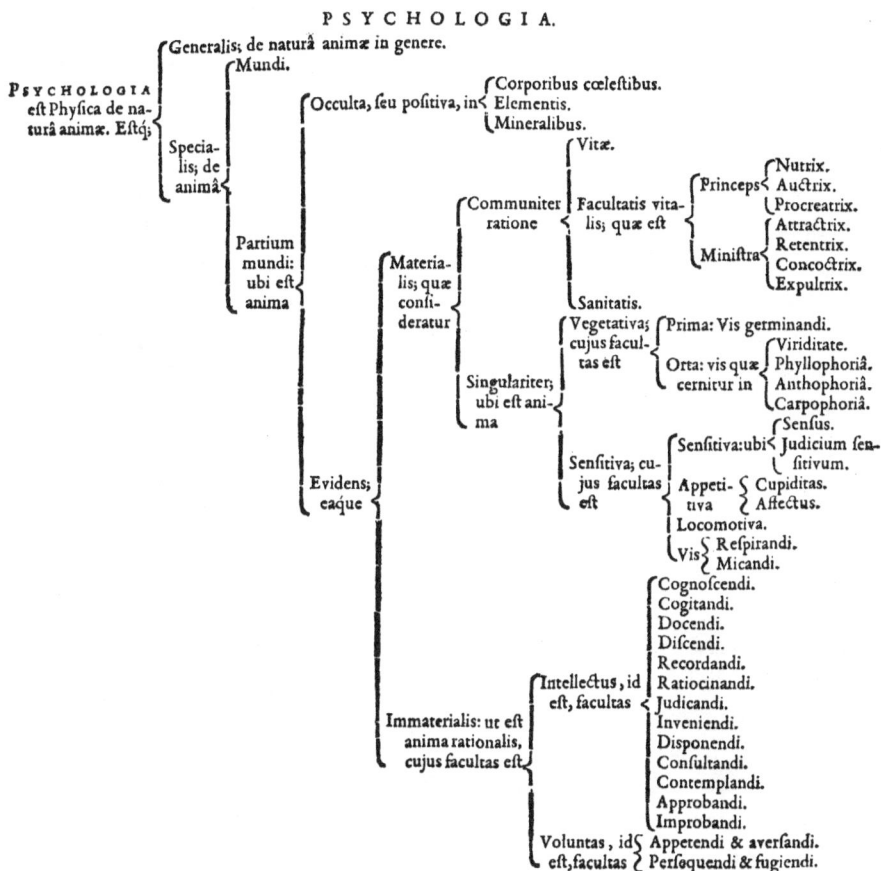

Fig. 3.3. The field of psychology. From Johann Heinrich Alsted, *Encyclopaedia* (Herborn, 1630), vol. 2, p. 796.

and sensitive life, and the "immaterial" rational soul endowed with the capacity of intellection and will. Once again, *psychologia* designated the generic science of the soul as it animates living and thinking beings. This generic science was distinct from pneumatology and referred principally to natural-philosophical disciplines. But, as already noted, Alsted's terminology was not consistent. The diagrams do much to restore order by getting rid of empsychology and extricating psychology from a pneumatology in which the human soul was examined alongside angels, demons, and God. Nevertheless, as du Pleix argued, although the soul, even when separated from the body, always maintains a relation to it—as form to matter—it can also be considered "as purely and simply a spirit [. . .] which can only be the object of Metaphysics."[18] The science of the soul was therefore grounded in different ontological orders, due to the very nature of the soul itself, and the articulation and divergence between *psychologia* and *empsychologia* was a fundamental and irreducible source of tension. This set of problems was hardly ever theorized, but it was explored by the Jesuit *doctor eximius* Francisco Suárez (1548–1617), one of the greatest figures of Renaissance Catholic Scholasticism, much read in Lutheran universities in the seventeenth century. He did not actually use the term "psychology," but what he wrote about the *scientia de anima* throws light on the epistemic autonomy of the science of the soul. Suárez argued that this science belonged to physics alone; du Pleix was explicitly targeting Suárez in maintaining the opposite.[19]

ON WHETHER *DE ANIMA* BOOKS CAN THEMSELVES CONSTITUTE A SCIENCE

In 1572, Suárez prepared a commentary on Aristotle's *De anima* for his university lectures. He began to look over these again toward the end of his life with a view to publication. The work, published posthumously in 1621 under the title *De anima*, is Suárez's most important text on psychology. In the systematic arrangement of Suárez's complete works, it was placed after books on spiritual creatures (angels) and the six days of Creation, and constituted the last book of three devoted to God as "author of all creatures." The treatises on God as Creator were intercalated between books about God *uno et trino* and the ends of man. Since the soul is what composite beings and purely spiritual entities have in common, the position of *De anima* reflects that of its object in the hierarchy of creatures.[20]

Suárez's *De anima* is divided into three parts: on the soul itself (book I, taking up a little less than a third of the treatise), its faculties (books II–V, a little less than two-thirds of the treatise), and the condition of the separated soul (book VI). The subject matter derives principally from Aristotle's work of the same name and was initially a commentary on it. The 1572 version, which was published for the first time only in 1978, differs considerably

18. Du Pleix, *Métaphysique*, pt. 2, chap. 1, § 5.

19. "En quoy qu'aucuns tiennent qu'en toutes façons elle est de l'objet de la Physique," with a reference to Suárez; ibid.

20. Francisco Suárez, *De anima*, in *Opera omnia* (Paris: L. Vivès, 1856), vol. 3. The works on Suárez's psychology which I have consulted do not address the question which concerns us here. See, however, Clare C. Riedl, "Suarez and the organization of learning," in Gerard Smith, ed., *Jesuit thinkers of the Renaissance* (Milwaukee: Marquette University Press, 1939).

from the 1621 version. It is of particular interest as regards the place and function of the *scientia de anima* in the organization of knowledge.[21] In the prooemium, Suárez defines the object of *de anima* books and sets out their position and rank within the sciences. His discussion shows us what is at stake in the distribution of psychological subject matter across different branches of knowledge, as well as the intellectual universe in which the question of the autonomy of the discourse on the soul took on meaning. His discussion leads to the issue of whether psychology is a physics or a metaphysics of the soul.

Suárez explained that physics is the science of sensible (observable) substances, some inanimate and some animate, some engendered (hence mortal) and others not. Physics must therefore include the study of the living being. But it is inconceivable, said Suárez, to separate a science of the parts from a science of the whole; the knowledge of the whole implies a knowledge of the parts, and setting out to know these is necessarily setting out to know the whole. Likewise, knowledge of the principles of the natural being (the object of physics) is knowledge of natural being itself. For Suárez, then, it followed that "these *de anima* books do not themselves constitute a science but are the introduction to the science of living beings," because they are concerned with the soul, which is the formal principle of all animate beings: "Ex quibus constat hos libros De anima non constituere per se scientiam, sed inchoare scientiam de viventibus" (§ 5).

It is clearly not that these books fail to define a body of knowledge but rather that they logically precede the sciences which investigate composite beings. However, some authors saw the living body, others, the soul, as the object of the *scientia de anima*. Suárez attempted to reconcile the two positions in arguing that an investigation into the soul must be the first part of a science of living beings, since it concerns their principle, and hence also their bodies. For Suárez, the *scientia de anima* can be considered a unified whole only insofar as it comprises three different sciences: the sciences of vegetative beings, of sentient beings, and of the thinking being. The *de anima* books, therefore, are only the beginning of a "triple science" (§ 9).

Suárez, moreover, questioned whether these books were properly part of physics. Since the existence of the rational soul demonstrates that the soul's operations are not entirely "physical" (§ 10), it would be justified to

21. Francisco Suárez, *Commentaria una cum quaestionibus in libros Aristotelis De anima* (1572), ed. and trans. Salvador Castellote et al. (Madrid: Sociedad de estudios y publicaciones, 1978–1981). I will cite the prooemium (vol. 1), referring to paragraph numbers in the text (§). Castellote's introduction compares the two versions; see also Rafael Gil Colomer, "El Tratado de 'anima' de Francisco Suárez," *Convivium*, 17–18, 1964, 128–141.

place the *scientia de anima* within metaphysics. The sensitive functions of
the soul are certainly material and performed by the body, but others, such
as the understanding of angels, are purely spiritual. Human understanding,
however, partakes of both materiality and spirituality. Though immaterial
in itself and able to do without a bodily medium, it depends to a certain
degree on the body. Man is therefore positioned between material and spiritual beings, and whereas investigation into the former is certainly part of
physics, the study of the latter transcends it.

As for man himself, it is hard to say unequivocally what science should
take him as its object, given the "difficulty" presented by the rational soul
(§ 12). One solution, Suárez explained, is to differentiate between two understandings of the rational soul: it can be conceived either as an immaterial substance, endowed with intellect and will, or as the form of the body
and as the source of material operations and of the understanding (which
is partly dependent on the senses). The study of man from the first perspective would belong to metaphysics; from the second, to physics. Both
approaches were justified, in Suárez's view, since, depending on the aspects
examined, the same object may be studied in metaphysics or in another
science. Nevertheless, Suárez concluded, "speaking in absolute terms," the
study of the rational soul should belong to physics. For matter and form are
incomplete when taken separately, and neglecting one or the other misses
the essence of the thing studied. Now the definition of the human being is
precisely that it is a composite of soul and body, and Suárez set out to determine the field to which it should belong.

Before returning to Suárez, let us briefly recall the Aristotelian understanding of such a definition. Take, for example, a house, a shelter which
protects against the elements. It is defined from a physical point of view by
its materials (stones, bricks, and wood), and from a dialectical point of view
by the form given to these materials in view of its finality. The naturalist
(*phusikos*) takes into account both matter and form. If he examines anger,
for example, he will do so materially (the boiling of the blood around the
heart) but also formally, as regards its cause and finality (the desire for re-
venge). Moreover, he treats them as motions of the body, or of a bodily part
or faculty, produced by a specific cause and oriented toward a specific goal.
The metaphysician (*philosophos*), by contrast, will study the form and the
properties separable from matter (*DA*, 403a). The definition of the human
being proceeds *per genus et differentiam*. For example, humans belong to
the category of living creatures; they have vegetative functions in common
with plants and animals, and sensitive functions in common with animals.
But the human being is the only animal to be endowed with reason: what

specifies him, therefore, is rationality, which is the *differentia* of the human species within the *genus* of living beings.

Suárez maintained that the *phusicus* treats the rational soul in terms of its ultimate difference, whereas the metaphysician does so in terms of its common predicates. In other words, the rational soul can be viewed either specifically, as the form of the human body, or generically, in terms of its immortality, immateriality, and indivisibility. It follows that the definition "Man is a composite of a body and a rational soul" comes from within physics, that the study of any being constituted of matter and form belongs unequivocally to natural philosophy, and that the "physician" cannot study one without the other (§ 13).

Suárez made other observations that, in his view, confirmed his conclusion (§ 14): the rational soul is essentially the form of the physical body[22] and its properties are partly dependent on the body; certain material affections, such as laughter, belong to man alone, and so necessarily derive from the rational soul. In short, the rational soul *is nature* since it is the intrinsic principle of the being it animates, and man is a natural being. For Suárez, it is basically mistaken to state that the rational soul belongs to physics as a principle of the vegetative and sensitive functions, but to metaphysics as a principle of intellection, since the intellective soul is the soul *of the body*, and the human being is constituted as human precisely by the soul (§ 15). Moreover, although the soul can exist without the body, it cannot do so *sine ordine ad corpus*, which is enough, for Suárez, to prove that it should be examined within the field of physics (§ 18). Hence the study of the soul in all its aspects belongs to natural philosophy. Natural philosophy does not address the understanding in itself (any more than any other *ens ut sic*) but rather the understanding as it operates in the body: "Non agit de intellectu ut omnino abstracto a corpore, sed ut intelligente in corpore" (§ 22).

Suárez did not question the legitimacy of discussing the disembodied soul within physics. He argued that, as Aristotle taught, when studying the rational soul, the naturalist is concerned with things that are separable from matter. But one cannot achieve "perfect knowledge" of the soul through one science alone, and since the soul is the bond between the intellectual and the sensible orders, its *perfecta cognitio* requires knowledge of the two (§ 24). Nevertheless, as Suárez stated again in his later *Disputationes metaphysicae* of 1597, the science of the soul belongs entirely to natural

22. We have already mentioned the doctrine that the rational soul is *per se et essentialiter* the form of the human body, decreed at the Council of Vienne in 1311 and reaffirmed at the Fifth Lateran Council in 1513.

philosophy. The *Disputationes* dealt with psychological issues only briefly, insofar as they were relevant to metaphysics. The rational soul was not discussed—whether as separate or united with the body—since Suárez maintained that treating it as though it were divided into several parts and linked to several sciences would be long-winded and cause confusion.[23] This stance was consistent with his notion of metaphysics as the science of being as such (*ut sic*), independent of theology.[24]

Suárez's arguments concerning the *scientia de anima* were closely linked to his analysis of the general question of *habitus* and the unity of a "science." Aristotle distinguished ten categories of being: substance, quality, quantity, place, time, relation, position, state (*habitus*), passion (affection or passivity), and act (action or activity). "Quality" defined that by virtue of which a person is said to be such and such, and *habitus* is a sort of quality (*Categories*, VIII). *Habitus* was defined as a lasting disposition (in which it differed from *dispositio*) which helped exercise a faculty or carry out an operation. Suárez explained that the faculties (*potentiae*) "capable of *habitus*" are those whose operations are by nature affected by a certain "indifference" or "indetermination."[25] The will was the most obvious one. But such was also the case for the intellect, which "partakes of indifference" when it must judge things of which it can have no clear knowledge (for instance, in matters of faith or opinion). *Habitus* helps the intellect demonstrate the truth more easily (1.12), for even after the understanding has received the "intelligible species" of the objects to be examined, it may encounter difficulties in reaching a judgment (4.3).[26]

The Scholastics established a list of *habitus intellectuales* modeled on the account of the "intellectual virtues" in the *Nicomachean Ethics*. It comprised *sapientia*, *intellectus* (or *intelligentia*), *scientia*, *prudentia*, and

---

23. Francisco Suárez, *Disputaciones metafísicas* (1597), ed. and trans. Sergio Rábade Romeo, Salvador Caballero Sánchez, and Antonio Puigcerver Zanón (Madrid: Gredos, 1960–1966), I.2.20.

24. Whereas Suárez defined *ens* as what can exist, Bonaventura and Aquinas defined the created being as an *esse*, a "being" only insofar as it "descends" from God and participates in the divine *esse*. For Suárez, God was not an extrinsic principle, the source of the subject of metaphysics, but part of the study of *ens ut sic*. This conception of metaphysics was later called "ontology." See Jean-François Courtine, "Le projet suarézien de la métaphysique," *Archives de philosophie*, 42, 1979, 235–274.

25. Suárez, disputation XLIV, "De habitibus," in *Disputaciones*, vol. 6, 1.11. In what follows I refer to this disputation by section and paragraph number.

26. A notion manifests itself in three ways: as *species*, aspect, and form. The intelligible species are abstracted from the images of things and specify what is knowable. They are to the intellect what sensible images are to the senses—not what the intellect knows but that by which it knows, just as images are not the objects perceived but that by which they are perceived. See, for example, Thomas Aquinas, *Summa theologica*, Ia, q. 85 art. 1–2.

ars. The virtues (but also the vices), as well as the different types of knowledge, are *habitus* which, for example, dispose the understanding *in ordine ad verum* or the will *in ordine ad bonum*. The difference between *intelligentia* and *scientia* was defined as follows: *intelligentia* is the *habitus* of the understanding by which we know principles or necessary and irrefutable propositions; *scientia* is a *habitus conclusionum*, by which we arrive, apodeictically, at necessary conclusions from necessary principles.[27]

In short, the *habitus* "incline" the intellect to accept an element of knowledge (*assensus scientificus;* 4.8). The problem of the unity of a science stemmed from the fact that, according to Aristotle, this unity depends on the unity of its object. An "indivisible" *habitus* may sometimes "incline" the understanding to a set of truths (for example, all the particular truths contained in a universal truth; 11.54), but other *habitus* are constituted by "collecting and subordinating" different qualities. This seemed to Suárez to be the case with the sciences. For example, we believe geometry to be a unified science, but its unity is not a "simple quality" but a "composition or coordination" of different qualities.

Similary, Suárez noted, health and beauty are, "according to the common manner of speaking and philosophizing," considered qualities; in fact, they result from "proportions" of several different qualities. Likewise *in habitibus scientiarum.* However, although the principles and conclusions of a science do not have the same degree of unity as a simple quality, one cannot, given the "connection or affinity" between the objects which a science explores, deny it all unity (11.55). The unity of the subject matter is therefore supplemented by a "coordination" between the simple *habitus* which make up a single science (11.63). While admitting that such arguments did not provide an entirely satisfactory solution, Suárez emphasized that he had used an analogous argument to explain the unity of metaphysics: "And I consider that this is how one must philosophize on other matters. The task will seem less difficult for one who bears in mind that the unity of a science is not exact and complete, but as if artificial" (11.69).

Suárez's examination of the *scientia de anima* also highlighted the variety of theories of the soul and their dispersion across the physics of the living body and the metaphysics of the rational soul. His conception of the science of the soul had some features in common with how Christophe de Savigny

27. "Intelligentia est habitus intellectus, quo principia cognoscimus. . . . Scientia est habitus apodicticus (qui comparatur per Demonstrationem) quo percipimus necessarias conclusiones e necessariis propositionibus"; Goclenius, *Lexicon,* s.v. "Habitus," p. 624. For an overview of the Scholastic treatment of *habitus,* see Mariano Fernández García, *Lexicon scholasticum philosophico-theologicum* (1910; Hildesheim: Olms, 1988), pp. 314–321.

at about the same time—or Alsted half a century later—characterized psychology: as a part of natural philosophy, though not a specific science in itself. Psychology functioned as a generic introduction to the threefold science of living beings. We have already noted, citing Micraelius's *Lexicon* (1662), that "psychology" tended to become the doctrine of the separable soul from the middle of the seventeenth century. The meaning of the term then evolved in two different directions. Sometimes the word designated a pneumatology, a metaphysics of the rational soul; more often, a physics of the mind, that is, an empirical investigation of the rational soul united with the body. The two meanings derived from debates within Aristotelianism (I mentioned Scipion du Pleix's critique of Suárez), debates which contributed to the collapse of Aristotelian frameworks and the emergence of new ways of thinking about the soul and its relation to the body.

FROM SOUL-FORM TO SOUL-MIND

After the soul ceased to be conceived as the principle of vegetative and sensitive life and became the human mind (*mens*), psychology could no longer be the generic science of living beings and became the science of the rational soul united with the body.[28] Hylomorphism persisted, however, particularly in the university. Its demise, due also to disputes within Aristotelianism itself, was neither rapid nor sudden, and it even lived on in certain anti-Scholastic *novatores*.[29] As far as the concept of soul was concerned, the "mechanized" vision of the universe and the explanation of physical phenomena in terms of the movements of matter played a major role in dislodging the Aristotelian definition.[30] The rejection of "qualities" and "forms" in

28. For overviews of the concept, see Helmut Holzhey, "Seele (§ 4. Neuzeit)," in Ritter et al., *Historisches Worterbuch*, vol. 9; and F. Vidal, "Soul," in Kors, *Encyclopedia*, vol. 4.

29. Christia Mercer, "The vitality and importance of early modern Aristotelianism" in Tom Sorell, ed., *The rise of modern philosophy: The tension between the new and traditional philosophies from Machiavelli to Leibniz* (Oxford: Clarendon Press, 1993); Christoph Lüthy and William R. Newman, "'Matter' and 'form': By way of a preface," *Early science and medicine*, 2, 1997, 215–226; and also Roger Ariew and Marjorie Greene, "The Cartesian destiny of form and matter," *Early science and medicine*, 2, 1997, 300–325.

30. For good recent introductions (but which do not examine the consequences of this revolution for psychology), see Steven Shapin, *The Scientific Revolution* (Chicago: University of Chicago Press, 1996); and Peter Dear, *Revolutionizing the sciences: European knowledge and its ambitions, 1500–1700* (Princeton: Princeton University Press, 2001). The historiographical situation sketched by Shapin has hardly changed: "Almost no historian of science has argued a case for early modern 'revolution' in the practices now known as psychology or sociology, and accordingly the historical literature on these subjects is sparse. On the other hand, the considerable *problems* posed by mechanism for philosophies of knowledge, mind, and moral conduct are standard topics

the natural realm entailed the rejection of the idea of souls in animals and plants. When René Descartes (1596–1650) banished the soul from nonhuman living beings, he performed one of the most radical acts to emerge from the mechanistic reform of natural philosophy in general, and physiology in particular.[31]

The name of Descartes is associated with "dualism" and with the origins of the mind-body problem.[32] The very fact that his name hardly appears in Enlightenment empirical psychologies shows to what extent these were already predicated on the transformation of the concept of soul for which he was largely responsible.

Descartes, in seeking to determine what could be known with certainty, took the path of radical doubt and imagined, in his First Meditation, a certain "evil genius" which endeavors to deceive him in his perceptions of the world and his own body. He decided to consider himself "as not having hands or eyes, or flesh, or blood or senses, but as falsely believing that I have all these things."[33] This brought him to the famous *cogito*.[34] In the Second Meditation, Descartes elaborated the thought experiment of disembodiment that made him assert that, since thought is the only attribute which cannot be detached from himself, he is nothing other than a *res cogitans*, a mind,

---

in the history of philosophy, though few practitioners in this idiom show a genuinely historical sensibility" (pp. 189–190). On the debates most relevant to psychology, see Daniel Garber, "Soul and mind: Life and thought in the seventeenth century"; Charles McCracken, "Knowledge of the soul"; and Gary Hatfield, "The cognitive faculties," in Daniel Garber and Michael Ayers, eds., *The Cambridge history of seventeenth-century philosophy* (Cambridge: Cambridge University Press, 1998).

31. Dennis Des Chene, *Spirits and clocks: Machine and organism in Descartes* (Ithaca: Cornell University Press, 2001).

32. See, in particular, Geneviève Rodis-Lewis, *Descartes: Textes et débats* (Paris: Livre de poche, 1984); John Cottingham, *A Descartes dictionary* (Oxford: Blackwell, 1993); and J. Cottingham, ed., *The Cambridge companion to Descartes* (New York: Cambridge University Press, 1992). Specifically on the physiology of the mind, see Richard B. Carter, *Descartes' medical philosophy: The organic solution to the mind-body problem* (Baltimore: Johns Hopkins University Press, 1983); G. A. Lindeboom, *Descartes and medicine* (Amsterdam: Rodopi, 1979); and Gary Hatfield, "Descartes' physiology and its relation to his psychology," in Cottingham, *The Cambridge companion*. See also Edward S. Reed, "Descartes' corporeal ideas hypothesis and the origin of scientific psychology," *Review of metaphysics*, 35, 1982, 731–752.

33. René Descartes, *Meditations on first philosophy*, in *The philosophical writings of Descartes*, vol. 2, trans. John Cottingham (1988; Cambridge: Cambridge University Press, 1996), First Meditation, p. 15.

34. "I think, therefore I am" appears at the beginning of pt. 4 of the *Discourse on Method* (1637) and in the *Principles of Philosophy* (pt. I, article 7, in the 1644 Latin edition). It is in the Second Meditation, however, that Descartes explains most clearly how he arrived at this single first certainty.

a rational soul, an intellect or reason (*mens, sive animus, sive intellectus, sive ratio*). The conclusion is grounded in the certainty that I am something that can be defined only by the fact it thinks; I am, insofar as I think. Now thinking is an activity of my mind, that is, of the *animus*, or rational soul. I am "a thing that thinks," that is to say, a thing "that doubts, understands, affirms, denies, is willing, is unwilling and also imagines and has sensory perceptions."[35] I am therefore not the other traditional "attributes of the soul"—nutrition, movement, and sensation. By the end of the *Meditations*, Descartes has demonstrated that thought is the intrinsic property of an immaterial and indivisible substance essentially distinct from any *res extensa*, from the material body in its extension and divisibility.

Henceforth, investigating the body could do without a science of the soul, and psychology would logically become a science of the rational soul in its union with the body. Why in its union with the body? Because the body is a machine, and animals, which are devoid of soul, are, literally, machines. The human being, by contrast, is made up of the substantial union of a soul and a body. Descartes's Sixth Meditation argues that feelings such as pain, hunger, or thirst teach us that the soul is not simply present in the body like a pilot in his ship, "but that I am very closely joined and, as it were, intermingled with it, so that I and the body form a unit."[36] This is the unity which Descartes assumed when studying the human being.

Descartes's *Treatise of Man*, published only in 1664, developed a mechanistic psycho-physiology which addressed not only sensation but also common sense, imagination, memory, the appetites, and the passions. At the end of the *Treatise*, Descartes emphasized that all these functions could be explained by the arrangement of the organs of the "machine," exactly as the functioning of a clock or an automaton is governed by the movements of its wheels and counterweights, such that it is not necessary "to conceive of any vegetative or sensitive soul or any other principle of movement and life than its blood and its spirits, agitated by the heat of the fire which burns continually in its heart and which is of no other nature than all those fires that occur in inanimate bodies."[37] In man, however, the body and the soul interacted at the pineal gland, which functioned by letting the animal spirits pass through it, moving them and, in turn, being moved by them. The pineal gland was the place where the impressions originating in

35. Descartes, *Meditations*, Second Meditation, p. 28.
36. Descartes, *Meditations*, Sixth Meditation, p. 56.
37. R. Descartes, *Treatise of man*, trans. Thomas Steele Hall (Cambridge MA: Harvard University Press, 1972), p, 113.

each of the double organs (eyes, etc.) were integrated. The images the rational soul contemplates when, joined to the "machine," it feels or imagines were inscribed there, as a widening of the "pores of the brain" by the passage of the animal spirits. Princess Elisabeth of Bohemia not only wanted Descartes to explain the "actions" and the "passions" of the soul in the body but also how an immaterial thing could move and be moved by the body.[38] In Descartes's most comprehensive response—*The Passions of the Soul* (1649)—he stated that while the soul is united with the entire body, it nonetheless "exercises its functions" in the pineal gland more particularly than elsewhere (art. 31).

In short, "Cartesian dualism" involved not only a body (which we know to have extension, a form, and the ability to move) and a soul (to which we attribute the property of thinking and willing) but also a third fundamental notion, that of the union of the two substances, which is the foundation of our ideas about the soul's capacity to move the body and the body's to arouse sensations and passions in the soul.[39] Investigating such interaction was wholly part of "physics." Pierre Sylvain Régis (1632–1707), author of a major Cartesian textbook, ranked knowledge of the soul as "the principal and most excellent part of metaphysics."[40] But in order to understand the soul, he went on, it is not enough to consider it as a thinking thing, since "experience shows us clearly that all the functions of the soul are completely dependent on the movements of the body with which it is united, such that knowledge of that union is absolutely indispensable."[41] It is indeed so necessary that Régis explained that he used the term "soul" to designate not the mind considered on its own but rather the "union of the mind with an organic body." For him, "the body and the mind taken separately do not make up a human being any more than the wick and the wax separately make up a candle; in other words, the human being and the candle depend absolutely on the union of the parts of which they are composed."[42]

---

38. Elisabeth to Descartes, 10/20 June 1643, letter CCCVIII, in Descartes, Œuvres, Correspondance, ed. Charles Adam and Paul Tannery (1897–1913; Paris: Vrin, 1982–1991), vol. 3, p. 685.

39. Descartes explained that in addition to such general "primitive notions" as number or duration, there is, for the body, only the notion of extension, for the soul, that of thought (including "the perceptions of the understanding and the inclinations of the will"), and "for the soul and the body together," only the notion of their union. Descartes to Elisabeth, 21 May 1643, letter CCCII, ibid., p. 665.

40. Pierre Sylvain Régis, *Système de philosophie, contenant la logique, métaphysique, physique et morale* (Lyon: chez Anisson, Posuel et Rigaud, 1691), vol. 1, p. 121.

41. Ibid., p. 122.

42. Ibid., p. 216.

Régis recast the Cartesian system in the terms of an Aristotelian-style empiricism.[43] He denied the existence of innate ideas and derived ideas from sensation, which he understood as a modification of the soul joined with the body. His definition of the human being as the union of two substances was nonetheless consistent with Cartesian anthropology.[44] Régis examined the understanding, then the will, and finally the state of the soul after death in his *Metaphysics*, but it was in his *Physics* (book VIII, "On Man and His Properties") that he examined—and, as he says, explained "physically"—the senses, the imagination, the judgment, reasoning, and the passions. Even the construction of syllogisms did not escape a physiological interpretation.[45] Régis thus undermined, at least methodologically, what Descartes in his Sixth Meditation called the "real distinction" between the body and the soul.

From an ontological point of view, however, this distinction persisted. The objects of Aristotelian and post-Aristotelian psychologies were different: the soul-form and the soul-mind, respectively. For post-Aristotelian psychology, the human being was indeed a union of two substances, but these substances were joined in a relation quite different from that of form to matter. The union of the body with the soul therefore emerged as problematic, beyond the terms of hylomorphism. In the seventeenth and eighteenth centuries, three theories informed the debates on this issue: Nicolas

43. Tad M. Schmaltz, *Radical Cartesianism: The French reception of Descartes* (New York: Cambridge University Press, 2002).

44. For example, "Asserimus enim hominem ex Corpore & Animâ componi, non per solam praesentiam, siue appropinquationem, vnius ad alterum, sed per veram vnionem substantialem." Descartes to Henricus Regius (a professor of medicine at the University of Utrecht), January 1642, letter no. CCLXVI, in Descartes, *Œuvres, Correspondance*, vol. 3, p. 508.

45. Take the syllogism "All men are animals; Peter is a man; therefore, Peter is an animal." Régis analyzes it as follows: "We hold for certain that the soul, once it has gained knowledge of the universal nature of humans and animals through the path which the spirits have taken from the trace of the human into that of the animal, sees clearly that the human species in its entirety is contained within the genus of animal. On the basis of that insight, the soul forms the first proposition of the argument, *All men are animals*. Following which, and having gained knowledge of Peter and of man in general through the habit which the spirits have of going from the trace of Peter to that of man, the soul again sees clearly that the human species contains Peter as one of its individuals, and from that it forms the second proposition, *Peter is a man*. Last, the animal spirits are moved by the same habit into the trace of Peter and then into that of animals, such that the soul sees clearly, in considering the whole of the human species, that Peter is included within it and, seeing this, understands that he is an animal, that is, forms the conclusion to the syllogism, *therefore, Peter is an animal*" (285–286). Régis, *Physique*, bk. 8, pt. 3, "Des causes physiques des fonctions de l'imagination, du jugement, de la raison, de la mémoire, et des habitudes corporelles et spirituelles," in *Système*, vol. 6, pp. 285–286.

Malebranche (1638–1715) advocated occasionalism, a Platonic version of Cartesianism in which God acted as a causal agent such that, for example, when the soul wished to move the body, God made it move; Gottfried Wilhelm Leibniz (1646–1716) put forward the theory of preestablished harmony, according to which the soul and the body functioned together like two perfectly synchronized clocks; the third theory was physical influence. The new intellectual framework did not exclude a purely pneumatological *psychologia* either, which was concerned with the soul-*mens* independently of its union with the body. This latter tendency contested Aristotelian Scholasticism[46] and thereafter contributed to the debates fueled by Cartesian philosophy.[47]

46. A good example of this orientation can be found in Johannes Schulerus, *Exercitationum philosophicarum liber IV: De anima rationali, sive Psychologia*, pp. 233–304, in *Philosophia novo methodo explicata: Cujus Pars prior continet Excercitationum philosophicarum Libros VI: Metaphysicam, Theologiam naturalem, Angelogra[p]hiam, Psychologiam, Logicam, Ethicam: In quibus, Assertis purioris Philosophiae principiis, Aristotelis Peripateticorumque errores passim refutantur* (The Hague: Ex Typographia Adriani Vlacq, 1663). Schulerus (1619–1674), a pastor and professor of philosophy at the Illustris Schola Auriaca in Breda in the Netherlands, illustrates the growing tendency in the seventeenth century to define psychology as the science of the mind: "Psychologiam non tantum originem, verum etiam omnia, quae ad mentis humanae cognitionem requiruntur, studiose investigat" ("Oratio inauguralis, de Philosophiae instaurandae necessitate, ejusque praestantia, & utilitate," in *Philosophia*, unpaginated). His arguments against the Aristotelian definition of the soul stress that the soul can exist independently of the body and that an immaterial entity can be neither the "formal cause" of a material object nor the source of its essence. If something were to be designated *forma hominis*, this could only be the union of the body with the soul, and not, as the Scholastics maintained, the *anima intellectiva* alone, since it is the body and the human soul which together constitute the specific difference of the human being (questio IV). Their union is neither a substance nor an accident, but simply the manner in which they are united to form a complete human being (questio XXXIV). Schulerus also argued for the autonomy of the soul by challenging Aristotle's tenet that one cannot think without images (questiones I and XXIII). Although the New Testament was sometimes cited, the analysis remained philosophical. For example, to the question of whether the return of the soul to the body is in itself natural, Schulerus replied that it is not, insofar as no *potentia creata* can unite two substances hypostatically; but also that it is, if by "natural" we mean that the soul's union with the body is, for the soul, a most *naturalis & conveniens* state, and that the separated soul is therefore attracted to its body (questio XLIII, the last in the *Psychologia*). Schulerus often targeted the *Conimbricenses*, the commentaries on Aristotle by the University of Coimbra Jesuits (*De anima* dates from 1592). He used Descartes eclectically, citing him in his *Psychologia* only in connection with the seat of the soul in the brain (questio VII). Descartes's influence was nevertheless clear from Schulerus's "dualism," and Descartes figured on the frontispiece of the *Philosophia*, while the "Imaginis titularis explicatio" (*in fine*) presented Descartes as upholding the "freedom to philosophize" (which was the official position of the Illustris Schola Auriaca).

47. A good example would be Johann Eberhard Schweling (1645–1714), a professor of law and later of "universal practical philosophy" at the University of Bremen and the author, among other texts, of *exercitationes* against Pierre-Daniel Huet, who contested Descartes. The proof that the essence of the mind (*mens*) is thought (*cogitatio*) led him to conclude that extension is that without

Eighteenth-century psychologists stressed again and again that what interested them was not the explanation of the union of the soul with the body but the empirical study of their interaction, or "commerce." In this they seemed to pursue the "physical" approach characteristic of the science of the soul-form. There was, however, an essential difference between these two worlds, namely, that the interaction between soul and body was unintelligible within the new philosophy.[48] The fact that Enlightenment psychologists presupposed but refused to deal with the union of the two substances, and limited themselves to phenomena which their own principles obliged them to regard as effects of the interaction of the body and the soul, is a sign of this fundamental difficulty.

In its focus on the interaction and mutual dependence of the body and the soul, rather than on the soul's nature, eighteenth-century psychology was heir to John Locke's critique of the notion of substance.[49] For Locke, the idea of substance was produced by combining simple ideas, gained through observation and experience, of the qualities of things (§ 3).[50] In itself, however, it was "nothing but the supposed, but unknown, support of those qualities we find existing" (§ 2). This was why Locke did not speak of substances, which in themselves are unknowable, but always confined himself to treating the *idea* of substance.

The concept of "body," as developed in the corpuscular theory of matter, replaced the notion of substance. The properties of bodies were the result of the "figure" and movement of corpuscles, through a link that was causal but no longer substantial. Thus,

---

which the mind can nonetheless know itself (§ 7). From this premise Schweling developed a criticism of the *veteres Magistri*, for whom the essence of the soul consists in qualities such as its spiritual nature, immateriality, incorruptibility, indivisibility, and impenetrability (§§ 9 et seq.). After recalling that thought is the essence of the soul, Schweling describes psychology (§§ 33–56) as bearing on the intellect and the will, then identified *mens* with *persona*, declared his opposition to traducianism and *psychopannychia*, demonstrated the immortality and incorruptibility of the soul, and rejected the doctrine of purgatory. References are to Schweling, *Philosophiae tomus pneumaticus: Continens Psychologiam, Theologiam naturalem, neque non Angelographiam* (Bremen: typis Hermanni Braueri, 1695).

48. Bernard Baertschi, *Les rapports de l'âme et du corps: Descartes, Diderot et Maine de Biran* (Paris: Vrin, 1992), p. 400.

49. See, in particular, R. W. Woolhouse, *Locke* (Minneapolis: University of Minnesota Press, 1983); John W. Yolton, *Locke: An introduction* (Oxford: Blackwell, 1985); Yolton, *A Locke dictionary* (Oxford: Blackwell, 1993); and Vere Chappell, ed., *The Cambridge companion to Locke* (New York: Cambridge University Press, 1994).

50. John Locke, *Essay concerning human understanding* (1690; 2nd ed., 1694), ed. Peter H. Nidditch (New York: Oxford University Press), bk. II, chap. 23, "Of our Complex Ideas of Substances"; henceforth referred to by paragraph number.

had we senses acute enough to discern the minute particles of bodies, and the real constitution on which their sensible qualities depend, I doubt not but they would produce quite different ideas in us: and that which is now the yellow colour of gold, would then disappear, and instead of it we should see an admirable texture of parts, of a certain size and figure. This microscopes plainly discover to us. (§ 11)

The properties we attribute to bodies are, in the last instance, a function of the limited nature of our sensory apparatus. Since we are unable to perceive "the bulk, texture, and figure of the minute parts of bodies, on which their real constitutions and differences depend," we characterize and differentiate bodies through their sensible or secondary qualities (§ 8). As regards the ultimate determinants of existence, particles have replaced substance.

Locke developed a similar argument concerning the notion of spiritual substance, except that he did not go so far as to attribute its operations to material corpuscles.[51] Since we see that thought or reasoning cannot subsist by themselves but, at the same time, we do not understand how they can be produced by a body, "we are apt to think these the actions of some other substance, which we call spirit" (§ 5). Just as the notion of matter refers merely to something in which "sensible qualities which affect our senses do subsist," so we know nothing of spirit except that it is that to which we attribute the function of substratum of our mental operations. But "from our not having any notion of the substance of spirit, we can no more conclude its non-existence, than we can, for the same reason, deny the existence of body" (§ 5). "Spirit," in sum, is the name for a complex idea formed by the combination of simple ideas from different mental faculties, and spiritual substance, just like material substance, is an abstraction that cannot be known in itself.

This kind of critique of the notion of substance, which was commonplace in eighteenth-century psychology, was related to another feature of Locke's thought which came to characterize the Enlightenment science of the soul. As stated above, psychologists did not seek to elucidate the union

51. In the chapter on "the extent of human knowledge" (*Essay*, IV.3, § 6), Locke maintained that we have ideas of matter and of thought, but that it is "impossible for us, by the contemplation of our own ideas, without revelation, to discover whether Omnipotency has not given to some systems of matter, fitly disposed, a power to perceive and think, or else joined and fixed to matter, so disposed, a thinking immaterial substance: it being, in respect of our notions, not much more remote from our comprehension to conceive that GOD can, if he pleases, superadd to matter a faculty of thinking, than that he should superadd to it another substance with a faculty of thinking; since we know not wherein thinking consists." On the ensuing debate, see John W. Yolton, *Thinking matter: Materialism in eighteenth-century Britain* (Oxford: Blackwell, 1984).

of the body with the soul, apart from some speculations on the physiological mechanisms by which ideas are formed. What they really sought to do was describe the cognitive faculties using a "historical, plain method."[52] Locke was classed as a psychologist in the eighteenth century, although his aim was not to write a psychology but to give an account "of the ways whereby our understandings come to attain those notions of things we have" and of the grounds for certain knowledge and men's diverse and contradictory "persuasions."[53] Thus, according to contemporary classifications of knowledge, the *Essay Concerning Human Understanding* belonged to logic. This logic was, however, based on the "historical" method of the Royal Society of London, with which Locke was associated, that is, on a descriptive approach calling on observation and experiment. This was the method advocated by eighteenth-century psychologists when they rejected working deductively from definitions which claimed to state the essence of things. The result was a new "physics" of the soul.

Exceptionally, someone would try to combine such empirical psychology with an Aristotelian perspective. In 1755, a doctor from Liège, Guillaume-Lambert Godart (1721–1794), published a *Physics of the Human Soul*. The title was appropriate, in his view, because he approached the soul "not so much in its substance as in the physical relation it has with a body."[54] Godart posited the "incomprehensibility of substances" and believed in "the real presence of the soul in the body" and the physical nature of the "reciprocal action" of the two substances.[55] In this, his *Physics* was an empirical psychology. Godart, however, was also a disciple of Aristotle, which is why he addressed not only the nature and seat of the soul, the "animal" and psychological functions (the sense of self, perception, imagination, judgment, the passions, memory, sleep, and dreams), as well as man's "metamorphosis" in the afterlife, but also the "vital" functions of nutrition and generation.

Godart interpreted the Aristotelian notion of soul-form as though it referred to the union of the two substances. He took the definition of the soul

---

52. Locke, *Essay*, introduction-chapter 1, § 2.

53. Ibid.

54. Guillaume-Lambert Godart, *La physique de l'âme humaine* (Berlin: Aux dépens de la Compagnie, 1755), pp. v–vi. Godart states that the work is a "sort of commentary" on his thesis "Specimen animasticae medicae," defended at Reims in 1745 (ibid., p. iii). On Godart, see Gaston R. Demarée, "Guillaume Lambert Godart: Médecin, philosophe et météorologiste: Un savant oublié du XVIIIᵉ siècle," *Ciel et Terre*, 109, 1993, 47–51; Godart and his *Physique* are discussed in Alphonse Le Roy, "La philosophie au pays de Liège (XVIIᵉ et XVIIIᵉ siècles)," *Bulletin de l'Institut archéologique liégeois*, 1860, 1–157, chap. 5.

55. Godart, *La physique*, pp. 21, 22, and 30.

as form to mean that the soul "has the capacity to be joined to the organized body, giving it life."[56] This capacity was a "relation" stemming from the nature of each of the substances, such that they could be joined to each other and exchange their impressions. Godart maintained that "we cannot doubt that Aristotle meant this mutual relation when he used the term *form*." As a consequence, the "rational" soul was not only the site of "thought, ideas, reasoning and the passions," but it additionally animated and "vivified" the body to which it was united.[57] In his attempt to reconcile Aristotle with his physics of the soul, Godart maintained that in the "vital and natural" organs, the soul performs only vegetative and sensitive functions, and as a rational principle, the soul is "only in the brain,"[58] even if it is also "wholly in all the body and wholly in each of its living parts." In this light, "our soul is what each of us feels to be the source of his actions."[59]

Despite Godart's particular brand of Aristotelianism, which I have singled out to illustrate a transitional phase which reached far into the eighteenth century, his *Physics* remains, even in its final speculations on the resurrection, exactly what the title suggests: a natural history of the post-Cartesian and post-Lockean soul.

PSYCHOLOGY AS A METAPHYSICS OF THE RATIONAL SOUL

In several respects Godart's *Physics* represented a typical Enlightenment empirical psychology, except that it was not *called* psychology. On the other hand, when the term *was* used, it could still refer, at the beginning of the eighteenth century, to something other than a physics of the soul-mind.

In 1702, in the context of a post-Lockean debate on the soul and matter, the physician William Coward (1657–1725) published his *Second thoughts concerning the human soul, demonstrating the notion of human soul, as believ'd to be a spiritual immortal substance, united to human body, to be a plain heathenish invention, and not consonant to the principles of philosophy, reason, or religion; but the ground only of many absurd, and superstitious opinions, abominable to the reformed churches, and derogatory in general to true Christianity.* Coward stated that, in contrast to

56. Ibid., p. 19.
57. Ibid., pp. 20 and 21.
58. Ibid., p. 73.
59. Ibid., p. 21.

the "psychomuthists" and their fabulous tales (*muthos*) about the soul, he could see nothing in the scriptures to support the idea of the human being as a composite of immortal and immaterial spirit, and mortal and material substance. Far from it, he argued, for in the Bible "human soul and life are the same thing."[60]

According to Coward, the Bible presents the soul as a breath infused by God into inert matter in order to give it life, sensation, and reason. The soul is life itself transforming matter into an "animal."[61] Since the soul can be reduced to the capacity for movement accompanied by sensation, it must be the whole person who dies (and not only the body) and who will be resurrected when God again breathes life into that person's material being. Human immortality began at that moment, and so had nothing to do with the continued existence of a substance separated from the body. Coward replied to his critics that he did not deny the human soul had consciousness, reason, and immortality, but he found that dividing the human being into two halves was contrary both to reason and to the principles of Christianity.[62]

Coward appealed to physiology and medicine to support his arguments. When the circulation of the blood slows down or is interrupted, symptoms appear which can affect even the understanding. On such occasions, the supposed "operations" of the soul are affected, whereas if the soul were an immaterial substance, a pure spirit, it would continue to function "freely and rationally" even in madmen. When the scriptures depict the difficulty of expelling spirits and demons from a body other than through a miracle of Christ, they confirm such observations.[63]

Two *Psychologies* figure among the objections to Coward, each with a different strategy. In one of them, the author maintained that, as demonstrated in scripture, in pagan literature (which illustrates the Greek and Roman belief in the soul and its survival), in the church fathers, and in

60. William Coward, *Second thoughts concerning the human soul, demonstrating the notion of human soul, as believ'd to be a spiritual immortal substance, united to human body, to be a plain heathenish invention, and not consonant to the principles of philosophy, reason, or religion; but the ground only of many absurd, and superstitious opinions, abominable to the reformed churches, and derogatory in general to true Christianity* (London: printed for R. Basset, 1702), p. 156. "Psychomuthists" on p. 46.

61. Ibid., p. 90.

62. William Coward, *The just scrutiny: or, a Serious enquiry into the modern notions of the soul, I, Consider'd as breath of life, or a power (not immaterial substance) united to body, according to the H. Scriptures, II, As a principle naturally mortal, but immortaliz'd by its union with the baptismal spirit, according to Platonisme lately Christianiz'd* . . . (London: printed for John Chantry, 1705), p. 96.

63. Matthew 17:14–18. Coward, *Second thoughts*, pp. 107–108.

various modern authors, an immortal soul is united with a human body and survives after death. Moreover, he claimed, our ignorance concerning the nature and substance of the soul does not imply it is mortal.[64] Like the rest of the work, which closed with "pious considerations" on death, judgment, heaven, and hell, his argument was composed mainly of a polemical medley of citations.

The other critique of Coward was philosophically more interesting, since it addressed the problem of the notion of substance, which was essential to Enlightenment psychology. For John Broughton, its author, "psychology" designated the "examination of the nature of the rational soul." The first part of his book set out to "demonstrate by rational argument the established doctrine of an immaterial and hence immortal substance united with the human body"; the second part argued for this doctrine against Coward's work.

Here, psychology meant a metaphysical discourse on the rational soul. Despite the clearly religious context and consequences of the debate, the issues were not ultimately theological.[65] Broughton's main argument against Coward was based on the soul's capacity for self-movement, as proved, in the author's view, by a large number of philosophers since antiquity. Broughton saw in the soul's capacity for self-movement a proof of the distinction between a spiritual principle of life and sensation, and a bodily principle of mechanical and local movement. He accused Coward of not even investigating the basis of his opponents' position—the idea of the immateriality of thought—with the result that, in dismissing a substance capable of subsisting

---

64. "Because the wisest of Men cannot tell *what* the Soul is, and *how* it is, in its *perfect Nature* and *Substance*, I hope it must no be allow'd upon that Ignorance, that the Soul is mortal." Alethius Phylopsyches, *ΨΥΧΗΛΟΓΙΑ; or Second Thoughts on Second Thoughts: Being a discourse fully proving from Scripture, the Writings of the Learned Ethnicks, Fathers of the Church, Philosophy, and the Dictates of right Reason, the separate existence of the Soul . . .* , (London: Printed and Sold by John Nutt, n.d. [1702 or 1703]), p. 84.

65. Of course, religion permeated all discourses so powerfully at the time that none can be termed wholly secular. It is nonetheless certain that the term "psychology" was not disseminated by religious debate. It figured in the title of two disputations on the condition of the separated soul after death, but to my mind this is an exceptional usage. Both disputations were chaired by the same professor, and they were in large part polemics against the doctrine of purgatory: Johannes Gerhard, praes., respond. Wolfgang Ernest Tüntzel, *Ψυχολογια generalis: H. e. Disquisitio de statu animarum post mortem* (Jena: Typis Ernesti Steinmanni, 1633); Johann Ernest Gerhard, praes., autor Iohannes Steinhusius, *Ψυχολογια, sive disquisitio de statu animae separatae* (Jena: literis Johannis Wertheri, 1663). J. Gerhard, a professor at Jena, was one of the foremost Lutheran theologians of his time; his son Johann Ernst Gerhard (1621–1668), also a theologian, was a distinguished specialist in Oriental languages.

separately from matter, he had established a "universal corporealism" which was nothing short of atheism.[66]

Broughton began by defining certain notions, among them "substance" and "human soul." Substance, for Broughton, was the nature or essence, known only to God, on which the existence of each and every thing depended. The human soul was a finite spirit so closely united with a body by its creator that it formed a single and complete nature and person.[67] Having set out these definitions, Broughton stated his intention of moving from effects to causes, following the analytic or a posteriori method appropriate to the pursuit of scientific knowledge.[68] Actually, however, Broughton proceeded from definitions, or, as he put it, from the ideas we have of things, such as matter or thought, to an analysis of the properties these must have in the light of the premises adopted.

Beyond Coward, Broughton's target was clearly Locke. As for many Christian divines, his disagreement hinged essentially on the notion of substance.[69] We have seen that Locke did not examine substances in themselves (he considered them to be unknowable) but rather the idea of substance. In his view, it was formed through a combination of simple ideas derived from the perceived qualities of things. "Had the poor Indian philosopher (who imagined that the earth also wanted something to bear it up) but thought of this word substance, he needed not to have been at the trouble to find an elephant to support it, and a tortoise to support his elephant: the word substance would have done it effectually" (*Essay*, II.13, § 19). For Broughton, by contrast, substance was the real *substratum* of things, their very essence and the necessary condition of their existence. He maintained that although substances could not be known in themselves, they could be through their accidents, and even that wherever there were accidents, there was necessarily a substance.

Broughton inferred from the existence of matter in movement that there was an increate and immaterial substance which was the first cause of the movement of created substances. Likewise, the existence of thought proved that of an incorporeal substance; thought was a substantial property of the mind, just as extension was a substantial property of bodies. Insofar as no bodily feature may be attributed to the mind, and vice versa, we must nec-

---

66. John Broughton, *Psychologia: or, An Account of the Nature of the Rational Soul . . .* (1703; Bristol: Thoemmes Press, 1990), preface, unpaginated.

67. Ibid., p. 3.

68. Ibid., p. 6.

69. Alan P. F. Sell, *John Locke and the eighteenth-century divines* (Cardiff: University of Wales Press, 1997), chap. 6.

essarily be dealing with two different essences. The fact that they are united in the human being only confirms the composite nature of man. Broughton thus adopted a position at the opposite extreme from Locke's supposition that God could give matter the capacity to think. He claimed that the incompatibility of the concepts of *thought* and *extension* was incontrovertible proof of the existence of an immaterial substance in human nature.[70] Even sensation proves this, said Broughton: we feel pleasure when we warm ourselves at the fire, but we will end up feeling pain if the heat increases. How could the same movement of matter, transmitted in the same way to the same organ, generate contradictory ideas?

Broughton believed this question remained mysterious within a materialist and mechanistic framework, but that all became clear if one accepted the union of the two substances: each is affected by the other, such that certain changes in the body produce pleasure in the mind whereas others cause pain. Starting from the premise that matter, due to its passive and divisible nature, is incapable of thought, Broughton observed that the hypothesis of thinking matter can be interpreted in only two ways: either thought is actually one of the unknown properties that God has bestowed on matter, or else matter does possess a hidden potential to think, which God could actualize if he so chose, "tho' it does not lye within the Compass of our Apprehension."[71]

The first possibility—that matter thinks—appeared to Broughton to be so absurd that it was not even worth rebutting. The second was in his view more problematic, since we do not know all the properties and potentialities of matter. One could indeed say that our idea of matter is incomplete and that God can act beyond our conception of things. Broughton's rebuttal was consistent with his method, in that he tested objections against his own definitions: of course, he conceded, God may transform every particle of matter in the universe into an immaterial thinking substance, but he cannot create "cogitative matter" without contradicting himself.[72]

---

70. This was a common argument in such debates. See Ben Lazare Mijuskovic, *The Achilles of rationalist arguments: The simplicity, unity, and identity of thought and soul from the Cambridge Platonists to Kant: A study in the history of an argument* (The Hague: Martinus Nijhoff, 1974).

71. Broughton, *Psychologia*, p. 57.

72. I have summarized the following sections from the first part of Broughton's *Psychologia*: IV, "Of extension and cogitation, as the attributes of body and mind, and how far their different ideas argues different substances"; V, "Of the repugnancy in our ideas, between cogitation and extension, and how far this proves an immaterial substance in human nature"; VI, "Of cogitation (as in man) requiring an immaterial substance"; VII, "Of physically extended substance, as in its nature utterly incapable of thought"; VIII, "That cogitation cannot be superadded to matter."

It followed that the union of the body with the soul did not destroy the distinction between the two natures. Broughton then asked how it is possible that bodily states have such an influence on the mind. We only ask such a question, he said, if we have forgotten that the human being is made up of the union of a body and a soul, and that we do not know how this union works. We do not think as immaterial beings alone but as immaterial beings united with the material world. Consequently, a quality which belongs to one part of our nature (the soul) and which, if separated, could not affect the other part (the body) can indeed affect our entire nature as composite beings.[73]

Anthony Collins described Broughton's *Psychologia* to his friend Locke as "a discourse on nothing, or . . . on something about which no one knows anything." In his reply, Locke admitted that merely reading the fifth section of the book, "Of the repugnancy in our ideas, between cogitation and extension, and how far this proves an immaterial substance in human nature," had convinced him that it was not worth reading the rest of a book by an author who argued so unconvincingly against his opponents but so convincingly against himself.[74] Broughton appealed to the ideas of "matter" and "thought" when he accused Locke of dishonestly hiding behind the idea of divine omnipotence in imputing to God the power of endowing matter with the faculty of thought. For if God can most certainly suspend the usual laws of the separation of substances in a particular case (such as when he once made a piece of iron float), it is impossible to imagine that he would permanently endow matter with thought. Given the essential characteristics of matter and thought, the association of the two terms would be simply unintelligible. Thought and extension are not compatible within one and the same substance, and this incompatibility is enough to prove that they belong to two distinct substances.[75]

Nothing could be further removed from Locke's "way of ideas," and the opposition between the two universes is clear cut. The *Psychologia*, as a metaphysics of the rational soul, can be viewed as a countermodel to the Lockean theories that turned the eighteenth century into the "century of psychology." The rejection of the term "psychology," and of what it designated, which we have examined above, should be understood in the light of Scholastic *Psychologies* and of a *Psychology* such as Broughton's. However,

73. Ibid., p. 101.
74. E. S. de Beer, ed., *The correspondence of John Locke* (Oxford: Clarendon Press, 1976–1989), vol. 8, letters 3311 and 3318.
75. Broughton, *Psychologia*, pp. 28–29.

in the eighteenth century, psychological subject matter occurred much less frequently in works of this type than in attempts to develop an empirical science of the soul.

THE NEW PSYCHOLOGY: CHRISTIAN WOLFF

Hegel stated that Christian Wolff had "entirely displaced the Aristotelian philosophy of the schools [i.e., Scholasticism], and made philosophy into an ordinary science pertaining to the German nation."[76] The development of a new psychology, complete with its terminology, was part of the process which Hegel was describing here. The presence of the term "psychology" in Wolff's vocabulary suggests a continuity between Wolff's vision, and the academic and conceptual fields within which the term was already circulating; at the same time, Wolff would redefine the term and reinvent the discipline it designated.[77]

Wolff, a Lutheran, studied mathematics and became a professor at Halle in 1707. He was accused by the Pietists of fatalism, Spinozism, and idealism and was exiled in 1723. He taught at Marburg before returning to Halle in 1740. Roughly from the death of Leibniz in 1716 to Kant's critical philosophy of the 1780s, Wolff was the leading philosopher, particularly in universities, for German-speaking Protestants and Catholics alike.[78] His vast work covers almost all areas of knowledge, first in German (by which he made a decisive contribution to the development of a vernacular philosophical language),

76. Georg Wilhelm Friedrich Hegel, *Lectures on the history of philosophy*, trans. Elizabeth S. Haldane and Francis H. Simson (London: Kegan Paul, Trench, Trübner, 1896), vol. 3, p. 353.

77. A facsimile edition of Wolff's works, including accompanying documentation, has been published by Jean École (general editor): Christian Wolff, *Gesammelte Werke*, ser. I, *Deutsche Schriften*; ser. II, *Lateinische Schriften*; ser. III, *Materialien und Dokumente* (Hildesheim: Olms, 1965–). We shall refer to it as follows: "Wolff, *G.W.*," followed by the series number (I, II, III) and volume number.

78. For summaries of Wolff's thought, it is worth consulting Beck, *Early German philosophy*, chaps. 11 and 12; J. École, "Wolffius redivivus," *Revue de synthèse*, no. 116, 1984, 483–501; Werner Schneiders, ed., *Christian Wolff, 1679–1754: Interpretationen zu seiner Philosophie und deren Wirkung* (Hamburg: F. Meiner Verlag, 1983); Marcel Thomman, "Wolff, Christian," in Denis Huisman, ed., *Dictionnaire des philosophes* (Paris: Presses Universitaires de France, 1984), vol. 2; and Giorgio Tonelli, "Wolff, Christian," in Paul Edwards, ed., *The encyclopedia of philosophy* (New York: Macmillan, 1967), vol. 8. For key works, see Mariano Campo, *Cristiano Wolff e il razionalismo precritico* (1939), in Wolff, *G.W.*, III.9; J. École, *La métaphysique de Christian Wolff* (1990), in Wolff, *G.W.*, III.12; G. Biller, "Die Wolff-Diskussion 1800 bis 1982: Eine Bibliographie," in Schneiders, *Christian Wolff*; J. École, *Index auctorum et locorum Scripturae Sacrae ad quos Wolffius in opere metaphysico et logico remittit* (1985), in Wolff, *G.W.*, III.10; and Dagmar von Wille, *Lessico filosofico della Frühaufklärung: Christian Thomasius, Christian Wolff, Johann Georg Walch* (Rome: Ed. dell'Ateneo, 1991).

then more extensively in Latin. Wolff sought to marry reason and experience (*connubium rationis et experientiae*):[79] in his view, this was the only
way for philosophy to arrive at the demonstrative certainties which until
then had been the preserve of mathematics, but which should characterize
the real *scientia*, defined as a *habitus asserta demonstrandi*, or the capacity
to draw conclusions deductively from certain and immutable principles.[80]

Despite the impression given by the rigidly syllogistic structure of the
Latin treatises—which Hegel deplored as a "barbarism of pedantry" or a
"pedantry of barbarism" and which was partly responsible, in his view, for
bringing Wolff's thought into discredit—Wolff's objective remained pragmatic.[81] Genuine virtue was for him the main fruit of philosophy.[82] His
most important German treatises, published between 1713 and 1726, were
called *Rational Thoughts on . . .*—an expression of Wolff's confidence in
reason and the powers of the human understanding, which he thought one
could know only by using these powers themselves. In his view, rationality was not purely speculative but was based on experience and should aim
to develop a practice, that is, a philosophy which all people could apply
in their lives.[83] This was the sense he gave to the *connubium rationis et
experientiae*.

Wolff distinguished a posteriori, empirical and historical knowledge of
things as they exist and take place, from a priori, rational and philosophical
knowledge of their reasons.[84] His philosophy was based on the principle of
noncontradiction (by which a thing cannot be at once affirmed and denied)
and of sufficient reason (by which everything which exists or occurs is as
it is and not otherwise). It was therefore concerned with things from the
perspective of their possibility or essence. By contrast, "history" dealt with
what actually existed and could be known through experience; it was not
strictly speaking a "science." Accordingly, Wolff divided every discipline

79. C. Wolff, *Psychologia empirica, methodo scientifica pertractata, quae ea, quae de anima
humana indubia experientiae fide constant, continentur et ad solidam universae philosophiae
practicae ac theologiae naturalis tractationem via sternitur* (1732),, ed. of 1738, in Wolff, *G.W.*,
II.5, § 497.

80. C. Wolff, *Philosophia rationalis sive Logica, methodo scientica pertractata et ad usum
scientiarum atque vitae aptata: Praemittitur Discursus praeliminaris de Philosophia in genere*
(1728), ed. of 1740, in Wolff, *G.W.*, II.1.1, § 30; C. Wolff, *Preliminary discourse on philosophy in
general*, trans. Richard J. Blackwell (New York: Bobbs-Merrill, 1963).

81. Hegel, *Lectures*, vol. 3, p. 356.

82. Wolff, *Philosophia rationalis . . . : Praemittitur Discursus praeliminaris*, § 112.

83. Nicolao Merker, "Cristiano Wolff e la metodologia del razionalismo," *Rivista critica di
storia della filosofia*, 22, 1967, 271–293; 23, 1968, 21–38.

84. Wolff, *Philosophia rationalis . . . : Praemittitur Discursus praeliminaris*, §§ 3 and 6.

into a "rational" part and an "empirical" part, although this division is not
as clear cut as might first appear:[85] empirical psychology, for example, is
"historical" but also belongs in philosophy. Be that as it may, Wolff maintained that what allows things to be understood or established as probable
must be examined first, with the result that the empirical sciences precede
the rational ones.[86]

Wolff defined philosophy as the "science of possibles" or of all possible
things, "so far as they can exist"[87]—that is, so far as they do not involve contradiction. Its major subjects were God, the human soul, and material bodies.
The discipline itself consisted of logic, metaphysics, practical philosophy,
physics, the philosophy of the arts (technology and the philosophy of the
liberal arts and of medicine), and the philosophy of law.[88] Metaphysics—the
science of being, the world in general, and spiritual substances—comprised
ontology, general cosmology, empirical psychology, rational psychology,
and natural theology.[89] Wolffian textbooks were organized according to this
structure, which was how post-Aristotelian psychology first came to be integrated into philosophy courses in Germany, and how it became a familiar
feature of classifications of knowledge.[90]

Since the definition of philosophy was a paradigm for that of other sciences, psychology became the science of things which are possible due to the
human soul ("scientia eorum, quae per animas humanas possibilia sunt").[91]
Psychology was viewed, like the other disciplines, from two perspectives,
empirical and rational. Wolff formulated his understanding of the distinction

---

85. Charles A. Corr, "Christian Wolff's distinction between empirical and rational psychology," *Studia leibnitiana supplementa*, 14, 1975, 195–215; J. École, "De quelques difficultés à
propos des notions d'a posteriori et d'a priori chez Wolff," *Teoresi*, no. 1–2, 1976, 25–34; École,
"De la notion de philosophie expérimentale chez Wolff," *Les études philosophiques*, no. 4, 1979,
397–406.

86. Wolff, *Philosophia rationalis . . . : Praemittitur Discursus praeliminaris*, § 132.

87. Ibid., § 29.

88. Ibid., chap. 3.

89. Ibid., § 79 for the definition of metaphysics.

90. Among the works from Wolff's time, see Georg Bernhard Bilfinger, *Dilucidationes philosophicae de Deo, anima humana, mundo, et generalibus rerum affectionis* (1725), in Wolff, *G.W.*,
III.18; Heinrich Adam Meißner, *Philosophisches Lexicon . . . aus . . . Christian Wolffens sämtlichen teutschen Schrifften . . .* (1737; Düsseldorf: Stern-Verlag Janssen, 1970); Ludwig Philipp
Thümmig, *Institutiones philosophiae wolfianae, in usos academicos adornatae* (1725–1726),
in Wolff, *G.W.*, III.19; and Alexander Gottlieb Baumgarten, *Metaphysica* (1739), ed. of 1779
(Hildesheim: Olms, 1982), German translation, 1776, 1783. Of Friedrich Christian Baumeister's
textbooks, all reprinted several times in the eighteenth century, see his *Philosophia definitiva
hoc est definitiones philosophicae ex systemate Lib. bar. a Wolf . . .* (1735), ed. of 1775, in Wolff,
G.W., III.7; §§ 689–976 contain the "Definitiones psychologicae."

91. Wolff, *Philosophia rationalis . . . : Praemittitur Discursus praeliminaris*, § 58.

between *Seelen-Geschichte*, the empirical history of the soul, and *Seelen-Wissenschaft*, the rational science of the soul, in his *Rational Thoughts on God, the World, and the Human Soul* (called the *German Metaphysics*), which was published in 1720, and later in two treatises in Latin, *Psychologia empirica* (1732) and *Psychologia rationalis* (1734).[92] Each had its own method:[93] rational psychology proceeded by deduction from definitions, irrefutable experience, axioms, and already proven propositions,[94] whereas empirical psychology was based on observation and even experimentation.

Wolff's empirical psychology is concerned with what we are aware to be occurring in our souls.[95] Its principal method is apperception, the voluntary act of bringing perceptions to consciousness, which Leibniz defined in his *Principles of Nature and Grace* (§ 4) as the consciousness or reflective knowledge of perception as an internal state.[96] Unlike so many later psychologies, Wolff's empirical psychology studied the inner life and faculties of the soul without drawing on physiology or neurology.[97] It aimed to

92. C. Wolff, *Vernünfftige Gedancken von Gott, der Welt und der Seelen des Menschen, auch allen Dingen überhaupt* (1720), in Wolff, *G.W.*, I.2. There were three accounts of Wolffian psychology in French: an anonymous summary published in 1745, and the whole of Wolff's system summarized by two Huguenot pastors from Berlin, Jean des Champs and Jean Henri Samuel Formey, in the 1740s. See J. Des Champs, *Cours abrégé de la philosophie wolffienne, en forme de lettres*, vol. 2, pt. 1, *Psychologie expérimentale*, pt. 2, *Psychologie raisonnée* (1747), in Wolff, *G.W.*, III.13.2.1–2; anonymous, *Psychologie ou Traité sur l'âme: contenant les connoissances, que nous en donne l'expérience* (1745), in Wolff, *G.W.*, III.46; and J. H. S. Formey, *La Belle Wolfienne*, vol. 5, *Psychologie expérimentale* (1753), in Wolff, *G.W.*, III.16.2. Des Champs's work is the better of the three; the anonymous work, which is also not bad, summarizes the first part of the *Psychologie expérimentale*. On Des Champs, see U. Janssens-Knorsch, "Jean Deschamps, Wolff-Übersetzer und 'Aléthophile français' am Hofe Friedrichs des Großen," in Schneiders, *Christian Wolff*; on Formey, see Eva Marcu, "Un encyclopédiste oublié: Formey," *Revue d'histoire littéraire de la France*, 53, 1953, 298–305.

93. A. M. Vittadello, "Expérience et raison dans la psychologie de Christian Wolff," *Revue philosophique de Louvain*, 71, 1973, 488–510.

94. C. Wolff, *Psychologia rationalis methodo scientifica pertractata, qua ea, quae de anima humana indubia experientiae fide innotescunt, per essentiam et naturam animae explicantur, et ad intimiorem naturae ejusque autoris cognitionem profutura proponitur* (1734), ed. of 1740, in Wolff, *G.W.*, II.6, § 3.

95. C. Wolff, *Psychologia empirica*, § 2. For the prolegomena to the *Psychologia rationalis* and the *Psychologia empirica* (§§ 1–10 and 1–9, respectively), see Robert J. Richards, "Christian Wolff's prolegomena to *Empirical* and *Rational Psychology*: Translation and commentary," *Proceedings of the American Philosophical Society*, 124, 1980, 227–239. Quotations from the prolegomena will generally be taken from Richards.

96. C. Wolff, *Psychologia empirica*, § 25 ("Menti tribuitur *Apperceptio*, quatenus perceptionis suae sibi conscia est"); see § 234 on the volitional nature of apperception.

97. J. École, "Des rapports de l'expérience et de la raison dans l'analyse de l'âme ou la *Psychologia empirica* de Christian Wolff," *Giornale di metafisica*, 21, 1966, 589–617; also in J. École, *Introduction à l'opus metaphysicum de Christian Wolff* (Paris: Vrin, 1985).

ground in experience the rational principles underlying what occurs in the human soul.[98] The first part of the *Psychologia empirica* discussed the human soul in general, and its sensitive and intellective faculties. The human soul's existence was proved by a procedure similar to that of the Cartesian *cogito* but formulated *more geometrico*: we have a sense of our own existence and are aware of ourselves and of things outside us; it follows that we exist; since the soul or mind (*mens*) is what, within us, has self-awareness and awareness of external things, we must be our soul and, consequently, the soul exists.[99]

Following these preliminaries, Wolff discussed the cognoscitive faculty, which he divided into upper and lower parts. The senses, the imagination as the capacity to produce images and to create them on the basis of other representations (*facultas fingendi*), memory, forgetting, and reminiscence belonged to the lower faculty; attention, reflection, the intellect, and different types of cognition and operations of the understanding belonged to the upper faculty. The second part of the *Psychologia empirica* dealt with the appetitive faculty (*facultas appetendi*), which was likewise divided into upper and lower parts. The lower part was associated with pleasure and aversion (*voluptas, taedium*), the notions of good and bad, sensitive appetite and aversion, and the affects; the upper part, with volition, nolition (*noluntas*), and freedom. The work ended on a chapter concerning the interaction (*commercium*) between the soul and the body. Wolff signaled that in this area he would consider only what was based on "irrefutable experience," namely, the "coexistence" of modifications of the soul and modifications of the sense organs, without examining the reasons for such a coexistence.[100] Since empirical psychology could demonstrate why certain actions were good or bad, bring to light the qualities and faculties of the soul, examine the inclination to virtue or vice, and describe cognitive operations, it also provided the foundations of natural law, natural theology, morals (practical philosophy), and logic.[101]

98. "*Psychologia empirica* est scientia stabiliendi principia per experientiam, unde ratio redditur eorum, quae in anima humana fiunt." Wolff, *Psychologia empirica*, § 1.

99. "Sumus enim nobis nostri rerumque aliarum extra nos conscii. Qui sui rerumque aliarum extra se conscius est, ille existit. Nos igitur existimus"; "Ens istud, quod in nobis sibi sui & aliarum rerum extra nos conscium est, *Anima* dicitur. . . . Etenim nos existimus, quatenus nobis nostria aliarumque rerum extra nos conscii sumus. Enimvero quatenus nobis nostri aliarumque rerum extra nos conscii sumus, anima sumus. Existit igitur anima nostra." Ibid., §§ 14 and 20–21. This passage is one of the very many which show that, as Hegel maintained, Wolff's definitions are purely nominal and ultimately based on everyday notions.

100. Ibid., §§ 947–949.

101. Ibid., §§ 6–10.

As regards the relation between empirical and rational psychology, Wolff argued that we have arrived at a truth if the results obtained a priori and a posteriori converge. If the conclusions of rational psychology do not concur with those of empirical psychology, then the former must be considered false; however, if a fact which has been demonstrated a priori has no equivalent a posteriori, then we must attempt to ascertain it empirically. That is why empirical psychology not only lends its principles to rational psychology but also "serves to examine and confirm discoveries made a priori concerning the human soul."[102]

The goal of rational psychology was therefore to give a priori explanations of the facts, axioms and proven propositions of empirical psychology, and thus to account for them in terms of the essence and nature of the soul.[103] In Wolff's view, the essence and nature of the soul is its capacity to represent the world to itself; the soul is a *vis repraesentativa universi*, and sensation is its first activity.[104] These principles underlay the *Psychologia rationalis*. It first examined the soul in general and its faculties, from perception to the intellect. It then addressed appetite and aversion (including the passions), and the interaction between the soul and the body. It discussed the correspondence between the operations of the soul and the movements of the body, the soul of animals, the nature of spiritual substances, the immaterial character of the soul, and its origin, immortality, and union with the body. Wolff explained that rational psychology could not provide new empirical knowledge but "increases our acumen for observing what occurs in our soul" and "discloses features of the soul which are closed to observation alone."[105]

Wolff and his disciples emphasized the novelty of their psychology. Wolff himself declared proudly that deriving a priori from the concept of *soul* everything that can be observed about the soul a posteriori was a new and bold undertaking which ran counter to previous conceptions.[106] The contents of Wolffian psychology, too, were considered innovative. Georg Bernhard Bilfinger (1693–1750), whose *Dilucidationes philosophicae* of 1725 were one of the best Wolffian textbooks, epitomized this innovation

---

102. Ibid., § 5.

103. J. École, "De la nature de l'âme, de la déduction de ses facultés, de ses rapports avec le corps, ou la *Psychologia rationalis* de Christian Wolff," *Giornale di metafisica*, 24, 1969, 499–531; also in École, *Introduction*.

104. Wolff, *Psychologia rationalis*, §§ 66–67.

105. Ibid., §§ 8–9.

106. "Novus cum sit ausus & praejudicatae opinioni adversus." Wolff, *Philosophia rationalis . . . : Praemittitur Discursus praeliminaris*, § 112 n.

when he contrasted the psychology of the ancients, which mixed physical and pneumatological observations in their discussions of the vegetative, sensitive, and rational soul, with psychology as a *science* of the human soul based on what can be known about the soul through experience and deduction.[107]

* * *

We have seen that the appearance of the word "psychology" in the sixteenth century cannot be taken as the sign of a new departure with respect to the *scientia de anima*; the authors using the term are not referring to a new enterprise which would mark a break with Aristotelian frames of reference. "Psychology" could designate the generic science of living beings, it could tend toward a "physical" or a metaphysical approach, presuppose or depend on, to a greater or lesser degree, the notion of the soul as entelechy, give precedence to investigating the body or the rational soul—but none of these different psychologies defined, whether in their form, contents or methods, a new science within natural philosophy.

This does not imply, however, that no development took place between the sixteenth and the eighteenth centuries. In the late sixteenth century, Francisco Suárez argued that *de anima* books did not constitute a separate science, and fifteen years later, Christophe de Savigny adopted the same position, but without the Scholastic apparatus. His *Tables of the Liberal Arts* (1587) represented the norm on the subject: since psychology was a theory of the soul-form, it was only a genus whose species were the sciences of living beings. Within this vast field, certain authors gave *psychologia* a narrower sense. In Goclenius's anthology, published three years after de Savigny's *Tables*, "psychology" referred to a body of treatises on the origin and transmission of the soul. Goclenius's usage certainly contributed to spreading the term within the academy—by means of an erudite typographical ornament in the style of the late Renaissance; Goclenius used "psychology" to designate a discourse on the origin and transmission of the soul, but his choice was more stylistic than conceptual. *Psychologia* in this period thus remained basically general and generic, denoting nothing more precise than the science of the soul.

---

107. "Est *nobis* hoc loco *Psychologia* scientia de Anima humana, quatenus ea, quae per experientiam de illa cognovimus, ex conceptu aliquo generali possunt legitime deduci & intelligi." Bilfinger, *Dilucidationes*, § CCXXXIII; § CCXXXII on the psychology of the ancients.

The Melanchthonian Snellius chose "psychology" to name the theory of man after the noblest part of the human being, which governs not only the body but also the will and the intellect. Here, as in most cases, the definition of the soul as *gubernatrix* took precedence over the concept of soul-form. When working from the Aristotelian definition, the science of the soul tended mainly to address the faculties (vegetative, sensitive, intellective) and, to a greater or lesser degree, how they functioned in and through the body. For Alsted, too, "psychology" was the same as the *scientia de anima*. Even in its first attested uses, the word was not explained but taken as though it were self-evident. It was sometimes defined, but only when it had to be, for example, in textbooks and lexicons. Until the eighteenth century, it was employed essentially in pedagogical and taxonomic contexts: it figured in diagrams of the organization of knowledge, in nomenclatures of the sciences, and in the title or body of texts designed for teaching, whose didactic goals required that the subjects to be communicated and learned should be labeled precisely. After it gained currency, however, it was widely used to refer to theories of the soul. But with the collapse of the Aristotelian worldview, the soul ceased to be the principle of vegetative and sensitive life and became the mind (*mens*); the notion of soul-form gave way to the concept of soul-mind. The anthropological function of the rational soul was fundamentally transformed by this, and a new psychology was called for. Since the name existed already, the task became to redefine the term and to use it to give impetus to the reform of the science of the soul.

René Descartes excluded the vegetative and sensitive soul from his explanation of vital processes, which he accounted for in mechanistic terms; the soul became a purely intellective entity. At the same time he insisted on the substantial union of the body with the soul, as the Scholastics had before him, a union constitutive of the human being. Thereafter, John Locke stressed that substances cannot be known in themselves and introduced the "way of ideas." This "way," which was both ontological and methodological, abandoned a priori definitions and syllogistic deduction in favor of observation and experiment, by which the origin of ideas in sensation was demonstrated. Descartes and Locke were not "psychologists." However, some of their philosophical principles formed the basis of psychology as it came to be reworked in the eighteenth century. At the same time, their thought displayed certain continuities with the *scientia de anima*: the soul was purely spiritual; in the living human being it was inseparable from the body; and in itself it was unknowable but could be known through experience, through which abstract concepts could be traced back to their origin in sensation. It was not until Christian Wolff in the 1720s that the *concept* of psychology was actually developed in a new, non-Aristotelian perspective.

Wolff accepted that knowledge originated in sensation. But this empirical fact had to be explained and confirmed a priori. Besides, for him, knowledge could not be arrived at by simple perception but required the active contribution of apperception. Far from rejecting syllogistic logic, as Locke did, Wolff made it the backbone of his method. At the opposite extreme from Locke, he sought to construct a system in which all knowledge would attain the demonstrative certainty of true "science," thanks precisely to the "synthetic" method, which proceeded by deduction from definitions and axioms. That is why his psychology had a dual structure, both empirical and rational. Each constituted a fully fledged discipline within philosophy and provided the grounds for natural law, natural theology, morals, and logic. The "age of psychology" proved nonetheless to be more Lockean than Wolffian. Although rational psychology remained in the curricula of German universities, those philosophers, naturalists, and physicians seeking to develop a science of the soul turned to empirical psychology. This was also true for Wolff's reception outside of German-speaking areas. His method was sometimes commended for the certainty it introduced into philosophy, but Enlightenment thinkers were attracted above all by the promise held out by empirical psychology. "Ce serait une belle chose de voir son âme" (What a fine thing it would be to see one's soul), wrote Voltaire at the beginning of the article "Soul" in his *Philosophical Dictionary*, underscoring that we can know neither the essence of the soul nor even the mechanisms of thought. But this did not prevent him from considering Locke the philosopher who had created the "history" and anatomy of the soul. Eighteenth-century psychology was characterized less by a uniform empirico-rational structure or a concern for systematicity than, on the one hand, by the fact that it was a discipline perceived as being in a process of renewal and revival, and, on the other, by its sensualist inclinations, its rejection of metaphysics as first science, its methodological self-awareness, and its notion of the human being as a composite of a soul and a body.

CHAPTER FOUR

Psychology in the Age of Enlightenment

In his lectures of the 1770s, Immanuel Kant (1724–1804) advocated integrating empirical psychology into university curricula as an autonomous discipline. He argued that psychology had remained subordinated to metaphysics because metaphysics had been wrongly construed and because psychology had until then been insufficiently systematic and its field of study too limited. But now the moment had come (as it had for anthropology) for it to acquire a place of its own alongside other university courses:

> The cause as to why empirical psychology has been placed in metaphysics is clearly this: one never really knew what metaphysics is, although it was expounded on for so long. One did not know how to determine its boundaries, therefore one placed much in it that did not belong there. . . The second cause was clearly this: the doctrine of experience of the appearances of the soul has not arrived at any system such that it could have constituted a separate academic discipline. Were it as large as empirical physics, then it would also have been separated from metaphysics by its vast extent. But because it is small and one did not want wholly to do away with it, one pushed it into metaphysics. . . . But now it has already become quite large, and it will attain amost as great a magnitude as empirical physics. It also deserves to be separately expounded, just as empirical phycics does; for the cognition of human beings is in no way inferior to the cognition of bodies; indeed, according to worth it is much to be preferred to the other. If it becomes an academic science, then it

is in the position to attain its full magnitude, for in the sciences an academic teacher has more practice than a scholar on his own.[1]

Kant considered that, given how far empirical psychology had evolved, it should be unequivocally acknowledged as autonomous, extricated from its dependence on other university disciplines, and given its own institutions.[2] This began to happen during Kant's lifetime. Since in the eighteenth century there were no psychology chairs and faculties, and since there were almost no single-discipline university careers, psychology could not be represented by "specialists." Yet not only were its theoretical conditions of possibility established, but it effectively existed and figured prominently in the organization of knowledge and in university curricula.[3]

We have already explored in what ways eighteenth-century psychology could be considered a discipline. But in order to explain Kant's confidence in psychology, we still need to understand the essential bond between psychology and anthropology, the forms taken by the psycho-anthropological field across Enlightenment Europe, and above all the methodological self-consciousness of empirical psychology, which was both a sign of the maturity Kant saw in this discipline and a crucial factor in its development.

PSYCHOLOGY, ANTHROPOLOGY, AND THE HUMAN SCIENCES

Eighteenth-century psychology cannot be considered independently of the ideal of a "science of man," to which many an intellectual enterprise of the

---

1. Immanuel Kant, *Lectures on metaphysics*, trans. Karl Ameriks and Steve Naragon (Cambridge: Cambridge University Press, 1977), p. 44; Kant, *Gesammelte Schriften* (Berlin: Walter de Gruyter, 1968), vol. 28, pp. 223-224.

2. In the *Critique of Pure Reason* (1781/1787, A848/B876), Kant nevertheless suggested that psychology should be placed within metaphysics.

3. We should mention in this connection Graham Richards's *Putting psychology in its place: An introduction from a critical historical perspective* (London: Routledge, 1996), chap. 2. Richards claims that psychology did not constitute a discipline before 1850 for four main reasons. First, Cartesian dualism excluded the soul from natural philosophy. Second, it was theologically dangerous to consider man a natural being. Third, the epistemology of the Scientific Revolution rejected metaphorical language and so discouraged the psychological exploitation of new scientific ideas. Last, the notion of the "individual mind," supposed to be one of the main objects of psychology, did not take shape in the West until around 1800. By the same author on the same question, see "The absence of psychology in the eighteenth century: A linguistic perspective," *Studies in history and philosophy of science*, 23, 1992, 195-211; as well as *Mental machinery: The origins and consequences of psychological ideas*, pt. 1, 1600-1850 (Baltimore: Johns Hopkins University Press, 1992). Kant's remarks above—and the present book in its entirety—suggest that these interpretations are simply not tenable.

Enlightenment aspired. "Man" was an essential preoccupation of the time. One of the period's major ambitions was, arguably, to comprehend him as an individual, social, and historical being and to create a general science, which came to be known as "anthropology," to include all knowledge pertaining to him, from anatomy to cultural history. This understanding of the place of "man" in the landscape of knowledge before the nineteenth century diverges sharply from that of Michel Foucault. In an influential passage from *The Order of Things*, Foucault declared that

> no philosophy, no political or moral option, no empirical science of any kind, no observation of the human body, no analysis of sensation, imagination, or the passions, had ever encountered, in the seventeenth or eighteenth century, anything like man; for man did not exist (any more than life, or language, or labor); and the human sciences . . . appeared when man constituted himself in Western culture as both that which must be conceived of and that which is to be known.[4]

According to Foucault, man taken individually or collectively became the object of a particular science "for the first time" in the nineteenth century. The concept of "man" was born, and the human sciences became possible only when the classical general theory of representation disappeared, and "the necessity of interrogating man's being as the foundation of all positivities was imposing itself in its place."[5]

If "man" as Foucault understands him corresponds to "man" as championed by eighteenth-century thought, then his arrival on the scene must be situated at least half a century earlier. But it is more likely that the "man" of the "archaeology of the human sciences" has nothing to do with its Enlightenment homonym, and that its meaning is decipherable only from within a Foucauldian framework. That said, Foucault's oracular pronouncement has been immensely influential in opening up the question of the "origins" of the human sciences and inspiring a fresh look at the continuities and discontinuities that might have existed between the human sciences in the nineteenth and twentieth centuries, and the earlier "science of man."[6] His-

4. Michel Foucault, *The order of things: An archaeology of the human sciences* (London: Routledge, 1997), pp. 344–345.

5. Ibid., p. 345.

6. Apart from Christopher Fox, Roy Porter, and Robert Wokler, eds., *Inventing human science: Eighteenth-century domains* (Berkeley: University of California Press, 1994); and Smith, "Does the history of psychology have a subject?," see Richard Olson, *The emergence of the social sciences, 1642–1792* (New York: Twayne, 1993); Olson, "The human sciences," in Roy Porter,

toriographical works drawing on Foucault have shown—sometimes against his own theses—that the history of psychology in the eighteenth century cannot be construed on the basis of psychological ideas chosen exclusively for their contemporary relevance; nor can psychology be treated as though it had always existed within certain discursive and ideological contexts (such as metaphysics, natural philosophy, or theology), but had simply been occluded or prevented from developing.

The concept of a "science of man" played a key role in the advent of the human sciences.[7] While the "sciences of man" exist today only in the plural, the "science of man," like "social science,"[8] started life in the singular. The idea of *one* science that would bring together and integrate all the disciplines concerning the human being had a fundamental regulatory function in the constitution of the modern field of the human sciences. As François Azouvi has noted, the history of the science of man has revolved around how far the scope of the different sciences which take man as their object may be extended, since it is "only their unity [which] constitutes the science of man in itself."[9] Despite much eighteenth-century reflection on the science of man, it was still commonplace, at the beginning of the nineteenth century, to maintain that this science had yet to be instituted. This was because all attempts at constituting it seemed to have got no further than juxtaposing disciplines, rather than integrating them in order to comprehend man "in every aspect."[10]

The most frequently cited form of this ideal unity is probably the introduction to *A Treatise of Human Nature* (1739), by David Hume (1711–1776). Since disciplines such as logic, morals, criticism and politics are based on human nature, their only valid foundation is a science of man based on experience and observation.[11] Developing just such a science was the central

---

ed., *The Cambridge history of science*, vol. 4, *Eighteenth-century science* (Cambridge: Cambridge University Press, 2003); Roger Smith, *The Fontana history of the human sciences* (London: Fontana Press, 1997); and Jan Goldstein, "Bringing the psyche into scientific focus," in Theodore M. Porter and Dorothy Ross, eds., *The Cambridge history of science*, vol. 7, *The modern social sciences* (New York: Cambridge University Press, 2003).

7. F. Vidal, "La 'science de l'homme': désirs d'unité et juxtapositions encyclopédiques," in Claude Blanckaert et al., eds., *L'histoire des sciences de l'homme: Trajectoire, enjeux et questions vives* (Paris: L'Harmattan, 1999).

8. Brian W. Head, "The origins of 'la science sociale' in France, 1770–1800," *Australian journal of French studies*, 19, 1982, 115–132.

9. François Azouvi, *Maine de Biran: La science de l'homme* (Paris: Vrin, 1995), p. 7.

10. Ibid, p. 10.

11. "There is no question of importance, whose decision is not compriz'd in the science of man; and there is none, which can be decided with any certainty, before we become acquainted with that science. In pretending therefor to explain the principles of human nature, we in effect

project of the Scottish Enlightenment, and it inspired many endeavors of
the time. The science of man was composed of a multiplicity of fields; but
these fields, which somehow had to be integrated, were themselves inspired
by an ideal of unity and interdependence. The science of man could not
exist without particular disciplines; the disciplines derived their meaning
and ultimate goals from the science as a whole. Psychology was frequently
seen as the foundational science of the edifice, in the first place because it
elucidated the principles and mechanisms of knowledge itself. In Hume,
for example, the principles governing the association of ideas (similarity,
contiguity in time or space, perception of causality) played the same fundamental role as Newton's laws of attraction and were endowed with the
same universal validity.[12] Second, psychology occupied a privileged position within the science of man because it focused on what makes humans
distinctive, namely, their composite nature and the constant interaction,
or "commerce," between soul and body. The very definition of the soul
sometimes echoed this focus and replaced its characterization as the power
of representation or that which thinks in man. Johann Georg Walch's influential philosophical dictionary of 1726, the first to be published in a vernacular language, defined the soul as the substance with which the body is
united, in such a way that the whole being of man results from this union.[13]
Instead of dwelling on the separable nature of the soul, Walch stressed its
union with the body to form the human being. This definition of the soul
more or less explicitly underpinned the anthropological discourse of the
Enlightenment.

In the sixteenth and seventeenth centuries, the discipline called *anthropology* focused on human physiology and, particularly, on human anatomy.
In the eighteenth century, anthropology became the general science of man,
comprising psychology, medicine, physiology, and philosophy in various
proportions—but usually not including the issues that would later be the fo-

---

propose a compleat system of the sciences." David Hume, *A Treatise of human nature: being an
attempt to introduce the experimental method of reasoning into moral subjects* (1739–1740), ed.
L. A. Selby-Bigge, rev. P. H. Nidditch (Oxford: Clarendon Press, 1978), p. xvi.

12. Hume, *Treatise,* pt. I, bk. I, sec. IV.

13. "Nach der gewöhnlichen Bedeutung dieses Worts ist die Seele diejenige geistliche Sub-
stanz, welche mit dem menschlichen Cörper vereiniget, so das durch diese Vereinigung der
Seelen und des Cörpers das volligen Wesen des Menschen entstehet." Johann Georg Walch, *Philosophisches Lexikon* (1726), ed. Justus Christian Hennings (1775; Hildesheim: Olms, 1968), art.
"Seele," 2, cols. 761–772. Walch (1693–1775) was consecutively professor of rhetoric, philosophy,
and theology at Iena.

cus of ethnography or physical, cultural, and social anthropology.[14] Its core subject remained man as a composite being. Like Montesquieu envisaging the soul in the body as "a spider in its web," the Enlightenment anthropologist and, a fortiori, the psychologist did not intend to reduce the soul to matter but to explore the mutual dependence of the two substances.[15] Three examples can be cited to illustrate this.

The first example is Alexandre-César Chavannes (1731–1800), a pastor and professor of dogmatic theology at the Academy of Lausanne. In 1788 he published an *Anthropology; or, General Science of Man, to Serve as an Introduction to the Study of Philosophy and Languages, and as a Guide in the Project of an Intellectual Education*.[16] This was one of the most ambitious works of its type. In it Chavannes outlined a number of sciences. The first was "anthropology proper." It considered man "as nature constituted him," that is, as a composite of body and soul, endowed with vegetative, animal, and intellectual life. Then came sciences that explore man from different angles: ethnology, which investigates the different nations and their origins; noology, the science of man as intelligent being, possessing a mind (*noos*); boulology, which approaches man as a being endowed with will (*boulē*), activity, and freedom. This science led to a discussion of natural moral laws and of the need for a revealed morality. There followed glossology, the science of man as a speaking subject; etymology, lexicology, and grammatology (the science of general grammar); and, last, mythology, "or the science of man as engaged at all epochs in inquiries into his origin and future destiny, but regrettably too inclined to go astray on this subject." These, for Chavannes, together made up the general science of man.

Chavannes's anthropology was representative in two respects. First of all, it outlined, on a significant scale, a discipline whose absence had been deplored at least since Hume. The Swiss pastor saw his project as filling this gap. He noted, for example, that although the classifications of knowledge published in encyclopedias demonstrate that the sciences "all relate more

---

14. Maretta Linden, *Untersuchungen zum Anthropologiebegriff des 18. Jahrhunderts* (Bern: Lang, 1976); Odo Marquard, "Anthropologie," in Ritter et al., *Historisches Wörterbuch*, vol. 1.

15. "L'âme est, dans notre corps, comme une araignée dans sa toile. Celle-ci ne peut remuer sans ébranler quelqu'un des fils qui sont étendus au loin, et, de même, on ne peut remuer un de ces fils sans la mouvoir." Charles de Secondat, baron de Montesquieu, "Essai sur les causes qui peuvent affecter les esprits et les caractères" (before 1748), in *Œuvres*, ed. Roger Caillois (Paris: Gallimard, 1951), vol. 2, p. 49.

16. Alexandre-César Chavannes, *Anthropologie ou Science générale de l'homme pour servir d'introduction à l'étude de la philosophie & des langues, & de guide dans le plan d'éducation intellectuelle* (Lausanne, 1788). See Gérald Berthoud, "Une 'science générale de l'homme': L'œuvre d'Alexandre-César Chavannes," *Annales Benjamin Constant*, no. 13, 1992, 29–41.

or less directly to man as their principal concern," they do not include a science of man:[17]

> Who would guess, on reading the authors who have written on different topics, or on reading an encyclopedia, that man is the center to which all sciences should lead and on which all rays of the light of the intellect should converge. The neglect here has been so striking that in the ordering of the sciences, in the genealogical tree that displays it and in the works intended for universal instruction, the idea of a *general science of man* has not even been conceived, or so much as mentioned, and the different parts of philosophy, psychology, logic, and morals have been treated without ever reflecting that they should be preceded by the science we call *anthropology*.[18]

Chavannes devoted all his efforts to developing this science, and his *Anthropology* of 1788 was only a summary of a much vaster work which remained unfinished and unpublished.[19] The first science, whose absence he deplored, was to be both propaedeutic, since it dealt with the essence of the objects of the specialized disciplines stemming from the nature of man, and synthetic, since it integrated the materials provided by these disciplines, confirmed their basis in observation and experience, and situated them in a Christian perspective. This "anthropology proper" was to precede the particular sciences that study man and that, when combined, would form anthropology as a general science.

Chavannes's anthropology sought to be exhaustive. Starting from physical man, it proceeded to man as a psychological and social being and assigned a central role to the study of language and to man's education, which relied ultimately on his intrinsic perfectibility to prepare him for life in this world and the next. Anthropology's "single goal" was to bring together three accounts, concerning, respectively, the progress of knowledge, the operations whereby humans strengthen their intellectual powers, and the origins and development of language.[20] In Chavannes's system, psychology

17. Alexandre-César Chavannes, *Essai sur l'éducation intellectuelle, avec le projet d'une science nouvelle* (Lausanne, 1787), p. 55.

18. Ibid., p. 56.

19. Alexandre-César Chavannes, *Anthropologie ou science générale de l'homme pour servir d'introduction à l'étude de la philosophie et des langues, et de guide dans le plan d'éducation intellectuelle*, 13 manuscript vols., undated, Département des manuscrits, Bibliothèque cantonale et universitaire, Lausanne.

20. Chavannes, *Essai sur l'éducation*, p. 144.

was not an autonomous science, and psychological subject matter was distributed across anthropology proper, boulology and noology.[21] Chavannes could nonetheless write that "if there is in anthropology an indisputable truth, it is the distinction between the two substances in man." The soul and the body are parts which are "essentially distinct but interlinked in the most intimate fashion, and subject to certain constant and invariable laws of union."[22] The soul cannot carry out an activity without the body being involved, and the body cannot undergo any change without this entailing a change in the soul: "The state of the soul ordinarily corresponds to that of the body."[23] This is, in condensed form, the fundamental premise of empirical psychology, which determines both its objects and its method.

Our second example is Jean-Paul Marat (1743–1793). In 1773, when Marat was still a doctor in London trying to make a reputation for himself through his "physical research" into fire, electricity, and light, he published *A Philosophical Essay on Man*, translated into French two years later. This was a far cry from the inflammatory newspaper he edited during the French Revolution, *L'Ami du Peuple*, which earned him his literary reputation. The *Philosophical Essay* was intended as a rejoinder to the materialism of Helvétius and La Mettrie, as well as to Buffon's anthropology. Marat's "attempt to investigate the principles and laws of the reciprocal influence of the soul on the body," as the subtitle reads, proclaims that to know man is to know the interaction of the two substances that constitute him. With the *philosophe*'s characteristic triumphalism, he claimed that we long remained in darkness and superstition concerning knowledge of man, but that the return to observation as a method will lead us out of ignorance.

Marat explained that the science of man had remained fragmented, its different parts isolated and disconnected from each other:

Anatomists examined the corporeal, and philosophers, the spiritual part. . . . Their observations were therefore more acute. Nevertheless, since they were carried out by different thinkers, by metaphysicians who were not anatomists, and by anatomists who were not metaphysicians, they remained disconnected, and the science of Man consisted wholly in

21. "Anthropology proper" investigated the mental faculties under the heading of "the life of the mind"; boulology included a "psychological theory" of "l'homme moral," which ranged from the impulsive causes of behavior to the desire to know one's origin and future destination; noology included psychology and logic, the former of which analyzed the intellectual faculties and operations.

22. Chavannes, *Anthropologie*, manuscript, vol. 1.2 ("Anthropologie"), p. 6r.

23. Ibid., p. 16r.

a number of scattered ideas. . . . The faculties of the soul, and the mechanism of the body were known, but not the whole Man as a composite
of both. No one had yet accounted for the singular relations between the
two substances which compose his being; scarce any one had noticed
their wonderful influence on each other. Man therefore was considered
as an enigma, an impenetrable mystery.[24]

The passage implicitly evokes Descartes, who, at the beginning of his *Treatise on Man*, said he would proceed by first describing the body and the soul
separately, and then showing how the two are united to form man. True
to this Cartesian inspiration, the proper goal of anthropology according to
Marat was to describe and explain the interaction between the soul and the
body, the mental and the physical, and thus arrive at a knowledge of man as
"compounded of both."

Marat divided his project into three parts. First, as an anatomist, he was
particularly interested in the nervous system and claimed to show that the
nervous fluid is the link between the body and the soul. He then investigated the faculties of the soul and observed their development from birth,
and from sensation through to the understanding.[25] Last, he focused on the
reciprocal influence of the body and the soul, concluding that their harmony
was entirely dependent on the "organization" of the body. Marat explained
that studying each substance in isolation provided only the basis for understanding the human being, and that subsequently one should study their
interaction, that is, "man himself." The focus on the relationship between
the soul and the body, once again, gave psychology a central position.

Our last example will be Ernst Platner (1744–1818), a contemporary of
Marat, who was professor of medicine and philosophy at Leipzig. In 1772
he published an *Anthropology for Physicians and Philosophers*. Here again
anthropology dealt with man from both a philosophical and a medical point

---

24. J[ean-]P[aul] Marat, *A philosophical essay on man: Being an attempt to investigate the
principles and laws of the reciprocal influence of the soul on the body* (London: printed for J. Ridley; and T. Payne,1773), 2 vols., 1, pp. vii, viii–ix. On Marat and his essay, see G. Matthias Tripp,
"Nachwort," in J.-P. Marat, *Über den Menschen oder die Prinzipien und Gesetze des Einflusses
der Seele auf den Körper und des Körpers auf die Seele*, trans. Joachim Wilke, ed. G. M. Tripp
(Weinheim: VCH, Acta Humaniora, 1992).

25. Marat had discussed these topics in the shorter *Essay on the Human Soul* (London:
printed for T. Becket, 1772), whose premises were the same as those of the *Essay on Man*. For
example: let the soul exist "before the body, to which it is united," and possess then "another
method of perceiving and understanding; yet it is certain that, being once subject to the laws of
this union, it no longer retains any thing of its former state, nor even the remembrance of a prior
existence" (p. 44).

of view, as a totality, with particular emphasis on the relations between the soul and the body (the title page features the portraits of Hippocrates and Plato side by side).[26] Platner thus gave pride of place to psycho-physical anthropology, a subject which by the mid-century had become one of the main research fields of the Faculty of Medicine at the University of Halle.[27]

Platner stated that he had as little intention of writing a psychology as Locke, Condillac, or Hume had had of writing an anthropology, and he did not wish to produce a treatise on anatomy or physiology either.[28] Rather, he sought to divide the knowledge of man into three sciences. First, anatomy and physiology, which would deal with the parts and functioning of the "machine" independently of their effects on the soul. Then, the capacities and properties of the soul would be studied independently of the body; this would be psychology or—for Platner it was the same thing—logic, aesthetics, and a large part of moral philosophy. Last, body and soul would be examined in their reciprocal relations, which was the task of anthropology proper.[29]

Platner maintained that his *Anthropology* was conceived more for those already initiated into philosophy and medicine than for students. It contained "more facts than speculation" (*mehr Fakta als Spekulationen*) which was why, Platner explained, it was written in an aphoristic style.[30] The point of departure was the idea that "sense of self" (*Selbstgefühl*) proves the truth of the immateriality of the soul. True to his own definition of anthropology,

---

26. Ernst Platner, *Anthropologie für Aerzte und Weltweise* (1772; Hildesheim: Olms, 1998). See Alexander Kosenina, "Nachwort," ed. G. M. Tripp,in Platner, *Anthropologie*; Kosenina, *Ernst Platners Anthropologie und Philosophie: Der philosophische Arzt und seine Wirkung auf Johann Karl Wezel und Jean Paul* (Würzburg: Königshausen & Neumann, 1989), chap. 1; and Harald Schöndorf, "Der Leib und sein Verhältnis zur Seele bei Ernst Platner," *Theologie und Philosophie*, 60, 1985, 77–87.

27. Carsten Zelle, ed., *"Vernünftige Ärzte": Hallesche Psychomedizinischer und die Anfänge der Anthropologie in der deutschsprachigen Frühaufklärung* (Tübingen: Max Niemeyer, 2001).

28. Platner, *Anthropologie*, pp. xv and xvii.

29. "Die Erkenntnis des Menschen wäre, wie mir dünkt, in drey Wissenschaften abzutheilen. Man kann erstlich die Theile und Geschäffte der Maschine allein betrachten, ohne dabey auf die Enschränkungen zu sehen, welche diese Bewegungen von der Seele empfangen, oder welche die Seele wiederum von der Maschine leidet; das ist Anatomie und Physiologie. Zweytens kann man auf eben diese Art die Kräfte und Eigenschaften der Seele untersuchen, ohne allezeit die Mitwirkung des Körpers oder die daraus in der Maschine erfolgenden Veränderungen in Betrachtung zu ziehen; das wäre Psychologie, oder welches einerley ist, Logik, Aesthetik und ein großer Theil der Moralphilosophie. . . . Endlich kann man Körper und Seele in ihren gegenseitigen Verhältnissen, Einschränkungen und Beziehungen zusammen betrachten, und das ist es, was ich Anthropologie nenne"; ibid., pp. xv–xvii.

30. Ibid., p. xviii.

however, Platner dealt mainly with psychological subjects, such as the generation of ideas, memory, the imagination, reason, and genius. He treated them in a psycho-physiological perspective, by situating the seat of the soul in the brain marrow, along with the mechanical and hydraulic processes which ensured the link between the body and the soul.[31] In 1790, Platner published a *New Anthropology for Physicians and Philosophers*. Despite its greater length and Platner's claim that its content was entirely original, it was not substantially different from the first version. It devoted more space to the human body as a whole and addressed anatomical issues, and particularly neurological ones, as well as the question of the seat of the soul. That accounted for only one-fifth of the book, and the remainder was devoted to the operations of the soul. It was to be followed by other volumes, which were never published, on mental pathology, the temperaments, the sexes, and the stages of life.[32]

Platner introduced a new idea that circumvented the perennial problem of soul-body interaction. He divided the "organ of the soul" into two and advocated *influxus physicus* over against occasionalism or preestablished harmony. His theses resembled those of the "animist" chemist and physician Georg Ernst Stahl (1660–1734), who, in his *Theoria medica vera* of 1707, rejected the view of the body as a simple machine and attributed its form and functions to an immaterial vital principle which he identified with the soul (*anima*) or nature. Platner thought that animal movement was caused by a soul power, a *Seelenkraft*, rather than by the irritability of muscular fibers (as Albrecht von Haller had maintained) (§ 270). For him, the human body was made up of an animal mass and a "nerve spirit" (*Nervengeist*), a subtle and invisible principle, different from the other fluids and from the solid parts of the body (§§ 1–3). This *Nervengeist* was the organ of the soul (§ 186). It followed that creatures without nerves, without an organ of the soul, and so without a soul, were not truly animals, but rather plants in animal form (§ 281). There were, moreover, two organs of the soul, a spiritual one (*das*

31. See also the aphorisms Platner published for his teaching. For example: "Alle bedeutende Zerrüttungen des Gehirns hindern überhaupt die Seelenwirkungen" (§ 66); "Die innere Seelenorgan, oder der Sitz der Seele, ist derjenige Theil des Körpers, in welchem alle Sinneneindrücke sich endigen, alle Bewegungen des Körpers sich anfangen, und alle Seelenwirkungen zunächst sich äußern" (§ 67). Ernst Platner, *Philosophische Aphorismen nebst einigen Anleitungen zur philosophischen Geschichte*, 2nd ed. (1793–1800; Brussels: Culture et Civilisation, 1970).

32. Ernst Platner, *Neue Anthropologie für Aerzte und Weltweise: Mit besonderer Rücksicht auf Physiologie, Pathologie, Moralphilosophie und Aesthetik*, vol. 1 (the only volume published) (Leipzig: bey Siegfried Lebrecht Crusius, 1790); referred to in the text by paragraph number (§). Platner frequently refers to his *Philosophische Aphorismen*. On the considerable influence of Platner's *Aphorisms* and anthropological works, see Kosenina, *Ernst Platners Anthropologie*.

geistige Seelenorgan) and an animal one (*das thierische Seelenorgan*), from which the higher and lower senses were derived (§§ 210–214). The human soul needed both, since if it had only the spiritual organ, it would be an angel, and if it had only the animal organ, it would be an animal. Man resulted from the union and joint influence of both.[33] Once again, only as a composite being can man be the proper object of anthropology; as Platner stated, "Man is the totality of the soul and the body, insofar as modifications in the soul are partly the effects and partly the causes of modifications in the body, such that through this mutual causal relation, body and soul are united in the most intimate and precise manner."[34]

In conclusion, if there is a leitmotiv in Enlightenment anthropology, it is certainly the idea that man as a composite totality of soul and body must be taken as an autonomous object of study. Despite their profound differences, Chavannes, Marat, and Platner all claimed that their anthropologies prefigured a new science, and they stressed the novelty of trying to understand the human being in accordance with his nature. This was probably not the human being Michel Foucault was referring to when he claimed that man did not exist in the eighteenth century; and yet, pace Foucault, this is the being who became, largely through empirical psychology, the "foundation of all positivities."

A REPUBLIC OF LETTERS

The development of psychology as a discipline, like the Enlightenment itself, displayed both considerable unity and national variations. Whatever their national origins, eighteenth-century psychologists championed empirical methods and shared a common Lockean heritage. But there were of course differences. Johann Nicolas Tetens (1736/38–1807), one of the major writers on psychology of the German Enlightenment, favored the British method of observation over against French speculative (*raisonnierende*) philosophy and the "geometric leanings" of the school of Leibniz and Wolff in his own

---

33. "Hätte die menschliche Seele nur allein das geistige Seelenorgan, so wäre sie Engel; nur allein das thierische, so wirkte sie als Thier. Durch die einige Verbindung und durch die Vermischten Einwirkungen beyder, wird das Mittelding zwischen Engel und Thier—der Mensch" (§ 215).

34. "Der Mensch ist in sofern das Ganze von Seele und Körper, wiefern die Veränderungen der Seele, theils Wirkungen, theils Ursachen von Veränderungen des Körpers, folglich Seele und Körper, durch dieses wechselseitige ursachliche Verhälhtniß, innigst und genau ve[r]bunden sind" (§ 175).

country.[35] He noted that this method—*die beobachtende*—which Locke, he said, was the first to use to study the understanding, was also employed in Germany by "our psychologists" in the field of empirical psychology *(Erfahrungs-Seelenlehre)*.[36] The issues which these psychologists addressed through observation were defined partly within the terms of Wolff's philosophy and partly in relation to French and French-speaking authors.

In practice, national trends, as far as they went, were rooted in institutional and socio-religious contexts that had specific characteristics but were often closely interwoven and quick to absorb other influences. The world in which psychology developed still belonged to a Republic of Letters. Works circulated, particularly from English and French to German and Italian,[37] and although Switzerland, Italy, and Germany were composed of autonomous states, they were viewed as single entities and as nations; Tetens spoke of "our psychologists," and travelers went "to Italy" or "to Switzerland."

In the eighteenth century, psychology blossomed particularly in Germany. This can be attributed to the history and pedagogical requirements of the Protestant universities, in which prevailed a tradition of metaphysical investigation: many German philosophers and psychologists, contrary to their counterparts in England and France, refused to reduce metaphysics to a sensualist theory of the origin of knowledge and pursued research into first principles. The deductive schemas and "geometric" method were abandoned, but not the imperative to arrive at fundamental universal concepts. Empirical psychology could provide the starting point for this kind of research, even if not its outcome.[38]

As we have seen, what was called "psychology" up to the end of the seventeenth century was generally based on the Aristotelian notion of the

35. Johann Nicolas Tetens, "Über die allgemeine speculativische Philosophie" (1775), in *Über die allgemeine speculativische Philosophie: Philosophische Versuche über die menschliche Natur und ihre Entwickelung*, ed. Wilhelm Uebele (Berlin: Verlag von Reuther & Reichard, 1913), p. 1 (I refer throughout to the original page numbers, which Uebele reproduces in the margin of his edition). Cf. Jeffrey Barnouw, "Psychologie empirique et épistémologie dans les *Philosophische Versuche* de Tetens," *Archives de philosophie*, 46, 1983, 271–289, p. 278. See also the useful introduction by Raffaele Ciafordone to translated extracts of the *Versuche* and *Über die allgemeine*: J. N. Tetens, *Saggi filosofici e scriti minori*, ed. Raffaele Ciafardone (L'Aquila: L. U. Japadre, 1983).

36. Tetens, *Philosophische Versuche* (1775), "Vorrede," p. iv.

37. See, for example, Manfred Kuehn, *Scottish common sense in Germany, 1768–1800: A contribution to the history of critical philosophy* (Kingston: McGill-Queen's University Press, 1987); on the dissemination of Bonnet's work, see Jacques Marx, *Charles Bonnet contre les Lumières 1738–1850* (Oxford: Voltaire Foundation, 1976).

38. See for example Michel Puech, "Tetens et la crise de la métaphysique allemande en 1775 (*Über die allgemeine speculativische Philosophie*)," *Revue philosophique*, no. 1, 1992, 3–29.

soul. Psychology was consequently the generic science of animate beings, serving as an introduction to the study of plants, animals, and man. When the Aristotelian frameworks disintegrated, by the 1720s at the latest, psychology became the science of the human mind. In university circles, it was Christian Wolff who gave this shift its most systematic form. Hegel mentions in his *Lectures on the History of Philosophy* that Wolff gave the discipline a systematic structure which had served as a standard "down to the present day," that is, until the 1820s.[39] This remark testifies to the immense influence of Wolff's thought in the university, as well as to its impact on philosophy and psychology. As explained above, Wolff developed two psychologies: one, "empirical," was a posteriori and based on experience; the other, "rational," was a priori and proceeded by deduction from definitions and axioms. Both were covered in the many Wolffian textbooks that remained in use until the end of the century.

Despite Wolff's role in refashioning German philosophy in non-Aristotelian terms, and the persistence of rational psychology in university curricula and philosophical lexicons,[40] developments in psychology essentially drew on the empirical approach. Johann Georg Sulzer's rallying cry was in this respect typical. In a 1759 account of "all the sciences and other branches of knowledge," he stated that one cannot altogether deny Wolff the credit for having invented and defined empirical psychology. Although he was not the first to carry out psychological observations, no one before him had collected the results into a system in which they were interlinked and organized on a conceptual basis. Nevertheless, said Sulzer, this part of philosophy can make even greater progress, and since the study of the human soul is the most noble of sciences, so empirical psychology is the most desirable of intellectual endeavors.

Sulzer wanted the "obscure objects of the soul," that is, the soul's behavior in cases of unclear knowledge or rapid judgment, to be investigated in preference to others.[41] He also thought it useful to collect extraordinary psychological cases which could not be explained by the soul's known properties, such as premonitions and the various forms of madness.[42] Sulzer considered that rational (*erklärende*) psychology was also of great value and deserved a prime position among the sciences, and that it was Wolff who

39. Hegel, *Lectures*, vol. 3, p. 353.

40. For example, Johann Christian Lossius, *Neues philosophisches allgemeins Real-Lexikon oder Wörterbuch der gesammten philosophischen Wissenschaften* (Erfurt: bei J. E. G. Rudolphi, 1805), art. "Psychologie."

41. Sulzer, *Kurzer Begriff*, § 206, pp. 158–159.

42. Ibid., § 207, pp. 159–160.

had developed it into a science, since no one before him had deduced logically (durch Vernunftschlüsse) the properties of the soul from its essence:
"But nobody could have undertaken this before empirical psychology became a well-grounded science."[43]

The numerous attempts to develop empirical psychology, as much by
university professors as by philosophers, doctors, or theologians without
links to the university, were closely related to the "anthropological turn" of
the first half of the eighteenth century.[44] While psychology in the German
Aufklärung did not form a homogeneous whole, it generally resisted Locke's
ascendancy.[45] The emphasis was metaphysical or psycho-physiological to
different degrees, more or less closely linked to philosophy or medicine, and
more or less concerned with elucidating the nature of the soul or applying
psychology to ethics, aesthetics, pedagogy, or morals. This profusion was in
evidence not only in curricula, university textbooks, and treatises, but also
in journals. From 1750 onward, for example, the Berlin Academy regularly
published psychological articles in its Memoirs, occasionally using "psychology" as a classificatory category.[46]

In the 1780s and 1790s the number of psychological and anthropological journals increased exponentially.[47] These journals, which were intended
for an enlightened audience, sought to develop psychology as an empirical
science with moral and "philanthropic" goals rather than as a university
discipline. The first and best-known one, the Magazin zur Erfahrungsseelenkunde (1783–1793), rejected the abstract and intellectualizing language of
the mental faculties in favor of observations which often took the form of
existential and personal narratives, based on the examination of the self.

---

43. "Dieses konnte auch niemand unternehmen, ehe die beobachtende Psychologie zu einer
ordentlichen Wissenschaft geworden," ibid., § 209, p. 162.

44. Apart from Dessoir, Geschichte, see Eckart Scheerer, "Die Berliner Psychologie zur Zeit
der Aufklärung (1764–1806)," in Lothar Sprung and Wolfgang Schönpflug, eds., Zur Geschichte
der Psychologie in Berlin (Frankfurt: Peter Lang, 1992); Georg Eckardt, Matthias John, Temilo van
Zantwijk, and Paul Ziche, eds., Anthropologie und empirische Psychologie um 1800 (Cologne:
Böhlau, 2001); and Zelle, "Vernünftige Ärzte." For the "anthropological turn" and its periodization, see Werner Krauss, Zur Anthropologie des 18. Jahrhunderts: Die Frühgeschichte der Menschheit im Blickpunkt der Aufklärung, ed. Hans Kortum and Christa Gohrisch (1978; Frankfurt:
Ullstein, 1987); and Carsten Zelle, "Sinnlichkeit und Therapie: Zur Gleichsprünglichkeit von
Ästhetik und Anthropologie um 1750," in Zelle, "Vernünftige Ärzte," pp. 5–10.

45. Klaus P. Fischer, "John Locke in the German Enlightenment: An interpretation," Journal
of the history of ideas, 35, 1975, 431–446.

46. The journal of the Berlin Royal Academy changed its name during the period we are
considering: History of the Royal Academy of Sciences and Arts (1745–1769), New Memoirs . . .
(1770–1786), Memoirs . . . (1787–1804).

47. Georg Eckardt and Matthias John, "Anthropologische und psychologische Zeitschriften
um 1800," in Eckardt et al., Anthropologie.

It advocated an empirical science of the soul (*Erfahrungsseelenkunde*) that emphasized *Erfahrung*, experience, not only as an epistemic category, but as including a personal experiential dimension, and its studies of both normal and pathological cases were supposed to contribute to a "moral medicine" capable of bettering humanity through self-knowledge.[48] The numerous accounts and discussions of dreams, for instance, highlight the role played by psychological material in the constitution of an identity and a "bourgeois experience of the self."[49]

Other journals followed in its footsteps. The *Psychological Journal* (1796–1798), edited by the Jena philosophy professor Carl Christian Erhard Schmid (1762–1812), was more interested in the systematic conceptual development of psychology but did not neglect the pragmatic or philanthropic aspects either. The *General Repertory of Empirical Psychology and Related Sciences*, by the theologian Immanuel David Mauchart (1764–1826), featured numerous accounts of pathological or altered states—drunkenness, dreams, madness, premonitions, visions, and split personality—as well as articles on morals, psychology in pedagogy and aesthetics, the history of psychology, and children (Mauchart published a diary of the "physical and psychological development" of his own daughter in 1798).[50] Mauchart, who advocated *Aufklärung* for the greatest number, sought to make psychology serve the general interest and be useful for all. In summary, empirical psychology after Wolff rapidly became a flourishing enterprise, seen as a most promising science and one of the principal tools of enlightenment.

It is well known, however, that Immanuel Kant refused to accept psychology as a science. In the light of Kant's notion of science, this is not as dramatic as it may seem. Indeed, science was for Kant, as it was for the Aristotelians and Wolffians, necessary knowledge, capable of being produced not only empirically, but also a priori, with apodeictic certainty. The

---

48. Karl Philipp Moritz, ed., ΓΝΩΘΙ ΣΑΥΤΟΝ *oder Magazin zur Erfahrungs-Seelenkunde als ein Lesebuch für Gelehrte und Ungelehrte* (1783–1793; Nördlingen: Franz Greno, 1986), 10 vols. The title *Gnothi sauton* means "know thyself."

49. Doris Kaufmann, *Aufklärung, bürgerliche Selbsterfahrung und die "Erfindung" der Psychiatrie in Deutschland 1770–1850* (Göttingen: Vandenhoeck & Ruprecht, 1995); and Kaufmann, "Dreams and self-consciousness: Mapping the mind in the late eighteenth and nineteenth centuries," in Lorraine Daston, ed., *Biographies of scientific objects* (Chicago: University of Chicago of Press, 2000).

50. *Allgemeines Repertorium für empirische Psychologie und verwandte Wissenschaften*, 1792–1793, 1798–1799, 1801); *Neues allgemeines Repertorium . . .* (1802–1803). On Mauchart's diary on his daughter (and other similar works), see Heidrun Diele, "'Man kann sich nun immer mehr mit ihr abgeben . . .': Tagebuch eines Vaters über seine 1794 geborene Tochter," *BIOS*, 13, 2000, 125–134.

Critique of Pure Reason (1780), demonstrates the impossibility of rational psychology: what can be said about the nature of the soul, starting from the proposition "I think," is either derived from experience (in which case it has to do with one's lived sense of self) or else applies to the unity of experience (in which case it is not knowledge but a general subjective form of representation). Psychology, therefore, neither uses nor provides knowledge gained a priori through pure concepts. The psychology to which Kant refused the rank of science in his critical writings was based on introspection or the sense of self (*Selbstgefühl*) and corresponded to Wolff's or Baumgarten's approach.[51] Thus, as he explained in his lectures on metaphysics from the 1770s, and later in his *Metaphysical Foundations of Natural Science* (1786), psychology can at best be a natural, "historical," and a posteriori description of the soul, a "physiology" of the internal sense and of thinking beings, as physics is the "physiology" of the external sense and material beings.

To the extent, therefore, that psychology lacked a "pure" element by virtue of which mathematics could be applied a priori to its object, the soul, Kant deemed that it could not be considered a science.[52] He extended this argument to chemistry but did not, however, exclude the application of mathematics to phenomena such as the intensity of sensation. Nineteenth-century German psychology developed partly in reaction to Kant's position.[53]

We should thus bear in mind Kant's notion of science in trying to to make sense of his critique of psychology. For, despite this critique, Kant remained very interested in psychology throughout his career and, as we have seen, thought it should become a university discipline. In his *Attempt to Introduce the Concept of Negative Magnitudes into Philosophy* (1763), he upheld the possibility of calculating our experience of pleasure and displeasure mathematically.[54] He used the textbook of the Wolffian Alexander Gottlieb Baumgarten (1714–1762) for his lectures on metaphysics, adopting

51. Thomas Sturm, "Kant on empirical psychology: How not to investigate the human mind," in Eric Watkins, ed., *Kant and the sciences* (New York: Oxford University Press, 2001).

52. Gary Hatfield, "Empirical, rational, and transcendental psychology: Psychology as science and as philosophy," in Paul Guyer, ed., *The Cambridge companion to Kant* (Cambridge: Cambridge University Press, 1992); Theodor Mischel, "Kant and the possibility of a science of psychology," *Monist*, 51, 1967, 599–622.

53. David Leary, "The philosophical development of the conception of psychology in Germany, 1780–1850," *Journal of the history of the behavioral sciences*, 14, 1978, 113–121; Leary, "Immanuel Kant and the development of modern psychology," in Mitchell G. Ash and William R. Woodward, eds., *The problematic science: Psychology in the nineteenth century* (New York: Praeger, 1982).

54. Wolf Feuerhahn, "Entre métaphysique, mathématique, optique et physiologie: La psychométrie au XVIIIᵉ siècle," *Revue philosophique*, 193, 2003, 279–292.

its structure and contents, including for psychology.[55] His *Anthropology*, published in 1798 and based on lectures given between 1772 and 1796, was in large part an empirical psychology, enhanced by material drawn from historical, biographical, and literary works, as well as travel writings. Its emphasis was "pragmatic" (rather than "physiological"); that is, it aimed to fashion conduct and character, and to show not what nature makes of man but "what *he*, as a free-acting being makes of himself, or can and should make of himself."[56] Like many of his contemporaries, Kant considered empirical psychology to be one of the essential tools of *Aufklärung*. Obeying the Enlightenment's prescription of *Sapere aude!* ("Dare to know") began with reflective knowledge of the self, which should then lead to a science of man.[57]

The Scottish Enlightenment's principal goal was just such a science of man, which was to include disciplines ranging from psychology and "animal economy" to moral philosophy and political economy. The watchword of its psychology was Hume's aspiration to apply "experimental philosophy to moral subjects."[58] But this project did not immediately include a *psychology*. For example, the thoroughly Lockean *Encyclopaedia Britannica*, printed in Edinburgh from 1768 to 1771, saw psychology as the metaphysical branch of anthropology. Anthropology was divided into the sciences of the human body and "a metaphysical examination of man, his existence,

55. Baumgarten, *Metaphysica* (1739; German translation 1776, 1783). Beck (*Early German philosophy*, p. 283) is probably quite right in stating that Baumgarten was "the most competent—and in the long run perhaps the only philosophically important—adherent of the Wolffian philosophy."

56. Immanuel Kant, *Anthropology from a pragmatic point of view*, trans. Robert B. Louden (Cambridge: Cambridge University Press, 2007), p. 231 (*Gesammelte Schriften*, vol. 7. p. 119). See Gary Hatfield, "Kant and empirical psychology in the 18th century," *Psychological science*, 9, 1998, 423–428; and Norbert Hinske, "Wolffs empirische Psychologie und Kants pragmatische Anthropologie: Zur Diskussion über die Anfänge der Anthropologie im 18. Jahrhundert," *Aufklärung*, 11, 1999, 97–107.

57. Kant made *Sapere aude!* into the watchword of the Enlightenment at the beginning of his text "Was ist Aufklärung?" (1784). It meant, for Kant, having the courage to use one's own reason. Yet the same injunction, taken from Horace's lesson to young Lollius on Stoic morality (*Epistles*, I.2.40), was also the motto of the Society of Alethophiles ("friends of truth"), a Berlin society named after the dedication in Wolff's *German Metaphysics* (1720); their emblem featured a bust of Minerva with *Sapere aude* and the portraits of Leibniz and Wolff engraved on her helmet.

58. Gladys Bryson's excellent book on the Scottish Enlightenment has stood the test of time: *Man and society: The Scottish inquiry of the eighteenth century* (Princeton: Princeton University Press, 1945). See also Peter Jones, ed., *Philosophy and science in the Scottish Enlightenment* (Edinburgh: John Donald, 1988); P. Jones, ed., *The "science of man" in the Scottish Enlightenment: Hume, Reid and their contemporaries* (Edinburgh: Edinburgh University Press, 1989); and Daniel Bruhlmeier, Helmut Holzhey, and Vilem Mudroch, eds., *Schottische Aufklärung: A hotbed of genius* (Berlin: Akademie Verlag, 1996).

his essence, his essential qualities and necessary attributes, all considered a priori"; this examination led to psychology as the theory of "the soul in general, and the soul of man in particular."[59] The *Encyclopaedia* assimilated psychology to older discussions on the soul's origins, essence, immateriality, and immortality—all topics on which, it observed, "the most subtle and abstract researches have been made, that the human reason is capable of producing; and concerning the substance of which, in spite of all these efforts, it is yet extremely difficult to assert any thing that is rational, and still less any thing that is positive and well supported."[60] For the *Encyclopaedia*, what went by the name of "psychology" was simply a remnant of pre-Lockean approaches. The true psychology was logic, which, insofar as it was based on the study of the mental faculties, explained "the nature of the human mind, and the proper manner of conducting its several powers."[61]

In its Scottish version, psychology was conceived as part of a unified science of man. For example, in the mid-eighteenth century, John Gregory (1724–1773), a professor of medicine at the University of Aberdeen, set to work on a natural history of the mind that was to examine all the faculties, from sensation to the understanding, and also the signs of the union between the soul and the body. Gregory's approach was historical and comparative and included research into animals and humans in different cultures and at different historical moments. He was opposed to the idea of the body as a simple machine (more specifically, to Hermann Boerhaave's iatromechanism), and aligned himself with a Stahlian animism, at least as regards his belief in a vital and sensitive principle uniting body and soul.[62] He explicitly positioned his work within the framework of a natural history of man and highlighted its continuity with a history of man as a social being. Another figure in this context was Thomas Reid (1710–1796), the principal philosopher of "common sense," a professor of philosophy at King's College, Aberdeen, and later of moral philosophy at the University of Glasgow. He lectured on "pneumatology"; what he taught under this name, however, was not the science of spiritual beings (the usual sense of the term) but a "history" of the human mind which explored, among other things, sensa-

59. William Smellie, ed., *Encyclopaedia Britannica; or, a Dictionary of Arts and Sciences* (Edinburgh: printed for A. Bell and C. Macfarquhar, 1768–1771), METAPHYSICS, 3, 174–203, p. 175.

60. Ibid.

61. *Encyclopaedia Britannica* (1768–1771), LOGIC, 2, 984–1003, p. 984.

62. John Gregory, *A comparative view of the state and faculties of man with those of the animal world* (1765). The work contains papers presented to the Philosophical Society of Aberdeen between 1758 and 1763. See Paul B. Wood, "The natural history of man in the Scottish Enlightenment," *History of science*, 28, 1990, 90–123.

tion, memory, the imagination, the understanding, judgment, and reasoning.[63] Reid is a good example of how a discipline such as "pneumatology" was not an end in itself but was conceived as the foundation for doctrines of knowledge, human freedom, and individual and social morality.

Reid did not mention "psychology," but the term was neither unknown in English nor always associated with Scholastic obscurantism. The English physician and philosopher David Hartley (1705–1757) placed psychology— or the "theory of the human mind, with that of the intellectual principles of brute animals"—within natural philosophy, alongside mechanics, hydrostatics, pneumatics, optics, astronomy, chemistry, medicine, and theories of arts and crafts.[64] In the 1760s, James Beattie (1735–1803), a professor of moral philosophy at the University of Aberdeen, lectured on "psychology, or science of the nature of the several powers or faculties of the human mind."[65] Although he considered psychology to be part of pneumatology, he treated it as an empirical science and did not address the immortality and immateriality of the soul.

Alexander Gerard (1728–1795), another Aberdeen professor, took a different approach. He divided his course on pneumatology into psychology and natural theology. Psychology was treated as natural philosophy but ended with a demonstration of the immateriality and immortality of the soul in order to lead on to natural theology. Gerard was particularly sensitive to the problems of the introspective method:

> Psychology is a Science of the most extensive utility, for all the other sciences are connected with it, and in a manner all our conclusions in it, as well as in Natural Philosophy, must be founded on Experiments and observations; Sometimes, the Phaenomena of the human mind is obvious to our consciences, at other times they must be learned by observing their Effects in Life. It is easy to make Experiments upon Body, but more

63. The basic principle of the philosophy of "common sense" is that there is a way of knowing and acting proper to human beings. Human nature is expressed in a common sense shared by all men, particularly in ordinary language and in grammatical elements (for example, the distinction between active and passive voice) which are present in all languages. Reid also thought that mental life cannot be reduced to sensation, and that the functioning of sensation presupposes certain "judgments" (such as causality or induction). See in particular Reid, *An inquiry into the human mind on the principles of common sense* (1764), *Essays on the intellectual powers of man* (1781), and *Essays on the active powers of man* (1788).

64. David Hartley, *Observations on man, his frame, his duty, and his expectations* (1749; London: printed for T. Tegg and Son, 1834), p. 223.

65. James Beattie, *Elements of moral science* (Edinburgh: printed for T. Cadell, 1790), vol. 1, p. 1.

difficult to form experiments concerning the human mind because it is very inconstant.[66]

In summary, Scottish Enlightenment works on what was called "psychology"—or, more commonly, something like the "philosophy of the human mind"[67]—adopted an empirical and naturalist approach. This approach was at first Lockean but was later dominated by the antiskepticism of the "common sense" philosophers. Psychology in this context formed the basis of logic, aesthetics, and moral and political philosophy and belonged in a science of man (conceived as an individual, social, and historical being) that became a key feature of curriculum reform in eighteenth-century Scotland.[68]

The *Encyclopaedia Britannica*'s objections to the word "psychology" were also present in the French context, but exacerbated by antireligious and anticlerical polemic. For the *philosophes*, the term smacked of Scholasticism, of the clergy's abstruse teachings, of unintelligible concepts and notions with no real referent. They therefore rejected it, as we saw above with Condillac, Destutt de Tracy, Garat, and Condorcet. The term occurred no more frequently, however, in the *antiphilosophe* authors who were in constant dialogue with their more famous opponents, and who would deserve (at least on those grounds) to be given proper acknowledgment within the Enlightenment movement.[69]

In their defense of Christianity against materialism, skepticism and fatalism, the *antiphilosophes* foregrounded the active nature of the soul, as well as the unity of the self and consciousness. They thus opposed the idea of a mechanically functioning organic machine which reacted more or less passively to sensory stimuli. The most remarkable of these thinkers, the Oratorian Joseph Adrien Lelarge de Lignac (1710–1762), attacked the "Optimist-Fatalists" in particular.[70] He accused them of subordinating God to neces-

---

66. Notes taken during Gerard's lectures on moral philosophy, cited in Paul B. Wood, *The Aberdeen Enlightenment: The arts curriculum in the 18th century* (Aberdeen: Aberdeen University Press, 1993), p. 110.

67. Dugald Stewart, *Elements of the philosophy of the human mind*, published in three volumes (1792–1827). Stewart, a disciple of Thomas Reid, was a professor of mathematics and later of moral philosophy at Edinburgh.

68. On this last point, see Wood, *The Aberdeen Enlightenment*, chaps. 2 and 5 (on the humanities, moral philosophy, the science of man, and the study of antiquity).

69. Robert R. Palmer, *Catholics and unbelievers in 18th-century France* (1939; New York: Cooper Square Publishers, 1961); Didier Masseau, *Les ennemis des philosophes: L'antiphilosophie au temps des Lumières* (Paris: Fayard, 2000).

70. Lelarge de Lignac's *Lettres à un Américain* (1751) criticized Buffon and championed the feeling of existence and individuality over against sensualism and the axiom that *nihil est in intellectu quod non prius fuerit in sensu*; he took a similar stance in his *Élémens de métaphysique*

sity and attributing an autonomous motor force to the soul (rather than to the laws governing the union of the two substances or, better still, to their Creator). As a disciple of Descartes and Malebranche, he countered the sensualists with the "sense of self" and experiences such as the limits of one's own will. He based the self on a "feeling of existence" irreducible to sensation and thought and railed against "our century's obsession" with the idea that we "do not know our own soul, although we can feel that we exist and can distinguish in ourselves what is active and what is passive, and cannot mistake ourselves for another."[71]

The "obsession" that characterized Lignac's opponents lay behind much of Enlightenment empirical psychology.[72] The critics of the notion of the soul as a spiritual substance without which no self is possible adopted a range of strategies.[73] The least risky involved describing the "movements" of the soul as manifested physiognomically or in the passions. A second strategy was to reformulate metaphysical concepts in psychological or physical terms: the *soul* gave way to the *mind*, which was then absorbed into the sphere of the mental (*le moral*). The third approach reduced mental operations to the functioning of matter, in line with various materialist doctrines, from La Mettrie's *The Natural History of the Soul* (1745) and *Man a*

---

tirés de l'expérience: ou Lettres à un matérialiste sur la nature de l'âme (1753). The most comprehensive study on Lignac remains F. Le Goff's thesis, *De la philosophie de l'abbé de Lignac* (Paris: Hachette, 1863). Maine de Biran was influenced by Lignac's work in his rejection of sensualism; see François Azouvi, "Genèse du corps propre chez Malebranche, Condillac, Lignac et Maine de Biran," *Archives de philosophie*, 45, 1982, 85–107.

71. Joseph Adrien Lelarge de Lignac, *Le témoignage du sens intime et de l'expérience, opposé à la foi profane et ridicule des fatalistes modernes* (Auxerre: François Fournier, 1760), vol. 2, p. 166. The work is essentially a critique of Bonnet's *Essai de psychologie*. Lignac claimed Bonnet himself had sent him the work via a third party (vol. 1, pp. 53–54); he knew the author's identity, but respected his anonymity by always referring to him as "the Psychologist." For Lignac's arguments against Bonnet, see *Le témoignage*, vol. 1, chap. 1; vol. 2, chaps. 5 and 6; vol. 3, chaps. 6 and 7. For Lignac's notion of the self, see Jean A. Perkins, *The concept of the self in the French Enlightenment* (Geneva: Droz, 1969), chap. 3.

72. I can give only the briefest account of what can be called current historiography here. See Sergio Moravia, *La scienza dell'uomo nel Settecento* (Bari: Laterza, 1978); Moravia, *Filosofia e scienze umane nell'età dei lumi* (Florence: Sansoni, 1982); Moravia, "The capture of the invisible: For a (pre)history of psychology in 18th-century France," *Journal of the history of the behavioral sciences*, 19, 1983, 370–378; and Moravia, *Il pensiero degli Idéologues: Scienza e filosofia in Francia (1780–1815)* (Florence: La nuova Italia, 1974).

73. Since Gabriel Bonno's *Les relations intellectuelles de Locke avec la France* (Berkeley: University of California Press, 1955), three important works have been devoted to Locke in France: John Yolton, *Locke and French materialism* (Oxford: Clarendon Press, 1991); Ross Hutchinson, *Locke in France, 1688–1734* (Oxford: Voltaire Foundation, 1991); and Jørn Schløsler, *John Locke et les philosophes français: La critique des idées innées en France au dix-huitième siècle* (Oxford: Voltaire Foundation, 1997).

Machine (1748) to d'Holbach's *The System of Nature* (1770) and Cabanis's
On the relations between the physical and moral aspects of man (1802).

In France as elsewhere in Europe, empiricism informed treatises on education and inspired programmes of pedagogical reform.[74] Toward the end
of the eighteenth century, psychology even existed briefly in institutional
form. The Idéologues integrated their "analysis of the understanding,"
which they saw as indispensable to the accomplishment of their political
ideals, into the *écoles centrales* (secondary schools), the École normale, and
the Institut, all institutions they had helped create during the French Revolution.[75] A certain François-Joseph Benoni Debrun (1765–1845), who was a
constitutional cleric before marrying and abandoning the priesthood, was
appointed grammar teacher at the *école centrale* of the Aisne département
in 1796. National directives prescribed that his lessons should include ideology, general grammar, French grammar, and logic. Debrun merged grammar,
logic, and morals into a "single corpus," preceded by a "psychography," or
"descriptive treatise on the soul." As he explained in the published version
of 1801, this arrangement of subjects formed "a course for which the title
Psychology is well suited, since its objects are the soul and its operations."[76]
As a sensualist, Debrun advocated "analysis," defined the soul as "the being
which has the potential to feel," and characterized thought as "any of the
soul's ways of being, which analysis can reduce to sensation."[77] He was not,
however, a materialist, maintaining that "my body is not my self, but a machine." His adherence to the immateriality of the soul was based on well-
worn arguments supposed to prove the heterogeneity of the two substances
("we see by means of the eye, but the eye itself does not see").[78]

Although Debrun's course was comparatively idiosyncratic, only its
title was really surprising, insofar as it mentions a term which the Idéologues tended to denounce and refused to use. Perhaps the continued use
by Debrun of the term "soul" ("psychography," "psychology") was a ves-

74. Marcel Grandière, "Le sensualisme dans les traités d'éducation à la fin du XVIII° siècle
(1760–1789): Quelques aspects," in Hubert Hannoun and Anne-Marie Drouin-Hans, eds., *Pour
une philosophie de l'éducation* (Dijon: CNDP, 1994); Grandière, *L'idéal pédagogique en France
au dix-huitième siècle* (Oxford: Voltaire Foundation, 1998), chaps. 5–7.

75. François Azouvi, ed., *L'institution de la raison: La révolution culturelle des Idéologues*
(Paris: Vrin, 1992). I touched upon Garat and his teachings in chap. 1.

76. François-Joseph Benoni Debrun, *Cours de psychologie: Traité de psychographie* (1801),
in Serge Nicolas, ed., *Un cours de psychologie durant la Révolution française de 1789: Le traité
de psychographie de Benoni Debrun* (Paris: L'Harmattan, 2003), p. vii. Page numbers in the text
refer to the original.

77. Ibid., § 10, p. 18.

78. Ibid., § 77 and 78, p. 88.

tige of his ecclesiastical past. The term entered academic curricula only after Napoleon had stifled the Idéologie movement, and principally after the Restoration. "Psychology" then brought the "sense of self" back into favor and was characterized, in the works of Maine de Biran (1766–1824), Pierre Laromiguière (1756–1837), Victor Cousin (1792–1867), and others, by a spiritualist eclecticism critical of Enlightenment sensualism and the physiological orientation of someone like Cabanis.

The Swiss scene was extremely varied, politically, religiously, and intellectually. Due to the large number of Swiss abroad, it was also influential beyond its borders.[79] Historians debate the existence of a "Swiss Enlightenment," but there is some consensus on Switzerland's role as an intermediary between the French *Lumières* and the German *Aufklärung*. What unified the Swiss Enlightenment was its adherence to Christianity understood as a religion compatible with reason and the scientific and epistemological developments of the century.[80] Eighteenth-century Swiss thinkers themselves characterized what was "Swiss" about their works and their country in terms of a middle ground between theological orthodoxy and atheism, a mystical appreciation of nature and a mechanistic vision of the world, and between theology and anthropology.[81] The connection between theology and anthropology, which we have already encountered in Chavannes, is particularly important: whereas in the French context the development of a natural history of man seemed to require an anti-Christian stance, as well as a critique of the concept of the soul and of the human being as a composite of two substances, in Switzerland such "dualism" and its Christian framework played a positive role in the advent of the human sciences. The significant impact of Charles Bonnet's work is another sign of this positive role, which will be further illustrated in chapters 7 and 8.

The Italian situation was as complex, but in a different way.[82] We will see below that, unlike its English, German, and French counterparts, the

79. The "Berlin Swiss" were particularly important in this respect. See Fontius and Holzhey, *Schweizer in Berlin*.

80. Ulrich im Hof, *Aufklärung in der Schweiz* (Bern: Francke, 1970); Samuel S. B. Taylor, "The Enlightenment in Switzerland," in Porter and Teich, *The Enlightenment*. It is symptomatic of this historiographical context that whereas im Hof attributes a certain role to Catholic thinkers, Taylor considers the Swiss Enlightenment to be essentially a Protestant phenomenon.

81. Helmut Holzhey and Simone Zurbuchen, "Die Schweiz zwischen deutscher und französischer Aufklärung," in Werner Schneiders, ed., *Aufklärung als Mission: Akzeptanzprobleme und Kommunikationsdefizite* (Marburg: Hitzeroth, 1993), p. 312.

82. For some of its features, see Luciano Mecacci, "Primi usi della parola 'psicologia' tra Settecento e Ottocento in Italia: Breve nota fino al 1830," *Teoria e modelli*, 8, no. 3, 2004, 31–39.

Italian encyclopedia, published in Venice between 1746 and 1751, had no article on psychology and did not treat the subject at all. While other Enlightenment encyclopedias reflected the development of psychological thought, this was not the case in Italy.

After publishing an antiskeptical essay on the human understanding, the great Modena historiographer Ludovico Antonio Muratori (1672–1750) published another text, on the imagination (1745).[83] This lively work, which was widely translated and often reprinted, mixed psychological views with theological, demonological, moral, and medical issues. While it explicitly abandoned the Scholastic strategy of setting opinions against each other and laying out questions, objections, and replies, it still belonged to a tradition of writings on the imagination which were psychological, even though they did not try to "do psychology."

Some of Muratori's contemporaries took his approach further. His friend Antonio Conti (1677–1749) returned to his native town of Padua after his ordination in Venice to devote himself to mathematics and work with the professor of "experimental philosophy," Antonio Vallisnieri (1661–1730). Conti also pursued his scientific, philosophical, and aesthetic interests in Paris and England. He met Newton, defended Homer in the disputes opposing Ancients and Moderns, was involved in philosophico-theological debates on Newtonianism, wrote on the nature of love and the system of monads, published poems and translations from English, French, Latin, and Greek, and remained in personal and epistolary contact with many scholars and philosophers. He is said to have corresponded with Christian Wolff.[84] In the preface to his works, Conti praised the precision and clarity of Wolff's ontology and empirical psychology. He was a staunch supporter of "experimental philosophy" and was convinced that the roots of aesthetic pleasure were to be found in sensation, the imagination, and the understanding. His essays on imitation, allegory, enthusiasm, and "poetic fantasies" were prefaced by a "short treatise on empirical psychology," which he claimed to be necessary for understanding the subjects broached. For, he said, without a theory of the soul, "one cannot investigate poetic doctrine in any depth."[85]

At his death, Conti left unpublished a voluminous *Treatise on the Soul*, which was most likely completed in 1745. Locke, Malebranche, Vico, and

83. Ludovico Antonio Muratori, *Della forza della fantasia umana* (1745), ed. Carlo Pogliano (Florence: Giunti, 1995).

84. See Giuseppe Toaldo's biographical note at the beginning of vol. 2 of Antonio Conti, *Prose e poesie* (Venice: presso Giambattista Pasquali, 1739–1756); also Nicola Badaloni, *Antonio Conti: Un abate libero pensatore tra Newton e Voltaire* (Milan: Feltrinelli, 1968).

85. Conti, *Prose*, vol. 1, "Prefazione," unpaginated, [3] and [24].

Wolff all left their mark on this work. Conti contested the Cartesian idea of transparent knowledge of the soul, maintaining that it can be known only through the body. He emphasized the inseparability of the two substances and went so far as to point out the extreme difficulty of proving the spirituality of the soul. Since these ideas could be seditious at the time, Conti protected himself by dividing his book into a historical part that reviewed different opinions on God, the human soul, and other spirits, and a psychological and gnoseological part which eschewed theological and metaphysical considerations.[86] Conti treated only the cognitive faculties, and in their most traditional form (secondo le ipotesi degli antichi), that is, the external senses, the common sense, imagination or fantasy, memory, judgment, and the understanding.[87] The first part of the treatise concerned the senses, the second part, the imagination, and the third part, the mind (mente).

In 1745, Conti wrote to the abbot Antonio Genovesi (1712–1769), praising his textbook on metaphysics and asking him to elucidate the nature and origin of ideas.[88] Genovesi, to whom we owe the blossoming of empiricism in the late eighteenth-century Neapolitan Enlightenment, was a professor of moral philosophy at the University of Naples, and later of "civil economy," a new discipline which won the philosopher numerous disciples and was extremely relevant in the context of the Bourbon reforms.[89] In his reply to Conti, Genovesi referred to some of his writings but above all stressed his ignorance of the nature and origin of perceptions and ideas; insofar as we do not and cannot know the essence of the soul, he stated, any inquiry into such questions is necessarily sterile.[90]

In his textbooks, Genovesi of course said nothing so radical. He stood by the principle that substances, essences, and ultimate causes were unknowable

86. An overview of the psychological part can be found in Conti, Prose, vol. 2, § XI, pp. 267–269 ("Trattato delle potenze conoscitive dell'anima"). The preface to the historical part ("Saggio di storia critica de' ragionamenti intorno Dio, l'anima umana e gli altri spiriti") and all the psychological part ("Trattato dell'anima umana") have been published in A. Conti, Scritti filosofici, ed. Nicola Badaloni (Naples: Fulvio Rossi, 1972).

87. Conti, "Trattato," in Scritti, p. 51.

88. Giovanni Gentile, Storia della filosofia italiana dal Genovesi al Gallupi (1929), in Opere (Florence: Sanzoni, 1935), vol. 18, p. 1.

89. Paola Zambelli, "Antonio Genovesi and eighteenth-century empiricism in Italy," Journal of the history of philosophy, 16, 1978, 195–208; Zambelli, La formazione filosofica di Antonio Genovesi (Naples: Morano, 1972), focuses on the intellectual context and theological dimension of Genovesi's work. On Genovesi's Newtonianism, including how it affected his psychology, see Vincenzo Ferrone, Scienza, natura, religione: Mondo newtoniano e cultura italiana nel primo settecento (Naples: Jovene, 1982), pp. 609–641.

90. Letter from Genovesi to Conti of 15 January 1746, cited in Gentile, Storia della filosofia italiana, p. 3.

but still entertained different ways of resolving the issue. Genovesi's Latin textbooks talked of "psychosophy";[91] one of his Italian textbooks for *giovanetti* defined psychology as the science of the soul, its faculties, instincts, passions, virtues, and vices, while another favored the term "anthropology" on the grounds that it was more comprehensive, common, and precise.[92] Despite the way in which the metaphysical elements were presented— *mathematicum in morem adornata*—the content was clearly Lockean. The count Lodovico Barbieri (1719–1791), a nobleman from Vicenza, criticized Genovesi roundly on this account. His 1756 *Treatise on Psychology* set out to refute the theory of innate ideas, Locke's reflections on thinking matter, and the systems of physical influence and preestablished harmony. He sought to show that the soul is an immaterial "active power," and that the mutual dependence of body and soul can only be explained by occasional causes. We are a far cry here from the contents and method of empirical psychology.[93]

Later in the century, the Milanese Cesare Beccaria (1738–1794) drew on associationist and sensualist psychology in his revolutionary work *On Crimes and Punishments* (1764).[94] Punishments, he explained, should be chosen for the lasting impression they make on people's minds, and for the least cruel impression they make on the culprit's body (§ XII). They are more useful when they are swift, because the less time elapses between the crime and the punishment, the stronger and more lasting is the association between the ideas of *crime* and of *punishment*, "such that imperceptibly one comes to be considered as the cause and the other as its necessary and inevitable effect" (§ XIX). Due to the way in which habits are formed and our sensibility functions, an execution is quickly forgotten and leaves no

91. Antonio Genovesi, *Elementa metaphysicae mathematicum in morem adornata*, 2nd ed. (1743; Naples: typis Benedicti, et I. Gessari, 1751), vol. 2, pt. 2 ("Qua continentur principia Psychesophiae"); Genovesi, *Disciplinarum metaphysicarum elementa mathematicum in morem adornata* (Venice: Remondini, 1779), vol. 3, *Psychesophia et Ethica*. The latter work is an *editio novissima* of the first *Elementa*.

92. A. Genovesi, *La logica per gli giovanetti* (1766; Naples: Stamperia Abbaziana, 1790), bk. V ("Dell'Ordinatrice"), chap. V ("Dell'Ordinamento delle nostre Idee"), p. 238; Genovesi, *Delle scienze metafisiche per gli giovanetti* (1766; Naples: Angelo Coda, 1791), pt. 3 ("Antropologia"), p. 315.

93. Lodovico Barbieri, *Trattato di psicologia nel quale si ragiona della natura dell'anime umane, e degli altri spiriti, della loro eccellenza sopra i corpi, della intelligenza, della volontà, della immortalità ec.* (Venice: Pietro Valvasenese, 1756).

94. Cesare Beccaria, *Dei delitti e delle pene* (1764), ed. Franco Venturi (1965; Turin: Einaudi, 1994), including a set of documents on the origins and dissemination of the work; referred to in the text by section number (§). The latest English version is *On crimes and punishments*, trans. Graeme R. Newman and Pietro Marongiu (New Brunswick, NJ: Transaction Publishers, 2009).

traces; life imprisonment is therefore more effective for crime prevention than capital punishment (§ XXVIII).

For Beccaria, morals, politics, and aesthetics had to be based on the study of human nature and hence on the science of man. His sensualist treatise on style sought to ground style in a "philosophy of mind" (*filosofia dell'animo*) which, he said, would be more appropriately called "psychology" or "metaphysics."[95] A similar approach was taken by his friend Pietro Verri (1728–1797), in a *Discourse on the Nature of Pleasure and Pain*, in which he attempted to break down different types of sensations into their components and discover the properties characterizing the painful and the pleasurable ones.[96] The essay drew on Locke and Condillac, but it was also linked to contemporary debates on the nature of individual and social happiness, to "economic" research into the possibility of calculating pleasure and pain, and to the development of a nonmechanistic physiology of sensibility. Here again, the ultimate goals of psychology lay beyond the discipline itself. As suggested by the fact that an anti-Wolffian polemic was published at the beginning of the 1780s in the Venetian *Giornale enciclopedico*, questions of method and the defense of "analysis" could reach a broad educated public.[97] These questions also influenced educational reform.[98] As in the rest of enlightened Europe, psychology in Italy was seen as the empirical basis for reforming minds and institutions, from logic and aesthetics to education and legislation.

95. C. Beccaria, *Ricerche intorno alla natura dello stile* (1770), in *Opere*, ed. Sergio Romagnoli (Florence: Sansoni, 1958), 2 vols., vol. 1, pp. 201–204.

96. Pietro Verri, *Discorso sull'indole del piacere e del dolore* (1773; expanded ed. 1781), ed. Silvia Contarini (Rome: Carocci, 2001), p. 58. For the context, see Contarini's introduction, "Una mappa della sensibilità."

97. Giovanni Scola, "Lettere ai seguaci del sistema sintetico" (1781), in Mario Berengo, ed., *Giornali veneziani del Settecento* (Milan: Feltrinelli, 1962). The letters criticized a Wolffian philosophy manual (*Institutiones philosophiae wolphianae*, by Ludwig Philipp Thümmig); they often used psychological arguments and sometimes referred to Charles Bonnet.

98. The figure of Father Francesco Soave (1743–1806), who finished his career as professor of "ideology" at the University of Pavia, is important in this respect. He placed psychology within metaphysics but discussed, with a typically *illuminista* critical scepticism, theories of the union of the two substances and the issue of the nature, spiritual character, immortality, essence, and origin of the soul. See F. Soave, *Istituzioni di metafisica*, parte prima ("Psicologia"), in *Istituzioni di logica, metafisica ed etica* (1804), "edizione corretta e accresciuta" (Venice: Nella Stamperia di Sebastiano Valle, 1813), vol. 3. On Soave's school textbooks, see Carlo Pancera, "L'importanza dei testi scolastici di Francesco Soave (1743–1806)," in Luciana Bellatalla, ed., *Maestri, didattica e dirigenza nell'Italia dell'Ottocento* (Ferrara: Tecomproject, 2000).

On a smaller scale, psychology in Spain was perceived as being able to play a similar enlightening role.[99] Partly thanks to Genovesi and the Portuguese Capuchin Luís António Verney (1713–1792), who wrote a *True Method of Study in Order to Be of Service to the Republic and the Church* (1746; translated into Spanish in 1760), the critique of Aristotle and Scholasticism came to be grounded in empirical, Lockean arguments rather than in Descartes and Gassendi.[100] Prior to Verney, the physician Andrés Piquer (1711–1772) published a *Modern Logic* that proposed to examine thought and the origin of error, not in terms of formal reasoning but of sensitive and intellective operations.[101] Even some Spanish Jesuits, such as the polymaths Juan Andrés and Antonio Eximeno y Pujades, who lived in exile in Italy after the expulsion of their order in 1767, declared themselves empiricist followers of Bacon and Locke and vigorously attacked Scholasticism.[102] Locke

99. Luis Rodríguez Aranda, "La recepción e influjo de la filosofía de Locke en España," *Revista de filosofía*, 14, 1955, 359–381; Francisco Sánchez-Blanco Parody, *Europa y el pensamiento español del siglo XVIII* (Madrid: Alianza, 1991), chap. 9; Antonio Jiménez García, "Las traducciones de Condillac y el desarrollo del sensismo en España," in Antonio Heredia, ed., *Actas del VI seminario de historia de la filosofía española e iberoamericana* (Salamanca: Ediciones Universidad de Salamanca, 1990). For a more general work, see Helio Carpintero, *Historia de la psicología en España* (Madrid: EUDEMA, 1994), pp. 67–79.

100. Luís António Verney, *Verdadeiro método de estudar para ser util à Republica, e à Igreja: proporcionado ao estilo, e necesidade de Portugal* (1746), ed. António Salgado Júnior (Lisbon: Livraria Sá da Costa, 1949–1952), vol. 3, letters 8–10. Having defined physics as the science which examines the nature of the body and the mind through their observable effects (p. 207), Verney placed the study of the soul within it. Physics first studied man "according to his organs and the machine of his body" (anatomy), then "according to the source of his passions and the force of his imagination." It thereafter examined spiritual substances, with a view to proving the soul's existence, freedom, and immateriality—all things, Verney claimed, familiar to each of us. Verney rejected Scholastic demonstrations and appealed to the readers' experience of their own understanding and will as proof of the distinction between the soul and the body. Similarly, when dealing with the union of the two substances, he refused to address Scholastic questions such as whether the soul should be called the "form" of the body or whether the body should be called *comparte* (participant) in the soul (p. 241).

101. Andrés Piquer, *Lógica moderna, o Arte de hallar la verdad, y perficionar la razón* (Valencia: En la Oficina de Joseph García, 1747).

102.   Adolfo Domínguez Molto, *El abate D. Juan Andrés Morell (Un erudito del siglo XVIII)* (Alicante: Instituto de estudios alicantinos, 1978), pp. 81–89 and 177–179. Mention should also be made of the Benedictine professor of theology at Oviedo, Benito Jerónimo Feijóo (1676-1764), who was a tireless proponent of the authority of "experience" and of experimental philosophy, and whose writings sometimes addressed psychological issues (for example, the imagination, memory, and physiognomy). While at times he used the vocabulary of "form" and "matter," Feijóo wrote of the reciprocal action of the soul and the body. Feijóo's anti-Aristotelianism comes clearly to the fore in his *Teatro crítico universal* (1726–1739) and its sequel, the *Cartas eruditas y curiosas* (1742–1760), where he discusses all sorts of subjects, including logic, metaphysics, and physics. For a good introduction, see I. L. McClelland, *Benito Jerónimo Feijóo* (New York: Twayne, 1969).

was banned in Spain—but only in 1804; Condillac's *Commerce and Government Considered in Their Mutual Relationship* (1776) and his *Logic; or The First Developments of the Art of Thinking* (1780), which were translated into Spanish in 1778–1780 and in 1784, had an enormous impact on Spanish political economy and logic, both before and after being banned in 1798 with the rest of the philosopher's works.[103]

In summary, then, the general science of man which emerged in the eighteenth century, whether penned by French *philosophes*, Italian priests, Spanish Jesuits, or German and Scottish academics, was essentially a natural and cultural history. The human being was treated as a composite of soul and body, and studying the interaction between these two substances was considered of prime importance. Psychology occupied a central position in this project. Authors were less concerned with the heterogeneity of the body and the soul than with its indissociability since, as many of them emphasized, they aimed to understand man as a totality. This position was consistent with the project of a natural history of man.[104] The weight of faith, theological doctrine, or apologetic goals varied according to individuals and particular contexts. Nevertheless, to the extent that the notion of man as a composite being subtended the empirical science of the soul, anthropology went hand in hand with psychology in a relationship of mutual implication and dependence.

METHODOLOGICAL DISCUSSIONS IN ENLIGHTENMENT PSYCHOLOGY

The historiographical neglect of eighteenth-century psychology contrasts sharply with Enlightenment psychologists' methodological awareness and with their claim to be following the methods of natural philosophy.[105] In

---

103. Jiménez García, "Las traducciones de Condillac"; Alain Guy, "Ramón Campos, disciple de Condillac," in José Luis Abellán et al., *Pensée hispanique et philosophie française des Lumières* (Toulouse: Association des publications de l'Université de Toulouse-Le Mirail, 1980).

104. Claude Blanckaert, "L'histoire naturelle de l'homme (XVIIIe–XIXe siècle)" (*Habilitation*, Paris, 1997). We are once again suggesting that the anthropological ideas of the Enlightenment were not based on materialism. But in order to see this, we need both to step outside the materialist tradition and to take seriously the authors who speak about the union of the body with the soul. For the opposite viewpoint, see Ann Thomson, "From *l'histoire naturelle de l'homme* to the natural history of mankind," *British journal for eighteenth-century studies*, 9, 1986, 73–80.

105. For a different approach, see Martin Müller, "Methoden psychologischer Forschung und Diagnostik im Deutschland des 18. Jahrhunderts," in Horst Gundlach, ed., *Arbeiten zur Psychologiegeschichte* (Göttingen: Hogrefe, 1994).

1786, for example, Jacob Friedrich Abel (1751–1829), at the time a professor of psychology and morals in a military academy in Stuttgart, published a work which he described as the first comprehensive textbook of empirical psychology.[106] For Abel, psychology was part of the science or theory of man (*Menschenlehre*), and its methodology was identical to that of the sciences of nature: one gathered data on specific phenomena, from which one derived general laws, which in turn served either to explain the phenomena or to find new rules applicable to new objects. It followed that psychologists worked like naturalists, except that they observed what went on within themselves.[107] Abel remarked that, since psychological notions cannot be tested as easily as those of the science of the body, becoming a psychologist is much harder than becoming a naturalist—"but also much more important."[108]

These sorts of ideas were widespread among psychologists and generated methodological debate, particularly on the empirical nature of the science of the soul, critiques of oversystematization, and the analysis and comparison of different methods.

Let us turn in the first instance to Wolff's statements in his *Empirical Psychology* on observation, experience, and experimentation.[109] His remarks are helpful in introducing vocabulary (including terms not immediately his) that can be found in later debates. They additionally show how empirical psychology was from the very start involved with the methodological questioning that was already present in logic and natural philosophy.

Wolff explained that the art of discovery (*ars inveniendi*), which involves deducing unknown truths from ones already known, can proceed either a priori or a posteriori (§ 454). In the latter case, which is the only one of interest to empirical psychology, findings are based on observation or experimentation (*ex experimentis*, § 457). Both are forms of "experience" (*experientia*), that is, of knowledge acquired by paying attention to our perceptions.[110] Observation involves no voluntary alteration of nature; experi-

---

106. Jacob Friedrich Abel, *Einleitung in die Seelenlehre* (1786; Hildesheim: Olms, 1985), p. vi, for the passage where he says he does not know of any other textbook, and p. vii for his claim that he will only deal with empirical psychology.

107. "Auch der Geist des Psychologen ist also überhaupt der Geist der Naturforscher," § IV, p. xxxi.

108. Ibid.

109. Wolff, *Psychologia empirica*; referred to in the text by paragraph number.

110. "*Experiri* dicimus, quicquid ad perceptiones nostras attenti cognoscimus. Ipsa vero horum cognitio, quae sola attentioni ad perceptiones nostras patent, *experientia* vocatur"; Wolff, *Philosophia rationalis*, in Wolff, *G.W.*, II.1.2, § 664. And: "Die Erfahrung (*experientia*) ist eine Erkenntniß, welche man durch die bloße Aufmerksamkeit auf eine Empfindung sich erwirbt,

mentation (*experimentum*), by contrast, requires it (§ 456). Watching the sky cloud over is an observation, whereas pumping air from a pneumatic machine is experimentation. The *ars observandi* used by physicists, doctors, and above all astronomers is the proper method of empirical psychology (§ 458). *Ars experimendi*, on the other hand, is used only by physicists—even if, Wolff suggested, it could be applied to the whole of philosophy and even to natural theology (§ 459).

Having established these key terms, we shall examine something which must appear improbable in the light of received historiography: in the eighteenth century several attempts were made to introduce measurement and quantification into the science of the soul.[111] In his *Empirical Psychology*, for example, Wolff defined pleasure (*voluptas*) and aversion (*taedium*) as the intuition or intuitive knowledge (*intuitus, seu cognitio intuitiva*) of, respectively, perfection and a true or apparent imperfection (§§ 511, 518). These definitions were to have a significant impact on aesthetics. In psychology, they inspired Wolff's two "theorems": one, that pleasure and aversion are proportional to the perfections and imperfections we are conscious of, and the other, that they are also proportional to the degree of certainty of our judgments about them (§ 522). Wolff situated these theorems in "the field of *psychometrics*, which yields a mathematical knowledge of the human mind, and which until now has remained but wishful thinking. The task [of psychometrics] is to teach us how to measure the magnitude of perfection and imperfection, and also the certainty of our judgments."[112] He went on to explain that such a science is possible because, when it comes to quantity,

---

z. E. durch Sehen, Hören, Riechen, Schmecken, Fühlen"; Walch, *Philosophisches Lexikon*, art. "Erfahrung," 1, cols. 1082–1084, col. 1083. This "modern" definition, which further specifies that experience includes perception, observation, and experimentation, is from the 1775 edition. The 1736 original defined experience as a knowledge we acquire over time by contact with all sorts of things in all sorts of states and roles ("Ist eine Erkenntniß, die durch Länge der Zeit aus allerhand specialen Fällen in allen und jeden Ständen und Aemtern erworben wird"; col. 1082). This definition belongs more to casuistry than to scientific theory.

111. Konstantin Ramul, "The problem of measurement in the psychology of the eighteenth century," *American psychologist*, 15, 1960, 256–265.

112. "Theoremata haec ad *Psycheometriam* pertinent, quae mentis humanae cognitionem mathematicam tradit & adhuc in desideratis est. In ea autem doceri debet, quomodo magnitudinem perfectionis ac imperfectionis nec non certitudinem judicii metiri debeamus" (§ 522). For various aspects of Wolff's project, see Alexandre Métraux, "An essay on the early beginnings of psychometrics," in Georg Eckardt and Lothar Sprung, eds., *Advances in historiography of psychology* (Berlin: VEB Deutscher Verlag der Wissenschaften, 1983); and particularly W. Feuerhahn's articles, which show how photometrics functioned as a model for psychometrics: "Entre métaphysique, mathématique, optique et physiologie" and "Die Wolffsche Psychometrie," in Oliver-Pierre Rudolph and Jean-François Goubet, eds., *Die Psychologie Christian Wolffs: Systematische und historische Untersuchungen* (Tübingen: Max Niemeyer Verlag, 2004).

the human soul obeys mathematical laws, and because, in the soul, arithmetic and geometric truths are mixed with contingency as much as they are in the material world.

Wolff did not pursue empirical psychometrics. His examples point the way to particular fields of inquiry, such as the measurement of "degrees of attention," that is, of an individual's capacity to remain focused in the midst of distractions or when studying long mathematical proofs (§ 243–248).[113] Wolff wanted his work to have the universal applicability of mathematics while also being socially useful. This was, in fact, how he conceived his philosophy as a whole. In his *Universal Practical Philosophy* he claimed that the freer an action, the more deliberate it is, and so the more it can be attributed to the individual who carried it out. In Wolff's view, this implied that a mathematical approach could be taken to the freedom of the soul and to the imputation of actions. Consequently psychometrics, used here to measure intentionality, could enable degrees of responsibility to be calculated, thus leading to a mathematically precise justice system.[114]

Other authors followed suit. One of the first was the Wolffian Gottlieb Friedrich Hagen (1710–1769), a philosophy teacher at the Bayreuth gymnasium. He imagined psychological experiments (*experimenta psychologica*) that would alter the soul, for example, by scaring people and then observing their reactions.[115] Such experiments could contribute significantly to self-knowledge, and those designed to produce knowledge of others would provide a firm basis for other observations.[116] Hagen also conceived a *dynametria* to measure the faculties (*dunamis*) of the soul, again within the framework of a sort of quantitative casuistry.[117] He argued that, like the "mechanical" faculties of the body, the "representative" faculties of the soul are finite in number; since they vary considerably from individual to individual, they may be compared quantitatively.[118]

113. See Wolff, *Philosophia rationalis . . . : Praemittitur Discursus praeliminaris*, § 17; Wolff, *Vernünfftige Gedancken*, §§ 268–272.

114. Wolff, *Philosophia practica universalis, methodo scientifica pertractata, Pars prior . . .* (1738), in Wolff, *G.W.*, II.10, §§ 606–608.

115. Gottlieb Friedrich Hagen, *Meditationes philosophicae de methodo mathematica . . .* (1734), in Wolff, *G.W.*, III.82, cap. III ("De experientiis"), § 37. Hagen (ibid., § 39) also conjectured that one could carry out theological experiments (on temptation and succumbing to sin) and philosophical experiments (particularly concerning different teaching methods).

116. Ibid., § 41.

117. G. F. Hagen, *Dissertatio mathematica de mensurandis viribus propriis atque alienis* (Giessen: Litteris Muellerianis, 1733). The term *dynametria* appears only at the end, p. 16, § 27.

118. Ibid., caput I ("De viribus in genere et humanis in specie"), particularly §§ 11–13.

Hagen wanted to be like the "geometer" who measures a circle through infinitely small angles. But what was to be measured in order to ascertain the power and degree of perfection of each faculty?[119] Attention would be gauged by the number of ideas which could be entertained concurrently, and by the time spent reflecting on and analyzing an idea; *ingenium*, by the speed at which hidden similarities were uncovered; and judgment, by the number of links made simultaneously between propositions. The will would be measured by examining the *modus appetitionis*, that is, how much one reflected before acting and to what extent one was guided by the "rational appetite" rather than the sensitive one. Even virtue could be computed, since it consists in the strength of one's desire for the good and for shunning evil.[120]

Not all faculties were equally measurable. Hagen explained that intellectual abilities may be natural or acquired. Through its natural abilities, the soul represents the world to itself. Acquired abilities result from repetition, observing rules or a knowledge of other disciplines (physicians, for example, must know mathematics, physics, anatomy, and *materia medica*, and metaphysics is a prerequisite for logic and the rest of philosophy).[121] Hagen commented on the difficulty of measuring acquired abilities, but he thought natural ones could be measured by the speed and number of acts carried out. For example, if one wanted to compare two boys' ability to learn history, the same historical event could be explained to them several times, until they had learned it by heart. Each time they reviewed their lesson, the points they remembered would be noted down. The more points they remembered and the less repetition needed, the faster they were learning. Natural intellectual powers, especially attention, could be measured in this way. The elements remembered each time would in addition indicate each subject's preferences, in line with his or her natural inclination.[122]

The psychometric ideal inspired both psychologists and philosophers. We have already mentioned Kant on the calculation of pleasure and displeasure. Charles Bonnet wondered whether the "number of correct consequences which different minds derive from the same principle" might

119. Ibid., caput II ("De mensura virium"), §§ 6–9.

120. Ibid., §§ 12–17.

121. G. F. Hagen, *Programma de mesurandis viribus intellectibus* (Halle: Litteris Hendelianis, 1734), p. 4.

122. Ibid., pp. 4–5. Hagen believed it possible to determine the natural aptitudes for studying philosophy. Since philosophy is the science of truth, he argued, the relevant faculties must be rational. The required "natural dispositions" could be classified (from the analytic to the synthetic) in degrees going from the capacity to discover and posit causes to the deduction of truths from principles and the combination of different principles; ibid., pp. 5 and 6–7.

not serve as the "basis for the construction of a *psychometer*: may we not suppose that one day we shall be able to measure minds as we measure bodies?"[123] Finding the appropriate unit of measurement remained the greatest challenge. The Swiss Johann Bernhard Merian (1723–1807), a prolific writer on psychology who taught philosophy at the Collège Français in Berlin and became in 1797 the secretary to the Berlin Academy, held that the intensity and duration of pleasure and pain "cannot be rigorously and arithmetically assessed."[124] However, he went on, since sensations differ in intensity and duration, we can examine "which of the two, pleasure or pain, is greater, . . . without attempting to assign a precise quantity to the greater one."[125] We learn from experience that pain is naturally associated with duration: we feel it "prolongs time," whereas pleasure shortens it.[126] Merian thought, however, that differences in intensity were impossible to quantify, even though we experience certain feelings as stronger than others: "Such knowledge belongs to a science which is still lacking, but which will be the masterpiece of the human mind if ever achieved, and which we will call *Psychometrics* when it is discovered."[127] In the meantime, the mind could be measured by the number of objects it could imagine, the clarity with which they were imagined, and the time needed to arrive at a clear image of them; sensations, in turn, could be measured by taking two individuals and comparing the distance at which each could still hear a sound distinctly or see an object and have a clear idea of it.[128]

Johann Gottlob Krüger (1715–1759), a professor of medicine and philosophy at Halle and then at Helmstädt, was interested in the question of sensation from very early on.[129] He was an admirer of Wolff, to whom he

123. Charles Bonnet, *Contemplation de la nature* (1764), 4th pt., chap. X ("Gradations de l'Humanité"), in *Œuvres*, vol. 4.1, pp. 132–133.

124. Johann Bernhard Merian, "Sur la durée et sur l'intensité du plaisir et de la peine," *Histoire de l'Académie Royale des Sciences et des Belles-Lettres de Berlin*, 1766, 381–400, p. 382. On Merian in Berlin, see Jens Häseler, "Johann Bernhard Merian—ein Schweizer Philosoph an der Berliner Akademie," and on a central aspect of his psychological thought, see Bernard Baertschi, "La conception de la conscience développée par Mérian," in Fontius and Holzhey, *Schweizer in Berlin*.

125. Merian, "Sur la durée," p. 382.

126. Ibid., p. 385.

127. Ibid, p. 390.

128. The examples can be found in Ramul, "The problem of measurement," pp. 258–259 and 262–263.

129. Ibid., pp. 261–262. Hatfield, "Remaking the science of mind," pp. 201–205; Carsten Zelle, "Experimentalseelenlehre und Ehrfahrungsseelenkunde: Zur Unterscheidung von Erfahrung, Beobachtung und Experiment bei Johann Gottlob Krüger und Karl Philipp Moritz," in Zelle, *"Vernünftige Ärzte"*; Zelle, "Experiment, experience and observation in eighteenth-century an-

dedicated his 1742 medical thesis on sensation. In his *Doctrine of Nature*, Krüger referred to Giorgio Baglivi's late seventeenth-century vivisection of dogs to support the idea that sensation resulted from the vibration of nerve fibers.[130] Having described the "nerve membranes" (*Nervenhäute*) as elastic and the nerve as a "taut rope," Krüger noted that the elasticity of the rope was proportional to the force with which it was slackened or stretched.[131] From this he formulated "mathematical laws" concerning the relation among the force with which an external object acts on the nerves (that is, affects their "tension"), the nervous activity thereby generated, and the vividness (*Lebhaftigkeit*) of the resulting sensation. He expressed these relations as proportions in the following formula: "$S:s = VT: vt$," where S and s represent the strength or intensity of two sensations, V and v the strength of the stimulation, and T and t the tension of each nerve fiber. It followed that if, for example, V was three times stronger than v and T was twice the tension of t, then the sensation S was six times more intense than s. It followed also that if $T = t$, then $S:s = V: v$. In other words, if the nerves had the same tension, then the strength of the sensations would be in the same proportional relation as that pertaining between the forces of the external objects that produced them. If all the sensory nerves had the same tension, then the increase in a sensation's strength would be proportional to the increase in the stimulus. Thus, the sensation produced by different instruments sounding with the same strength was proportional to the number of instruments; and the sensation felt by the skin following the impact of an object was proportional to the distance covered by the object or to its weight (if the weight differed while the distance remained the same).[132]

Experimentation, which took pride of place in Krüger's *Doctrine of Nature*, was introduced into psychology some years later. In 1756, Krüger published a mostly programmatic and methodological *Essay on Experimental*

---

thropology and psychology: The examples of Krüger's *Experimentalseelenlehre* and Moritz's *Erfahrungsseelenkunde*," *Orbis litterarum*, 56, 2001, 93–105.

130. "Die Empfindung geschieht durch eine zitternde Bewegung." Johann Gottlob Krüger, *Naturlehre: Zweyter Theil, welcher die Physiologie oder Lehre von dem Leben und der Gesundheit der Menschen in sich fasset* (Halle: Carl Hermann Hemmerde, 1743), chap. 6 ("Von der Empfindung überhaupt"), § 315, p. 568.

131. Ibid., §§ 315–316.

132. Ibid., § 317 ("Mathematische Gesetze der Empfindung"). Further on (§§ 324–327), Krüger reformulated the doctrine of the temperaments (choleric, melancholic, sanguine, and phlegmatic) in the light of his doctrine of the nerves—thus confirming, in his view, its medical usefulness. His theory of pathology also presupposed his doctrine of the nerves. See *Naturlehre: Dritter Theil, welcher die Pathologie, oder Lehre von den Kranckheiten in sich fasset* (1749; Halle: Carl Hermann Hemmerde, 1755), chap. 11 ("Von allzuheftigen Seelenwürckungen").

Psychology. As in Wolff, both observation and experimentation were forms of "experience,"[133] but only experimentation transformed nature. Krüger's aim was to demonstrate that it was possible to do experimental psychology without using the tools of experimental physics. For him, psychological experimentation was based on the interaction of the body and the soul and involved causing extraordinary changes in the former in order to produce in the latter changes that would not normally come about in nature.[134] Experiments could be carried out on criminals and animals; mostly, however, they were furnished by medical case studies. It is not unreasonable, said Krüger, to consider such case studies experiments, in which an extraordinary change in the body produces an equally unusual state in the soul; thanks to such cases, he added, we have access to psychological experiments without getting human blood on our hands.[135] The value Krüger attributed to such pathological *Experimente* is embodied in the three-hundred-odd pages of medical observations he appended to his *Essay*.[136] Krüger's project nevertheless remained basically an experimental psychology.[137]

Enlightenment psychologists were therefore generally convinced that psychometrics was the ideal approach to empirical psychology, and that experimentation would allow this science to come into its own. But their methodological focus was principally on a critical and comparative appraisal of the methods best suited to the science of the soul.[138]

In 1770, Christian Gottfried Schütz (1747–1832), a philosophy professor at Jena and subsequently one of the first Kantians, published a translation

133. "So wohl Wahrnehmungen als Versuche (*observationes et experimenta*) sind Erfahrungen." J. G. Krüger, *Versuch einer Experimental-Seelenlehre* (Halle: Carl Hermann Hemmerde, 1756), § 6, p. 14.

134. "Man durch ausserordentliche Veränderungen, die man mit dem Leibe vornimmt, Veränderungen in der Seele zuwege bringen könne, die sich sonst nach dem gewöhnlichen Lauffe der Natur bey ihr nicht gezeigt haben würden; und dieses heißt nichts anders, als daß es möglich sey, Experimente mit der Seele zu machen"; ibid., § 7, p. 18.

135. Ibid., p. 20.

136. Johann Gottlob Krüger, *Anhang verschiedener Wahrnehmungen, welche zur Erläuterung der Seelenlehre dienen*, printed as an appendix to the *Versuch*.

137. Zelle ("Experimentalseelenlehre" and "Experiment") emphasizes Krüger's preference for medical observation, arguing that it prefigures the interest in individual cases which characterized the novels of the second half of the century. In this respect, Krüger's experimental psychology would be closer to Moritz's *Erfahrungsseelenkunde* than may first appear.

138. Among the works of eighteenth-century quantitative psychology, the longest has yet to be studied in detail: Christian Albrecht Körber, *Versuch einer Ausmessung menschlicher Seelen und aller einfachen endlichen Dingen überhaupt, wie solche der innern Beschaffenheit derselben, gemäß ins Werck zu richten ist, wenn man ihre Kräffte, Vermögen und Würckungen recht will kennen lernen* (Halle: in der Lüderwaldischen Buchhandlung, 1746). Ramul touches on it briefly ("The problem of measurement," pp. 262–263).

of Charles Bonnet's *Analytical Essay on the Faculties of the Soul.*[139] He appended to it not only an annotated synopsis of Condillac's *Treatise on the Sensations* but also his own "Considerations on the different methods in psychology"—one of the discipline's first methodological texts.[140] Methodological reflection was for Schütz a crucial way of identifying obstacles to progress in psychology and discovering their causes. Mere "conjectures and hypotheses" were futile; what was needed was to observe the soul, but that was precisely the problem (190). Aware of his pioneering role, Schütz conceded that one could find in classical and modern authors remarks relevant to psychological method; methodology, however, still had to be addressed in a more comprehensive and coherent fashion. We will then have a better idea of the value of research already carried out, and will avoid dead ends in the future (190).

According to Schütz, "psychological difficulties" fell into three categories, relating to the knowing subject (*das erkennende Subject*), the object observed, and the way in which the subject treated it (191). The "spectacle" or "theatre" (*Schauspiel*) of the soul was extremely varied and uneven. But who watched the show? The soul itself, at once actor and spectator. This was the most significant source of methodological difficulty. Some of the problems were due to the soul itself, as it observed itself by means of "cognitive powers" (*Denkungskräfte*) more naturally directed toward the external world. Other difficulties stemmed from the fact that the soul was the object of study, and from the way research was carried out (192).

The first category of methodological problems, deriving from the fact that the soul was the observing subject (*das betrachtende Subject*), included several types of difficulty. First of all, there was the problem of getting beyond sensory impressions. Our attention is accustomed to being directed outward, and the soul, unaccustomed to dispensing with sensible images. That is why, Schütz maintained, philosophers made the soul into something material. When the "geometer" despairs of rectifying the circle (that is, calculating the length it would have as a straight line), he makes it into an infinite polygon. Likewise, incapable of conceptualizing a purely spiritual

---

139. On the methodological thought of Schütz and Tetens (discussed below) see Thomas Sturm, *Kant und die Wissenschaften vom Menschen* (Hamburg: Mentis Verlag, 2009), chap. 2.

140. Christian Gottfried Schütz, "Betrachtungen über die verschiednen Methoden der Psychologie; nebst einem kritischen Auszuge aus des Hrn. Abt von Condillac *Traité des sensations*," in Charles Bonnet, *Analytischer Versuch über die Seelenkräfte*, trans. C. G. Schütz (Bremen: Johann Henrich Cramer, 1770), vol. 1, pp. 187–273. I will concentrate here on §§ 1–5 and 12–15 (erroneously numbered 14); §§ 6–11 deal with Condillac's *Treatise*. Referred to in the text by page number.

nature, psychologists have imagined the soul as made up of the most subtle matter. Such "philosophical approximations" were not, in Schütz's view, as anodine and practical as the geometrical ones; depicting the soul as a breath or a subtle flame does not contribute to psychology's advancement (194). It would be worthwhile, Schütz concluded, to write a history of the influence of sensory notions (sinnliche Begriffe) on philosophical opinions concerning the soul, as a way of ascertaining whether there is anything of real value in these approximations (196).

The soul, Schütz continued, turns out to be a less acute observer of itself than of other objects. This is because of the "astonishing speed" with which representations succeed one another—and this brings us to the second difficulty arising from the soul as observing subject. Our attention cannot follow these representations as easily as it follows bodies. Besides, contemplating spiritual matters leads to exaltation and enthusiasm (Schwärmerey) more readily than does observing physical objects, which is more laborious and concentrated (197).[141] Finally, the speed at which the soul hastens from thought to thought makes observing ephemeral states and continuous processes very difficult. It follows, for Schütz, that few people are capable of grasping the development of genius or the of the child's psychological faculties (198).

A third difficulty arising from the soul as observing subject occurs when obscure representations (dunkle Vorstellungen) eclipse clear ones, such that the soul ceases to be aware of itself. Certain states such as deep sleep, fainting, or madness defy psychological observation (198). It would be useful for the "art of discovery" if a genius could note down how he works; but just as a soldier thrown into the thick of battle is unable to give a full account of it, so the genius, in his fervor, is incapable of being fully conscious of his method (199).

The second category of methodological problems stemmed from the fact that the soul is itself the object of inquiry. Unlike the eye, which does not see itself when it sees, the soul must observe itself in order to know itself. But, said Schütz, since the soul is endowed with consciousness, it thinks it knows itself already. That is why the inquiry into the nature of the soul is sometimes reduced to internal sensation (200). The conviction that we know the soul so well already that we do not need to study it is a common error of judgment which hinders the progress of psychology (204). But

---

141. Schwärmerey: the term occurs in the philosophico-religious context of the critique of fanaticism, superstition, and other delusions, which were often attributed to the powers of the imagination.

even without this hindrance, observing oneself remains a major challenge. While the understanding is busy examining its own processes, it neglects the task at hand or changes direction; orators, for example, often lose their thread in following their own thoughts. Worse still are the exceptional and overwhelming psychological states which make us incapable of reflection; a painter, for example, is unable to execute a self-portrait if overcome by the deepest sorrow (200). One could object that it is always possible to observe someone else. But, Schütz warned, knowing someone else is as difficult as knowing oneself, since each of us wears a mask, and even in the absence of hypocrisy many things will always remain hidden from us (201).

The problem, therefore, lay in the structure of the soul, which is both varied and uniform. It would, however, be an error to multiply the faculties to help account for the complexity of certain psychological phenomena—that would be no better, Schütz stated, than multiplying or changing names. Conversely, it would be equally erroneous to believe that what appears simple really is so, since complexity (*Zusammensetzung*) may be well concealed, and much attention and circumspection are needed to ascertain whether a phenomenon has simple or complex causes. For instance, one and the same tendency may have several causes (202). The properties and functioning of the soul are of "astonishing complexity," but in addition, the soul is like a "rapid current" that seems to remain the same while it is constantly changing. Sometimes the changes are imperceptible, at others a chance event will overturn a whole system of thoughts and tendencies. Schütz found an example in those artists whose style completely changed after they saw the canvas of some great master. We cannot but wonder, he wrote, at this kind of impact, which is so powerful that years may pass without a similar one occurring, and we are led to ask how many rapid changes the soul can indeed tolerate (203). Schütz called for the collaboration of teachers, whose observations, he declared, could be most useful to psychology (204).

The last category of methodological difficulties related directly to the methods used in psychology: the empirical, the analytic, and the synthetic (or systematic). Each had strong points as well as failings. The empirical method appeared to be the most reliable, since it restricted itself to observation and experimentation. But it was also the least likely to make significant progress and was still subject to error (through poorly conducted observation, drawing false conclusions, and confusing the contingent with the essential). The analytic method—which Schütz called the "method of arbitrary connection" (*Methode der willkürlichen Verbindung*)—consisted of selecting certain cases and making analogies between them and other

cases (205). The "analyst" risked "tearing the fabric" by dividing phenomena up too much, when they should be taken as wholes. The "synthesist"
(Synthetiker), on the other hand, presupposed certain principles and general
propositions from which he deduced conclusions which he then sought to
"square" with the "real phenomena" of the soul (205). He thus risked inventing concepts that had nothing to do with reality, getting lost in useless
speculation, and constructing a system devoid of content (206).

Schütz left little doubt about his preference for the empirical method
and set out three rules for it. The first was to carry out a large number of
observations but without forgetting that they are merely "individual cases"
to be generalized with the utmost circumspection (264). The second rule
was to describe the phenomena together with their surrounding circumstances (Nebenumstände). The reasons for such a rule were that, when isolated from their context, phenomena appear different, and their different
aspects may seem contradictory to each other. The third rule of the empirical method was to take into account every minutest detail, since great
events may stem from modest causes, just as a serene affection may grow
into an intense passion (265).

There were also some sources and procedures which Schütz wished to
see more widely used in psychology (266–268). These were child development, to elucidate the origin of children's abilities and inclinations; extraordinary phenomena such as madness, visions, and somnambulism, in order
to determine how the body behaves during these states and what role the
soul plays; proceeding by analogy, which could help discover what really
occurs in the case of "obscure representations," during sleep, for example;
conceiving the "so-called faculties of the soul" in relation to each other
rather than in isolation, in order to have a better understanding of why they
differ so greatly between individuals; and, last, asking that psychology focus
more on man as a whole—not only on the soul but on the soul in its union
with the body. In summary, Schütz declared, it would be preferable for psychology "to be treated more as an anthropology."[142]

Although Schütz's proposed anthropological inflection of psychology
would principally use the empirical method, all methods seemed to have
a valid contribution to make. Analysis supplemented the limits of experience, by breaking down facts and reducing the results obtained to simple
principles. This required a lively but orderly imagination, and the formula-

---

142. "Endlich sollte man auch mehr auf den ganzen Menschen, nicht bloß auf die Seele,
sondern auf die Seele, mit dem Körper verbunden, Achtung geben. Es wäre überhaupt vortheil-
hafter, wenn die Psychologie mehr als Anthropologie behandelt würde" (268).

tion of a small number of principles (269). This is where, in Schütz's view, Condillac and Bonnet went wrong. The *Treatise on the Sensations* (1754) and the *Analytical Essay on the Faculties of the Soul* (1760) used the fiction of a hollow and inanimate statue which is first exposed to isolated sensory stimuli (starting with the scent of a flower) and ends up becoming a wholly sentient and thinking being (fig. 4.1).

In both Bonnet and Condillac, the awakening of the senses activated all the faculties, even abstract thinking. But how, Schütz asked, could the statue react to scent if it were not already endowed with sensibility? (269–270). In spite of its shortcomings, the synthetic method, which Schütz accused of having been most prejudicial to psychology, could in his view be useful. It could help organize and clarify the observations made, and even contribute to their precision: in grouping observations together, it could enable gaps, contradictions, and elements out of order to be identified. All in all, Schütz concluded, each method is imperfect in itself, but excellent results can be achieved when all are used together (273).

Johann Nicolas Tetens put more emphasis than Schütz on introspection as the appropriate procedure for empirical psychology. Not only did he apply it in his voluminous *Philosophical Essays on Human Nature and Its Development*, but he also subjected it to an in-depth critical appraisal which bears the mark of his eclecticism, his knowledge of the sciences, and his pragmatic outlook.[143] As mentioned above, Tetens began by espousing the method of "observation" he ascribed to Locke and German empirical psychology, which he characterized in the following terms:

> Take the modifications of the soul as they are known by the sense of self; observe them again carefully and in different circumstances, noting how they arise and the laws governing the action of the forces

---

143. Tetens taught physics, mathematics, and philosophy at different times in his career and, apart from his work on psychology, published on subjects as diverse as meteorology, electricity, etymology, school curricula, the application of mathematics to retirement pensions, the hydraulic technology of dams, the laws of armed conflict, the concept of the divinity, and the proofs of the existence of God. He left university teaching in 1789 to become a high-ranking civil servant in Denmark's justice and finance departments. For overviews of Tetens's work, see Barnouw, "Psychologie empirique"; and Barnouw, "The philosophical achievement and historical significance of Johann Nicolas Tetens," *Studies in eighteenth-century culture*, 9, 1979, 301–335. Tetens's ideas on psychological development proved of interest to late twentieth-century psychologists: Marianne Müller-Brettel and Roger A. Dixon, "Johann Nicolas Tetens: A forgotten father of developmental psychology?," *International journal of behavioral development*, 13, no. 2, 1990, 215–230; Ulman Lindenberger and Paul B. Baltes, "Lifespan psychology: In honor of Johann Nicolaus Tetens (1736–1807)," *Zeitschrift für Psychologie*, 207, 1999, 299–323.

Fig. 4.1. The psychologist holding out a flower for the statue to smell. Frontispiece to Jean Henri Samuel Formey, *Entretiens psychologiques, tirés de l'Essai analytique sur les facultés de l'ame, de Mr. Bonnet* (Berlin: chez Joachim Pauli, 1769). (Bibliothèque Publique et Universitaire de Genève. Photo: Jean Marc Meylan.)

that produce them; then compare and break down the observations into their elements, and in this way find the simplest properties and modes of operation, as well as their interrelations. These are the essential procedures for a psychological analysis of the soul grounded in experience.[144]

According to Tetens, this was the method to study nature—and the only method capable of showing the operations of the soul "as they really are" and of enabling us to reach principles from which we may reliably derive causes, such that we end up with certainties, rather than simple conjectures, concerning the nature of the soul (iv).

Tetens set this method against the "analytic," or "anthropological," which reduced psychological phenomena to states of the brain (iv). Since these states cannot be observed, such methods should be called "metaphysical." They simply reduce changes in the immaterial self to changes in the brain; the immaterial self, however, participates as an active motor force in these changes and can itself be transformed at the same time as the brain (v). For Tetens, the organ of thought was merely a machine driven by the soul.[145] There was no question of denying, for example, that memory processes leave traces in the brain (vii); the problem is that authors such as Bonnet and Hartley added to this general and persuasive model more precise but unfounded hypotheses concerning the way in which changes in the brain produce representations (vii). Those hypotheses on the "physics of the brain," such as Hartley's theory of nerve vibrations, were purely conjectural, and it was therefore useless to try to reach any conclusions as long as the brain remained so little understood (xiii).

Tetens's critique did not exclude brain physiology from psychology but rather assigned it a different position. If one conceded that Bonnet's and Hartley's "metaphysical analyses" could teach us something, they should come at the end and not the beginning of psychology (xiii–xiv): "*Psychological* analysis must take priority" (Die *psychologische* Auflösung muß vorgehen; xiv). And even if we made do with "metaphysical psychology," its

144. "Die Modifikationen der Seele so nehmen, wie sie durch das Selbstgefühl erkannt werden; diese sorgfältig wiederholt, und mit Abänderung der Umstände gewahrnehmen, beobachten, ihre Entstehungsart und die Wirkungsgesetze der Kräfte, die sie hervorbringen, bemerken; alsdenn die Beobachtungen vergleichen, auflösen, und daraus die einfachsten Vermögen und Wirkungsarten und deren Beziehung auf einander aufsuchen; dieß sind die wesentlichsten Verrichtungen bey der psychologischen Analysis der Seele, die auf Erfahrungen beruhet. Diese Methode ist die Methode der Naturlehre." Tetens, *Philosophische Versuche*, "Vorrede," p. iv, henceforth referred to in the text by page number.

145 "Das Denkorgan ist eine Maschine, wozu die Seele die bewegende Kraft ist," (v).

theses should always be tested against observational knowledge (*Beobach-tungskenntnisse*; xiv). The desire to interpret states of the soul as changes in the brain had made even some very good observers overlook psychological facts (xv). In psychology as in other sciences, Tetens continued, it was both natural and expedient to invent hypotheses when experience and rational approaches were lacking; for the creator of hypotheses (*Hypothesendichter*), the observer (*Beobachter*) and the "airy creator of systems" (*der luftige Systemmacher*) all contribute something to knowledge—but within the limits of their approaches.

Tetens recognized that self-observation was far from straightforward but emphasized the operations which made it valuable to the psychologist. These were the observation of individual phenomena, their dissection or analysis (*Zergliederung*), and comparison between various findings (*Vergleichung*); thanks to these procedures, assertions concerning individual facts (*einzelne Sätze*) could lead to general propositions (*Allgemeinsätze*: xvi–xvii). Each operation had specific pitfalls. Tetens noted, for example, how easy it was to trick oneself into believing one was reproducing an observation. Whoever cannot distinguish a true perception from what is said about it is not qualified to become an "observer of the soul" (xvii); in this the imagination (*Phantasie*) is a major source of error (xvii–xviii).

Since each observation is singular, generalization, which is achieved by comparison, constitutes one of the major steps of the observational method (xix). But in order not to mistake what is probable for what is certain, what might be for what is, the comparison must be real and not the mere possibility of an analogy (xx). Only a large number of observations can form the basis of concepts and principles (xxii). Yet, Tetens continued, in order to go beyond simply accumulating material and isolated observations, one needs a "general philosophy" of the relations between the different states of things, in order to determine, for example, to what extent similar effects result from similar causes (xxiii).

Tetens's methodology was, in the last instance, based not only on the practice of observation, but also on an epistemic and almost moral distinction between hypotheses and empirical propositions (*Erfahrungssätzen*; xxix). In this, it modeled itself on natural philosophy. Psychology, however, was the site of particular problems. It was more difficult to distinguish arguments from accounts of experiments in psychology than in other sciences, in which a pure enumeration of events (*Aufzählungen der Begebenheiten*) was more common (xxix). Reasoning and observing should go together, untainted by imagination or oversystematization. That is why he himself,

Tetens claimed, never extended his arguments beyond what he considered to be indisputably confirmed by observation (xxx).

"THE BEST WAY TO PERFECT THIS FINE SCIENCE"

Charles Bonnet's contribution to the methodology of Enlightenment psychology went well beyond the fact that he was criticized by Schütz and Tetens. His philosophical and psychological texts abound with methodological comment, particularly on the "art of observing," which he considered to be the right method not only for physics but also for metaphysics. Bonnet was the inspiration behind Jean Senebier (1742–1809) in his *Essay on the Art of Observing and Performing Experiments*, which has been described as the century's first attempt at systematizing fundamental rules for investigating nature.[146] The *Essay* was written for a competition organized by the Dutch Society of Sciences on a subject Bonnet himself proposed.[147] Senebier, who translated Spallanzani and carried out research into vegetable physiology, said his idea of reflecting on the art of observing was inspired by "reading Bonnet again and again," notably his *Analytical Essay on the Faculties of the Soul* and *Considerations on Organized Bodies*. Senebier took Bonnet's approach as a model of the analytic method and proceeded through a sort of *mise en abîme* in which he "analyzed the important links in the chain" of analysis itself.[148]

The competition's prizewinner was the pastor Benjamin Carrard (1730–1789), from the canton of Vaud. He thought, like Senebier, that one of the best ways of elucidating the rules of the art of observation was to examine how the best observers practiced it. Bonnet again provided the model for the qualities and procedures required for successful observation (although in a less exclusive way than Senebier and without Senebier's enthused tone): concrete details, Carrard confirmed, are a more precious contribution "to

---

146. Marx, *Charles Bonnet*, pp. 172–173; for a comparison of Bonnet and Senebier on methodological issues, see pp. 171–178.

147. Bonnet, who was admitted to the society in 1768, proposed "this most important subject: to what extent may the art of observing contribute to the perfection of the mind." Cited in Marx, *Charles Bonnet*, p. 416.

148. Jean Senebier, *Essai sur l'art d'observer*, 2nd ed. (Geneva: J. J. Paschoud, 1802), vol. 2, p. 288. Senebier was awarded the silver medal (*accessit*). His text was published in the *Memoirs* of the society in 1772, and then in Geneva in 1775. The *Essay* became known in its significantly reworked and enlarged version of 1802.

the great art of observing [. . .] than precepts, which . . . always have something vague and unspecific about them."[149]

In 1765, as soon as he was inducted as a foreign member into the Dutch
Society of Sciences, Bonnet assumed a leading role in defining the society's
competition subjects.[150] After the competition on the art of observing, he
proposed another question which was "in some sense" to be its "application or elaboration":

> What is the usefulness of psychological science for the education and
> guidance of man and for the happiness of societies, and what would be
> the best way to perfect this fine Science and further its progress?[151]

Bonnet criticized educators who were satisfied with general principles,
when, to be of any value, knowledge of man required empirical research.
He argued that since societies are made up of individuals, politics is underpinned by psychology, and since psychology also underpins education, it
becomes the cognitive source of the social happiness which politics should
have as its goal. Given this role of psychology in education and politics,
Bonnet argued, we must ensure its progress, and do so on the basis of the
reciprocal influence of physical and mental phenomena.

The competition prize went to Jean Trembley (1749–1811), nephew of
the famous naturalist Abraham Trembley (1710–1784). Although Trembley
was certainly a minor figure, he was one of the first authors to address methodological issues specifically in psychology. As a polymath, a great traveler
(unlike the sedentary Bonnet) moving intellectually and geographically between the French and German domains, and having no stable profession, he
was one of the many agents of cultural dissemination characteristic of the
Republic of Letters and the Swiss Enlightenment.[152] He also contributed in a

149. Benjamin Carrard, *Essai . . . sur cette Question: Qu'est-ce qui est requis dans l'Art
d'Observer; & jusques-où cet Art contribue-t-il à perfectionner l'Entendement?* (Amsterdam:
chez Marc-Michel Rey, 1777), p. 72.

150. Apart from the art of observing and the usefulness of psychology, Bonnet proposed two
other prize topics: "What are the foundations and the characteristics of analogy, and how should a
philosopher use analogy in his search for physical and mental truths?" and "What are we to think
of the gradations which several philosophers, both ancient and modern, have supposed to exist
between natural beings, and to what extent can we ascertain the reality of these gradations and of
the order which nature observes through them?" See Marx, *Charles Bonnet*, pp. 417–418.

151. Charles Bonnet to the Dutch Society of Sciences, Genthod, 13 June 1770, Archives of
the Hollandsche Maatschappij der Wetenschappen, Rijksarchief in Noord-Holland, Haarlem.

152. François Rosset, "La vie littéraire et intellectuelle en pays romand au XVIIIe siècle,"
in Roger Francillon, ed., *Histoire littéraire de la Suisse romande*, vol. 1, *Du moyen âge à 1815*
(Lausanne: Payot, 1996); Taylor, "The Enlightenment in Switzerland."

minor way to the eighteenth-century blossoming of the sciences in Geneva and was for a few years one of the adopted Swiss of the Berlin *Aufklärung*.[153] Trembley's varied interests—mathematics, physics and psychology, aesthetics, theology, and politics—shared methodological and epistemological convictions concerning the preeminence of psychology and the moral and cognitive value of empirical fact. These convictions also informed his political commitments, which focused on the fight against "prejudices," including those which, to his mind, threatened Christianity.[154]

In the mid-1760s, the young Trembley studied privately with Bonnet. Bonnet wrote for him a vast guide to reading the *Analytical Essay on the Faculties of the Soul*, comprising 2,045 questions, beginning with "What is the source from which all our ideas of reflection arise?" and ending on "What is the best way to read the *Analytical Essay*, and what benefit can one hope to derive from it?."[155] While the terminus a quo of the guide was an empirical question, the terminus ad quem related to the method and morals of the *Essay*. Many other questions were also designed to make the young Trembley aware of methodological issues.[156] The overall aim of the guide was to help him not only learn the contents of the *Essay*, but also explore and develop the principles it expounded: "What memory has retained should be implemented by judgment and wisely embellished by imagination."[157] The guide mirrored the preeminent role for Bonnet of attention in the acquisition of knowledge in general and psychological knowledge in particular: attention, he explained, enables us to make abstractions, and that is why it is "the mother of genius."[158] The guide as a whole can be seen as an affirmative answer to one of its own questions, "Can the principles expounded by our analyst concerning attention give rise to useful logical applications?" (no. 642). It also put into practice Bonnet's conviction that a

---

153. On Geneva, see René Sigrist, *L'essor de la science moderne à Genève* (Lausanne: Presses polytechniques et universitaires romandes, 2004).

154. F. Vidal, "Psychologie empirique et méthodologie des sciences au siècle des Lumières: L'exemple de Jean Trembley," *Archives des sciences*, 57, no. 1, 2004, 17–39.

155. "Questions sur l'Essai Analitique," Ms. Bonnet 13, notebook of 225 sheets, dated from 1 February 1768 to 21 March 1775, Department of Manuscripts, Bibliothèque publique et universitaire, Geneva (henceforth BPU).

156. For example: "How should the true psychologist proceed?" (no. 15), "What are the principal difficulties of the analytic method?" (no. 17), "Why does the author begin his work with the surprising expression 'I suppose'?" (no. 59), "Does language provide many examples of terms taken from matter and transferred to the mental?" (no. 540), and "What practical consequences should be drawn from our profound ignorance of the real essence of things?" (no. 1115); ibid.

157. Ibid., "Avertissement," fol. 1r.

158. Bonnet, *Essai analytique sur les facultés de l'ame* (1760), § 530, in *Œuvres*, vol. 13.

"history of attention" (in the sense of a natural history), if well done, would "surpass any *Logic*" since it would be "logic in action."[159]

Bonnet's methodological vocabulary does not differ fundamentally from that of his critics. Like these, he recommends amassing facts on the basis of repeated observations and experiments.[160] Observation and experimentation must be supplemented by analysis and synthesis. The scientist combines the collected facts into groups, derives principles and laws by induction, and from these laws draws conclusions which he tests empirically. Since similar effects may have the same cause, it is legitimate to use analogy to formulate a hypothesis which may inspire fresh observations and experiments. Such was, in Bonnet's view, the method of the natural sciences, which should also be used in metaphysics. The philosopher considered his own metaphysics to be "almost entirely physical."[161]

Observation and experimentation did not exhaust methododology. Imagination and curiosity, conjecture and hypothesis also had a part to play, as long as they were monitored by reason and the spirit of observation:

> One can never formulate too many hypotheses on an obscure subject. Hypotheses are the threads that may lead us to the truth along varied paths, or enable us to discover new lands. Conjecture is the spark which good physics uses to ignite the torch of experience. . . . Let the imagination roam; but let reason always hold the reins of this dangerous steed. Let us leave no stone unturned, invent new suppositions and create new hypotheses; but let us never forget that they are but conjecture, and hypothesis, and that they should never take the place of fact.[162]

"I will construct hypotheses," Bonnet boldly declared in the opening pages of his *Analytical Essay*, "and I will make sure that these hypotheses are based on facts and appear to derive naturally from them."[163] Although Schütz and Tetens thought he had failed in this, his insistence on the positive role of hypothesis and conjecture is methodologically significant.

159. Bonnet, "Analyse abrégée de l'*Essai analytique*," § XX, in *Œuvres*, vol. 7.

160. On Bonnet's methodological thought, see Raymond Savioz, *La philosophie de Charles Bonnet de Genève* (Paris: Vrin, 1948), chap. 14; also Marc Ratcliff, "Une métaphysique de la méthode chez Charles Bonnet," in Marino Buscaglia et al., eds., *Charles Bonnet savant et philosophe (1720–1793)* (Geneva: Editions Passé Présent, 1994).

161. Bonnet, *Œuvres*, vol. 1, p. vii.

162. Bonnet, *Considérations sur les Corps organisés* (1776), chap. 3, § 24, in *Œuvres*, vol. 3, p. 11.

163. Bonnet, *Essai analytique*, p. 2.

Jean Trembley, for one, accepted it wholeheartedly. In 1773, while temporarily replacing Horace-Bénédict de Saussure (1740–1799) on his course in logic at the Geneva Academy, he referred to the discovery of the aberration of light to argue that hypotheses are as necessary and useful in psychology, natural law, and politics as they are in experimental physics and mathematics.[164] When Abraham Trembley accused Bonnet of being content with hypotheses and of following his own principles too systematically, Jean Trembley replied:

> If hypotheses were to be banished from physics, what would become of them? Not only would the hypotheses physics contains be excluded, but also everything that has since been confirmed by experience and calculation, and which previously was only hypothetical.[165]

The arguments and calculations of celestial mechanics, for example, clearly presuppose factual observation. But, Trembley noted, when we are talking about light emissions, for example, "since we have no idea of their nature or of the laws obeyed by the force that causes them, we can base our calculations only on hypotheses which do not stem from the nature of things."[166] Basically, our knowledge of nature advances by testing hypotheses against experience. Such a test is vital if a fact is to be accepted as such, rather than as part of some synthetic and deductive "system." These principles, which Trembley expressed already in his youth, remained the guiding thread of his epistemology and methodology, and manifested themselves clearly in the many physical "observations" he published, mostly in the *Memoirs* of the Berlin Academy.

Jean Trembley's reply to the question on the usefulness of psychology and the way to ensure its progress seems to have been the first metapsychological work in the French language.[167] In it psychology was treated as an autonomous discipline. Trembley developed Bonnet's methodological

---

164. J. Trembley, *Cours de logique*, chap. 8 ("De hypothesi"), fol. 104. Notes taken by Georges-Constantin Naville (BPU, Ms. Cours univ. 779).

165. J. Trembley, "Examen des Remarques sur l'*Essai Analytique*: Par Mr. Jean Trembley," fol. 5 bis recto (BPU, Fonds Trembley 33).

166. J. Trembley, "Observations sur l'attraction & l'équilibre des Sphéroïdes," *Mémoires de l'Académie Royale des Sciences et Belles-Lettres de Berlin*, 1799–1800, 68–109, pp. 91–92.

167. J. Trembley, "Réponse à la question, proposée par la Société de Harlem: Quelle est l'Utilité de la Science Psychologique dans l'éducation & la direction de l'Homme, & relativement au bonheur des Sociétés? Et quelle serait la meilleure maniere de perfectionner cette belle Science, & d'accroitre ses progrès?," *Verhandelingen, uitgegeeven door de Hollandsche Maatschappye der Weetenschappen, te Haarlem*, 20, no. 1, 1781, 1–310; referred to in the text by page

reflections further by addressing his vision of psychology's progress while unobtrusively introducing the German discussions and contributions to the discipline. Conversely, the work of Tetens and Schütz was closely linked to the *Essay on Psychology* and the *Analytical Essay*, which were translated into German at the beginning of the 1770s; in 1777, Jean Trembley could report to Bonnet that the latter's ideas were thriving in Germany.[168]

Trembley divided his prize essay into three parts. It was first necessary to "have an accurate idea of what psychology should be," in order to avoid both "dreaming up systems" and accumulating detail or "minor isolated facts" which one might then treat as principles (239–240). This implied above all characterizing psychological methodology. Trembley explained that psychology should involve both analysis and synthesis. Analysis establishes the most general facts, for example, that all humans have a love of happiness and of themselves, or that the soul experiences pleasure at carrying out its activities. Synthesis then demonstrates these facts logically: someone who did not love himself would not seek self-preservation, and an active being who derived no pleasure from his activities would not be acting according to the principle of happiness (245). All the other facts are causally and logically dependent on these general ones, which are considered ratified.

Trembley also explained that we cannot attain knowledge of primary causes and first principles but only of effects: "A body that falls is not gravity but is a result of gravity" (261). Similarly, since we know the soul only through the body, we must approach it by making a large number of "well-analyzed physical observations" (276). In short, psychology should imitate physics. But in doing so, it is all the more necessary to "stick close to the facts . . . since we cannot apply a mathematical analysis to them and our reasoning is forced to do without signs, such that we run a much greater risk of going astray" (282–283). Since the object of psychology is "within ourselves" (61), we must get used to "sustained self-observation and awareness of what is going on within us" (285–286). At the same time, we should turn to the type of history pursued by the unnamed authors Trembley characterized as the "good historians who are working at present." Such historians (discussed in chap. 6) do not limit themselves to the "tedious details" of sieges and military camps but map the "course of the human mind," study habits and customs, and thereby shed light on human nature.

---

number. The French original occupies the lower third of each page, the rest being reserved for the Dutch translation.

168. Letter from Jean Trembley to C. Bonnet, 14 July 1777, cited in Marx, *Charles Bonnet*, pp. 419–420. See also Dessoir, *Geschichte*, pp. 130–131.

For Trembley, psychology, like physics, had only made progress "since it [had become] experimental and started calculating effects without being encumbered by causes" (274). Its internal organization should likewise mirror that of physics. Psychologists' desire to create a "total system" had prevented them from studying anything in depth; physics began to make progress only when physicists devoted themselves to "particular branches." Psychology should follow its example:

> Why do some psychologists not undertake a psychological examination of the senses and the way in which we come to know and judge objects, while others could address the psychological history of children and also of animals, which, since they too are sensible beings, have several types of relation to us. Others could look at the history of the passions or the different classes of action of men in society; still others at the philosophy of history, etc. Psychology would then become an equally vast and illuminating science, which would accumulate new riches every day, such that one day we may be able to bring together all these different treasures and dare to undertake a theory of man. This is without doubt the principal way to perfect this science. (292–293)

Specialization is presented here as an essential condition of the progress which psychology has yet to achieve, after aligning itself methodologically and epistemologically with physics. Once it has become an empirical science based on the description of phenomena and the inductive formulation of general laws, it should abandon the idea of systematicity and the search for causes and divide itself into specialist areas. The history of physics shows us that only "particular questions" are able to enrich the sciences (30–31). Methodological principles thus operate also at the level of the contents and organization of the discipline. Just as the goal of physics is the discovery of simple and general laws which can unite disparate phenomena, so the regulatory ideal of an empirical and specialized psychology takes the form of a science of man unified around laws of the same type.

Such a methodology had at once an epistemic and a moral dimension. While most of Jean Trembley's works were concerned with questions of analytic geometry and its application to problems in physics and mechanics, he did not suggest, unlike Hagen or Krüger, that psychology should *apply* mathematical analysis. Rather, he sought to bring both algebra and psychology, along with aesthetic judgment, into an arena defined by common methodological rules which drew their legitimacy from psychological analysis. Since psychology demonstrates how the mind works, and since

valid methods are to be based on knowledge of the mind, psychology is the
only science whose method derives from its own discoveries. This was why
Trembley felt psychology could underpin a cognitive morals, valid for all
forms of knowledge, and grounded in experience, the mechanisms of habit
formation and the association of ideas, and the controlled use of curiosity,
attention, and imagination. For Trembley, as for many contemporaries, the
fact that psychology could be the foundation of an epistemic ethic placed it
at the apex of the hierarchy of the sciences. And from these heights psychology could envisage the ultimate ends of the human being.

★ ★ ★

For Carrard and Senebier, who wrote on the "art of observing," but also for
Abraham Trembley and Horace-Bénédict de Saussure, Bonnet's method was
the very incarnation of scientific logic. It was embodied in Bonnet's works
on natural history and also addressed directly, in psychological terms, in
his *Analytical Essay*. Not only could this method lead to probable truths
in natural philosophy, but it could also prove the existence of God, the immortality of the soul, and the teleological character of the world.[169] Methodology was both an ethics of inquiry and an inquiry into ethics, and this
dual aspect, far from being secondary, was an integral part of the discourses
on method of the time. Tetens, for example, despite being a tireless critic of
Bonnet, ended his *Philosophical Essays* with a discussion of human perfectibility, of how to realize "the most noble project of which man is capable,"
that of "working toward the betterment of humanity." To this end, Tetens
said, we need to be guided by "enlightening reason" (*aufklärende Vernunft*)
which—even if Tetens does not say so in so many words—finds an essential
expression in sound empirical methods.[170]

The originality of these discourses did not lie in the conviction that
thinking well and acting well go together, or in how method was conceived.
Most of the thinkers mentioned here saw themselves as followers of Bacon,
who, as Garat put it, is the "universal legislator" of the knowledge enterprise.[171] Central to that "legislation" is the Baconian fact—not the fact of
the Aristotelian and Scholastic tradition, limited to what takes place ha-
bitually and is part of common experience, but extended to events that are

---

169. Jacques Marx, "L'art d'observer au XVIIIᵉ siècle: Jean Senebier et Charles Bonnet,"
Janus, 61, 1974, 201–220.

170. Tetens, *Philosophische Versuche*, vol. 1, p. xxxvi.

171. Garat, "Analyse de l'entendement," p. 155.

particular and singular, even exceptional and strange. The Aristotelian approach aimed at general propositions on the way things normally occur, from which the causes of other facts could be deduced. In the Baconian perspective, inductive reasoning from facts did not aspire to demonstrative certainty, and only if observations and experiments were repeatable were they deemed capable of leading to "physical" and "moral" certainty.[172]

Isaac Newton (1642–1727) too played a tutelary role in psychological methodology. His presence, however, was far from monolithic, and he was relatively absent from the German psychological scene.[173] That said, psychological terminology often reflected "Newtonian" methodology. A case in point is the analytic method, as defined in question 31 of his *Optics* (1704). It involves carrying out experiments and observations, drawing general conclusions by induction from facts, and accepting as counterarguments only reasoning established in a similar fashion. Analysis was then followed by synthesis or "composition," which consisted of taking already validated causes as principles, using the latter to explain phenomena derived from these causes, and presenting the explanations demonstratively. Psychologists echoed these rules, along with those from the third book of Newton's *Mathematical Principles of Natural Philosophy* (1687; edition of 1713), namely, that only causes of natural phenomena which are true and sufficient to explain their appearance should be accepted; that as far as possible the same effects should be attributed the same causes; and that, despite hypotheses to the contrary, propositions arrived at by induction should be considered exactly or nearly true until other phenomena make it necessary to modify them. The famous *hypotheses non fingo* of the "General Scholium" (*Principles*, edition of 1713, bk. III, *in fine*), in which hypothesis is defined as that which is not deduced from phenomena, was also evoked, as was the exclusion of hypothesis from "experimental philosophy," developed in question 31 of the *Optics*. Abraham Trembley's criticism of Bonnet and Jean Trembley's reply are one of the many echoes of this issue which preoccupied scholars. All in all, despite the variety of readings of Bacon and Newton, Enlightenment psychologists often declared their adherence to principles which they identified with the natural philosophy of these thinkers.

172. Lorraine Daston, "Baconian facts, academic civility, and the prehistory of objectivity," *Annals of scholarship*, 8, 1991, 337–363.

173. Paul Wood, "Science, philosophy, and the mind," in Porter, *The Cambridge history of science*, vol. 4.

Beyond these principles, however, methodological reflection contributed to psychology mainly due to the relations it established between investigative procedures and psychological objects.

"Attention" became central to methodology when it developed into a major theme in psychology and attracted a large number of studies.[174] Only Bonnet, to my knowledge, went so far as to claim that the empirical study ("a history") of attention could replace logic, but the idea was widespread. The methodologies of the sciences were based on attention. "Observation," says the Encyclopédie, "is the attention of the soul turned toward objects proposed by nature. Experimentation is the same attention, but directed toward artificially produced phenomena."[175] Sensation is no more than a mediocre, if necessary, point of departure. One need only read Senebier to realize how far his notion of observation is from the passivity often associated with sensualism, and how firmly it is linked to psychology as the science of the soul's activity.

Although Senebier could write that the observer "merely makes himself receptive to the sensations which external objects arouse," observation remained for him "the reflective act by which the soul views the objects which concern it, by means of the senses."[176] His Essay highlighted how faculties and qualities which go well beyond sensory precision and intensity contribute to observation, for example, attention, the coup d'oeil, or ability to understand at a glance, perceptiveness, discernment, intelligence, imagination, and judgment.[177] Observation, he stated, "presupposes reasoning, without which we would perceive without noticing and view without seeing."[178] This is why we have to analyze the subject before observing it, and plan our observations so as to generate ideas and establish links which would not occur to us while we were engrossed in looking. "When we observe without prior preparation," says Senebier, "we see only that which strikes the senses; we do not get beyond the outer shell of objects and everything beneath remains hidden to even the most perspicacious observer."[179] In

---

174. David Braunschweiger, Die Lehre von der Aufmerksamkeit in der Psychologie des 18. Jahrhunderts (Leipzig: Hermann Haacke, 1899); Gary Hatfield, "Attention in early scientific psychology," in Richard D. Wright, ed., Visual attention (New York: Oxford University Press, 1998).

175. Paris Encyclopédie, OBSERVATION (Grammaire, Physique, Médecine), 13, 313–321, p. 313.

176. Senebier, Essai, vol. 1, pp. 27 and 24.

177. Ibid., vol. 1, pt. 2 ("De l'observateur pendant qu'il observe"), chap. 1 ("L'observateur doit avoir du génie").

178. Ibid., vol. 1, p. 147.

179. Ibid., vol. 1, p. 294.

short, the goal of the art of observing is not descriptive precision and accuracy. Rather, it "consists of discerning the qualities of the beings we are studying, tracing their effects, grasping their similarities and differences, discovering their mutual relations, and inferring, whenever possible, from an effect . . . to its cause."[180]

What holds for the naturalist is even truer for the psychologist, who must direct the attention of his soul to the operations it carries out in order to observe itself. Psychological observation—which, like all observation, is an "attention of the soul"—is based on apperception, a voluntary effort to become aware of what is happening in one's soul. And this operation, which itself belongs to the soul, is also what makes us ourselves. For Condillac, "from the very first scent our statue's capacity for having sensations is wholly devoted to the impression made on its sense organ. This is what I call attention."[181] For Bonnet, the statue's sense of smell enables it to begin "enjoying existence," but not to know that it exists, for as long as the statue does not practice apperception and separate perception from the perceiving subject, it has no "I." And if the statue has only one sensation, it will have no attention, because it has no choice. Attention, says Bonnet, "seems to presuppose the presence of different ideas, one of which the soul selects."[182] Either way, attention is not simply the beginning but the very condition of psychic existence, and even sensualism's detractors are more or less in agreement on this point. The feeling of existing or sense of self is not a passive perception but an inner experience of growing awareness, which is achieved only when attention is directed *toward oneself*.

This ontological appreciation of attention coincided with its promotion as an epistemic and social value in the second half of the eighteenth century. In pedagogy, attention, along with experience and memory, was conceived as the basis of the learning process. In Germany, attention directed at oneself, one's feelings and thoughts, was a means for the enlightened bourgeoisie to affirm its sense of self. For experiential psychologists such as Karl Philipp Moritz (1756–1793), who edited the *Magazin zur Erfahrungs-seelenkunde* and wrote the famous "psychological novel" *Anton Reiser,* no detail was so insignificant as to be passed over.[183] Attention became a

---

180. Ibid., vol. 1, p. 33.

181. Condillac, *Treatise on the sensations,* trans. Geraldine Carr (London: Favil Press, 1930), p. 4.

182. Bonnet, *Essai analytique,* § 47.

183. Michael Hagner, "Toward a history of attention in culture and science," *Modern language notes,* 118, 2003, 670–687, pp. 673–679.

powerful tool for training and controlling the self and others.[184] It prescribed an extremely strict framework for those wishing to become observers or carry out experiments, and a discipline which ended up taking over the mental and physical existence of the naturalist, giving him as much work as pleasure.[185] Psychology displayed a unique reflexivity, since it revealed the function and significance of attention through the practice of attention itself; this alone justified its status as first science.

A further way in which methodological debate contributed to shape psychology is a consequence of this first point. Since methodological inquiry focused on the epistemological procedures and psychic acts needed to obtain psychological knowledge, it entailed a critique of faculty psychology. Schütz referred to the "so-called" faculties of the soul, saying that it was better to try to understand them in their interaction than in their static organization. Tetens contested the apparently automatic character of the processes which, in Condillac or Bonnet, seemed to lead from sensation to knowledge and chose to focus on the soul's autonomous activity (*Selbsttätigkeit*), which he made into the fundamental psychological faculty (*Grundkraft*).[186] Bonnet himself admitted that he did not know if the soul produced sensation actively or passively, but defined it as "a *force*, a *power*, capable of acting or producing *certain effects*."[187] As we shall see below (chap. 8), the notion of the soul as an active force inspired the reformulation of metaphysics, logic, and morals in psychological terms. It also contributed to making the system of faculties dynamic, and even to breaking it apart, as it was replaced by a science seeking to grasp mental "acts" and what "was going on" in the soul.

184. For example, during the antimasturbation campaign that began in the 1770s, particularly in Germany, educators stressed the importance of paying constant attention to schoolchildren's behavior and physiognomy and proposed various attentional techniques to monitor young people and make them avoid or admit to onanism. These methods were advocated particularly in texts published by the *Allgemeine Revision des gesamten Schul- und Erziehungswesens* (1785–1792), a periodical for educational reform edited by the Philanthropinist leader Joachim Heinrich Campe (1746–1818). See Christa Kersting, *Die Genese der Pädagogik im 18. Jahrhundert: Campes "Allgemeine Revision" im Kontext der neuzeitlichen Wissenschaft* (Weinheim: Deutscher Studien-Verlag, 1992).

185. Lorraine Daston, *Eine kurze Geschichte der wissenschaftlichen Aufmerksamkeit* (Munich: Carl-Friedrich-von-Siemens-Stiftung, 2001); and Daston, "Attention and the values of nature in the Enlightenment," in L. Daston and F. Vidal, eds., *The moral authority of nature* (Chicago: University of Chicago Press, 2004).

186. Tetens, "Ueber die Grundkraft der menschlichen Seele und den Charakter der Menschheit," eleventh essay in the *Philosophische Versuche*. See Barnouw, "Psychologie empirique."

187. Bonnet, *Essai analytique*, § 46.

The methodological discussions were emblematic of the science to which they belonged: despite its "disciplinary" consolidation, eighteenth-century psychology did not become a closed system. On the contrary, different plans for what it should look like, different research programs and different visions of the discipline were developed in different places, with a wealth of links to medicine, theology, and philosophy. When Kant expressed the desire to see psychology become an academic discipline, he was not thinking of exactly the same contents and methods as the Scottish philosophers, the German Wolffians, the Swiss naturalists, or the French encyclopedists. Although the "age of psychology" was inspired by a widely felt confidence in the value of the empirical study of the soul, it presented a varied landscape. But its protagonists agreed on the disagreements which it was legitimate to have—concerning how dependent knowledge was on sensation, for example, or the number of faculties and their mode of inter-relation—and they were, last, united by a common project, to be realized by "paying attention to what is going on in the soul."

CHAPTER FIVE

Historicizing Psychology

Post-1960 critiques of the historiography of psychology suggest that the retrospective construction of the discipline within a "long" history and on a triumphant course toward scientificity is a late nineteenth-century invention generated by the wish to legitimate a supposedly young science.[1] Although such critiques are valid, the retrospective construction of psychology dates in fact from the eighteenth century, when it emerged in the context of the history of philosophy, and contributed to bringing about psychology as an autonomous academic field. Historical narratives helped fashion the discipline by situating the study of the soul in a temporal perspective, presenting empirical psychology as an Enlightenment achievement, and transforming psychological discourse into a transhistorical entity. These developments were not initially due to individuals setting out to "do psychology," but rather the work of bibliographers, encyclopedists, historians of philosophy and *polyhistors*, scholars of universal learning who sought to synthesize knowledge and spread it far and wide across the

---

1. George W. Stocking, "On the limits of 'presentism' and 'historicism' in the historiography of the behavioral sciences," *Journal of the history of the behavioral sciences*, 1, 1965, 211–217; Robert M. Young, "Scholarship and the history of the behavioral sciences," *History of Science*, 5, 1–51, 1966. See also Mitchell G. Ash, "The self presentation of a discipline: History of psychology in the United States between pedagogy and scholarship," in L. Graham, W. Lepenies, and P. Weingart, eds., *Functions and uses of disciplinary histories* (Dordrecht: Reidel, 1983). For an overview of historiographical issues, see Ernest R. Hilgard, David E. Leary, and Gregory R. McGuire, "The history of psychology: A survey and critical assessment," *Annual Review of Psychology*, 42, 1991, 79–107. On the relations between history and legitimation, see Nicholas Jardine, *The scenes of inquiry: On the reality of questions in the sciences* (Oxford: Clarendon Press, 1991), chap. 6.

Republic of Letters.[2] In discussing writings on the soul, *polyhistors* defined the contents, scope, and limits of psychology, as well as its relation to other disciplines. They were concerned with the history and organization of the totality of scholarly knowledge (*Gelehrsamkeit*), understood as mobilizing not only the senses, memory, and *ingenium* for a "common" and descriptive understanding of things but also the judgment needed to grasp underlying principles.[3]

INVENTING A BIBLIOGRAPHIC TRADITION

Establishing empirical psychology as an academic specialism was made possible partly through the creation of a body of authoritative authors, that is, a bibliography which defined the borders of the discipline. As early as the sixteenth century, major bibliographic catalogs of philosophy were being published, reflecting the structure of textbooks and teachings, and including *de anima* writings.[4] It was a long-established tradition to consider writings on the soul as one thematic corpus, sometimes divided up internally. Thanks to that tradition, historians benefit from an exceptional collection of catalogs, ranging from those of the prolific bibliographer Martin Lipenius, who, in his *Bibliotheca philosophica* (1682), classified works on the soul alphabetically from Aristotle to contemporary *Psychologiae*,[5] via Johann Georg Theodor Gräße's *Bibliotheca psychologica*[6] and the monumental catalogs of the American theologian Ezra Abbot or the German psychiatrist Heinrich

2. *Polyhistor* is close to the English "polymath" and the French *polygraphe*. For German usage in the eighteenth century, see Johann Heinrich Zedler, *Grosses vollständiges Universal-Lexicon aller Wissenschaften und Künste* (1732–1750; Graz: Akademische Druck- und Verlags-Anstalt, 1993), s.v. "Polyhistor" and "Polyhistorie." Walch, *Philsophisches Lexicon*, s.v. "Polyhistorie," warns against overuse of the term. For the polymathic project and its most famous representative, Daniel Morhof, see Françoise Waquet, ed., *Mapping the world of learning: The Polyhistor of Daniel Georg Morhof* (Wiesbaden: Harrassowitz Verlag, 2000). For a more general overview, see Anthony Grafton, "The world of the polyhistors: Humanism and encyclopedism" (1985), in *Bring out the dead: The past as revelation* (Cambridge, MA: Harvard University Press, 2001).

3. Walch, *Philosophisches Lexicon*, s.v. "Gelehrsamkeit"; Johann Andreas Fabricius, *Abriß einer allgemeinen Historie der Gelehrsamkeit* (1752–1754; Hildesheim: Olms, 1978), vol. 1, pp. 3–4.

4. Michael Jasenas, *A history of the bibliography of philosophy* (Hildesheim: Olms, 1973).

5. Martin Lipenius, *Bibliotheca realis philosophica omnium materiarum, rerum, & titulorum, in universo totius philosophiae ambitu occurentium* (1682; Hildesheim: Olms, 1967), s.v. "Anima."

6. Johann Georg Theodor Gräße, *Bibliotheca psychologica oder Verzeichniß der wichtigsten über das Wesen der Menschen- und Thierseelen und die Unsterblichkeitslehre handelnden Schriftsteller älterer und neurere Zeit* . . . (1845; Amsterdam: E. J. Bonset, 1968).

Laehr,[7] to those of Hermann Schüling and Wilhelm Risse.[8] While the term "psychology" can give thematic homogeneity and temporal continuity to the subject matter, it cannot by itself define the contents of a discipline. Yet in the eighteenth century bibliographic activity sometimes played precisely this instituting role, insofar as it separated psychology, as the empirical science of the soul, from pneumatology, the metaphysical science of spirits.

Burkhard Gotthelf Struve (1671–1738), a professor of history and later of law at Jena, discussed the *scriptores pneumaticos* in his *Bibliotheca philosophica*. He explained that he would not comment on debates about the transmission of the soul, or on Scholastic or Platonic writers, because they were "outmoded."[9] The considerably enlarged edition of 1740 did not create a special category for "psychological" authors but presented Wolff's concept of empirical psychology and devoted considerable space to the Wolffian school.[10] Erudition was not a goal in itself but was, in a manner typical of the early *Aufklärung*, subordinated to the goal of improving human life. Authors such as Struve were in general anti-Scholastic and anti-Cartesian but saw themselves as eclectic and searched for truth in the most varied sources. It was this sort of pragmatic philosophy—whether *aulica* ("princely") in the case of Christian Thomasius or *instrumentalis* in the case of Johann Franz Budde—that the historians and bibliographers of the first half of the eighteenth century sought to further. Struve's attitude to pneumatological literature illustrates this trend and highlights the growing dissociation of pneumatology from psychology, henceforth singled out for its cognitive and moral usefulness and for the role it assigned to empirical knowledge. Within their field, then, bibliographers of philosophy gave impetus to a movement of disciplinary differentiation.

An example on a grand scale of this movement is provided by a history of scholarly literature in the liberal arts and philosophy, published in 1718 by Gottlieb Stolle (1673–1744), a professor at Jena who also wrote similar works for law, medicine, and theology. It was a *historia literaria*, a commentary on books presented chronologically and thematically. The chapter

7. Ezra Abbot, "Literature of the doctrine of a future life: or, A catalogue of works relating to the nature, origin, and destiny of the soul," in William Rounseville Alger, *A critical history of the doctrine of a future life* (Philadelphia: George W. Childs, 1864); Heinrich Laehr, *Die Literatur der Psychiatrie, Neurologie und Psychologie von 1459–1799* (Berlin: Reimer, 1900), 4 vols.

8. Schüling, *Bibliographisches Handbuch* and *Bibliographie der psychologischen Literatur*; Risse, *Bibliographia philosophica*.

9. Burkhard Gotthelf Struve, *Bibliotheca philosophica in suas classes distributa* (1704; Jena: apud Ern. Claudium Baillar, 1707), chap. 5, § IX.

10. Ludwig Martin Kahle, *Bibliothecae philosophicae Struvianae emendatae, continuatae atque ultra dimidiam partem* (Göttingen: impensis Vandhoek et Cunonis, 1740), chap. 4, § XII.

on pneumatology, like the other chapters, covered its subject from antiquity up to the most recent debates between Wolffians and anti-Wolffians. The result was a kind of historicized bibliography somewhere between the annotated book list and the history of ideas. The "history of pneumatology," said Stolle, makes it clear that we have generally confused the dictates of Revelation and the teachings of Nature.[11] Implicit in this remark is the conviction that psychology should be differentiated from pneumatology. This conviction prompted Stolle's praise for Jean Le Clerc (1657–1769), a Protestant theologian and friend of John Locke (as well as his first biographer). In limiting the scope of his pneumatology, Le Clerc had created, in Stolle's opinion, a more judicious work than any of his predecessors.[12]

Around the midcentury, Johann Andreas Fabricius (1696–1769), author of philosophical, moral, educational, and popularizing works (including on the art of eloquence in galantry), published a vast *Outline of a General History of Learning*. In it he clearly distinguished pneumatology from psychology, which he traced back to antiquity. Pneumatology, he explained, had been treated both as as autonomous discipline and as belonging to metaphysics, natural philosophy, ontology, or natural theology; it included, in addition to the soul, the existence of benign and evil spirits, phantoms and ghosts, and magic and sorcery. Psychology, by contrast, focused on the human soul and dealt with its nature, its union with the body, its immortality, origin, migration, and state after death, its seat in the body, and its presence in animals.[13] It therefore encompassed every discourse on the human soul.

11. Gottlieb Stolle, *Anleitung zur Historie der Gelehrheit, denen zum besten, so den Freyen-Künsten und der Philosophie obliegen* (1718; Jena: in Verlegung Johannes Meyers seel. Erben, 1736), p. 494.

12. Ibid., p. 500. Le Clerc, who taught at the Remonstrant seminary in Amsterdam, devoted only 160 pages to his pneumatology out of about 1,500 pages covering logic, ontology, a history of Eastern philosophy taken from Thomas Stanley, physics, botany, zoology, and a *de corpore* treatise. In the first edition of his *Opera philosophica* (which was actually a school textbook), Le Clerc dedicated his *Logic* to Robert Boyle (who died before he could see it) and his *Pneumatology* to Locke (to whom he also dedicated the *Logic* in subsequent editions). Despite their traditional form and terminology, the *Opera* were clearly indebted to Locke. In his pneumatology, Le Clerc defined the mind (*mens*) as a substance capable of thinking, while declaring at the same time that its "intimate nature" is unkowable. He maintained that pneumatology would be useless if it were reduced to mere speculative contemplation of the *animus* and stressed the moral and cognitive importance of exploring the mind empirically. Jean Le Clerc, *Pneumatologia*, in *Opera philosophica* (1692; Amsterdam: apud Joan. Ludov. de Lorme, 1710), vol. 2, esp. sec. I, chaps. 1 and 2. See Maria-Cristina Pitassi, "Jean Le Clerc bon tâcheron de la philosophie: L'enseignement philosophique à la fin du XVIIᵉ siècle," *Lias*, 10, 1983, 105–122.

13. Fabricius, *Abriß*, vol. 1, p. 4. Hauptstück, § XXXXVI. The *Abriß* was more of a *bibliographie raisonnée* than a history book in the strict sense. It was composed principally of lists of authors and works arranged chronologicaly and thematically, with short commentaries.

Whereas the range and contents of psychology as outlined by Fabricius corresponded to the academic definition of the discipline as it had existed since the late sixteenth century, by the second half of the eighteenth century psychology's focus had moved toward examining the operations, especially cognitive operations, of the soul joined to the body. Bibliographic works were instrumental in this development.

The first bibliographic catalog in which psychology was assigned a chapter of its own was probably Michael Hißmann's *Guide to Knowledge of the Best Writings from All the Branches of Philosophy* (1778). Alongside psychology were various disciplines traditionally included in this type of work (metaphysics, natural theology, morals, natural law, and politics) as well as new ones, namely, pedagogy and aesthetics, the latter appearing for the first time as a bibliographic category.[14]

Michael Hißmann (1752–1784), a professor of philosophy at Göttingen, was a committed sensualist and materialist.[15] He translated Condillac's *Essay on the Origin of Human Knowledge*, speculated on the links between the brain, the nerves and the functions of the soul, related psychology to anatomy, physiology, and medicine, and criticized those who reduced psychology to a doctrine of the faculties of a thinking substance. He was convinced that the theory of the association of ideas had great explanatory power, and, in his contempt for metaphysics and dogmatic theology, he sought to make empirical psychology into the foundation of philosophy. As for pneumatology, Hißmann observed that neither the external senses nor any of the internal senses could yield knowledge of spiritual substances.[16] "Metaphysical psychology" therefore dealt only with what we do not understand or remains doubtful. But, he added ironically, it may be useful to identify what we do not or cannot know, since "tracing the limits of human knowledge is no less the work of genius than advancing up to those limits."[17]

"Hütet euch, Weltweisen, Seelensklaven zu werden!" (Beware, philosophers, lest you become slaves of the soul!). This warning, taken from Hißmann's *Philosophical Essays*, was the expression of a materialist and

14. Michael Hißmann, *Anleitung zur Kenntniss der auserlesenen Litteratur in allen Theilen der Philosophie* (Göttingen: im Verlage der Meyerschen Buchhandlung, 1778), §§ 71–107. See Jasenas, *History of the bibliography of philosophy*, pp. 69–77.

15. Otto Finger, *Von der Materialität der Seele: Beitrag zur Geschichte des Materialismus und Atheismus in Deutschland der zweiten Hälfte des 18. Jahrhunderts* (Berlin: Akademie-Verlag, 1961).

16. Hißmann, *Anleitung*, § 128.

17. Ibid., § 130.

mechanistic vision which was not typical of German psychology at the time.[18] While Hißmann's polemical energy put the issue of the legitimate limits of psychology into particularly sharp relief, the issue was widely addressed, and not necessarily from a materialist standpoint. Hißmann edited a journal "for philosophy and its history," reprinting texts published in the proceedings of various academies.[19] The prominent place it gave to psychology emphasized Hißmann's conviction that philosophy should become psychological. The same objective informed his bibliographic *Guide*, whose longest chapter was devoted to the writings "of psychology or logic"—with the conjunction strongly suggesting equivalence.[20]

The chapter on "psychology or logic" comprised annotated bibliographies on psychology in general, logic, sensation and the senses (including their laws and the "physics" of their operation), the history of the doctrine of ideas, and various psychological phenomena and faculties, including memory, imagination, enthusiasm (*Schwärmerei*), sleep, dreaming, somnambulism, the understanding and reason (*Verstand* and *Vernunft*, which, according to the author, were the most important terms in all psychology, § 84), folly and fury, syllogistic reasoning (whose importance, given its artificiality and "false subtleties," Hißmann considered overstated, § 86), genius, consciousness and the sense of self, language, and, last, various "logical" subjects such as uncertainty, "rational skepticism," idealism, the rejection of skepticism, proof and probability, practical logic, discovery and interpretation, documentary critique, and the art of observation.

Hißmann claimed to categorize psychological writings according to their principal contents (§ 71, *in fine*). This he actually did, but in such a way as to reformulate the whole of philosophy in psychological terms. Such large-scale psychological transformation began with logic. For, Hißmann explained, "insofar as logic consists of a valid set of rules for the best possible use of the powers of the human soul, the true rational philosophy [*Vernunftslehre*], or logic, belongs to psychology" (§ 4).[21] Moreover,

since psychology, as a theory of human understanding, constitutes half of all philosophy, and since the considerations in the other half also

18. Michael Hißmann, *Psychologische Versuche, ein Beytrag zur esoterischen Logik* (Frankfurt, 1777), p. 279.

19. Michael Hißmann, ed., *Magazin für die Philosophie und ihre Geschichte* (Göttingen: im Verlage der Meyerschen Buchhandlung, 1778–1783).

20. Hißmann, *Anleitung*, henceforth referred to in the text by paragraph number.

21. For the terminology, see Zedler, *Grosses vollständiges Universal-Lexicon*, s.v. "Logicke"; Walch, *Philosophisches Lexicon*, s.v. "Logik" and "Vernunftlehre."

refer to it, the content of the writings on the human understanding also spread over, in different ways, into the branches of psychology [*Seelenlehre*] whose proper object is the human will. That is why these writings can also be put to good use in practical subjects.

Besides, there exist psychological writings which explore a subject matter that in fact belongs to general practical philosophy, and in the course of this exploration, many observations most pertinent to theoretical philosophy are also made. That is why aesthetic writings, among others, are turned to excellent account by moralists [*praktische Philosophen*] as well as by aesthetic theorists [*theoretische Aesthetiker*]. (§ 71)

The assimilation of psychology to logic, the differences Hißmann emphasizes between psychology and *metaphysische Seelenlehre*, and his remarks on syllogistic reasoning all indicate his epistemological preferences and explain his choices regarding "the best writings from all the branches of philosophy." His vision of the value of empirical psychology could, however, be found in authors far removed from his mechanistic materialism, and the breadth he gave to psychology through his headings and references prefigured the didactic tools which would be used to fashion the new discipline.

By the end of the century, then, a canon restricted to Malebranche, Locke, Wolff, and Condillac—the authors mentioned in the brief Wolffian article on psychology in Diderot and d'Alembert's *Encyclopédie*—had grown to a substantial bibliography of core works. For example, the philosophy course taught at Lucerne by the Capuchin friar Heinrich Walser included a sixty-six-item bibliography of famous psychological authors.[22] Around two-thirds of the authors were Germans writing in the second half of the eighteenth century. The list also included journals (the *Magazin zur Erfahrungsseelenkunde* among others) and such works as Locke's *Essay Concerning Human Understanding* and Leibniz's response in his *New Essays*, Albrecht von Haller's physiology, texts by francophone authors (Condillac, Diderot, Helvétius, but also Malebranche, Bonnet, Tissot on nerve illnesses, and a treatise on magnetic somnambulism), as well as English and Scottish writers (Thomas Arnold on madness, Alexander Gerard on genius, Beattie, Hartley, Hume, Reid, Priestley), and Muratori's book on the imagination.[23]

22. Heinrich Walser, "Syllabus scriptorum in materia psychologica celebrium," in *Institutiones philosophicae*, Liber III, *Psychologia* (Augsburg: M. Rieger, 1791), pp. vi–ix.

23. See also the list drawn up by the physician Alexander Crichton: it includes "British psychologists" (Locke, Hartley, Reid, Stewart, Priestley, Kames, and others), as well as German and French authors, among them the abbé de Condillac. A. Crichton, *An Inquiry into the Nature and Origin of Mental Derangement* (London: printed for T. Cadell, 1798), vol. 1, p. xxvii.

Walser's course included a "universal pneumatology" but excluded rational psychology. His treatment of empirical psychology showed how the field had grown by assimilating subjects which until then had belonged to logic or metaphysics. Despite their differences, the Catholic Walser and the materialist Hißmann contributed to defining a new field of knowledge by establishing lists of authors, and references.

CONSTRUCTING A HISTORY FOR PSYCHOLOGY

Applying the term "psychology" retrospectively, and to any psychological subject matter whatsoever, is rightly open to criticism. The retrospective approach did, however, produce two features which were crucial to the discipline's identity: a bibliographic corpus and a historical narrative. The two genres were mutually determining, since the authors in the bibliography were also the actors of the historical narrative; the sequence of their writings revealed the dynamics of history and above all its progress. This was behind Hißmann's declared preference for a chronological order, a simple "genealogy of psychological systems" which could show how each author exploited the ideas of his predecessors.[24] Such a historicizing project was carried out principally in Germany in the eighteenth century, and was inspired by developments in the historiography of philosophy.

It is important in this context to distinguish polemical or apologetic narratives from those with real historical intent. For example, it is sometimes assumed that the historiography of psychology originated with book I of *De anima*; that book, however, discusses earlier theories of the soul in order to criticize them and justify Aristotle's conception. Christian Wolff, who, as we have seen, qualified his treatment of the soul as new, bold, and flying in the face of received opinion, behaved as though no psychology at all had existed before him: "Pars philosophiae quae de anima agit Psychologia a me appellari solet."[25] His disciple Bilfinger likewise dismissed the psychology of the ancients, holding that the distinction among vegetative, sensitive, and rational souls involved an incomprehensible mix of physical and pneumatological considerations. There is no attempt at historical understanding here, merely a judgment designed to define psychology in such a way as to admit only Wolff's vision. Similarly, in controversies around materialism, chronological accounts of theories of the nature, origin, and fate of the soul served to support particular positions. For example, *The history of the*

---

24. Hißmann, *Anleitung*, § 73.
25. Wolff, *Philosophia rationalis . . . : Praemittitur Discursus praeliminaris*, § 58.

philosophical doctrine concerning the origin of the soul, and the nature of matter, by Joseph Priestley (1733–1804), sought primarily to prove that the notions of an immortal and immaterial soul, and of the person as composed of two substances, were rooted in Eastern pagan beliefs and were consistent neither with scripture and Christian Revelation nor with the natural sciences.[26] While the argument might appear historical, its explicit goal was to support Priestley's Unitarian theology.

Of course, no history is entirely disinterested, but not every history is polemical. In the eighteenth century, the history of philosophy, from which the history of psychology developed, became a way of philosophizing and a philosophical discipline in itself. Every sort of text, from insignificant essays—and there were dozens of them—to major textbooks, pinpointed and denounced the errors of the past. Their perspective was basically philosophico-literary or historico-critical, aiming to integrate historical material and the lessons of history into philosophical activity.

A significant part of psychology had long been considered philosophical. This remained the case in the eighteenth century, but the subject matter came to be perceived as belonging to a new science. This shift was inseparable from the adoption of a historical approach, as the historiography of psychology developed hand in hand with the discipline whose history was being written.

The eighteenth century witnessed the publication of a great many histories of the sciences, both general and concerning particular fields (natural history, botany, chemistry, mathematics, physics, astronomy, medicine, and literature).[27] These recounted the steady progress of reason as manifested in the growing emphasis on observation and the creation of "philosophical and reasoned" systems that combined theory with experience. Even when these histories went back as far as biblical times, the sciences were still deemed to originate in ancient Greece, where historians sometimes found (for example, in Hippocrates) empirical observations they considered valid. By demonstrating the conditions of intellectual progress, historiographical

26. Joseph Priestley, *The history of the philosophical doctrine concerning the origin of the soul, and the nature of matter* (1777; New York: Garland, 1976) (follows the *Disquisitions relating to matter and spirit*).

27. Giovanni Getto, *Storia delle storie letterarie* (Florence: Sansoni, 1969); Rachel Laudan, "Histories of the sciences and their uses: A review to 1913," *History of science,* 31, 1993, 1–21; Dietrich von Engelhardt, *Historisches Bewußtsein in der Naturwissenschaft von der Aufklärungs bis zum Positivismus* (Freiburg: Karl Alber, 1979); Jost Weyer, *Chemiegeschichtsschreibung von Wiegleb (1790) bis Partington (1970)* (Hildesheim: Gerstenberg, 1974).

activity was viewed as sustaining it and so became a condition of the advancement of learning. The history of philosophy played a similar role when it established itself as a philosophical discipline in the eighteenth century. It developed, particularly in Germany, from a field of largely philological erudition, via an eclectic pragmatic history and the conception of history as progress, to the assertion of the superiority of Kantian criticism.[28]

Toward the end of the century, the history of philosophy in Germany, like biblical history or classical philology, was transformed into a hermeneutics and harnessed to the project of a cultural history of humanity. It interpreted its sources as the manifestation of cultures that should be understood both comparatively and for themselves, in order to reveal the stages by which human perfectibility was realized.[29] Whereas the history of most sciences was not incorporated into scientific practice, the history of philosophy became a way of philosophizing. The same would be true of the historiography of psychology when it came to be seen as furthering the psychological history of humanity.

"PSYCHOLOGIAE HISTORICO-CRITICAE SPECIMINAE"

When, in 1736, Gottlieb Stolle noted the absence of a history of pneumatology, the history of philosophy already had behind it twenty years of significant theorization and methodological reflection and was developing separate histories of different philosophical disciplines. The work of Christoph August Heumann (1681–1764), a renowned polymath and professor at the University of Göttingen, epitomized the move from a "practice" to a "discipline" conscious of its own concept and procedures. Heumann's *Acta philosophorum* (1715–1726) was the first journal to specialize in the history of philosophy and one of the first to reflect on the theory and method of this new specialism.

In 1715 Heumann explained that the history of philosophy could be approached in two ways: either by philosophical discipline, or according to a

---

28. Lucien Braun, *Histoire de l'histoire de la philosophie* (Paris: Ophrys, 1973); Maria Assunta Del Torre, *Le origine moderne della storiografia filosofica* (Florence: La nuova Italia, 1976); Mario Longo, *Historia philosophiae philosophica: Teorie e metodi della storia della filosofia tra Seicento e Settecento* (Verona: IPL, 1986); Giovanni Santinello, *Storia delle storie generali della filosofia*, vol. 2, *Dall'età cartesiana a Brucker*; vol. 3, *Il secondo illuminismo e l'età kantiana* (Brescia: La Scuola,1981; Padua: Antenore, 1988).

29. See, for example, Hans Erich Bödeker et al., eds., *Aufklärung und Geschichte: Studien zur deutschen Geschichtswissenschaft im 18. Jahrhundert* (Göttingen: Vandenhoeck & Ruprecht, 1992).

geographic and chronological order from the origins to the present, in the different regions of the world. In his view,

> this second method is the better one since it is universal, and it is somewhat easier to extract from a history of philosophy done in this way the special history [*Special-Historie*] of a particular discipline, and to present it separately. History by this method is like the history of the church. Once it has been written in a chronological and geographic order, little effort is required to do the special history of councils, heresies, popes, etc.[30]

Heumann's disciples went on to write universal, chronologically and geographically structured histories. Internally, philosophies—*barbarica* of east or west, *hebraeorum*, *exotica* or *graecanica*, pre-Christian or Christian—were arranged by author, school, or "sect" and sometimes contained headings corresponding to the disciplines which could become the object of particular histories. Heumann taught that particular histories could be geographic (arranged by nation, country, or region), topographical (by institution or town), chronological, "technological" (by "art"), disciplinary, "anthropological or biographical," or bibliographic.[31]

Heumann did not invent the history of disciplines, but he thematized its method. He devoted to it an entire chapter of his *Conspectus reipublicae literariae* and furnished lists of histories, including, for the sciences, works on mathematics, astronomy, arithmetic, and medicine. His bibliographies mention not only histories in the strict sense but also overviews and glossaries, in which the historical dimension was always present, at least in the form of a chronological arrangement. Heumann's prestigious disciple Johann Jacob Brucker (1696–1770) likewise distinguished two types of *historia philosophica*: the most important in his view was "broad and general," encompassing the history of ideas, as well as the history of authors and their contexts since the beginning of the world; the other, "narrower" *historia dogmatica* was concerned with the development of doctrines within particular disciplines.[32]

---

30. Christoph August Heumann, "Eintheilung der Historiae Philosophicae," *Acta philosophorum, sive gründliche Nachrichten aus der Historia philosophica,* 3. Stück, 1715, 462–472, pp. 462–463.

31. Christoph August Heumann, *Conspectus reipublicae literariae, sive Via ad historiam literariam iuventuti studiosae aperta* (1718; Hanover: apud haeredes Nic. Foerster et filii, 1763), chap. 1, § 7.

32. Johann Jacob Brucker, *Kurtze Fragen aus der Philosophischen Historie . . .* (Ulm: bey Daniel Bartholomäi und Sohn, 1731–35), Theil I, § VII.

The Augsburg pastor Brucker was the author of an immense and innovative body of work.[33] Like Heumann, he argued from the premise of the historical and geographic unity of human reason and human nature that the history of philosophy and the history of the intellect were one and the same. His account combined three key features: it was eclectic, in its choice of the opinions to retain and in its refusal of authority; it was critical, in its adherence to reason as the basis for judgments on the past; and it was pragmatic, in its desire to find in history truths that are both useful and grounded in reason.[34] Brucker's history was also informed by a theory of knowledge derived from the "experimental" philosophy and inductive logic advocated by Boyle and Locke. He set his vision of the progress of philosophy against the notion of *prisca sapientia*. Rather than base history on the idea of *philosophia perennis*, he favored the eclectic presentation of different schools of thought. Brucker made a sharp distinction between, on the one hand, the *sapientes* (such as Adam or Moses), who possessed revealed wisdom, or the mathematicians, astronomers, and doctors of antiquity, who worked in a purely empirical and practical manner, and, on the other hand, the *philosophi* who proceeded systematically, inductively, and by rational argument.[35]

Brucker's *Critical History of Philosophy* (1742–1747) was "universal" and was, as Heumann had proposed, constructed so as to allow readers to extract from it particular histories.[36] The preeminent position and rank of psychology within it and among the philosophical sciences was determined by a long tradition: "Post Deum non nobilius in pneumatologia metaphysicae iuncta argumentum occurrit, quam quod de anima tractat, et speciali nomine psychologia dicitur" (IV.2, 706). On the one hand, Brucker used the term "psychology" to refer to the natural history of the human soul joined to the body, the metaphysics of the separate or separable soul, and sometimes both at once; in this sense, *all* the discourses on the human soul

33. For biographical studies, see Wilhelm Schmidt-Biggemann and Theo Stammen, eds., *Jacob Brucker (1696–1770): Philosoph und Historiker der europäischen Aufklärung* (Berlin: Akademie Verlag, 1998), 1st pt.

34. For a concise formulation of this approach, see Brucker, *Kurtze Fragen*, pp. 15–16.

35. For Brucker's theory of knowledge and its sources (including, in the field of the history of science, Hermann Conring's *De hermetica medicina* [1667] and Daniel Le Clerc's *Histoire de la médecine* [1702]), see Constance Blackwell, "Thales Philosophus: The beginning of philosophy as a discipline," in Kelley, *History and the disciplines*; Blackwell, "Jacob Brucker's theory of knowledge and the history of natural philosophy," in Schmidt-Biggemann and Stammen, *Jacob Brucker*; and W. Schmidt-Biggemann, "Jacob Bruckers philosophiegeschichtliches Konzept," in Schmidt-Biggemann and Stammen, *Jacob Brucker*.

36. Johann Jacob Brucker, *Historia critica philosophiae a mundi incunabilis usque ad nostram aetatem deducta*, 4 vols. and appendix (1742–47; Hildesheim: Olms, 1975), henceforth referred to in the text by volume and page number.

which Brucker examined were "psychological." For more recent periods, on the other hand, he tended to apply the term only to the science of the embodied mind.

At the beginning of his presentation of recent psychological discussions ("psychologiae recentioris fata"; IV.2, 706–710), Brucker stated that anyone writing a "special history of the discipline" would have to speculate on obscure and controversial issues. This, however, was not what he intended to do. Instead of intervening in past debates, he wished to contribute positively to philosophy by presenting the doctrines such as to draw out their meaning and interconnections. This method, Brucker explained, would enable the reader to identify in his *History of Philosophy* materials for a critical history of psychology: "Quo pacto multa eiusmodi psychologiae historico-criticae speciminae in hoc nostro opere historico-critico et philosophico reperientur" (appendix, 924).

Brucker first of all examined how past philosophers had approached "psychology." In Plato, he singled out the *Timaeus* (because of its focus on the embodied soul) rather than the *Phaedo* (which concerns the soul's immortality). To link psychology to natural philosophy, Brucker explicitly chose what he considered most intelligible and regularly emphasized the obscurity and difficulty of the subject (I, 712). For example, in his summary of the *psychologia aristotelica*, he remarked upon the futility of trying to reduce it to clear concepts, especially with regard to such disputed issues as the immortality of the soul and the agent intellect (I, 821).

Similarly, Brucker accused Plotinus of treating psychology "more iterum suo obscurissime" (II, 411). Yet he still considered the *Enneads* essential reading as the foundation of a cosmology and an ethics. Here we see a second feature of Brucker's approach: the moral interest. For example, when entering the "labyrinth" (II, 1049) of Cabalistic psychology, he warns against overuse of allegory and evokes the risk that it may lead to deism and enthusiasm (II, 1038). He later discards Christian Thomasius's pneumatology in favor of the latter's moral philosophy (IV.2, 496–498).

As for John Locke, we have seen that the *philosophes* thought he reduced metaphysics to an "experimental physics of the soul." Nineteenth-century psychologists would single him out as their predecessor, even if, as Théodule Ribot noted, the Lockean revolution did not immediately produce a corresponding transformation in scientific and philosophical terminology.[37] For James Mark Baldwin, the problem of knowledge was displaced in

---

37. Théodule Ribot, *La psychologie anglaise contemporaine (école expérimentale)*, 2nd ed. (Paris: Baillière, 1875), pp. 33–34.

Locke "from metaphysics to facts,"[38] and this interpretation has been generally adopted by later histories of psychology.[39]

For Brucker and his contemporaries, however, Locke's revolution took place in logic.[40] Brucker referred to the wealth of psychological material to be found in the *Essay on Human Understanding* (although "psychology" and its cognates are not mentioned by name) and stressed that they were derived "non ex libris sed experientia interna" (IV.2, 609). He placed Locke in the same category as Ramus, Malebranche, and Tschirnhaus, whom he described as Germany's Malebranche; these thinkers had all recently carried out reforms in logic, a "science which guides the cognitive faculty toward discovery of the truth."[41] But since it was deemed essential for logic to be grounded in a natural history of the understanding, Locke's work came to be classified as psychological, and logic as a branch of psychology. Hißmann, for

38. James Mark Baldwin, *History of psychology: A sketch and an interpretation* (New York: G. P. Putnam's Sons, 1913), vol. 2, pp. 1–2.

39. Even Klemm, who did not think Locke had contributed much of value to psychology, considered him to be writing from the viewpoint of empirical psychology. See Otto Klemm, *A history of psychology*, trans. Emil Carl Wilm and Rudolf Pintner (1911; New York: Charles Scribner's Sons, 1914), pp. 58–59.

40. The opinion of the preacher and pedagogue Johann Christoph Stockhausen, who thought Locke's *Essay* had more to do with metaphysics than with logic, seems atypical. See J. C. Stockhausen, *Critischer Entwurf einer auserlesenen Bibliothek für die Liebhaber der Philosophie und schönen Wissenschaften* (1751; Berlin: bey Haude und Spener, 1771), p. 25.

41. Fabricius (*Abriß*, I, 362) placed Locke in the field of logic, following on from Ehrenfried Walther von Tschirnhaus (1651–1708). Tschirnhaus was a German nobleman who, alongside his research into mathematical and technical subjects (particularly mirrors and "burning lenses," or solar furnaces, and porcelain production), set out to find a method to "perfect as best can be our understanding, insofar as this is possible by natural means" (p. xi). The first edition of his *Medicina mentis* (1686, dated 1687) was subtitled *Tentamen genuinae Logicae, ubi disseritur de methodo detegendi incognitas veritates*; the second (and definitive) edition, from 1695, characterized the "medicine of the mind" as *artis inveniendi praecepta generalia*. One need only look at the titles, then, to see the connection between Tschirnhaus's *Medicina*, on the one hand, and, on the other, Descartes's *Rules for the Direction of the Mind* and Spinoza's *On the Improvement of the Understanding*. Tschirnhaus rejected syllogistic reasoning, which can discover only known truths, and looked for a method by which the true could be distinguished with certainty from the false, error be avoided, and new truths be discovered. He criticized Descartes's rules for laying down "*what* should be observed, when searching for the truth of things unknown," rather than showing "*how* to obtain what should be observed" (p. 158). Accordingly, Tschirnhaus wanted, for example, to replace nominal and real definitions with definitions containing the efficient cause of the thing defined, its "generation" or "first mode of formation" (pp. 67–68). On Tschirnhaus, his works and his context, see Eduard Winter et al., eds., *Ehrenfried Walter von Tschirnhaus und die Frühaufklärung in Mittel- und Osteuropa* (Berlin: Akademie Verlag, 1961). The following editions and translations have helpful introductions and notes: Johannes Haussleiter et al., *Medicina mentis* (Leipzig: J. A. Barth, 1963); Lucio Pepe and Manuela Sanna, *Medicina mentis* (Naples: Guida, 1987); and Jean-Paul Wurtz, *Médecine de l'esprit ou préceptes généraux de l'art de découvrir* (Paris: Ophrys, 1980). Page numbers given here refer to the 1695 Latin edition.

example, did not include Locke's *Essay* among works of logic but placed it after Malebranche's *Search after Truth* in the bibliography on psychology.[42]

In Brucker's view, however, Locke's work still belonged to logic, and the psychological content of recent works could be attributed to transformations in metaphysics and pneumatology (IV. 2, 706–710). This content was always polemical, particularly on the issue of materialism. Brucker mentioned Hobbes and Spinoza, as well as English and, to a lesser extent, German polemics on the immortality and immateriality of the soul. The debates on the faculties of the soul seemed to him to be more elusive and confused. Moreover, the problem of the nature of the soul and its union with the body was linked to the problem of freedom. There were further controversies around the distinction between *spiritus* or *mens* and *anima*, the origins and transmission of the soul, and, last, the "enigma" of its interaction with the body. Rarely, Brucker confided, have more problematic and doubtful controversies been witnessed in ancient and modern philosophy than these psychological ones, "of which one can truly say that no two opinions are alike" (appendix, 924). Faced with such "inextricable difficulties," the wisest course was to accept ignorance (IV. 2, 709–710)—a conclusion consistent with Brucker's pragmatic and eclectic position.

What is significant here is the presence, even if only virtually, of a history of psychology. The headings Brucker used to punctuate his narrative from antediluvian times to the Enlightenment introduced thematic threads which became the backbone of special histories. But headings such as "Ethica," "Metaphysica," "Physica," "Pneumatologia," or "Cosmologia" had less power to shape the disciplines they designated than did "Psychologia." When Brucker explained that his presentation of past philosophies contained "multa psychologiae historico-criticae specimina," he laid bare the constitutive function of historical discourse: by linking together scattered fragments, his critical history was actually contributing to fashioning a field of knowledge.

The historiography of aesthetics provides a good comparative example. As mentioned above, aesthetics figured for the first time under its own name in a bibliography of philosophy in Hißmann's *Anleitung* (1778). In his *Reflections on Poetry* (1735), and later in his *Metaphysics* (1739) and *Aesthetics* (1750), the Wolffian philosopher Alexander Gottlieb Baumgarten defined aesthetics as the *scientia cognitionis sensitiva* and as the counterpart

---

42. Hißmann, *Anleitung*, §§ 73–74.

of logic or the theory of the understanding.[43] The term eventually came to designate a new discipline. But in 1756, Georg Andreas Will (1727–1798), a professor of philosophy at Altdorf at the time (and later of poetics, history, and logic), made a "solemn speech on the aesthetics of the ancients." Appealing to *historia philosophica*, he argued that the antiquity of metaphysics proved that of aesthetics: Aristotle, Cicero, Quintilian, and Horace were *aesthetici*, and just as Homer came before poetics and Plato before logic, so the *aestheticus* existed before Baumgarten.[44] Will thus formulated a premise of later histories, namely, that aesthetic questions were ancient, and that, as illustrated by Baumgarten's frequent references to Horace's *The Art of Poetry*, the discipline formed in the eighteenth century merely assembled elements until then scattered across various philosophical fields, as well as in poetics and rhetoric. At the same time, by incorporating aesthetics into a tradition, he endowed it with a prestigious history which would henceforth form part of the discipline and strengthen its status as a legitimate intellectual enterprise.

THE HISTORY OF THE "THEORY OF IDEAS"

In 1723, the young Jacob Brucker published a *Philosophical History of the Theory of Ideas*. The work extended from the Greeks to the moderns, with Greek and Roman antiquity defining the first period. The second period extended from the first Christian philosophers, via the church fathers and the medieval debates on nominalism and universals, to the Renaissance and the "Scholastics' verbiage." In a third period, Locke rejected Cartesian innate ideas and explained, correctly in Brucker's view, that all ideas are acquired *per experientiam*.[45] In Brucker's view, the theory of ideas had over the centuries grown into a philosophical *arena maxima* in which the moderns still displayed their talents.[46] The questions debated within it were metaphysical (as in Platonism), logical (for instance, the debates on the categories and universals, from Aristotle to Ramus), and psychological (innatism versus

---

43. Alexander Gottlieb Baumgarten, *Aesthetica*, Latin-German ed. and trans. Dagmar Mirbach (1750; Hamburg: Felix Meiner, 2007), § I.

44. Georg Andreas Will, *Oratio sollemnis de Aesthetica veterum* (Altdorf: sumtibus Laurentii Schupfelii, 1756).

45. Johann Jacob Brucker, *Historia philosophica doctrinae de ideis qua tum veterum imprimis graecorum tum recentiorum philosophorum placita enarrantur* (Augsburg: apud Dav. Raym. Mertz, et I. Iac. Mayer, 1723), p. 240.

46. Ibid., p. 301.

the role of sensory experience). The theory of ideas was therefore perfect as
the subject of a *historia dogmatica*: it possessed a strong thematic unity and
ran through the whole of philosophy, outside a history of the disciplines. It
was, however, most connected to the history of logic; Brucker showed how
the notion of *idea* had left metaphysics and natural philosophy and become
a logical concept.[47] When he wrote his history, it was hardly conceivable
that a half century later the theory of ideas would be predominantly psychological. The publication of Hißmann's *History of the Doctrine of the Association of Ideas* in 1777 testifies to the transformation of a fundamental
philosophical issue into a psychological topic. This does not mean that the
theory of ideas stopped being relevant to philosophy but that philosophy
itself was becoming psychological. It was partly through this process that
psychology emerged, and that its *Special-Historie* could be written.

The association of ideas would be a central theme in nineteenth-
century experimental psychology and was an important topic in various
early twentieth-century schools, from psychoanalysis to behaviorism. Histories of psychology written by psychologists have invariably focused on
the long-standing and uninterrupted presence of this theme, either with
a view to blaming nineteenth-century psychology for not getting beyond
eighteenth-century associationists[48] or in order to situate modern psychology within a longer line[49] of predecessors, founders and successors.[50] It is
only recently that attempts have been made to restore the historical complexity of the debates—which touched on fields as varied as theology and
physiology—instead of treating the subject purely in the light of the contemporary concerns of psychology.[51]

Hißmann saw himself both as continuing the work of Brucker and as a
historian of psychology. He explained in his *Guide* to philosophical writings that psychological texts could be classified either by dividing authors

47. C. Blackwell, "Epicurus and Boyle, Le Clerc and Locke: 'Ideas' and their redefinition in
Jacob Brucker's *Historia philosophica doctrinae de ideis*," in Marta Fattori, ed., *Il vocabolario
della République des Lettres: Terminologia filosofica e storia della filosofia: Problemi di metodo*
(Florence: Olschki, 1997).

48. Walter Popp, *Kritische Bemerkungen zur Associationstheorie*, 1. Theil, *Kritische Entwickelung des Associationsproblems* (Leipzig: Johann Ambrosius Barth, 1913).

49. Édouard Claparède, *L'association des idées* (Paris: Octave Doin, 1903); H. C. Warren, *A
history of association psychology* (London: Constable, 1921).

50. Luigi Ferri, *La psychologie de l'association depuis Hobbes jusqu'à nos jours* (Paris:
Germer Baillière, 1883).

51. Chiara Giuntini, *La chimica della mente: Associazione delle idee e scienza della natura umana da Locke a Spencer* (Florence: Le Lettere, 1995). See also Robert M. Young, "Association of ideas," in Philip P. Wiener, *Dictionary of the history of ideas* (New York: Scribner's,
1973–1974), vol. 1.

into those who discuss the nature of the human soul and those who "have philosophized on the soul in a more materialist manner and have preferred observation to speculation" or by following a purely chronological order. He favored the second method because it allowed one to include the "genealogy of psychological systems," and to assume that at least the authors who were not "partisan" had been able to benefit from the works of their predecessors.[52] "All valid philosophy," Hißmann declared, "is a reasoned history," whose concepts must be derived from "true things known by observation and experience." There is, then, a "true history" to be written, capable of "laying the foundations of the true philosophy" by revealing how the direct examination of phenomena enabled knowledge to advance. The *History of the Doctrine of the Association of Ideas* was informed by the two major principles it was designed to illustrate: "Die ganze brauchbare Philosophie ist eine räsonnierende Geschichte; Die wahre Geschichte ist die Grundfeste von der wahren Philosophie."[53]

Hißmann warned that it would be unjust to compare his work to Brucker's without taking into account the time at which each was writing. Fifty years earlier, he explained, the laws governing the association of ideas were not yet known. Hißmann aimed to pick up the story where his predecessor left off.[54] For him, as for Brucker, the histories of the general organization of the sciences, of particular disciplines, and of theories and concepts were closely related. Since the concept of *idea*, like logic itself, had become psychological, Hißmann's field of inquiry had become the history of psychological theory, which included "the most remarkable operations of the soul" [1].

Hißmann aimed to investigate "if and how psychologists have applied and used" the doctrine of the association of ideas in order to explain the manifestations of the human soul [2]. His approach was both historical and anthropological: "Men have always entertained complex notions [*zusammengesetzte Begriffe*], from time immemorial, ever since they were men. But did they know the laws governing the association of ideas on the basis of which they pondered and dreamed?—that is the question which this *History* asks" [10]. The way the question was formulated implied that the laws of association (similarity, spatial or temporal contiguity, and causality)

---

52. Hißmann, *Anleitung*, § 73.
53. Ibid., § 41.
54. Michael Hißmann, *Geschichte der Lehre der Association der Ideen, nebst einem Anhang vom Unterschied unter associierten und zusammtgesezten Begriffen, und den Ideenreyhen* (Göttingen: im Verlag Victorinus Voßpiegel und Sohn, 1777), [5]; referred to in the text by page number; numbers in square brackets refer to the unpaginated "Vorbericht."

were valid, and that the theory's progress consisted in an increasing awareness of these laws.

Hißmann's history emphasized continuity, the absence of revolutionary periods, and the importance of rediscoveries. What was needed, in his
view, was a new way of reading the past. The case of Hobbes seemed to him
exemplary: while Hobbes was a central figure for the history of morals and
natural law, psychology had paid less attention to him than it should have
[6], and historians of philosophy should henceforth attend to those figures
who risked falling into oblivion [7]. Hißmann's reading of the moderns was
similar. Hume, for example, claimed that no one before him had treated
in detail the laws of association, and on reading Priestley's presentation of
Hartley's theory one would think, Hißmann observed, that it was Locke
who had discovered association [3–5]. Yet the moderns sometimes created
entire doctrines on the basis of a few paragraphs from ancient philosophy
[3], and neither Locke nor any other recent philosopher could claim the
honor of having discovered the laws of association and the "attraction" of
ideas [7]. On the contrary, the ancients clearly knew this doctrine, even if
they did not understand its laws "as accurately as today" (6) and did not
make these into "general laws" (7). Moreover,

> if the ancient naturalists [Physiker] had been as attentive to the mani-
> festations of the world of bodies as the ancient psychologists were to
> the operations of the soul and the laws governing them, Newton would
> never have been the first to discover that powerful force of nature which
> maintains the vast bodies of the universe in their eternal paths where
> they have been revolving for thousands of years. (7)

A century after Hißmann, the Scottish philosopher William Hamilton
(1788–1856) explained that the laws of gravitation and of association were
certainly analogous, but that the former had been discovered thanks to
the work of several generations while the latter had been defined, applied,
and generalized by a single ancient philosopher (Aristotle), who should be
viewed as the Copernicus, Kepler, and Newton of the intellectual world.[55]
Clearly, whether knowledge of the association of ideas was the product of a
collective effort or of individual genius, its long history lay beyond doubt.

---

55. William Hamilton, "Contribution towards a history of the doctrine of mental suggestion
or association," in W. Hamilton, ed., The works of Thomas Reid (Edinburgh: Maclachlan and
Stewart, 1863), vol. 2.

The undisputed superiority of the "most recent centuries of philosophy" resided, for Hißmann, in having gone beyond description to develop
principles and theories from which it was possible to "deduce all phenomena without difficulty" and assemble them into a totality (*Allgemeinheit*),
such that only really extraordinary facts were now considered exceptions
(8). Hißmann's *History* enacted this vision of the progress of science, bringing into psychology subjects hitherto classed in philosophy. For example,
leaving aside the issue of universals, which had so preoccupied Brucker, he
highlighted Aristotle's rare "mechanistic explanations," which had been
neglected by his disciples (13–16). And, whereas Brucker disregarded Roman
rhetoric and oratory, Hißmann was interested in their relation to the arts of
memory, since, as he explained, those who had developed them "were psychologists and were doing no more than what we are doing today, namely,
gathering psychological means of assisting memory" (18). While Brucker
expressed his aversion to Scholasticism, but made room for the nominalist
controversy, Hißmann only disparaged it. In his view, the church fathers
simply grafted their Christian faith onto Plato and Aristotle, while their
followers paraded "a barbarous-scholastic countenance (the most frightful
of all monsters)" in their attempts to make philosophy conform to their
"theological chimeras" (27). What was new here was not a well-worn anti-
Scholasticism, but the insistence on empirical psychology.

For Hißmann, Malebranche's renewal of logic on the basis of the laws
governing ideas represented a turning point in its history. Malebranche was
also the first, in Hißmann's view, to anticipate the laws of association and
the way in which ideas are linked to traces in the brain (35–43). Locke had
not, when all was said and done, advanced the issue, despite the other services he had rendered to philosophy, since he had simply named it "association of ideas" and cited "curious facts," but not explored laws already
known (43–44). Similarly for Condillac. It was not until Hartley that, in
Hißmann's view, the association of ideas was elevated into the key subject
matter of morals, aesthetics, and almost all psychology (54). The lineage
went from one "mechanistic" thinker to another:

> Malebranche was perhaps the first thinker after Aristotle to meet the
> challenge of elaborating a mechanistic psychology, which Hartley ex-
> amined in depth before it was quite openly developed by the excellent
> author of the *Essay on Psychology*. (63)

This "excellent author," anonymous at the time, was Charles Bonnet, in whose work Hißmann found important indications concerning the

neurological mechanisms of association (67). After reviewing the treatment
of association by the Scotsmen Henry Home and Alexander Gerard, whom
he greatly respected (67–81), Hißmann summarized the examples given by
Ernst Platner in his recent *Anthropology for Physicians and Philosophers*
(1772) and expressed the hope that the anatomical bases of association
would one day be elucidated (84–85).

History, for Hißmann, blazed the trail. Theorization followed observation and gave rise to laws which were proved, but which should be explored
and generalized further. Thus, in two essays placed after his historical narrative, Hißmann sketched some of the future work to be undertaken in the
light of the *History*: the first was a short account of the "dogmatic results"
of the historical narrative (that is, its consequences for the theory), and the
other an essay on the "difference between associated and complex notions,
and series of ideas." There were two laws which Hißmann deemed irrefutable: coexistence or simultaneity, and similarity. He suggested adding a
"law of the physical connections between our internal organs" (86), called
for by the psychology of sensation and the passions. This law and its illustration (the effects of music) show how Hißmann was attempting to extend
the domain of psychology beyond the problem of knowledge, to aesthetics
and the study of feeling.

Hißmann was both a psychologist and a historian, aware that he was
writing history so as to shape a new discipline or help reform an established
one. His history of the theory of the association of ideas illustrates how
psychology comes into being through the transformation of logic (which
was why it could claim direct descent from Brucker). But it did not simply
reflect an existing situation. Although Hißmann maintained that history
was "progressing" toward his own philosophical and scientific position,
his historiography played a more constitutive role: the bibliographic and
narrative construction of a historical topic helped define an entire field of
knowledge.

PHILOSOPHERS WRITE THE HISTORY OF PSYCHOLOGY

In the wake of Heumann and Brucker, psychology became a standard topic
within the history of philosophy. I have already mentioned Fabricius's *Outline of a General History of Learning* (1752–1754), which took Brucker as a
major authority. But there were other figures.

The Italian priest Appiano Buonafede (1716–1793), who wrote under the
name of Agatopisto Cromaziano on the "restoration" of philosophy, com-

plained that psychology—which he considered the most important of the philosophical disciplines—had a history rife with dangerous topics of inquiry. Before discussing the early eighteenth-century English debates on the soul, he addressed "the psychological discoveries, truths, errors, and sometimes also fantasies of the followers of Descartes, Malebranche, Locke, and Leibniz, as well as of the Wolffian reformers."[56] Although the work often smacked of Catholic propaganda, it was translated and expanded in German by Karl Heinrich Heydenreich (1764–1801), who was one of the first to grasp the impact of Kantian criticism on the historiography of philosophy.

Dietrich Tiedemann (1748–1803), who became a professor at Marburg in 1786, has gone down in the history of psychology as the author of the first systematic observations to be published on the development of children's "mental faculties."[57] But at the time of its publication, his six-volume *Spirit of Speculative Philosophy* was the most important history of philosophy to appear since Brucker. Tiedemann, unlike the Kantians, did not think it possible to define philosophy other than historically, and he did not believe that any single system could make sense of history and reveal the errors and truths of the past. Reason advanced and systems were perfected, but this occurred over time and relative to other systems rather than to a single philosophy transformed into an absolute. Precisely in order to compare systems, Tiedemann made "cross sections" to bring out philosophers' views on God, the soul, and the world (Wolff's subjects of philosophy), without, he explained, examining the connection between those views in each individual author.[58] Psychology (*Seelenlehre*) was one of the main headings in his work and was used so much more systematically than Brucker used *psychologia* that Tiedemann's alphabetical index made it possible to extract from the general history of philosophy a special history of psychology from Pythagoras to Berkeley.

Tiedemann's successor at Marburg was Wilhelm Gottlieb Tenneman (1761–1819), whose *History of Philosophy* would be translated into French by the spiritualist philosopher Victor Cousin. Tenneman was a Kantian who took critical philosophy as his model and wrote history to demonstrate how

---

56. Agatopisto Cromaziano, *Kritische Geschichte der Revolutionen der Philosophie in den drey letzten Jahrhunderten* (1766–72), trans. K. H. Heidenreich, ed. of 1791 (Brussels: Culture et civilisation, 1968), p. 192.

57. Juan Delval and Juan Carlos Gómez, "Dietrich Tiedemann: La psicología del niño hace doscientos años," *Infancia y Aprendizaje*, 41, 1988, 9–30.

58. Dietrich Tiedemann, *Geist der spekulativen Philosophie* (1791–1797; Brussels: Culture et civilisation, 1969), I, p. vii.

philosophy had constituted itself as the science of the ultimate principles of nature and freedom. He saw in psychology a way of combating empiricism. Incapable of solving the problems of philosophy, empiricism seemed in addition to prove metaphysics superfluous, which led to skepticism. Psychology countered empiricism by revealing the internal rules of the human mind and carrying out a "profounder exploration of the foundations and laws of the understanding."[59]

Georg Gustav Fülleborn (1769–1803), another Kantian, emphasized the unity and "spirit" of systems, the way in which the very structure of the understanding informed the questions posed, and the procedures through which philosophy achieved systematic unity. History in this light was more of a series of snapshots than a narrative of how systems developed or were generated from each other. It nonetheless situated authors on a temporal scale, and divided historical material so as to enable particular histories to be written.

Fülleborn, for whom philosophy comprised both rational and empirical psychology,[60] explained that history could be written in two different ways:[61] either by reconstructing the philosophical disciplines of logic, psychology, morals, and others, examining their treatment by individual authors, and leaving the headings blank if one found nothing; or by writing a *Psychology of the Ancients*, a *Logic of the Ancients*, and so forth, in which each philosopher's thought would be treated as a totality.

Fülleborn illustrated the first approach. For example, he reconstituted an empirical psychology by binding "into a whole" Aristotle's ideas on the concept of the soul, sensation, memory, and the internal and external senses.[62] Compared with Brucker's *psychologia aristotelica*, Fülleborn's followed more closely the conceptual divisions of the psychology of his time and discussed neither the immortality of the soul nor the agent intellect. Fülleborn applied the same method to the controversial Dominican Tommaso Campanella (1568–1639), whose cosmology, pneumatology, and psychology he reconstituted. While he noted many omissions which a "critical philosophy" could not simply pass over, he also found in Campanella ac-

59. Wilhelm Gottlieb Tenneman, *Geschichte der Philosophie* (Leipzig: bei Johann Ambrosius Bart, 1798–1819), XI, pp. 298, 518–519.

60. Georg Gustav Fülleborn, "Versuch einer Uebersicht der neuesten Entdekungen in der Philosophie," in G. G. Fülleborn, ed., *Beyträge zur Geschichte der Philosophie* (1791–1796; Brussels: Culture et civilisation, 1968), 2. Stück (1792), 102–135.

61. G. G. Fülleborn, "Vermischte Bemerkungen zur Geschichte der Philosophie (I)," in Fülleborn, *Beyträge*, 7. Stück (1796), 173–188, p. 175.

62. Ibid., pp. 177–188.

curate observations, for instance, on sensation as the basis of the link between ourselves and objects.[63] The result was a history of philosophy, but constructed by writing special histories and analyzing topics as defined by their modern disciplinary boundaries.

The issue of philosophy as a profession was linked to that of particular histories. Fülleborn noted, for instance, that the first Greek scholars were not professional philosophers and did not develop coherent systems. Philosophy as yet had no institutional basis and so had not yet become a "livelihood" ("Die Philosophie war keine Facultäts-, keine Brodwissenschaft").[64] In Fülleborn's opinion, institutionalization had negative effects: it called for systems and textbooks, and textbooks set artificial limits on a science, encouraged bigotry, and sometimes reduced philosophy to a sort of preparatory course for other professions.[65]

Once disciplines had been established, however, and integrated into courses and institutions, there was no question of rejecting them, and it was finally disciplinary demarcations which provided the framework through which Fülleborn conceived and presented the various thinkers he discussed.[66] In his "Brief History of Philosophy," for instance, he placed metaphysics in speculative philosophy and divided it into psychology, cosmology, and theology, with various topics and authors treated by discipline. Thus, under psychology, Fülleborn contrasted the position he attributed to Locke, that "there is only one science concerned with the observation of the external manifestations of the soul," with Hume's belief that "such a science is impossible," and with Wolff's claim that we know the nature of the soul a priori. He also reviewed different opinions on the soul itself (for example, Is it material or a simple substance?) and its link to the body.[67] Fülleborn's scheme, which corresponded to Wolff's tripartite division (God, the world, and the soul), reflected the structure of university textbooks. The history of philosophical or psychological specialisms could be written from within this framework. For example, Fülleborn outlined a history of physiognomy and defended it not only because of its capacity to account for men's character

63. G. G. Fülleborn, "Thomas Campanella ueber die menschliche Erkenntniss," in Fülleborn, Beyträge, 6. Stück (1795), 124–162, pp. 142–143.

64. G. G. Fülleborn, "Über den Einflus anderer Wissenschaften und äuserer Verhältnisse auf die Philosophie und dieser auf jene," in Fülleborn, Beyträge, 3. Stück (1793), 52–69, p. 58.

65. Ibid., p. 67.

66. See the example of Aristotle in G. G. Fülleborn, "Kurze Geschichte der Philosophie," in Fülleborn, Beyträge, 3. Stück (1793), 3–51, pp. 20–21.

67. Ibid., pp. 45–50.

but also because of its position in relation to other disciplines: physiognomy was part of empirical psychology, which itself belonged in philosophy.[68]

We have seen that the history of psychology initially took shape within the history of philosophy. Brucker, in his treatment of the *psychologia aristotelica*, no less than Fülleborn in his summary of Aristotle's empirical psychology, explicitly intended to write histories of philosophy. However, the two historiographies could change places, signaling the emergence of an autonomous history of psychology. For example, in 1790 the *Psychologisches Magazin* published a long anonymous article on "the first psychological investigations of the Greeks," particularly Plato and Aristotle. The author warned that one should not expect to find there a fully fledged empirical psychology, since the Greeks were still too involved in developing the idea of philosophy itself. But they nonetheless developed the "first idea of a psychology."[69] The earliest period, up to Socrates, produced only speculations and hypotheses. During the second period, which ended with Aristotle, the notion of a systematic psychology emerged at the same time as sophisticated discussions on certain phenomena. The period after Aristotle saw the first complete outlines of a psychology as well as isolated contributions which were at the time attributed to other disciplines.[70]

According to the anonymous author, the real beginnings of psychology dated from the time of Socrates, a period during which the Delphic "Know thyself" was first given "sense and content."[71] Plato was the key figure both for philosophy and for psychology. Although he did not treat psychology scientifically, or approach it as a specific object, he gave significant space to psychological considerations. Aristotle, by contrast, was "the first to recognize psychology as a particular science."[72] But, as the author explained, since the Greeks did not distinguish between empirical and rational psychology, they did not create a psychology based on experience and on observing the manifestations of the internal sense.[73]

---

68. G. G. Fülleborn, "Abriss einer Geschichte und Literatur der Physiognomik," in Fülleborn, *Beyträge*, 8. Stück (1797), 1–180.

69. Anonymous, "Ueber die ersten psychologischen Versuche bei den Griechen, vorzüglich des Plato und Aristoteles," *Psychologisches Magazin*, 1, 1790, 298–388, pp. 300–301. The author might have been Friedrich August Carus, discussed in chap. 6.

70. Ibid., pp. 301–302.

71. Ibid., pp. 317–318.

72. "Er war der erste, der Psychologie als eine besondre Wissenschaft erkannte." Ibid., p. 322.

73. Ibid., p. 322–323.

Nevertheless, Aristotle made important contributions to the study of the imagination, memory, and reminiscence.[74] But that was not all: he almost arrived at the Kantian discovery that time and space constitute "the forms of all representation."[75] All in all, the author concluded, if one were to piece together everything Aristotle wrote on the objects of psychology (*Gegenstände der Psychologie*), one would have "the first complete outline of a science," which still lacked sound empirical and theoretical foundations, but which could be expected to advance.[76] The purpose of such a narrative was not to reconstitute a *psychologia aristotelica* which would remain part of Aristotle's philosophy. Rather, it was to show how a particular discipline (psychology), which was closely linked to the other branches of philosophy, could be construed as an autonomous entity—and how it actually came to be so in the course of its history. This presupposition of continuity and teleology implied identifying an origin (Aristotle as the first philosopher to make psychology into a particular science) and a terminus ad quem (eighteenth-century empirical psychology) but nevertheless leaving the latter open to the future. A "special" history of psychology was thus created that was inseparable from the constitution of the specialized discipline that the history described.

* * *

The works examined in this chapter demonstrate the role of the historian of philosophy in creating the history of psychology. As the discipline began to take on the characteristics of a profession, people involved in it undertook to write its history—a process similar to that occurring in the natural sciences. Thus, Hißmann, who wrote works on psychology, also wrote the history of a psychological doctrine. In the historical writings of Tiedemann, who was also a psychologist, the history of psychology figured prominently. The growing importance of psychology, and the debates and questions it raised in late eighteenth-century German universities, more than justified its position in Fülleborn's history of philosophy. Yet even an author as exhaustive as Tiedemann omitted central figures of the psychology

74. Ibid., p. 362.
75. Ibid., p. 372.
76. "Wenn man alles, was er [Aristote] gedacht und gesammelt hat, zusammenfaßt, so hat man einen ersten vollständigen Grundzüge einer Wissenschaft, welche freylich nur durch vervielfältige, geprüfte und vielseitige bearbeitete Beobachtungen, wenn sie nach bestimmten leitenden Principien angestellt und geordnet werden, ihrer Vervollkommnung entgegen sehen kann." Ibid., pp. 386–387.

of his time, for example, Bonnet, despite the fact that the latter was well known in Germany. This was not due to oversight but because these figures did not fit into his schema: for him, the history of philosophy could not be equated with the history of philosophical disciplines. In certain cases, the specificity of the history of philosophy even meant neglecting psychology altogether. For example, as late as 1807, an encyclopedic textbook on the history of philosophy omitted psychology and classified Hißmann's work on the theory of association alongside other histories of "philosophical theories."[77]

In summary, the elaboration of an autonomous history of psychology went hand in hand with developments which led to giving disciplinary form, and institutional and intellectual autonomy, to the very object of the historical narrative. Each process sustained the other. Their constitutive bond was represented on a grand scale by the groundbreaking history of psychology by the Leipzig philosophy professor Friedrich August Carus (1770–1807). Published posthumously in 1808, this 770-page history would have been dubbed "special" by Heumann or Brucker. It was certainly anchored in the history of philosophy, but it also drew on the cultural and anthropological "history of humankind," which we discuss in the next chapter.

Carus's work seems to have been ignored. Nevertheless, after the publication of his *History*, textbooks appeared which included historical narratives playing much the same role as they do today, that of a short propaedeutic account of the origins and progress of the discipline.[78] While fragmentary histories of psychology existed before Carus, his *History* represents the advent not only of a specialized historiographical genre but also that of its subject matter. Psychology already had its own journals and was the object of publications and teachings. It would henceforth have its own history.

The *Elements of General Psychology*, by Franz Anton Nüsslein (1776–1832), is a good example of an early nineteenth-century textbook. Nüsslein, who taught philosophy and natural history at the lyceum of Dillingen, had a romantic conception of life, in which the soul, an immaterial spirit, was the incarnation of nature ("Die Seele ist auch Natur"), its exact likeness ("Gleichbild"), and a microcosm which condensed in itself the whole of the

---

77. Johann Heinrich Martin Ernesti, *Enzyklopädisches Handbuch einer allgemeinen Geschichte der Philosophie und ihrer Literatur* (1807; Düsseldorf: Stern-Verlag Janssen, 1972).

78. Richard A. Littman considers Carus's *History* "premonitory because it antedated psychology as an empirical, research discipline by around sixty years"; Littman, "Psychology's histories: Some new ones and a bit about their predecessors: An essay review," *Journal of the history of the behavioral sciences*, 17, 1981, 516–532, p. 517. Such an interpretation overlooks the role of historical narrative in the constitution of the discipline.

universe.[79] For him, science was an investigation into first principles, as was psychology, which was consequently a theory of the nature of the soul, as opposed to some purely empirical description (§§ 2, 3).

Nüsslein adopted the classical division of the mental faculties into knowledge, feeling and the will and added intuition, prophetic dreams, and natural and artificial somnambulism to the list of internal and external senses. He also discussed reason (*Vernunft*) in connection with the "senses" of truth, beauty, morality, and religion. His historical narrative began with Aristotle (§§ 26–34), in whose works, Nüsslein declared, psychology first figured as an "autonomous science." The philosopher's ideas were then not taken up until the sixteenth century, when Melanchthon, Vives, and others revived psychology. After the collapse of Scholasticism, rationalism and empiricism breathed new life into psychology, but also revealed its multiplicity in the form of opposing systems. Cartesian dualism enlarged the scope of psychology by integrating into it the theory of the harmony between the soul and the body. Spinoza (whom Nüsslein rated highly; § 8) treated the soul in an original way, but his ideas never took root in psychology. Leibniz's monadology paved the way for the reworking of psychology by Christian Wolff, "that Homer of Leibnizian philosophy" (§ 31). Locke's empiricism generated Hume's skepticism, which in turn led to Kantian criticism. Kant posited that noumena were unknowable and so disallowed reflections on the *an sich* of the soul; he thus transferred the soul from metaphysics to physics and demoted it from its rank of science to that of an observational natural history. Nüsslein believed that, in his own times, psychology was reawakening and that once again thinkers could dare to speak about the nature of the soul (§ 33).

Nüsslein aimed to show the damage caused by empiricism and criticism, and to defend romantic psychology as a genuine science of the soul. As we have seen, Hißmann and Carus likewise described the progress of psychology, Hißmann up to the materialist doctrines of sensualism, and Carus up to a general science of man tinged with Kantianism. Although each narrative defended the validity of its author's psychological and methodological positions, neither can be reduced simply to a strategy of legitimation.

This is because these narratives also produced a new object: the past as psychology. Psychology became a set of questions that existed *a mundi incunabulis* but could be further explored and developed. Historiography

---

79. Franz Anton Nüsslein, *Grundlinien der allgemeinen Psychologie zum Gebrauche der Vorlesungen* (Mainz: bey Florian Kupferberg, 1821), §§ 18, 23, 19, henceforth referred to in the text by paragraph number.

proved that there existed a science in progress which could achieve intellectual and institutional autonomy through its own internal movement.
Its function was therefore genuinely formative: the creation of psychology
as a historical entity was one of the mechanisms whereby psychology was
fashioned as a new discipline. The historical narrative did not simply justify
particular points of view but reinforced the autonomy of its object, gave
form to its contents, and sketched in its borders. For example, Brucker classified the issue of the agent intellect and the immortality of soul under
psychologia aristotelica; by the time of Hißmann and Fülleborn, this issue
was no longer part of psychology. And by the end of the eighteenth century,
"rational psychology" was no longer included in the field which would
henceforth be termed "psychology."

   This process of exclusion, which is vital for defining the borders of a field
of knowledge, goes hand in hand with strategies of incorporation and expan-
sion. We have seen this at work in the example of the history of the association of ideas. Historiographical activity did not, however, confine itself to
defining limits and contents. It was integrated into the very subject matter
it treated, becoming a sign of how far humanity had evolved in its psychological thought and practice. Through bibliographic works, the history of
psychology could be staked out both synchronically and diachronically, and
the discipline characterized by a genealogy of questions and protagonists.
Since the historiographical approach also implied a critique of the past, it
showed how far knowledge of the soul and of humanity had progressed, and
in so doing established criteria by which the present and above all future
legitimacy of psychology would be judged.

   The history of psychology during the *Aufklärung* was as much a narrative about the past as the formulation of a program for the future. Historical
sensibility went hand in hand with a reflexive awareness of the constitutive role of history and historiography, of the mutually determining links
between the narrative and the narrated object. A new aspiration emerged,
beyond that of distinguishing psychology from pneumatology and nature
from Revelation: to ground a science of man in the psychological and cultural history of humanity.

CHAPTER SIX

Psychology and the History of Humankind

"Aristotle studied nature in Homer"
—Hugh Blair, *A Critical Dissertation on the Poems Of Ossian* (1765)

W e have seen that the history of psychology emerged as a genre from within the history of philosophy, and that it was not simply a narrative about the past. Just as the history of philosophy was a way of philosophizing, so the history of psychology became a way of doing psychology, and of bringing together the study of the individual mind with that of humankind. In the perspective of late eighteenth-century anthropology and cultural history, psychological ideas expressed collective mentalities which were themselves the result of physical, social, and political circumstances. Writing the history of psychology thus went beyond its initial function of defining "psychological" themes and authors to become *the* method of empirical psychology on the scale of the history of humanity.

This transformation, which extended empirical psychology to past civilizations and "primitive" peoples, and reaffirmed its break with rational psychology, occurred in a context in which the science of antiquity (*Altertumswissenschaft*) and the cultural and anthropological history of humanity were beginning to take shape. Philology was establishing itself as the quintessential science for understanding human culture, while Homer was raised to a unique position and came to dominate the Enlightenment's image of the ancient world.[1] For Johann Joachim Winckelmann (1717–1768), Greek art was the highest form of naturalism; it may well have idealized nature, but in so doing it revealed her true designs. Likewise, Homer not

---

1. See for example Hugh Honour, *Neo-classicism* (Middlesex: Penguin Books, 1979), chap. 2.

only possessed vast and accurate knowledge—something with which he had been credited since antiquity—but he additionally expressed it in a language and with a sensibility that were still true to nature, uncorrupted by civilization. It was just such an ideal of primeval purity that John Flaxman sought to capture in the famous outline drawings with which he illustrated the *Iliad* and *Odyssey*.

Homer was portrayed as the inventor of poetry, and as a man of unparalleled breadth of knowledge and observational precision. The *Iliad* and the *Odyssey* were read as texts based on the observation of nature and mankind. Such readings broke with a long tradition of allegorical interpretation, which nonetheless had also taken Homeric poetry as a reliable source of empirical facts about geography and natural history, religion, customs, and human psychology. In the framework of nonallegorical interpretations, psychology posed particular problems, since it could not be examined in the same way as, for example, geography. Written descriptions of places and landscapes could be checked in situ. In psychology, however, since not only the express content but also the form, style, and vocabulary helped forge a mentality to which one had access only through the text itself, hardly any access to extratextual sources was possible. If, as Herder maintained, words are inseparable from thought (rather than being simply its vehicle and representation), then the analysis of language provided the foundations for a psychological and cultural history, and classical and biblical philology had to become disciplines of cultural interpretation.

FRIEDRICH AUGUST CARUS AND THE "HISTORY OF HUMANITY"

Like Hißmann and Tiedemann, Carus was at once a psychologist, a philosopher, and a historian.[2] After studying theology, ancient languages, philology, and history at Leipzig and Göttingen, he became a professor of philosophy at the University of Leipzig in 1797. He left behind him a significant body of

---

2. Friedrich August Carus, *Geschichte der Psychologie* (1808; Berlin: Springer Verlag, 1990), introduction by Rolf Jeschonnek. For aspects of Carus's thought relevant to this chapter, see Jeschonnek's introduction; and Jörn Garber, "Von der 'anthropologischen Geschichte des philosophierenden Geistes' zur *Geschichte der Menschheit* (Friedrich August Carus)," in Garber and Thoma, *Zwischen Empirisierung und Konstruktionsleistung*. Carus's *Geschichte* was originally published as vol. 3 of his *Nachgelassene Werke*, ed. Ferdinand Hand (Leipzig: bei Iohann Ambrosius Bart und Paul Gotthelf Kummer, 1808–1810). Hand provides biographical information in his "Vorrede" to *Nachgelassene Werke*, vol. 7.

work comprising, alongside the *History of Psychology*, a *Psychology*, *Ideas on the History of Philosophy* and *Ideas on the History of Humanity*, a *Psychology of the Hebrews*, and several *Philosophies* of morals and religion. He also wrote about the sources of Anaxagoras's "cosmotheology" and published in Fülleborn's journal of the history of philosophy. Carus was a direct heir of those who maintained that writing a discipline's history was vital for the discipline itself, but he inflected this principle toward reconstituting the cultural and psychological history of humanity. His principal goal, during his brief academic career, was to develop a "history of humanity." Such a history, in his view, presupposed psychology and was a prerequisite for any history of the arts and sciences, particularly of philosophy.[3] During his years at Leipzig, Carus placed increasing emphasis on the history of psychology and on empirical psychology, going so far as to offer practical exercises in psychological observation. For him, however, psychology was above all the basis for, and an integral part of, a history of humanity. Carus pursued a combined history of the science of man and of psychology (*Geschichte der Menschenkunde und der Seelenlehre*) whose ultimate object was a totality in which the history of ideas and disciplines would be inseparable from the psychological history of humankind.

Carus's history thus depicted the progression from myths about the soul to the empirical psychology of his time. When conceived "in its most elevated and comprehensive" form, such a history concerned nothing less than the history of humanity's increasing consciousness of its own "spiritual nature.[4] Since psychology was one of the fundamental forms of this realization, narrating its history would contribute to furthering human self-consciousness. Carus insisted on the difference between his approach and a natural history of external and internal facts: in psychology, he declared, "observation remains the least important moment unless it is cognized and repeated by self-consciousness and secured through further elaboration."[5] Likewise, the *Geschichte der Menschheit* should be neither a "purely empirical description of the external characteristics, expressions, transformations, and differences between men" nor a history of the "human mind,

3. F. A. Carus, *Ideen zur Geschichte der Menschheit*, in *Nachgelassene Werke*, vol. 6, pp. 74–75.

4. "Geschichte der allmäligen Klarheit des Selbstbewußtseins der geistigen Natur." Carus, *Geschichte*, p. 4.

5. "Die Beobachtung bleibt der kleinste Moment, wenn sie nicht durch das Selbstbewußtsein erkannt und wiederholt und durch die weitere Bearbeitung fixiert wird." F. A. Carus, *Psychologie*, in *Nachgelassene Werke*, vol. 1, p. 27.

heart, and will" confined to a purely empirical psychology, a "natural description of the faculties of the soul."[6]

For Carus, then, a work such as the *Outline of the History of Humanity* (1785), by the Göttingen professor Christoph Meiners (1747–1810), in which the physical features and cultural practices of different peoples were compared, was merely an *äussere Naturbeschreibung*, a natural history of the external aspects of man.[7] He set against this a history that would be characterized by a "cosmopolitan sense" (*Weltbürgersinn*).[8] Carus's outlook and vocabulary may arguably reflect Kant's "cosmopolitan point of view" (*weltbürgerliche Absicht*), with its vision of a universal political order in which the original capacities of the human race would be able to develop.[9] It most certainly expresses the "pragmatic" viewpoint of the Göttingen historians with whom Carus would have been familiar—"pragmatic" meaning, in this context, a history which looks for causes in the conditions and development of a people's material and intellectual culture.[10]

Carus believed that the various objects of the history and historiography of humanity, from the poems and fables of antiquity to Enlightenment anthropologies and "histories of the human mind," contributed to the psychological progress of humanity just as much as the history and historiography of psychology.[11] As Georges Gusdorf has rightly observed, at a time when psychology was acquiring a "regulatory function" with respect to the other human sciences, Carus expanded its scope so that it became the core of cultural anthropology.[12]

Some of Carus's works are entitled *Ideas on the History of . . .* This suggests an affinity with Johann Gottfried Herder (1744–1803), and the vocabulary and style of the two authors indeed show their spiritual kinship. Although Carus cited Herder only occasionally, his vision of the history of humanity agreed with the fundamental tenet of Herder's *Reflections on the*

---

6. Carus, *Ideen zur Geschichte der Menschheit*, pp. 75, 76.

7. Ibid., p. 27. Cf. Christoph Meiners, *Grundriß der Geschichte der Menschheit*, ed. of 1793 (1785; Königstein: Scriptor, 1981).

8. Ibid., p. 76.

9. I. Kant, *Idea for a Universal History from a Cosmopolitan Point of View* (1784).

10. See Norbert Waszek, "Le cadre européen de l'historiographie allemande à l'époque des Lumières et la philosophie de l'histoire de Kant," in Myriam Bienenstock, ed., *La philosophie de l'histoire: Héritage des Lumières dans l'idéalisme allemand?* (Tours: Université François Rabelais, 2001). For further details, see Gudrun Kühne-Bertram, "Aspekte der Geschichte und der Bedeutungen des Begriffs 'Pragmatisch' in den Philosophischen Wissenschaften des ausgehenden 18. und des 19. Jahrhunderts," *Archiv für Begriffsgeschichte*, 27, 1983, 158–186.

11. See Carus, "Geschichte der Geschichte der Menschheit," in *Ideen zur Geschichte der Menschheit*, pp. 6–32.

12. Gusdorf, *L'avènement des sciences humaines*, pp. 90–91.

philosophy of the history of mankind (1784–1791), namely, that humanity (*Humanität*) is the goal of human nature, and that this goal is revealed only in history. If there is a god in nature, said Herder, then there is also a god in history; since man is part of creation, even in his worst wrongdoings, he obeys laws that are as excellent as those which govern terrestrial and celestial bodies. Even amid the errors and horrors of nations, the supreme law of nature remained "Let man be man."[13]

According to Herder, humanity came into being through the development and use of reason (*Vernunft*), which in turn could take place only through language (*Sprache*).[14] Without language, the operations of the soul are neutralized and the subtle structure of the brain is useless. No idea could exist without its corresponding verbal expression, and hence no reason could exist without language: "Eine reine Vernunft ohne Sprache ist auf Erden ein utopisches Land." Language alone has made man human.[15] Since the education (*Bildung*) of mankind was realized by means of language, Herder considered the "philosophical comparison of languages" the best way to investigate a people's mind and heart.[16]

For Herder, the education of humanity involved passing through a series of cultural forms corresponding to specific times and places, which is why the historian should judge each culture dispassionately and impartially.[17] The value of a culture or nation depends on what humanity could achieve at that particular stage. But since Herder's philosophy of history was at once teleological and naturalistic, cultures could be characterized by their level of development, as well as by the extent to which they realized the human being's humanity.

For all its originality of form and content, Herder's philosophy was indebted to the cultural history that was developing at the time in Germany, especially at the University of Göttingen. His treatment of ancient Greece makes this particularly clear. For instance, Herder's presentation of the sources of Greek mythology—regional legends, popular beliefs, traditional tales, and primitive attempts to explain the world and create a society—drew on the writings on myth of his friend Christian Gottlob Heyne.

13. "Selbst bei allen Gräueln und Fehlern der Nationen blieb das Hauptgesetz der Natur känntlich: 'der Mensch sei Mensch!'" J. G. Herder, *Ideen zur Philosophie der Geschichte der Menschheit* (bk. 15, chap. 1), in *Herders sämmtliche Werke*, ed. Bernhard Suphan (1887–1913; Hildesheim: Olms, 1994), vol. 14, pp. 207 and 209. All references to Herder's works are to this edition (by volume and page number).

14. Ibid. (bk. 9, chap. 2), 13, p. 355.

15. "Nur die Sprache hat den Menschen menschlich gemacht." Ibid., p. 357.

16. Ibid. pp. 363 and 364.

17. Ibid. (bk. 12, chap. 6), 14, p. 85.

Moreover, it was to Christoph Meiners, a student and later colleague of Heyne's, that Herder referred for the history of the Pythagorean school after explaining that the Greeks' dominating interest in moral philosophy and human nature and customs set the tone for their poetry, history, and political organization.[18]

THE PRIMITIVES AND THE ANCIENTS

The history of psychology was the means by which Carus placed psychology at the center of cultural anthropology. His approach was modeled on that of the University of Göttingen, where he had studied. The university was named "Georgia Augusta" after its founder, George II of England (in his capacity as elector of Hanover), and from its very foundation in 1737, it was conceived as an Enlightenment institution centered on the natural and human sciences rather than theology. Particularly after 1770, the creation of scholarly journals and academies, a museum, and a botanical garden turned Göttingen into one of the most influential and progressive universities in Europe.[19]

Among the most distinguished scholars at Göttingen was Christian Gottlob Heyne (1729–1812), a professor of poetics and rhetoric from 1763 to his death. He was a philologist and an archaeologist, the author of numerous editions of Greek and Latin texts, the director of the University library, which he aspired to make universal and expanded to some two hundred thousand volumes, and the editor of the *Göttingische Anzeigen von gelehrten Sachen*, to which he contributed no fewer than seven thousand reviews.[20] He is of interest to us here because of his conception of the study of antiquity as the quest to understand humanity itself in ancient times, which was how his student (and critic) Friedrich August Wolf (1759–1824), a professor at Halle, characterized this science.[21] For some educational purposes,

18. Ibid. (bk. 13, chaps. 2 and 5), 14, pp. 100 and 125–126.

19. Luigi Marino, *Praeceptores Germaniae: Göttingen 1770–1820*, trans. Brigitte Szabó-Bechstein (Göttingen: Vandenhoeck & Ruprecht, 1995), which is a revised and enlarged edition of Marino, *I maestri della Germania: Göttingen 1770–1820* (Turin: Einaudi, 1975); Michael Carhart, *The science of culture in Enlightenment Germany* (Cambridge, MA: Harvard University Press, 2008). I can do no more than touch on this context here.

20. Apart from Marino and Carhart, see Gioachino Chiarini, "Ch. G. Heyne e gli inizi dello studio scientifico della mitologia," *Lares*, 55, 1989, 317–331; also Dieter Burdorf and Wolfgang Schweickard, with the collaboration of Annette Gerstenberg, eds., *Die schöne Verwirrung der Phantasie: Antike Mythologie und Kunst um 1800* (Tübingen: Francke Verlag, 1998).

21. "Die Kenntniß der alterthümlichen Menschheit selbst." Friedrich August Wolf, *Darstellung der Altertumswissenschaft nach Begriff, Umfang, Zweck und Wert* (1807; Berlin:

it may be sufficient to study the classical texts, but the science of antiquity, wrote Wolf, pursues empirical knowledge of human nature and of man's original faculties and inclinations, as well as of the ways they act on each other and are affected by external circumstances.[22]

Wolf's science of antiquity aimed to investigate nations and cultures of antiquity through the traces they left behind. Philology and textual analysis were no longer ends in themselves but tools of anthropological and historical research. The new philology no longer conceived language from a sensualist perspective, as a system of arbitrary signs designed to express ideas, but rather as something essential to humanity's development.[23] Philology thus joined the historiographical paradigm which had abandoned the biblical narrative as a privileged source about the origins of humanity in favor of a resolutely historicist and anthropological approach.[24]

An example of such a stance is provided by the Basel philosopher Isaak Iselin (1728–1782), one of the main figures of the Swiss Enlightenment. In *On the History of Humanity* (1764) he defined three successive states— the state of nature, the savage state (*Wildheit*), and the state of civilization (*gesitteter Stand*)—which he related, respectively, to sensibility (*Sinnlichkeit*), the imagination (*Einbildung*), and reason (*Vernunft*), as well as to the three ages of man (childhood, youth, and adulthood).[25] In accordance with this scheme, the entire first book of the *History* was devoted to individual

---

Akademie-Verlag, 1985), pp. 124–125. The Italian version contains a useful introduction by the translator, Salvatore Cerasuolo: F. A. Wolf, *Esposizione della scienza dell'antichità secondo concetto, estensione, scopo e valore* (Naples: Bibliopolis, 1999). For Wolf's debt to Heyne and to his friend Wilhelm von Humboldt (another student of Heyne's and the author, in a similar vein, of the essay "Über das Studium des Alterthums, und des griechischen inbesondere" [1795]), see Anthony Grafton, "Prolegomena to Friedrich August Wolf" (1981), in Grafton, *Defenders of the text: The traditions of scholarship in an age of science,* 1450–1800 (Cambridge, MA: Harvard University Press, 1991). See also Sotera Fornaro, "Lo 'studio degli antichi,' 1793–1807," *Quaderni di storia,* 22, 1996, 109–124.

22. Wolf, *Darstellung,* p. 126.

23. This also explains the importance of philology for politics and education, as Robert S. Leventhal has shown in "The emergence of philological discourse in the German states, 1770–1810," *Isis,* 77, 1986, 243–260.

24. Gérard Laudin, "Changements de paradigme dans l'historiographie allemande: Les origines de l'histoire de l'humanité dans les 'histoires universelles' des années 1760–1820," in Chantal Grell and Jean-Michel Dufays, eds., *Pratiques et concepts de l'histoire en Europe XVIe–XVIIIe siècles* (Paris: Presses de l'Université de Paris-Sorbonne, 1990).

25. Isaak Iselin, *Über die Geschichte der Menschheit* (1786; Hildesheim: Olms, 1976). This work, which was reprinted several times, was first published in 1764 under the title of *Philosophische Muthmaßungen über die Geschichte der Menschheit.* Simone Zurbuchen, "Iselin, Isaak," in Knud Haakonssen, ed., *The Cambridge history of eighteenth-century philosophy* (New York: Cambridge University Press, 2006), vol. 2.50.

psychology. Another example is Johann Christoph Adelung (1732–1806), a librarian from Dresden known for his lexicographical work. In his *Essay on the History of Human Culture*, he systematized the correspondence between the ages of humanity and those of the individual.[26] The Pentateuch, for example, was in his view composed of sensory images drawn from the physical world and expressed in a childlike language. Christoph Meiners, in his *Outline of the History of Humanity* and in the numerous articles he contributed to the *Göttingisches historisches Magazin* (1787–1794), concentrated on the present. He used writings by travelers and explorers to describe the environments, physical features, and material cultures of a wide range of peoples, as well as their political organization and laws, sociocultural practices, customs, and production, as well as their mental capacities and characteristics. For Meiners, these elements were interconnected, and their description resulted in a "history of humanity" that made intercultural comparison possible.[27] The goal of all these authors, despite the different ways each combined metaphysics, morals, history, and ethnography, remained the reconstitution of the different phases of a total history of mankind.

A similar approach was also adopted in the field of Bible studies. In Göttingen, Johann David Michaelis (1717–1791) linked biblical exegesis not only to philology but also to geography, history, and archaeology and compared the "fables" of the Hebrews with those of Greek mythology.[28] His disciple and successor Johann Gottfried Eichhorn (1752–1827), who was also a student of Heyne's, criticized those who read the Pentateuch as though it were a modern text without taking into account its age and language, and he saw it as a particularly comprehensive description of the "history of the civilization and education" (Geschichte der Cultur und Aufklärung) of an ancient people.[29] In a different vein, Herder examined the "spirit of Hebrew poetry," emphasizing its origin in sensation and sensory images, and its relation to the structure of pastoral society. For him, the societal setting distinguished the poetry of the Hebrews from anything "savage" peoples

---

26. Johann Christoph Adelung, *Versuch einer Geschichte der Cultur des menschlichen Geschlechts* (Leipzig: Hertel, 1782).

27. In addition to Marino and Carhart, see Britta Rupp-Eisenreich, "Des choses occultes en histoire des sciences humaines: Le destin de la 'science nouvelle' de Christoph Meiners," in B. Rupp-Eisenreich and Patrick Menget, eds., "L'anthropologie: Points d'histoire," special issue, *L'ethnographie*, 79, no. 90–91, 1983, 131–183.

28. Anna-Ruth Löwenbrück, "Johann David Michaelis et les débuts de la critique biblique," in Yvon Belaval and Dominique Bourel, eds., *Le siècle des Lumières et la Bible* (Paris: Beauchesne, 1986).

29. Johann Gottfried Eichhorn, *Einleitung in das Alte Testament* (1780–1783; Leipzig: Weidmannische Buchhandlung, 1803), vol. 1, p. v (from the preface to the 2nd ed., 1787).

could achieve; no Iroquois warrior or Huron hunter had ever produced anything like it.[30]

In Herder's view, Hebrew poetry was a purely human creation.[31] It followed that a people or a period should not be judged against another people or period, and that Old Testament poetry should be analyzed and evaluated in terms of the nature of man as it developed historically.[32] In such a perspective, poetic legends and fictions were not considered to be the fruit of an uncivilized people's untamed imagination, but based on an empirical reality, a *Naturwahrheit*. For example, representations of the kingdom of the dead derived from burial practices and ignorance of the separation of body and soul. The fact that, in the East, tombs were vast caves containing many corpses must have led to the idea of an underground kingdom peopled by helpless shades, while the sight of decomposing corpses gave rise to the belief in a monster which destroyed their substance. For Herder, such explanations of myth were as natural as the effects of climate and geography and could therefore be found among all ancient peoples.[33]

The classical and biblical fields shared not only a historicist and anthropologizing hermeneutics, but also a method. In his most famous work, the *Prolegomena ad Homerum* (1795), Friedrich August Wolf, a former student of Michaelis at Göttingen, claimed that the originals of the *Iliad* and the *Odyssey* were lost forever. Homer, he argued, did not know how to write and could not have composed such long texts. These had been altered in the course of their oral transmission, before being written down in the Hellenistic period. The original compositions therefore had no artistic unity and were not all by the same author. Such ideas were controversial but not new; Wolf's originality lay in the way he appropriated Eichhorn's model of historico-textual analysis.

---

30. "Kein kriegender Irokese, kein jagender Hurone dichtete so." J. G. Herder, *Vom Geist der Ebräischen Poesie: Eine Einleitung für die Liebhaber derselben, und der ältesten Geschichte des menschlichen Geschlechts* (1782–1783), in *Herders sämmtliche Werke*, vols. 11–12, 12, p. 30. On the question which concerns us here, see especially Herder, *Vom Geist*, pt. 2, chap. 1 ("Vom Ursprung und Wesen der Ebräischen Poesie").

31. Ibid., p. 7. Herder (ibid., p. 31) believed that Hebrew, in contrast to Greek and Latin, had remained a *Kindheitssprache*. I have mentioned only Göttingen scholars, but the crucial author for Herder was the Leipzig professor Johann August Ernesti (1707–1781). Herder admired and drew inspiration from the *Institutio interpretis Novi Testamenti* (1761), in which Ernesti established the principles for a historico-critical and purely secular approach to the scriptures.

32. James L. Kugel, *The idea of biblical poetry: Parallelism and its history* (New Haven: Yale University Press, 1981), stresses the formal and above all metrical aspects of the Old Testament. On the way in which the Bible became a literary work, and the emergence of an aesthetic approach that placed the Bible alongside Greek and Latin classics, see chap. 8.

33. Ibid., pp. 17–19.

Eichhorn had studied the methods of the Masoretes, the Middle Eastern Jewish scholars who devised systems to fix the pronunciation and paragraph and verse divisions of the Bible during the first millenium (the term *masorah* became connected with the idea of transmission and acquired the meaning of "tradition"). The Hebrew alphabet itself has no vowels, but the Masoretes developed a vowel notation system, as well as punctuation and marginal annotations. At stake was nothing less than the authentic letter of the biblical text and its original meaning. In his *Critical History of the Old Testament* (censured upon publication in 1678), the French Oratorian Richard Simon (1638–1712) explained that the original texts and meanings had been lost due to the many additions and changes made over centuries by necessarily fallible scholars. Wolf transposed this debate and its underlying methods to ancient Greece, viewing the Hellenistic commentators and grammarians as the Masoretes of the Homeric texts. He was concerned, however, with rediscovering the methods used rather than reconstituting a single original text.[34]

Beyond these specialized issues was the project of creating a science of antiquity that would give a total picture of humanity in ancient times. Two new disciplines were deemed essential to the success of the new *Altertumswissenschaft*: the study of myth and cultural anthropology. Heyne was once again central to this initiative. He did not view mythology as a "mass of chimeras, reveries and absurdities," as Fontenelle put it in *Of the Origin of Fables* (1724), as a hodgepodge preserved out of "blind respect for antiquity" and ignorance about primitive man. On the contrary, myth revealed ancient peoples' philosophy and history, while betraying their incapacity for abstract thought. Comparing humanity in antiquity to children and "primitives" was no longer pejorative but illuminating. For Heyne, myths had historical as well as moral or physical contents. They should be studied in relation to the traditions, times, and places in which they were produced and understood in the light of their creators' conceptions of faith, truth, authority, and judgment.[35]

34. Grafton, "Prolegomena." See also the introduction to F. A. Wolf, *Prolegomena to Homer* (1795), trans., introduction, and notes by A. Grafton, Glenn W. Most, and James E. G. Zetzel (Princeton: Princeton University Press, 1985).

35. C. G. Heyne, "An interpretation of the language of myths or symbols traced to their reasons and causes and thence to forms and rules," extracts translated from "Sermonis mythici seu symbolici interpretatio ad causas et rationes ductasque inde regulas revocata," *Commentationes Societatis Regiae scientiarum Gottingensis*, 1807, in Burton Feldman and Robert D. Richardson, *The rise of modern mythology* (Bloomington: Indiana University Press, 1972). Heyne often stressed this historicist approach, for example, at the beginning of "De origine et causis

Heyne's goal was to develop a "physics of ancient myth," that is, an explanation of myth in terms of natural causes, humanity's infancy, its degree of linguistic development, and the sensory origin of thought. For Heyne, since primitive man was limited to sensation, he had no access to the causes of phenomena, which he therefore feared; from his fear myths were born. Since primitive languages tended to express abstractions concretely, thoughts were formulated as though they were actions, and this also gave rise to myths. Instead of naming nature, man described it using the names of gods. Modern travelers and philosophers talked of primitive peoples' "religion," "deities," and "God," but these terms, Heyne wrote, were not appropriate in connection with beings incapable of elaborating abstract concepts and metaphysical notions. To say that, from the psychological and linguistic point of view, the humanity which created myths was still in its infancy was not a suggestive analogy or a contemptuous metaphor but a fact that could be proved by empirical examples.[36]

Herder, for his part, believed that the nations still in a state of historical infancy (im Kindheitszustande der Welt) should be left to their simple images, since "that is how children see, speak, and feel."[37] And if the epic poem is indeed "the living word" and "the voice" of prehistory, if by its very essence it belongs to "the infancy of the world," then Homer is its poet par excellence.[38] This conception seemed valid even if Homer had not existed as an individual.

Half a century before Herder, Giambattista Vico (1668–1744) wrote that Homer was the Greek people or, more exactly, that he "was an idea or heroic archetype of the Greeks who recounted their history in song."[39] The Iliad

fabularum homericarum commentatio," Novi commentarii Societatis Regiae scientiarum Gottingensis, 8, 1777, 34–58.

36. C. G. Heyne, "Inquiry into the causes of fables or the physics of ancient myths" (1764), in Feldman and Richardson, The rise of modern mythology (including extracts translated from "Quaestio de causis fabularum seu mythorum veterum physicis," in Heyne, Opuscula academica collecta et animadversionibus locupletata [1785–1812; Hildesheim: Olms, 1997], vol. 1).

37. Herder, Vom Geist, in Herders sämmtliche Werke, vol. 12, p. 9.

38. "Epos war das lebendige Wort, die Stimme der Vorwelt"; "Das Epos gehört in der Kindheit der Welt." Herder, "Homer, ein Günstling der Zeit" (1795), in Herders sämmtliche Werke, vol. 18, pp. 443 and 444.

39. Giambattista Vico, New Science: Principles of the new science concerning the common nature of nations, 3rd ed. (1744), trans. David Marsh, intro. Anthony Grafton (London: Penguin Books, 2001), § 873, p. 381. See bk. 3, as well as the frontispiece and its "Explanation." The image shows Metaphysics standing on the globe of nature. On her breast, she wears a jewel, convex so that the rays of Providence can be "reflected and refracted." This shows that metaphysics "recognizes God's providence in public moral institutions" rather than seek "the private illumination of intellectual virtues as a guide to private morality." The ray of Providence is then reflected onto a statue of Homer, "the earliest pagan author to come down to us," symbolically conveying the

represented Greece in its youth, "burning with sublime passions, like pride, anger, and thirst for vengeance" and therefore glorifying Achilles, "the hero of violence"; the *Odyssey*, by contrast, represented an age "when the spirits of Greece had been somewhat cooled by reflection, which is the mother of prudence," such that it admired Ulysses, "the hero of wisdom."[40] The Homeric epics did not formulate a philosophy but rather a "poetic wisdom," the fruit of memory and imagination (*fantasia*) rather than of the reflexion (*riflessione*) on which philosophers would later base their works: "Among the pagans, wisdom began with the Muse."[41] In the "world's childhood" (*mondo fanciullo*), therefore, men "were by nature sublime poets," while the first "wise men of the Greek world were theological poets."[42] Thus, the childhood of the world and the world of childhood were reflections of each other.

"Childhood" here meant the "image of the past in the present."[43] Nevertheless, while the "childhood of humanity" was a commonplace, comparisons with real childhood were not. A certain Johann Nast, who taught at the Stuttgart gymnasium, took the metaphor literally and examined Homer's language from the point of view of its similarities with popular language and children's speech. The fact that Greek in Homer's time was not written showed that it was in its infancy. To prove his point, Nast explored in detail the modes of expression, noting the figures that were employed spontaneously, the use of conjunctions and verbal forms, how the manifestations of the soul were conveyed, and even remarked on the verbosity of Homer's

---

discovery of the true Homer. For directly relevant studies, see Paolo Cristofolini, *Vico et l'histoire* (Paris: Presses Universitaires de France, 1995); and Olivier Remaud, *Les archives de l'humanité: Essai sur la philosophie de Vico* (Paris: Seuil, 2004).

40. Vico, *New Science*, § 879, p. 383. Some scholars argue that Vico influenced Heyne, Wolf, and Herder. Vico, however, is not even indirectly present in the context that concerns us here. See Silvia Caianiello, "La lecture de Vico dans l'historicisme allemand," *L'art de comprendre*, no. 7, 1998, 139–167. Only much later did F. A. Wolf summarize bk. 3 of the *Scienza nuova*—and in unambiguously critical terms: Vico, "der lebhaft umherspringende Ragionatore," certainly had better ideas than more recent authors, but they were swamped by "strange and erroneous notions" (bei manchen wunderlichen und irrigen Begriffen). He was a visionary totally lacking in historical rigor: "Historische Strenge ist zwar nirgends in diesem Räsonnement; kaum scheint Vico davon eine Idee gehabt zu haben. Alles hat eher das Ansehen von Visionen: doch nähern sich solche Visionen oft der Wahrheit mehr und haben grössern Werth, als die ebenso unbewiesene Wiederholung des gemeinen beweislosen Glaubens, wie sie sich bis zu einer gewissen Zeit so allgemein in den Schriften unserer Gelehrten und schönen Wissenschafter fand." F. A. Wolf, "Giambattista Vico über den Homer," *Museum der Alterthums-Wissenschaft*, 1, 1807, 555–570, pp. 557 and 569–570.

41. Vico, *New Science*, § 365, p. 137.

42. Ibid., §§ 187 and 199, pp. 89 and 91.

43. Remaud, *Les archives de l'humanité*, p. 244.

heroes, similar in this respect to "charming infants" and "innocent primitives" (*Naturmenschen*).[44]

The analogy between primitive and ancient peoples was elaborated more often than that between primitives and children. It became commonplace to claim that knowledge of the customs and psychology of "savage" peoples provided insight into those of the ancients. These insights, of course, were not always as disinterested as Heyne would have wished. Thus, in 1724, on the basis of a five-year stay in Canada and numerous accounts by seventeenth-century missionaries, the French Jesuit Joseph-François Lafitau (1681–1746) compared the customs of the American Indians to those of primitive times. As the frontispiece of his work illustrated, Lafitau sought to prove that the American Indians, the Huron and Iroquois, had a religion, and that it manifested a primitive consciousness of Revelation.

Lafitau was less concerned with the specificities of primitive culture than with detecting elements of Christianity, not only in clearly religious forms but also in customs and traditions as a whole. Moreover, rather than considering myths and symbols cryptic descriptions and explanations of the physical world (and therefore, liable to encourage atheism), he referred them to the principles of Christianity. Just like the scholar on his frontispiece (fig. 6.1), Lafitau gathered material from the four corners of the earth and established comparisons by which he discovered "vestiges of the mystery of the Very Holy Trinity" in the mysteries of Isis, the works of Plato, the religions of the East and West Indies, Japan, and Mexico, as well as in pagan mythology.[45] His extensive survey constituted an apologetics in the form of an ethnography and a cultural history aimed at proving the religious unity of humanity. Voltaire heaped irony on the Jesuit.[46] From a different perspective,

44. Johann Jakob Heinrich Nast, *Ueber Homers Sprache aus dem Gesichtspunkt ihrer Analogie mit der allgemeinen Kinder- und Volks-Sprache* (Stuttgart: bei Johann Benedikt Metzler, 1801). Almost two centuries later, the psychologist David R. Olson claimed that writing transformed certain linguistic objects into objects of consciousness (including the idea of the inner self); these, as we shall see, are said to be lacking in Homer's heroes. See Olson, *The world on paper: The conceptual and cognitive implications of writing and reading* (New York: Cambridge University Press, 1994), particularly chap. 11.

45. Joseph-François Lafitau, *Customs of the American Indians compared with the Customs of Primitive Times* (1724), ed. and trans. William N. Fenton and Elizabeth L. Moore (Toronto: Champlain Society, 1974), vol.1, p. 31.

46. Lafitau's claim that "the Americans are descended from the ancient Greeks" prompted Voltaire to remark, "These are his [Lafitau's] reasons: the ancient Greeks hunted, so do the Americans; the Greeks had fables, some Americans do too; the ancient Greeks had oracles, the Americans have witches. There was festive dancing in Greece, there is dancing in America. One must admit that these reasons are indeed convincing." Voltaire, *Essai sur les mœurs et l'esprit des nations* (1756), chap. 8 ("De l'Amérique"), in *Œuvres* (Paris: Armand-Aubrée, 1829–1831),

Fig. 6.1. Frontispiece to Joseph-François Lafitau, *Mœurs des sauvages amériquains, comparées aux mœurs des premiers temps* (Paris, 1724). "The frontispiece depicts a woman in the attitude of writing, and presently making comparisons between several monuments of antiquity: pyramids, obelisks, representations, pantheon statues, medals, works of ancient authors, as well as between travel Relations, maps, travel books and other curiosities from America in the midst of which she is seated. Two genii bring those monuments close to each other, and help her compare, by making her see the relationships they might have between them. But time, whose function is to make all things known and in the end to reveal all things, is making these connections still clearer by recalling her to the source of all being, and making her touch with her finger the connection between all these monuments and the origin of man, the heart of our religion and the entire doctrine revealed to our first fathers after their sin, which he shows to her in a kind of mysterious vision" (Joseph-François Lafitau, *Customs of the American Indians compared with the Customs of Primitive Times* (1724), ed. and trans. William N. Fenton and Elizabeth L. Moore [Toronto: Champlain Society, 1974], vol. 1, p. 7).

however, the premise that one could compare the customs of the earliest peoples with those of the "primitives" known to the eighteenth century would later enjoy enormous popularity. As Constantin François de Volney (1757–1820) would ironize, intercultural and transhistorical comparisons demonstrated that the Greeks were true savages, like those of America, and that in honoring the Greeks, we honor the spirit and customs of a barbarous era.[47]

Barely a generation after Lafitau, the comparison of the past with the present appeared self-evident, as suggested by Antoine-Yves Goguet's *The Origin Of Laws, Arts, and Sciences, And Their Progress Among The Most Ancient of Nations*, published just before the author's death in 1758. The work was soon available in German, English, and Italian and was extremely influential among historians of "the human mind." Goguet explained his approach unambiguously:

When I found myself quite destitute of facts and historical monuments, particularly in the first ages, I consulted what has been said, both by ancient and modern writers, on the manners of savage nations. I imagined, that the conduct of these nations would give us pretty clear and just ideas of the state of the first wandering colonies, immediately after the confusion of tongues and dispersion of families. We may collect, both from ancient and modern relations of this kind, several points of comparison, capable of removing many doubts which might arise about certain extraordinary facts, which I have thought proper to build upon. The relations concerning America in particular, have been extremely

---

vol. 12, p. 31. Voltaire's derision (which is also apparent in the articles "America" and "Population" of his *Philosophical Dictionary*) may have gone too far. But it is also true that Lafitau concluded, on the strength of long and erudite comparisons, that "the largest number of American peoples came originally from those barbarians who occupied the continent and islands of Greece" (*Customs*, vol. 1, p. 79–80, and chap. 2 in general).

47. Volney was not referring to Lafitau but to the beneficial nature of a work that would examine antiquity in the light of primitive peoples: "It would dissipate a great number of illusions, by which our judgement is misled in the ordinary modes of education. We should be enabled to form just notions of the golden age, when men roamed about naked, in the woods of Thessaly and Hellas, feeding upon herbs and acorns. We should see, in the early Greeks, just such savages as those of America, placed in a similar country: for Greece, when over spread with trees and bogs, was much colder than at present. We should learn that the Pelasgi, a race dispersed from the Alps to Taurica, were merely the primitive wild hordes, wandering and hunting for their bread, like Hurons and Algonquins of the present age, or the Celts and Germans of old." Constantin François de Volney, *A View of the Soil and Climate of the United States of America* (1804), trans. C. B. Brown (New York: Hafner, 1968), pp. 413–414. More generally, see Giuseppe Pucci, *Il passato prossimo: La scienza dell'antichità alle origine della cultura moderna* (Rome: La Nuova Italia Scientifica, 1993), chap. 9.

useful to me on this article. We may judge of the state of the ancient world, for some time after the deluge, by the condition of the greatest part of the new world when it was first discovered. In comparing what the first adventurers have told us concerning America, with what antiquity has transmitted to us concerning the manner in which the inhabitants of our continent lived in those times which were reckoned the first ages of the world, we cannot but perceive the most evident and striking resemblance and conformity. I have, therefore, pretty often compared the relations of modern travellers with those of ancient historians, and intermingled their narrations, with a design to support the testimony of ancient writers, to shew the possibility, and even reality of certain facts which they relate, and certain customs which they mention. These different passages thus compared and brought together, mutually support each other, and lay a solid foundation for everything I have said concerning the progress of the human understanding, in its improvements and discoveries, which I date from the deluge.[48]

Unlike Lafitau, who had lived with American Indians and compared Huron and Iroquois customs with those of the ancients, Goguet collected textual material and sought to corroborate Herodotus, Strabo, Pliny, and Pausanias using modern travelogues. The primitives who entered the comparisons were invariably Nordic, because, since they lived neither in temperate climates nor in lands of plenty, they were obliged, like archaic men, to fight for their survival.[49] The present, then, was the key to the past, and

48. Antoine-Yves Goguet, *The Origin Of Laws, Arts, and Sciences, And Their Progress Among The Most Ancient of Nations*, trans. from the French (1758; Edinburgh: Printed for George Robinson . . . , 1775), 3 vols., 1, pp. xiv–xv.

49. The argument from climate was widely used, though often implicitly. In Winckelmann's *History of the Art of Antiquity* (1764), artistic excellence was attributed to a temperate climate. The Scotsman James Burnett, Lord Monboddo (1714–1799), author of the famous *Of the Origin and Progress of Language*, wrote an *Antient metaphysics* designed to revive "antient theism," an antimechanist and antimaterialist philosophy in which mind was responsible for the movement of matter. In it, he insisted on the difference between North and South Americans. Due to the climate, the latter "will have little spirit or understanding, and will be a dull, sluggish, unwarlike people." The former, however, "could not subsist without the invention and practice of the arts of hunting and fishing, nor without much toil and labor"; moreover, since they were always fighting one another, they developed into a "noble race of men" similar to those of Homer's heroic age. Monboddo, *Antient metaphysics*, vol. 3, *Containing the history and philosophy of men: With a preface containing the history of antient philosophy, both in antient and later times . . .* (Edinburgh: Cadell, 1784), pp. 206 and (on Homer) 106. Monboddo exemplified a genre that was typical of the Scottish Enlightenment and well known to German historians, namely, "conjectural history." See, in addition to Wood's "The natural history of man," Paulette Carrive, "L'idée d'"histoire naturelle de l'humanité' chez les philosophes écossais du XVIIIe siècle," in Olivier

the past enhanced the value and knowledge of the present. The ethnography of the time corroborated classical texts, which in turn demonstrated the similarity between primitive peoples and humanity in its infancy. These suppositions informed Winckelmann's vision to such an extent that despite never having seen an Indian hunt, he could find in the native American's movements those of Achilles.[50]

Christian Gottlob Heyne was one of Goguet's many readers, and his approach can be understood as a historicist extension of psychology to cultural anthropology. In 1779, Heyne wrote two papers for academic ceremonies, one on "the life of the most ancient peoples, particularly in Greece" and the other on "the life of ancient Greece"—both illustrated "by comparisons with savage and barbarous peoples."[51] These were not scholarly studies but rather statements in support of a research program.

For Heyne, travel writings not only confirmed the reliability (*fides*) of ancient history but also proved that a common nature underlies the immense variety of human thought, customs, and institutions. The peoples described by the ancient historians were mentally and physically similar to those of eigtheenth-century America, Asia and Africa. The Hebrew Bible described populations that lived at the beginnings of human society, were nomadic, or ranked as primitive. To judge their customs and beliefs by our own thinking and habits (*e nostro sensu ac more*) would lead to erroneous interpretations. They should be examined in relation to nations that subsist under comparable conditions of existence.[52] The same can be said, continued Heyne, of certain characteristics of ancient Greek life and customs

Bloch, Bernard Balan, and P. Carrive, eds., *Entre forme et histoire: La formation de la notion de développement à l'âge classique* (Paris: Méridiens Klincksieck, 1988); Annette Meyer, "Das Projekt einer 'Natural history of man' in der schottischen Aufklärung," *Storia della storiografia*, 39, 2001, 93–102; and Meyer, "The experience of human diversity and the search for unity: Concepts of mankind in the late Enlightenment," *Studi settecenteschi*, 21, 2001, 245–264.

50. "Behold the swift Indian outstripping in pursuit the hart: how briskly his juices circulate! how flexible, how elastic his nerves and muscles! how easy his whole frame! Thus Homer draws his heroes, and his Achilles he eminently marks for being 'swift of foot.'" Johann Joachim Winckelmann, *Reflections on the Painting and Sculpture of the Greeks: with Instructions for the Connoisseur, and an Essay on Grace in Works of Art*, trans. Henry Fusseli (London, printed for the Translator, and sold by A. Millar, 1765), p. 6.

51. C. G. Heyne, "Vita antiquissimorum hominum, Graeciae maxime, ex ferorum et barbarorum populorum comparatione illustrata: Commentatio I" and "Vita antiquioris Graeciae ex ferorum et barbarorum populorum comparatione illustrata: Commentatio II" (1779), in Heyne, *Opuscula*, vol. 3. For an Italian translation see Claudia Pandolfi, "Civiltà antiche e selvaggi moderni: Due dissertazioni di Christian Gottlob Heyne," *I castelli di Yale*, 2, 1997, 253–287.

52. "Falsa scilicet persuasione ex suo more et ritu inter Christianos et Europaeos recepto iudicant de barbarorum sensibus et iudiciis: de quibus tamen, nisi ad ipsorum mentem et intelligentiam accommodate, quod probabile sit, statui nequit. . . . Adumbrata est haec opinio e nostro

which resemble those of the "savage and barbarous" peoples described by travel writers. Some features, such as food, clothing, bodily care, or type of habitation, are directly linked to the natural environment. Others, such as family structures, children's education, and community rules, belong to the social realm, while yet others relate to beliefs, prejudices, and explanations of natural phenomena. Heyne found it easy to link sun and star worship to agricultural societies, which is why he insisted we should suspend our notions of God and religious worship and attempt to understand primitive peoples in terms of their own customs and systems of thought.

Transhistorical comparison seemed helpful in this connection. For example, the Greek practice of attributing a divinity to a certain place (particularly mountains, forests, rivers, and springs) could be found among primitives, as could the custom of personifying nature in the form of gods endowed with human characteristics, vices, and virtues. "And still today," Heyne noted, barbarous peoples are convinced, as were the ancients long before them, that illnesses, epidemics, famines, and disasters of all sorts are sent by wrathful gods whose anger has more or less the same causes as in humans. Ancient Greeks and Canadian aboriginals share an absolute reverence for certain animals or places, and both punish transgression severely: none of Ulysses's companions survived killing and eating Apollo's sacred oxen, and no native dares visit certain sacred islands of Lake Huron and Lake Superior.

Heyne criticized modern accounts of the "history of mankind." In his view, the words used by modern authors to describe the customs of primitive peoples and translate what they said were not to be taken in their usual sense. To do so would be to imagine, erroneously, that the same words designate the same concepts. Heyne emphasized the need to grasp how primitive peoples themselves viewed things, to reach "ad *sensus* istorum barbarorum," and to understand them in the light of their way of life, their particular circumstances, and their level of development.[53] His examples (the terms "mystery," "magus," "priest," "seer," "worship") indicated the domain in which he thought the "nature of the human mind" and its dependence on culture and developmental level were most likely to be revealed. This was the area of religious expression, and of beliefs about the causes and

---

sensu ac more; non vero ex sensu ac iudicio hominum rudium et a nostris religionibus et institutis alienorum." Heyne, "Vita antiquissimorum," p. 10.

53. "Epimetrum," ibid., p. 31.

origins of marvels, symbols, and myths—another field which was common to the study of the ancients and of primitive nations.

Within such a project for a comparative cultural anthropology of primitive and ancient man, Homer's position was unparalleled. In the Quarrel of the Ancients and the Moderns, supporters of the latter denounced what Charles Perrault had, in his *Parallel of the Ancients and the Moderns* (1688–1693), characterized as Homer's coarseness, childishness, and tendency to excess.[54] Nonetheless, both Homer's advocates and his detractors agreed that the poet depicted landscapes and culture as he saw them. His poems were believed to describe ancient customs and human nature in their native simplicity, and therefore to be a reliable source of information in every domain, from the science of nature to the science of human conduct, from metallurgy to politics, and from medicine to morals.[55] The shield Hephaestos forged for Achilles, and which was the subject of a famous *ekphrasis* in book XVIII of the *Iliad*, came to represent the entire circle of sciences and arts known at the time.[56]

54. Donald M. Foerster, *Homer in English criticism: The historical approach in the eighteenth century* (New Haven: Yale University Press, 1947); Chantal Grell, *Le dix-huitième siècle et l'antiquité en France, 1680–1789* (Oxford: Voltaire Foundation, 1995); Noémie Hepp, *Homère en France au XVIIe siècle* (Paris: Klincksieck, 1968); Anne-Marie Lecoq, ed., *La Querelle des Anciens et des Modernes, XVIIe–XVIIIe siècles* (Paris: Gallimard, 2001); Françoise Letoublon and Catherine Volpilhac-Auger, with the collaboration of Daniel Sangsue, eds., *Homère en France après la Querelle (1715–1900)* (Paris: Champion, 1999); Kirsti Simonsuuri, *Homer's original genius: Eighteenth-century notions of the early Greek epic (1688–1798)* (New York: Cambridge University Press, 1979), particularly chap. 10 ("The primitivists and the primitive bard"); Sotera Fornaro, "Homer in den deutschen und französischen Aufklärung," in Veit Elm, Günther Lothes, and Vanessa de Senarclens, eds., *Die Antike der Moderne: Von Umgang mit der Antike im Europa des 18. Jahrhunderts* (Hanover: Wehrhahn, 2009). On the modern reception of Homer, see Georg Finsler, *Homer in der Neuzeit von Dante bis Goethe: Italien, Frankreich, England, Deutschland* (Leipzig: Teubner, 1912).

55. Alexander Pope was not particularly concerned with the accurate information which the *Iliad* was supposed to contain, but his translation includes an "index of arts and sciences." Seventy years later, the Italian poet Melchiore Cesarotti (1730–1808), whose translations of Ossian were held in higher esteem than those of Homer, derided with a wealth of examples the "literary obsession" of taking Homer as a source of information. See his *L'Iliade di Omero*, pt. 1, *Ragionamento preliminare storico-critico* (pt. I, sec. IV), in *Opere* (Pisa: dalla Tipografia della Società letteraria, 1802), vol. 6.

56. "In the sculpture of Achilles's shield alone, he [Homer] has included what may be reckoned the circle of the sciences and the arts, which were then known in Greece." Walter Anderson, *The philosophy of ancient Greece investigated, in its origin and progress, to the aeras of its greatest celebrity, in the Ionian, Italic, and Athenian schools . . .* (Edinburgh: printed by Smellie, 1791), p. 12. In this work, Anderson (1723–1800), a Protestant minister, criticized the allegorical interpretatons we discuss below.

Homer had been credited with accurate depiction already in antiquity. However, largely to absolve him of the charge of blasphemy, early commentators identified in his texts physical, moral, or theological allegories.[57] Yet they believed they were simply letting the text speak for itself. For example, Porphyry, in his conclusion to a commentary on the eleven lines describing the grotto of the nymphs at Ithaca (*Od.* XIII.102–112), denied making forced interpretations or fanciful speculations:

> This sort of exegesis should not be considered forced, nor should it be equated with the sort of thing fanciful interpreters try to render plausible. When one takes into consideration the ancient wisdom and the vast intelligence of Homer, along with his perfection in every virtue, one cannot reject the idea that he has hinted at images of more divine things in molding his little story.[58]

Porphyry here betrays the influence of his master Plotinus, but, apart from the reference to "more divine things," his view was widely held.

The finest psychological example is provided by the scene in which Athena prevents Achilles from killing Agamemnon. When Agamemnon was forced to give up his captive Chryseis, daughter of Apollo's Trojan priest, he announced insolently to Achilles that he would simply take Achilles's captive Briseis instead. As the hero drew his sword, Athena (sent by Hera), who was standing behind him, grabbed him by the hair. Achilles, dumbfounded, turned around and recognized the goddess. She then commanded him to control his anger and end the quarrel, and Achilles obeyed (*Il.* I.193–214). Heraclitus, the author of *Homeric Problems*, interpreted this passage in the light of the Platonic theory of the soul and even claimed that Plato had plagiarized Homer. In this scene, he argued, Achilles's reason was eclipsed by anger, and Plato (e.g., *Timaeus*, 90a) located reason, or the *logisticon*, in the head, just as did Homer, for whom the head "occupies the most important position in the body." The hero regained his composure with the help of rea-

---

57. On this vast subject, see Félix Buffière, *Les mythes d'Homère et la pensée grecque* (Paris: Belles Lettres, 1956); Robert Lamberton, *Homer the theologian: Neoplatonist allegorical reading and the growth of the epic tradition* (Berkeley: University of California Press, 1986); R. Lamberton and John J. Keaney, eds., *Homer's ancient readers: The hermeneutics of Greek epic's earliest exegetes* (Princeton: Princeton University Press, 1992); Don Cameron Allen, *Mysteriously meant: The rediscovery of pagan symbolism and allegorical interpretation in the Renaissance* (Baltimore: Johns Hopkins Press, 1970), chap. 4.

58. Porphyry, *On the cave of the nymphs*, trans. and with introduction by Robert Lamberton (Barrytown, NY: Station Hill Press, 1983), § 36, p. 40.

son: "This change of heart due to sane thinking is very properly identified in the poem with Athena," who incarnates wisdom.[59] For Heraclitus, the passage revealed psychological truths: a goddess would have calmed Achilles down completely, whereas when Athena advised him to insult rather than physically attack Agamemnon, she allowed him to keep traces of anger. She thus represented a "human [rather than a divine] reason" (logismos anthropinos): "The Episode of Athena, whom Homer represents as the mediator in Achilles's anger against Agamemnon, may thus be seen to merit an allegorized interpretation."[60] The passage was read as though it conveyed an empirical truth or a naturalistic observation expressed in the cognitive and linguistic terms that were at humanity's disposal in Homer's time.[61]

This type of interpretation did not entirely disappear in the eighteenth century, and it even reached new heights. The first volume of the vast but unfinished *Primitive World, Analyzed and Compared with the Modern World* (1775), by Antoine Court de Gébelin (1728–1784), was devoted to the "allegorical genius" of the primitive world. For example, Court de Gébelin analyzed the legend of Hercules as an allegory of agriculture. The hero's labors represented the clearing and cultivation of land; they were precisely called *labors* "because the activities of the countryside are the real labors of man, those on which societies and empires are built."[62] The forest in which the lion of Nemea dwelled was the forest to be cleared; the hydra of Lerna

59. Heraclitus, *Homeric problems*, ed. and trans. Donald A. Russell and David Konstan (Atlanta: Society of Biblical Literature, 2005), 19.1 and 7. The work was first printed at the beginning of the sixteenth century and attributed at the time to the Pythagorean Heraclitus of Pontus (fourth century BC); it probably dates from the first century AD.

60. Ibid., 20.9–12.

61. The psychological and moral dimension of these readings of Homer comes to the fore in certain translations of the *Iliad*. For example, George Chapman's memorable version (1598–1611) fleshes out Athena's speech to make Achilles into a righteous person capable of autonomous moral decisions: "I come from heaven to see / Thy anger settled, *if thy soul will use her sovereignty / In fit reflection*. . . . / And cease contention; draw no sword: use words, and such as may / *Be bitter to his pride, but just*; . . . / . . . throw / Reins on thy passions, and serve us" (*Il.* I.207–214; Chapman's additions are in italics). In Alexander Pope's equally famous translation of the *Iliad* (1715–1720), Athena says: "To calm thy fury I forsake the skies: / Let great Achilles, to the Gods resign'd, / To reason yield the empire o'er his mind / . . . / Command thy passions, and the Gods obey" (*Il.* I.275–286 of the translation). No such moralizations can be found in the magnificent Italian and German neoclassical translations by Vincenzo Monti (1810) and Johann Heinrich Voß (1793).

62. Antoine Court de Gébelin, *Monde primitif, analysé et comparé avec le monde moderne, considéré dans son génie allégorique et dans les allégories auxquelles conduisit ce génie*, new ed. (Paris: chez l'Auteur . . . , 1777), p. 199. See Anne-Marie Mercier-Faivre, *Un supplément à "l'Encyclopédie": Le "Monde primitif" d'Antoine Court de Gébelin: Suivi d'une édition du "Génie allégorique et symbolique de l'antiquité," extrait du "Monde primitif,"* 1773 (Paris: Champion, 1999).

was the harvest, reaping the ears of corn; cleaning out the stables of Augias was the rainy weather of winter, during which the plowman repaired and washed out the cowsheds—and so forth. Once again, allegory was thought to be the only mode of expression possible in archaic times, but it was also supposed to conceal accurate observations and descriptions, which were there to be deciphered.

The dominant mode of interpretation in the eighteenth century, however, was empiricist and realist and was applied to any text believed to be primitive. For example, the poems of Ossian, the Gaelic bard said to have lived in the third century, were considered equivalent to Homer's epics; when published in the early 1760s, they were acclaimed as priceless expressions of humanity in its infancy. The first poem, *Fingal*, became particularly popular. They were all, in fact, compositions (or, as recent scholarship would have it, imaginative translations or creative reconstructions) by the Scotsman James Macpherson (1736–1796). The debate on their authenticity, which broke out immediately, did not affect their alleged value as the music of a language and a nation, the voice of a people from a lyrical era.

The type of philosophy of humanity that inspired Carus and other anthropologizing readers can be found in the enormously influential Ossianist Hugh Blair (1718–1800), a professor of rhetoric and belles lettres at the University of Edinburgh, and Ossian's most zealous defender. For Blair, since primitive language was poetic, one could expect to find poems among the ancestral treasures of every nation. As the human mind progressed, the understanding came to prevail over the imagination. The history of humanity was in this respect similar to individual development, which was why poetry, as the "offspring of the imagination," was so impassioned and vigorous in young nations.

Blair noted that Ossian's mythology was concerned above all with apparitions of spirits of the dead. In keeping with the notions of a primitive or "rude" age, these spirits were not purely immaterial, but "thin airy forms."[63] The same ideas could be found in Homer. For Blair, however, Ossian depicted ghosts with greater imagination, solemnity, dignity, and decorum than Homer. His ghosts, in short, were sublime and thus reflected an increasingly fashionable aesthetic value, influenced by Macpherson's poems themselves. The superiority of Ossian's mythology, in Blair's view, came

---

63. Hugh Blair, *A critical dissertation on the poems of Ossian, the son of Fingal*, 2nd ed. (1765; New York: Garland, 1970), p. 57. Also in James Macpherson, *The poems of Ossian and related works*, ed. Howard Gaskill, introduction by Fiona Stafford (Edinburgh: Edinburgh University Press, 1996).

from the fact that it was not "local and temporary" but belonged to human nature itself and was shared by all periods, countries, and religions.[64] For the philosopher Christoph Gottfried Bardili (1761–1808), Ossian's description of *Volksideen* suggested the "natural" character of the notion of divinity. This seemed to be confirmed by the customs of recently discovered islanders.[65] The idea of immortality, Bardili claimed, would never have appeared among primitive peoples had they not felt love and pity, or experienced pain at the loss of beings whose presence they wished to perpetuate. And how do we know this? he asked: "Man lese den Ossian!"[66]

It was Homer, however, who received most praise for his capacity to observe and reproduce reality faithfully. There are countless examples of this from the eighteenth century onward. Thus was Heinrich Schliemann (1822–1890) inspired to search for Troy in the place where he did, and to explore Ithaca for the cave of the nymphs which, he reported, was "very spacious" and corresponded exactly to Homer's description.[67] A generation later, faith in Homer's empirical accuracy motivated the French Hellenist Victor Bérard (1864–1931) to retrace Ulysses's sea voyage, and to conclude that the 1,200 negatives taken by his travel companion, the great Swiss photographer Fred Boissonas, enabled him to "go back over all [of Homer's] descriptions and depictions, word for word" and correct his earlier identifications

64. Ibid., p. 63.

65. Christoph Gottfried Bardili, *Epochen der vorzüglichen Philosophischen Begriffe, nebst den nöthigen Beylagen*, erster Theil, *Epochen der Ideen von einem Geist, von Gott und der menschlichen Seele: System und Aechtheit der beiden Pythagoreer, Ocellus und Timäus* (1788; Brussels: Culture et civilisation, 1970), p. 10.

66. Ibid., pp. 120–121. The value attached to Ossian coincided with the increasing popularity of myth and the decreasing value of fables. The late eighteenth-century discovery of myth corresponded to the wish to return to the origins of humanity and recapture the plenitude of ancient times, since mythic narratives were deemed to have an energy that civilized tongues had lost. On this development, see Jean Starobinski, "Fable and mythology in the seventeenth and eighteenth centuries" (1981), in *Blessings in disguise; or, The morality of evil*, trans. Arthur Goldhammer (Cambridge, MA: Harvard University Press, 1993).

67. Heinrich Schliemann, *Ilios: the city and country of the Trojans: The results of researches and discoveries on the site of Troy and throughout the Troad in the years 1871, 72, 73, 78, 79: Including an autobiography of the author* (London: John Murray, 1880), p. 49. The issue of Homer's accuracy is still a live one. While some deny that the town discovered by Schliemann was indeed Troy, others consider that the late Manfred Korfmann's excavations, in the 1980s, confirm the epic tradition as well as Schliemann's intuitions. Predictably enough, Korfmann's conclusions sparked a controversy. See http://www.uni-tuebingen.de/troia/deu/kontroverse .html; for Korfmann's debated reconstructions, see the first of his chapters in Korfmann et al., *Troia: Traum und Wirklichkeit* (Stuttgart: Konrad Theiss Verlag, 2001). For the question prior to Schliemann, see Chantal Grell, "Troie et la Troade de la Renaissance à Schliemann," *Journal des savants*, 1981, 47–76.

of places.[68] Bérard's boundless confidence in the objectivity of photography proved no less excessive than the most extravagant allegorical readings.

There was, of course, no photography in the Enlightenment. Evidence was provided by explorers' reports, and Lafitau was simply stating a commonplace when he wrote in 1724 that

> the science of the manners and customs of different peoples has some quality so useful and interesting that Homer thought he ought to make it the subject of an entire poem [the *Odyssey*]. Its aim is to set forth the wisdom of Ulysses.[69]

Homer again provided the model, which the gentleman traveler Robert Wood (1716–1777) would follow. His 1769 *An Essay On The Original Genius and Writings of Homer: With A Comparative State Of The Ancient And The Present State Of The Troade* was extremely influential, especially in Germany. Heyne read it with enthusiasm and declared, "Whoever knows no other men than his compatriots, us Europeans, should not read Homer and certainly not judge him. It is from travel writings and descriptions of the lands of the savages and other peoples living in societies and political orders which are still primitive that one learns most for Homer."[70] Wood examined Homer from the point of view of his "mimetic powers." He stressed the poet's closeness to and respect for nature and declared that his great merit was to have made a "faithful transcript, or . . . a correct abstract of human nature, impartially exhibited under the circumstances which belonged to his period of society, as far as his experience and observation went."[71] When evaluating Homer's descriptions of the customs and particularities of different nations, he emphasized the bard's experience of travel and claimed that even if a culture had entirely vanished or left barely visible traces, one could still discern general similarities between a people's ancient and present mores.

68. Victor Bérard, *Les navigations d'Ulysse I: Itaque et la Grèce des Achéens* (Paris: Armand Colin, 1927), p. 20.

69. Lafitau, *Customs*, 1, p. 29.

70. "Aus Reise- und Länderbeschreibungen der Wilden und andere Völker, die in einer noch ungebildeten Gesellschaft und Staats-verfassung leben, lernt man das meiste für den Homer." C. G. Heyne, review of *An essay on the original genius of Homer*, by R. Wood, *Göttingische Anzeigen von gelehrten Sachen*, 32. Stück, 1770, 257–270, pp. 258–259.

71. Robert Wood, *An Essay On The Original Genius and Writings of Homer: With A Comparative State Of The Ancient And The Present State Of The Troade* (1769; Washington, DC: McGrath Publishing Co., 1973), pp. viii and xiii.

Wood, who had journeyed extensively in Palestine, Egypt, and Asia Minor, and published sumptuous works on the ruins of Palmyra and Balbec, detected the customs of the *Iliad* in certain Eastern regions, while the American Indians he read about seemed to him to corroborate Homer's depiction of human nature.[72] In short, Homer "always remained faithful to nature and never based his work on a fictional ideal. Of course, it would have been perfectly easy for him to alter characters for the better or the worse, had he wanted to. But why would he have drawn on the powers of the imagination, when nature offered him realities in such abundance?"[73] This belief, formulated by this anonymous encyclopedist of the 1770s, was unanimously accepted at the time; the opposition between imagination and reality was much less significant than the assertion of Homer's fidelity to nature. This conviction, combined with a belief in the immanent connection between language and thought, accounts for the status Homer had in Carus's outline of the history of humanity and, as part of it (in Carus's view), in the history of psychology.

TOWARD A TOTAL HISTORY OF PSYCHOLOGY

For Carus, since the history of mankind determined the questions facing the history of psychology, these questions were extremely wide ranging.[74] How, in bygone eras, did humans think and speak, and how did psychological ideas affect language and the understanding of experience? How did people unknowingly develop psychological theories? Such questions, Carus maintained, revealed parallels between diverse objects and fields of history and historiography: psychology and languages, biography, religion, theology, natural law, morals, and the natural sciences. Other questions concerned human experience directly: What did past humans seek to know about themselves and others? What was their self-image? Which powers did they believe in and how did they honor them? How did they behave toward each other? How did they form the notions of animal, child, spirit, angel, god? How did the common people, writers, artists, physicians, educators, and leaders view the soul across the ages and under different political regimes? How did man gradually achieve a higher spiritual life and a more enlightened understanding?

---

72. Ibid., p. 155, n.
73. HOMERE (*Histoire littéraire*), in Yverdon *Encyclopédie*, ed. de Felice, 23, 369a–372a, p. 370b.
74. Carus, *Geschichte*, p. 5, henceforth referred to by page number in the text.

As we have seen, Carus conceived the history of psychology and of the science of man as a single totality and, in line with a schema inherited from the historiography of philosophy, he divided it into universal and particular. Universal history concerned the natural science of man (*Menschen-Naturkunde*). The historical progress "of psychological observation and thought" could be tracked through the development of the spirit of observation (*Geist der Beobachtung*) and of the scientific elaboration of observed facts (7–8). The rest fell into four major parts. The first was a "psychological history of the psychological sense"; this sense was a sort of "instinct" Carus described as being similar to the physiognomist's. It included the history of the spirit of observation (its scope, objects, and fields of application) and an investigation into the development of self-consciousness. The second part comprised a "history of psychological education," including a "history of the theoretical art of observation" and its rules (8). There followed a "philosophical history of anthropology and psychology" (*Menschen- und Seelenkunde*). In its "analytic" version, it was concerned with the psychological concepts, theories, dogmas, basic notions, and principles present in various domains (for example, in theology as regards the doctrines of original sin and divine inspiration). "Analytic history" dealt with the gradual separation of physics and metaphysics. It also concerned the emergence of notions about the soul, life, being, appearance, substance, law, psychic health and illness, and psychological phenomena and faculties. In its "synthetic" form, the history of anthropology and psychology described systems and systematic projects, including rational and empirical psychology and their relation to philosophy.

A third form of the history of psychology and the science of man consisted of a "hermeneutic" history of the grounds of explanation in psychology as well as of psychological explanations used in other sciences. The last part concerned the "aesthetic history of the forms of representation [*Darstellungsformen*] of psychological observations and facts," namely myth, allegory, fable, and all other written or spoken forms, from children's books to academic textbooks.

By contrast, particular history (*Specialgeschichte*) concentrated on the contents of psychological discourses, that is, the findings of the "spirit of observation and inquiry," in a range of individuals, nations, and periods, including, as Carus stated, their digressions, erratic ramblings, and even downright aberrations (12). It consisted of a chronological presentation of isolated observations and of the "gradual accumulation of psychological material": hypotheses and explanations, the "wealth of psychological experiences" contained in languages, and how "humanity manifests itself partly in the common people and partly in the cultivated people of the same nation" (12).

Carus noted that since psychological doctrines were often implicit, a wide range of sources had to be mobilized to grasp them. Moreover, since ancient and uncivilized (*ungebildet*) peoples did not yet possess self-consciousness, it was particularly difficult to formulate their anthropological ideas without denaturing them (14–15). Hence the challenge Carus had to face, given that the final object of his investigation was the human being's humanity, rather than the soul itself, the concepts used to designate it, or psychology as a science (16–17). Such an exploration involved a great many subjects and sources: the development of "degrees of consciousness"; how people acted in different circumstances; languages (Carus considered that the history of a psychological word could be more useful than the history of language in general, 21); poets and orators; historians and biographers; the empirical and practical sciences of a period, even if they provided "views" rather than systematic discourses; the writings, often in the form of thoughts and maxims, of moralists and authors experienced in human affairs; and philosophical works based on observing the "inner man."

Initially, Carus explained, the history of psychology had dogmatic and polemical goals and referred more frequently to the metaphysical psychology of the past than to the "new physical psychology" (30). It later took a more genuinely historical turn, while remaining nonetheless closer to a literary compilation than to a critical and "pragmatic" narrative. The "authentically historical," *das Aechthistorische*, appeared only in the last quarter of the eighteenth century, and even then, Carus observed, a work like Tennemann's was dependent on the history of philosophy (31). Significantly, Carus included in his historiographical bibliography works on the theory of the soul, such as the largely "metaphysical" *History of the Human and Animal Souls*, by Justus Christian Hennings.[75] It was nonetheless clear

---

75. Carus criticized Hennings (1731–1815), a professor of logic and metaphysics at Jena, for defending the soul's immateriality after having admitted that one can know almost nothing about the soul: "Die Seelenlehre ist die rechte Residenz der Hypothesen"; ibid., p. 669. Hennings's ultimate goal was to combat materialism. Thus, even though he discussed psychological methodology and the structure of the mental faculties, he devoted most of his book to issues such as the simplicity, immortality, immateriality, and origin of the soul, its seat, its harmony with the body, and the debate on the soul of animals. See J. C. Hennings, *Geschichte von den Seelen der Menschen und Thiere: Pragmatisch entworfen* (Halle: bey J. J. Gebauers Witwe und J. Jac. Gebauer, 1774).

that, in his opinion, a "truly historical" history of psychology had yet to be written.[76]

Thinking perhaps of Hißmann's history of the theory of the association of ideas, Carus found that even the most "truly historical" histories available were limited. A comprehensive vision of the whole would distinguish the new from the old; it would show the persistence of certain necessary ideas (for example, that humans possess something above their "animal" nature); it would clarify the present from an intellectual, terminological, and anthropological point of view; and it would enrich *Menschenkunde* with a wealth of psychological materials (including those provided by the "science of man in antiquity"). Such a total history would encourage a "true psychology" and stimulate progress by demonstrating its possibility (38). This was clearly Carus's ideal—which explains why, despite its seven hundred pages, he saw the *History of Psychology* merely as a programmatic sketch.

The first part of the *History of Psychology* outlined a "universal history of the science of man" (*Universalgeschichte der Menschenkunde*, 39–87), in which Carus identified three lines of development. First, the human being ceased to be an object and became a subject. Second, rather than simply observing external phenomena, he began registering what went on inside himself. Last, the science of individual man (*Individual-Menschenkunde*), which simply assembled disparate cases, gave way to a systematic universal psychology. Carus thus described a progress from a dogmatic to a critical perspective (from metaphysics to physics, and from rational to empirical psychology), from a panoply of causes to a few simple predictive laws, and from realism to idealism, that is, "from the mechanical explanatory principles of a general materialism to the teleological principles of an elevated spiritualism" (44).

Carus's account was organized into four "epochs" (85–87). Initially, humans were not differentiated from other living beings, whether animals or gods. The supposed lack of distinction between subject and object led Carus to conclude that, as far as observation was concerned, there was no human being. When initially identified as such, humans were considered purely material, and that is why the doctrine of the soul began as a doctrine of the body; observations were carried out, but they were not self-reflexive. The emergence of the distinction between internal and external, or soul and body, led to the first forms of pneumatology, or metaphysical psychology, as well as to a rudimentary empirical psychology. Last, the two natures

---

76. Carus, *Geschichte*, pp. 30–33, for the list of the "Hülfsschriften" in the section "Geschichte der bisherigen Bearbeitung oder der Vorarbeitung zur Geschichte der Psychologie."

of man were conceptually united—subject and object, nature and freedom, individuality and universality. The universal history of the science of man thus situated the concept of the soul within natural and social contexts and demonstrated how its evolution was a vital factor in the constitution of human subjectivity.

In Carus's view, the "point of departure" for an *Universalgeschichte* could not be inquiries into the origin of the soul, or a contemporary "system of psychology," or the emergence of the scientific spirit of inquiry (which would mean omitting Pythagoras and Socrates), or even the first stages of embryogenesis or child development. Rather, one should ask, How did the intuition of a not-I (*NichtIch*) make its appearance, followed by the intuition of souls or of something soul-like (*etwas Seelenartiges*, 45)?

Carus therefore embarked on a history of the concept of the soul. Initially, "mythico-psychological" concepts mixed "imaginary representations" with sensory data. Man's keen imagination made him mistake these representations for real objects and believe that some of them were spirits. Since the divine was placed within the orbit of man's external senses, the soul was primitively treated as a "half fetish" (49), and since the soul did not entirely belong to the individual, one could lose it as one could lose one's fetish. This mental universe allowed for "sympathetic or psychological cures" by priests, as well as magical procedures to recover lost souls. The idea that the soul was external, perceptible, and substantial defined a first form of materialism. As we shall see, abandoning such materialism was, for Carus, a prerequisite for the emergence of a scientific psychology.

Such was the psycho-cultural framework in which the first notions of divinity and spirit appeared. Gods and spirit had in common a principle of movement. This was for Carus the only form of monotheism to exist in the ancient world. But it was purely nominal, since every being endowed with movement had its own spirit. Air was thought to be the first means by which invisible beings were transported; most savages considered the soul a sort of air, visible but without flesh. The philosophers, Pythagoras for example, used this imaginary representation (*Phantasiebild*) to describe the soul's substance (52–53). Subsequently, as these ethereal forms were given matter, divinity was endowed with breath and a body of flesh. Thus originated the idea of the soul as breath and exhalation, which in turn gave rise to the belief that it was the source of these bodily functions. Revelation, at this stage, did not concern a doctrine of the soul but the soul itself—supposed to have been received from the gods as a divine breath, with the human body as its instrument.

The observation of dead bodies—rigid, immobile, cold, and not breathing—suggested that the soul was not merely the source of breath and movement, but also of life. Yet since the soul still remained a "gift from elsewhere," the gods who bestowed it could choose to give several souls to the same person. This would explain why certain primitive peoples from Greenland, Canada, and the West Indies imagined a multiplicity of souls, some of which could fly outside the body. The belief in a soul which resembled spirits and gods meant either that souls and spirits possessed, or were in themselves, bodies or that terrestrial bodies were distinct from the supernatural bodies of spirits.

Carus saw in this distinction between a flesh-body and a soul-body a first differentiation between the two substances (63). Such a primitive notion of a dual nature could be found among the most diverse peoples, from China to Greenland and from North America to India, and it often included transmigration as one of the forms of the soul's activity. Everything that possessed movement, breath, and life—gods, men, animals, and plants—was endowed with soul, which is why the expressions first used to designate inner movement were almost identical to those designating the soul, and they were used interchangeably. The first psychology was in fact a "psychotheology," and therefore, Carus maintained, it worked to a certain extent against its own materialism (66).

Another concept emerged when the soul was thought to be the seat of the qualities of personified spirits. Poets, for example, did not themselves compose but rather were inspired, possessed by spirit (Be-geisterte). Carus noted that what in the eighteenth century was considered the result of chance or outstanding talent was in archaic times revered as divine (68–68). Likewise, exceptional sensibility or intelligence was not viewed as a natural faculty but as a gift from the gods, as a wondrous power of supernatural origin. The "poetic psychology" that conveyed such beliefs could not be systematic. It was practical rather than theoretical, closer to religious psychagogy than empirical psychology, and less a genuine science than a means of spiritual guidance and elevation ("mehr Seelenleitung und Seelenerhebung als Seelenkunde," 72).

When dreams and poems began to be derived from an internal sense, the notion of the soul was internalized. Wonders (Wunder) were no longer felt to be supernatural events but only "supra-terrestrial"—enigmatic and incomprehensible. The understanding (Verstand), which was still purely practical, became an inner power, such that man no longer existed "solely in another but also in himself. He will henceforth talk of his spirit" (76). In what would not have seemed out of place in a "history of humankind" or in Heyne's work on mythology, Carus speculated on how animal domestication furthered the sense of self (*Selbstgefühl*), for example in the case of shepherds, whose activity increased their sense of being human. Sedentarization was accompanied by greater attention to the weak, the tempering of brute appetite, and a greater capacity to know and master the world. Agriculture speeded up the process of humanization, as man learned to detect regularity in nature. While fulfilling their pastoral duties, priests gathered many "remarkable psychological and anthropological observations," and trade furthered both the speculative spirit and an awareness of human weaknesses. "Since to this very day," Carus wrote, "the technical world of craftsmen has been better able to develop a spirit of observation than the entire world of thinkers [*Denkerwelt*] . . . , the technical arts have rapidly led to philosophy, as practice led to theory. This was how a special psychology [*Special-Seelenkunde*] was conceived and prepared" (79).

The soul was thereafter understood as a substance endowed with internal faculties. At the same time, a *psychologische Kunstsprache* (80), a specialized psychological language, was invented. The first universal psychology viewed the soul as an immaterial substance endowed with faculties. The soul's transformation into a hypothetical "substratum" paved the way for metaphysical psychology and the distinction between a science of the soul and a science of the body, Cartesian dualism, and the Wolffian division of empirical and rational psychology. Various camps formed: the realists and the idealists, the materialists and the spiritualists.

The last concept of the soul in Carus's outline was the idea of man as a totality uniting reason and freedom. "Not only does this concept cover the most general feature of humanity but also the universality of man's natural constitution and destination [*Naturanlage und Naturbestimmung des Menschen*], in interaction with his individuality." Such a concept enables a "most profound self-knowledge." What emerges is not so much an anthropology as an "anthroponomy" linked to philosophy in general and to the philosophy of nature in particular.[77]

> Henceforth, as they perfect themselves, all the philosophical sciences
> sustain each other mutually and are perfected with and by each other.
> While one (analytic philosophy) defines and secures the method of the
> psychologist, another (moral philosophy) assigns it a lofty goal and its

77. Carus's choice of "anthroponomy" highlights the aim of arriving at a law (*nomos*) rather than at a simple description of human nature and human history and suggests the philosophical (rather than merely natural-historical) significance of *Menschenkunde*.

most exalted meaning, while a third one (the philosophy of nature)
places it in the perspective of the totality and the full truth. (84)

In short, the "universal history of the science of man" is the history of humanization itself and of how the human being became a subject for himself.

Particular history, on the other hand, which took up the largest part of
Carus's work, focused on successive beliefs, doctrines, and theories of the
soul. For the earliest periods it also described peoples' psychology (in the
sense of their "mentality"). Carus organized his narrative chronologically
by author and country, as a kind of chapter within universal history, whose
broad outlines it followed. After a sketch of the Eastern world, he turned
to Greece, dividing his account into seven periods. The first ran from the
reign of the imagination to that of practical intelligence (from Homer to the
moral maxims of gnomic poetry). There followed the ascendancy of practical intelligence (up to Socrates). Then, under the influence of Socrates, came
the integration of the first psychological investigations into philosophy, the
first "scientific treatment" of the concept of soul, and the seeds of a systematic psychology (from Plato to the Roman Stoics). The period "from a new
metaphysical rationalism to a naturalist empiricism" went as far as Bacon,
Montaigne, and the Protestant Scholastics. The ensuing three periods were
organized by country, since Carus maintained that once a people became a
nation, psychological writings were marked by national temperament and
character. There followed the systematization of psychological observations (*Seelenerfahrungen*), from Descartes to the German post-Wolffians of
the 1790s.

Carus both followed and went beyond Heumann's or Brucker's indications concerning the *Special-Historie* of disciplines. He followed them insofar as his history of psychology demonstrated that a particular history of
the science of the soul as a branch of philosophy could indeed be written;
but he went beyond them insofar as his *History* combined all the sciences it
drew on—mythography, anthropology, philology, the science of antiquity,
and the philosophy of language—into a history of humankind. For Carus,
the object of the history of psychology was not the fashioning of a discipline, since philosophical concepts and practices were made possible and
determined by self-consciousness, as well as by a certain idea of the human being, and these also belonged to psychology. Carus emphasized the
split between pneumatology or metaphysical and rational psychology, on
the one hand, and empirical psychology, on the other. Empirical psychology was the terminus ad quem of the historical inquiry; at the same time,
it opened onto the future and was partially constituted by the narration of

its own history. Carus's brief general conclusions highlighted the program-matic character of his historiography:

> If we turn again to the overview we have given of this field, on which thinkers of different periods and nations have worked so zealously, it becomes clear to us that there were many systems, but only one science. System building comes and goes, but the truth remains. No system has been entirely infallible, but each has been valuable, and the invention of new words brought with it the invention of new truths. We have a wealth of observed facts but even more opinions. It was generally indi-viduals who were observed, in particular circumstances, and not always in a universal and impartial fashion. But a superior point of view can be claimed for psychology, from which it can be established as a philosophi-cal science. Phenomena are linked together due to the unity of nature: to find this unity is to arrive at the principle that underlies psychology. (759–760)

These conclusions foreground Carus's philosophical position, while also em-phasizing how the history of humanity (in the two senses of historiography and anthropology), psychology, the philosophy of nature, and the histori-cal accounts of the sciences of man and the soul fit together coherently. "Psychology, or the theory of human subjectivity, is as much a unity as the human subject."[78] The premise of the unity of nature and of human-kind grounded the possibility of a universal *Menschenkunde* which would include the history of psychological concepts and practices. In short, the history of psychology appeared to Carus to be a crucial means for advancing the history of the science of man, and this history itself, to be an aspect of the evolution of human self-consciousness.

THE PSYCHOLOGY OF THE HEBREWS

Carus's book on the psychology of the Hebrews "based on their sacred books" is exemplary in this respect.[79] From its title alone—if not from its bulk, some 450 pages—we can tell that it was not the first of its kind. "Bib-lical psychology" was considered one of the oldest sciences of the church, going back to Tertullian and Melito of Sardis's lost treatise on the soul and

---

78. "Die Psychologie oder menschliche Subjectivitätslehre ist in sich selbst nur Eine, wie das menschliche Subject." Carus, *Psychologie*, p. 23.

79. F. A. Carus, *Psychologie der Hebräer*, in *Nachgelassene Werke*, vol. 5.

the body. In the mid-nineteenth century, the Erlangen theologian Franz Delitzsch described it as comprising writings that use the Bible to examine questions relating to the soul's origin, nature, faculties, and fate after death.[80] He found a first non-Scholastic example in a "Guide to a True Psychology and Anthropology Contained in the Holy Scriptures," by the Danish anatomist Caspar Bartholin (1585–1629). This text, published around 1620, starts out from the book of Genesis—"And the Lord God formed man of the dust of the ground, and breathed into his nostrils the breath of life; and man became a living soul" (Genesis 2:7)—and gives brief accounts of subjects ranging from the senses to the state of the soul after death.[81]

Bartholin thus sketched the project of a scripturally based psychoanthropology.[82] In contrast to Bartholin's biblicism, the Lutheran theologian Jacob Carpov (1699–1768) adopted a rationalist approach and used Wolff's faculty psychology to compare Christ's soul and that of human beings.[83] A Psychologia Salomonis of 1686 criticized atheist readings of Ecclesiastes and Proverbs from yet other points of view, and a Psychologia rabbinica of 1719 examined ideas on the soul's survival after death and tried to demonstrate the Koran's debt to Jewish commentaries.[84]

Studying the psychology of the Hebrews effectively meant reconstructing their doctrines of the soul, which naturally fell within the history of philosophy. Johann Franz Budde (1667–1729), a Lutheran eclectic and professor of theology at Halle, discussed pneumatology in his history of Hebraic

80. Franz Delitzsch, System der biblischen Psychologie (1855; Leipzig: Dörffling und Franke, 1861).

81. Caspar Bartholin, "Manuductio ad psychologiam veram adeoque anthropologiam ex sacris literis exstruendam" (ca. 1618–1619), in Delitzsch, System, "Anhang."

82. An extreme form of this project can be found in the Pietist theologian Johann Tobias Beck's Umriss der biblischen Seelenlehre: Ein Versuch (1843; Stuttgart: J. F. Steinkopf, 1871).

83. Jacob Carpov, Psychologia sacratissima: Hoc est de ANIMA CHRISTI hominis in se spectata commentatio theologico-philosophica (Frankfurt: Io. Adam. Melchior, 1738).

84. Franz Wörger, Psychologia Salomonis: Post tot interpretum sinistras detorsiones à Varenio dextrè explicata, & adversos Spinosa aliorumque Atheorum inanes cavillationes adhuc uberius defensa (Hamburg: apud Gothofredum Schultzen, 1686). (August Varenius, the author of Gemmae Salomonis [1659], was a professor of theology and Hebrew at Rostock). Johann Egger, [Sefer Ha-Nefesh] Sive Psychologia rabbinica, quae agit de mentis humanae natura & praecipue ejus extremis, ex mente magistrorum Judaeorum accedit hinc inde sententia Mohammedis & Arabum (Basel: Typis Joh. Ludov. Brandmülleri, 1719). Egger sought to prove such theses as "The intellectual soul survives after death and preserves its individuality after leaving the body"; "The souls of the just return to God and their reward is to apprehend their first cause and enjoy voluptas divina"; and "All souls continue to exist after death, and the souls of the wicked are made to suffer not by being deprived of the beatific vision but by extraordinary punishments and eternal pain."

philosophy.[85] His disciple Johann Jacob Brucker did likewise, but without treating the pneumatology or psychology of the Hebrews separately.[86] In his *Foundations of Psychology Drawn from the Scriptures* (1769), the Pietist theologian Magnus Friedrich Roos (1727–1803) examined the Hebrew terms corresponding to *psukhē* and *pneuma*, *anima* and *spiritus*, dwelled at length on the Old Testament notion of "heart," discussed the cognitive faculties, the will, and the senses, and examined the various states of the soul, from its transmission (which he was unable to explain) to its reunion with the resurrected body.[87]

Unlike the biblicists, who believed the Bible literally contained a revealed psychology, Roos thought that scripture provided the basis on which to elaborate a psychology consistent with Revelation. Like Roos's *Fundamenta*, the more scholarly *Observations on Sacred Psychology*, by Georg Friedrich Seiler (1733–1807), a professor of theology at Erlangen, read the Bible as a psychological work.[88] Seiler emphasized the importance of not interpreting biblical terms through later, particularly Greek and Roman, terminology.[89] He also assumed that Old Testament authors were far removed from the "metaphysical subtlety" of modern philosophers as regards the human soul and mind, which they considered to be pure, subtle, invisible, but nonetheless material substances.[90] In this Seiler turned his back on the allegorical tradition of Philo of Alexandria, the Jewish Hellenistic philosopher who saw in the biblical terms *ru'ah* (spirit), *nefesh* (the soul, sometimes

85. Johann Franz Buddeus, *Introductio ad historiam philosophiam ebraeorum* (1702; Halle: Impensis Orphanotrophei, 1720).

86. Brucker, *Historia critica philosophiae*, vol. 1, bk. 1 ("De philosophia barbarica post diluvium"), chap. 1 ("De philosophia veterum hebraeorum"); vol. 2, bk. 2 ("De philosophia iudaeorum").

87. Magnus Friedrich Roos, *Fundamenta psychologiae ex Sacra Scriptura sic collecta, ut dicta eius De Anima eiusque facultatibus agentia collecta, digesta atque explicata sint* (Tübingen: sumptibus Lud. Friedr. Fuessi, 1769).

88. Among other works, Georg Friedrich Seiler published books of religious instruction and religious popularization (including some for children and young people), a defense of the harmony of reason and Revelation against Thomas Paine, and a *Biblical Hermeneutics* which, while advocating a moral reading of the Bible and stressing that Revelation could be received and interpreted only through the universal principles of reason, accused Kant of having gone too far in that direction. See his *Biblische Hermeneutik, oder Grundsätze und Regeln zur Erklärung der heiligen Schrift des Alten und Neuen Testaments* (Erlangen: Bibelanstalt, 1800). Thomas Paine, in *The Age of Reason: being an investigation of true and fabulous theology* (1795), proclaimed himself a Deist, before systematically attacking every Christian dogma. Seiler replied in his *Das Zeitalter der Harmonie, der Vernunft und der biblischen Religion* . . . (1802).

89. G. F. Seiler, *Animadversiones ad psychologiam sacram* (Erlangen: Typis Cammererianis, 1778–1787), I, pp. 3–4.

90. Ibid., I, p. 4.

synonymous with *ru'ah*), and *neshamah* (the breath of life) a confirmation
of Plato's tripartite division of the soul. Instead, he examined the real usages of these terms and of others, such as *lev* (heart) and *kelayot* (kidneys).
On the basis of historical and linguistic considerations he concluded that
faith in the resurrection of the body existed among the Israelites before the
Babylonian exile and was consequently not derived from other traditions.[91]
In short, Roos and Seiler believed Old Testament psychology to be based
both on empirical observation (concerning the bodily seat of the emotions,
for example), and on revealed dogma (God breathed life into man, the body
will be resurrected, and so forth).

A less theologizing example can be found in "Ideas for a psychology of
the Bible," probably written by Immanuel David Mauchart. It took scripture
as a source of psychological observations and insights, with a view to establishing what the Bible teaches about the human soul.[92] Mauchart discussed
three possible methods for biblical psychology: one applied a preconceived
system to biblical psychological statements, another brought together and
systematized psychological material scattered throughout the Bible (along
the lines of what was done for a "sacred" natural history, physics, zoology,
and numismatics), and a third method approached scripture *mit psycholo-
gischem Forschungsgeiste*, that is, with a psychological spirit of inquiry,
in order to identify material suitable for inclusion in a "psychology of the
Bible."[93]

For Mauchart, Roos and Seiler applied the second method.[94] He preferred
the third, which consisted of identifying in the Bible statements and observations to be empirically tested or even used to extend and perfect the
discipline. Nonetheless, Mauchart explained, since the scriptures are not a
"psychological treatise," they cannot explain the empirical manifestations
of the soul.[95] For example, when Herod heard about Jesus's miracles, he
thought John the Baptist had risen from the dead (Matthew 14:1–2, Mark
6:14–16). But within the terms of his own faith, Herod could not have believed in the resurrection of the dead. The fact that he held this belief on
that occasion, Mauchart concluded, confirmed a fact already established by

91. Ibid., I, p. 26.
92. Immanuel David Mauchart, "Ideen zu einer Psychologie der Bibel," *Allgemeines Repertorium für empirische Psychologie und verwandte Wissenschaften*, 6, 1801, 3–41, p. 3.
93. Ibid., p. 5.
94. Ibid., p. 11.
95. Ibid., pp. 11 and 25–26.

empirical psychology, namely, that fear makes one superstitious and that the passions distort rational judgment.[96]

In Carus's view, the very idea that by proceeding in this fashion one could arrive at a biblical psychology was like believing one could reconstruct Homeric psychology by doing a character study of Achilles.[97] Carus's principal source naturally remained the Old Testament, but his approach to the past and other cultures substituted a biblical or sacred psychology for an anthropologizing and historicist hermeneutics. As he understood it, "biblical psychology" represented "a description of the Hebrews' science of man (*Menschenkunde*) and its various ramifications in different countries" (5). Such a psychology could take several forms. In one, the scriptures were read through a preconceived rational psychology and thus resulted in "a psychological dogma where one simply asks, What can the Bible teach us about the soul?" (6). A biblical psychology was thus obtained through or with the Bible (*durch die Bibel oder mit ihr*). A second approach, illustrated by Mauchart, did not investigate concepts alone but also gathered psychological facts and insights dispersed throughout the sacred text, thus giving rise to a psychology "by the Bible." For Carus, such a psychology should be accompanied by a "psychological critique" of biblical modes of observation (including style and language), as well as by a "historico-psychological interpretation" of the chosen passages (7–8).

The third method, which Carus advocated, proceeded from and according to the Bible (*aus und nach der Bibel*), and ultimately opened onto

the history of [ancient Hebrew] psychological culture [*psychologische Cultur*] and concepts—hence not a purely historical development of the concepts of the *soul* and its faculties but also the history of the spirit of observation [*Beobachtungsgeist*], of the states of the soul, and so forth. Hence above all (a) the psychology of both the common people (popular beliefs and ways of speaking) and the educated [*Gebildete*] class; (b) the psychology of the nation as a whole, as well as of individual authors, who can then be examined one by one. (8)

Underpinned by philology—"Der Psycholog hält sich überall an den ursprünglich[en], etymologischen Sinn der Worte" (8)—Carus's biblical psychology took the form of a "general history of the psychological development [*psychologische Bildung*] of the Hebrew and Jewish nation" (10),

96. Ibid., pp. 29–34.
97. Carus, *Psychologie der Hebräer*, p. 7, henceforth referred to by page number in the text.

which covered three different areas. The first was a history of the "psychological sense" and of degrees of self-consciousness (*Selbstbewußtseyn*), the second a history of psychological culture, and the third a history of the "science of the soul" (*Seelenkunde*), that is, of the psychological explanations and principles present in the Bible itself. These histories were to draw on the most diverse sources: traditions and customs; language and ways of defining and naming (*Bezeichnungsarte*), which express a nation's intellect; religion, which manifests its "degree of feeling"; the eventful history of the Israelites as a reflection of their environment; their political and religious leaders; and the Hebrew national character and temperament as a result of circumstances (11–12). Carus set himself the "historico-hermeneutic" task of assessing how insightful the scriptural writers were concerning "human nature." He noted that while they lagged far behind the *Seelenforscher* and *Seelenerklärer* "we today call psychologists," they were at least their equals as "observers of men" (*Menschenbeobachtern*, 23–24).

Carus divided the history of the Hebrews' "psychological culture" into periods going from Abraham to the Roman destruction of the Temple of Jerusalem. He proceeded principally by selecting and analyzing concepts, drew up bilingual "psychological lexicons" (in Hebrew and Latin) based on several Old Testament books, and devoted thirty-odd pages to Hellenistic psychological vocabulary. Such a lexical and grammatical approach was totally attuned to the rejection of allegorism and was consistent with the most advanced methods of biblical interpretation. Carus's general approach was historical and anthropological, and he adopted the vision which Herder, in particular, had of language: if there was no thought without language, and if the limits of thought were set by language, then it was vital to identify the usage and different senses of words. Carus also offered sociohistorical considerations on the evolution of the Israelite people and included chapters on the "science of man" in Jesus, the Apostles, and Philo of Alexandria, followed by appendixes on maladies of the soul and on supernatural explanations of psychological phenomena as they appear in the Old Testament.

Carus's treatment of Jesus is a good illustration of his approach. Instead of reviewing Jesus's statements on the soul or its faculties, Carus considered Jesus as a *Menschenkenner* and examined his insights into human nature. He thus highlighted Jesus's interest in human beings and his talent for observing himself and others, as attested, for example, in John 2:23–25: "Now when he was in Jerusalem at the passover, in the feast day, many believed in his name, when they saw the miracles which he did. But Jesus did not commit himself unto them, because he knew all men, and needed not that any should testify of man; for he knew what was in man."

While the *Psychology of the Hebrews* was partly a history of psychologi-cal concepts, it was also the backbone of Carus's social, cultural, religious, and psychological history of Israel, which was itself a chapter in the history of humanity. Psychology did not appear as a discipline, however, and it is significant that there is no "Hebrew psychology" in the *History of Psychol-ogy*. This is probably because of the predominant role Greece was thought to have played in the emergence of Western culture. However, Carus was not as radical as Friedrich August Wolf, who excluded from the *Altertum-swissenschaft* the Hebrews, along with the Egyptians, the Persians, and all ancient Mediterranean civilizations on the grounds that they exhibited "civilization" or *Policirung*, but had nothing akin to genuine (Greco-Roman) "culture." He claimed that the progress of human consciousness, which in his view was inseparable from the development of empirical psychology as a discipline, only got underway with the Greeks, in a world which dawned with Homeric poetry.

HOMERIC PSYCHOLOGY

The expression "Homeric psychology" referred to two complementary do-mains. The first was the "mentality," sense of self, and psychic life of the humans depicted in Homer; it was the basis for a psycho-anthropology of ancient Greece. The second was composed of Homeric ideas and beliefs on the soul's essence, properties, structure and functioning, and fate after death. These two interdependent domains were both grounded in an analy-sis of Homer's language, particularly of the words designating psychological faculties and processes. Modern-day research in this field, which began with Erwin Rohde's great work of 1894, *Psyche: The Cult of Souls and the Belief in Immortality among the Greeks*, is a direct descendant of late eighteenth-century inquiries like those of Carus.[98]

Rohde (1845–1898) was a philologist, and a friend of Nietzsche who de-fended the *Birth of Tragedy* against its critics. On the basis of linguistic analysis, as well as archaeological and ethnological findings, Rohde showed that the self in Homer was both a material body and the *psukhē* which de-parted from it to survive after death as a shade in human form. In the wake of Herbert Spencer (1820–1903), he noted that this idea persisted among so-called savage peoples, and he applied to the Homeric *psukhē* Spencer's theory that, for primitive peoples, the soul was a sort of animist double.

---

98. Erwin Rohde, *Psyche: The Cult of Souls and the Belief in Immortality among the Greeks* (1894), trans. W. B. Hillis (London: Routledge & Kegan Paul, 1925; reprint, Routledge, 2000).

The English philosopher explained that primitive man conceived of his soul as his double because he saw himself in dreams. Similarly, Rohde argued that Homer did not represent the soul as the site of psychic unity, because he considered psychological activity an emanation of the body's internal organs.

Although the *psukhē* in Homer was man's "second self," Rohde maintained that some expressions prefigured the dematerialization of the concept and its transformation into an abstract notion of life. His ideas gained wide acceptance. Wilhelm Wundt, for example, admired his extension of animism from primitives (*Naturvölker*) to civilized peoples (*Kulturvölker*).[99] Rohde inspired scholars who went beyond his focus on *psukhē* in their analyses of Homeric vocabulary.

By the 1920s it had became customary to compare Homer's representation of psychological phenomena with the beliefs and ideas of "primitives." The comparison invariably confirmed that Homer was incapable of abstract analysis and synthesis, lacked any notion of "soul" or "spirit," and confined himself to concrete description.[100] In the tradition of late-Enlightenment "histories of mankind," authors contrasted Homer's "primitive" psychology with the post-Cartesian and post-Lockean notion of a subjective and individualistic self, seen as the natural outcome of the progress of Western civilization.[101]

Such viewpoints culminated in *The Discovery of the Mind*, by Bruno Snell (1896–1998), a professor of classical philology at Hamburg. First published in 1946, and including essays written between 1929 and the mid-1940s, it has gone through several new editions and has often been translated and reprinted. For Snell, Homer's characters did not apprehend their actions as stemming from a soul, or their bodies as a unity. *Psukhē* kept the

99. Wilhelm Wundt, *Völkerpsychologie: Eine Untersuchung der Entwicklungsgesetze von Sprache, Mythus und Sitte*, vol. 4, pt. 1, "Mythus und Religion" (1905; Leipzig: Alfred Kröner Verlag, 1920), pp. 24–25. The concept of "animism" had been invented by the Englishman Edward B. Tylor in his *Primitive Culture: Researches into the Development of Mythology, Philosophy, Religion, Art and Custom* (1871) to characterize the first stage in the development of religion. Tylor argued that the experience of dreams, trances, and witnessing death led primitive peoples to imagine the existence of souls and spirits, and then to project these onto the natural world.

100. See, for example, Joachim Böhme, *Die Seele und das Ich im Homerischen Epos* (Leipzig: B. G. Teubner, 1929), who criticized Rohde's animist interpretation.

101. Christopher Gill argues that the Greek notion of what we would call a "person" is not a primitive form of the modern "subjective-individualistic" notion but the expression of an "objective-participative" personhood, in which the human being is defined more by shared behavior and beliefs within a community than by autonomy and a unified will and consciousness. See C. Gill, *Personality in Greek epic, tragedy, and philosophy: The self in dialogue* (Oxford: Clarendon Press, 1996).

human being alive, but it had in Homer a primarily eschatological value and lacked the psychological functions it gained in later philosophies. Moreover, it offered no organic link with other "psychological" terms: *thumos* and *noos*, for example, did not designate parts of a unified psychic entity, but rather autonomous functions or their organic seat.

Snell argued that an analogous fragmentation affected the Homeric representation of the body. For instance, he identified several verbs conveying different modalities and ways of seeing—but no term for vision. Similarly, the visual arts depicted the body as dislocated, as a collection of disjointed limbs.[102] Since Snell's *Discovery*, and the equally acclaimed books by Richard Onians and E. R. Dodds (respectively, *The Origins of European Thought* and *The Greeks and the Irrational*), many works have refined their analyses with philological, philosophical, psychological, or anthropological insights. Snell's intepretations have also been challenged, particularly regarding the absence in Homer of any notion or consciousness of the self.[103] They nonetheless remain the object of a wide consensus.[104]

The verdict, then, has been that Homer's epic poetry made no distinction between the soul and the body and embodied a "psychology without a *psukhē*," which was more concerned with behavior than with an inner world.[105] Herder anticipated such a judgment when he summarized the difference between Homer and Ossian as that between a purely objective and a purely subjective poetry.[106] In Homer, *thumos*—breath, "the organic air, hot, humid, in motion"—was the seat of inner life, including thought and sensation; it was the seat of subjectivity because we are particularly aware

102. Bruno Snell, *The discovery of the mind in Greek philosophy and literature* (2nd ed., 1948), trans. T. G. Rosenmeyer (New York: Harper, 1960), chap. 1.

103. See, for example, Shirley Darcus Sullivan, *Psychological activity in Homer: A study of phrēn* (Ottawa: Carleton University Press, 1988), chap. 1.

104. Richard Broxton Onians, *The origins of European thought about the body, the mind, the soul, the world, time and fate* (Cambridge: Cambridge University Press, 1951), chap. 1; Jan N. Bremmer, *The early Greek concept of the soul* (1983; Princeton: Princeton University Press, 1993).

105. Jacqueline de Romilly, *"Patience, mon cœur": L'essor de la psychologie dans la littéra-ture grecque classique* (Paris: Belles Lettres, 1984), pp. 44–45 (pp. 23–45 on Homer). The following works provide useful summaries and analyses: Hartmut Erbse, "Nachlese zur homerischen Psychologie," *Hermes*, 118, 1990, 1–17; E. L. Harrison, "Notes on Homeric psychology," *Phoenix*, 14, 1960, 63–80; James Redfield, "Le sentiment homérique du moi," *Le genre humain*, no. 12, 1985, 93–111; Shirley Darcus Sullivan, "A multi-faceted term: *Psyche* in Homer, the *Homeric Hymns*, and Hesiod," *SFIC*, 6, 1988, 151–180.

106. "Schon das unterscheidet Homer von Ossian ganz und gar, daß Jener, wenn ich so sagen darf, rein-objektiv, dieser rein-subjektif dichtete." Herder, "Homer und Ossian" (1795), in *Herders sämmtliche Werke*, 18, p. 453. Bardili attributed the psychological origins of the belief in immortality to Ossian's force of inner feeling.

of changes in breathing, and these changes markedly accompany subjective experience.[107]

Actions we would associate with thought or emotion were rendered by means of a soliloquy or physical reactions. Since Homer's characters conceived their psychological experience in bodily terms, "psychological" phenomena were depicted as, literally, embodied. As James Redfield notes, Homer's heroes, too, saw themselves as endowed with feelings, thoughts, passions, plans, hopes, fears and imagination. Simply, the space in which their subjectivity unfolded was corporeal: "The inner self [was] none other than the organic self." The Homeric hero addressed parts of his body because he was particularly aware of them. For instance, when Ulysses felt the urge to kill the servants who had slept with Penelope's suitors, he asked his heart to be patient (Od. XX.6–30).[108]

Processes that seem unequivocally internal, such as decision making, were externalized as divine interventions. An example of this is the scene mentioned above, in which Athena prevents Achilles from killing Agamemnon (Il. I.193–200). Heraclitus interpreted the episode as an allegory of the Platonic theory of the soul. Contemporary scholars, by contrast, focus on deriving experiential knowledge and theories from textual analysis. For them, the scene shows that decision making, as a psychological process, comes from the outside, which in turn confirms that "man in Homer was not yet understood as the source of his own actions."[109]

The "exteriority" of the Homeric self in Homer has been explained on the basis of the "oral theory" proposed by Milman Parry and Albert Lord.[110] Parry, Lord, and later scholars argue that the procedures used by bards from the Balkans and the south of the former Yugoslavia, namely, oral composition in the course of the "performance" itself, and with the help of traditional formulas, show how the Iliad and the Odyssey were created. Homeric psychology, in this interpretation, would correspond to the process of production of oral poetry, that is, to a composition-in-performance.[111] For example, just as the actions of Homer's heroes are often described as

---

107. Redfield, "Le sentiment," pp. 99 and 100.

108. Ibid., pp. 99, 100, 103, 108.

109. Romilly, Patience, p. 36.

110. Milman Parry, "The traditional epithet in Homer" (1928), in The making of Homeric verse: The collected papers of Milman Parry, ed. and trans. Adam Parry (Oxford: Clarendon Press, 1971); Albert Lord, The singer of tales (Cambridge, MA: Harvard University Press, 1960).

111. Bennett Simon and Herbert Weiner, "Models of the mind and mental illness in ancient Greece," pt. 1, "The Homeric model of mind," Journal of the history of the behavioral sciences, 2, 1966, 303–314; Joseph Russo and B. Simon, "Homeric psychology and the oral epic tradition," Journal of the history of ideas, 29, 1968, 483–498.

determined from the outside, so the *aoidos*, or singer of epics, declares in all sincerity that he merely reproduces an ancient source. Both the archaic psychological structure and the oral poetic process seem rooted in the structure of a society that values community and public recognition and privileges shame over guilt and honor over integrity.[112] The Homeric hero's lack of self-consciousness, as well as the absence of a model of the mind in Homeric epics, has also been attributed to the absence of writing, since writing (so the argument goes) leads to an awareness of words and makes it possible to distinguish between them and ideas.[113]

In contrast to allegorical readings, these sorts of explanation are based on a textual archaeology that places Homeric poems in their original contexts and in a comparative perspective. The Enlightenment historians of the psyche also combined philology, historico-critical exegesis, and cross-cultural comparison, but they examined primitive mentalities less in their own right than as an early stage in the progress of the mind. In their opinion, Homeric poetry reflected humanity's childhood, an early phase in mankind's march toward the intellectual and social ascendancy of reason and reflexivity.

It should therefore come as no surprise that Carus, following in the footsteps of the historians of civilization and philosophy, believed that the imagination dominated the first period in the history of psychology. His vision was no different from Bardili's, except that while Bardili did not even mention Homer, the Greek bard was Carus's main source. Homer's cardinal position in his history of psychology illustrates the Enlightenment reinterpretation and reevaluation of the poet. Johann Jacob Brucker, in his *Critical History of Philosophy*, had chosen the camp of Charles Perrault and the moderns and rejected the way in which allegorical interpretation celebrated the Greek poet's supposed wisdom. He refused to attribute to Homer a *philosophia perfectissima*, and to consider him a moral, theological, or scientific authority.[114] By the second half of the eighteenth century, the cultural history of humankind had overturned this position: the *Iliad* and the *Odyssey* were precious testimonies to humanity's history, the purest and most ancient expressions of poetry and language, and a means of gaining insight into the human mind in its infancy.

---

112. On this last point, see E. R. Dodds, *The Greeks and the irrational* (Boston: Beacon Press, 1951), chap. 1.

113. Olson, *The world on paper*, chap. 11.

114. Brucker, *Historia critica philosophiae*, pt. II ("De philosophia graecorum"), bk. I ("De infantia philosophiae graecae"), chap. I ("De philosophia graecorum fabulari"), §§ XXX–XXXIV.

Carus's views on these subjects, as his bibliographies make clear, owed much to the philological, mythological, and ethnological research published in Germany in the 1780s and 1790s.[115] However, Carus additionally made Homer and his universe part of the history of psychology itself, thus extending the limits of the discipline back to an epoch prior to Aristotle, Plato, and the Pre-Socratics. With Homer's inclusion, the history of psychology was still the history of a discipline, but it additionally championed the unity of psychology and its history, and their constitutive link to anthropology and cultural history. Such a transformation required believing that the psychic life of primitive man was wholly under the sway of sensation. Carus may have derived this conviction from a text by the very young Schelling on archaic myths, legends, and philosophical ideas—a text evocative of Heyne that perfectly instantiates how the childhood of humankind was conceived at the time:

> Nowhere is the entirely natural source of mythical representations easier to apprehend than in *psychological* myths. When sensory [*sinnlich*] man wants to represent thoughts, feelings, or sensations, he must, as it were, represent himself [*sich selbst darstellen*]. He has no names for the objects of the internal senses other than those derived from the objects of the external senses. For him, the analogy between the two is all the easier to make because in all humans, and especially in primitives [*Söhnen der Natur*], intense sensation is linked to bodily sensation. The latter [the "sons of nature"] never learned to restrain their sensations, they had not heard enough descriptions and analyses of sensation before these sensations arose in them, and so they were incapable of conducting psychological experiments [*keine psychologische Experimente*] on their sensations.[116]

Poetry, Carus explained, had two essential qualities which enabled it to reveal man's being (21–22). First (as Aristotle had stated in *Poetics* IX), it was more philosophical than history because it described events which could have happened; as such, its scope was infinite. Second—a thesis reminiscent of Herder—poets apprehended "total impressions." Instead of analyzing and observing, they intuitively grasped nature as a whole. Carus thought these

---

115. Carus, *Geschichte*, pp. 57–59 for works on the history of the notions of soul and mind, and pp. 128–129 for works on "Homeric psychology."

116. Friedrich Wilhelm Joseph Schelling, "Ueber Mythen, historische Sagen und Philosopheme der ältesten Welt" (1793), in *Werke* 1, ed. Wilhelm G. Jacobs et al. (Stuttgart: Frommann-Holzboog, 1976), vol. 1, pp. 54–55 of the original, pp. 235–236 of this edition.

features led to the emergence of the "analogy with nature" (the animal was assimilated to the human, the inanimate to the living), as well as to the development of an aspiration toward the "purely human." Early poetry was a prime source of psychological insights; primitive man was incapable of distancing himself sufficiently from his own sensations to describe and analyze them.

Each type of poetry showed a different aspect of man: elegy, idyll, and erotic poetry revealed his sensual desires, tragedy his noble side, and comedy his weaknesses. In the epic, the poet tended to be self-effacing and give center stage to other characters (24). Carus's observations on poetry were consistent with his idea of what the science of antiquity could contribute to the science of man. He thought that impressions and intuitions formulated in ancient times could be more illuminating than modern psychologies, that the *Menschenkunde* of antiquity contained more factual observations than opinions, and that it approached man as a totality endowed with few distinct faculties, but with a particularly "concentrated and energetic" force (35). That is why, said Carus, it was important to develop "not only a psychology based on the ancient philosophers . . . but also a code of the psychological wisdom of the ancients, in entirely historical form, or else a science of man derived from the most ancient masterpieces of humanity, particularly therefore the Old Testament and Homer" (36).

The concern to avoid anachronistic readings, which Carus shared with Heyne and Herder, went together with a faith in the greater psychological accessibility of the remotest times: "What we can barely glimpse from within our narrow bourgeois sphere is given to us by the wealth of intuitions and assertions contained in the psychological instincts of the primitive world [*Urwelt*]" (37).[117] Primitive man's inability to develop systems and theories even implied that one could have a privileged access to his psyche and the mentality of his era, which were conveyed most explicitly through language. We have already seen that, for Carus, the "poetic psychology of the ancients" or the psychology of "the most ancient childlike poetry" could not be systematized. And just as well, Carus added, because had it been, "it would have remained forever, as Herder correctly intuited in *The Spirit of Hebrew Poetry*, a labyrinth of lifeless and empty rules" (72).

---

117. See Redfield's judgment two centuries later: Homer's epic poetry "enables strata of experience to resurface which have been repressed by more modern modes of conceptualization. The epic poem restitutes a language which allows us to speak about ourselves in non-philosophical terms" ("Le sentiment," p. 109).

Carus argued that the mythology of primitive peoples enclosed a "pure science of man" (eine reine Menschenkunde), and that early poets created a first "history of man" which was more psychological than anthropological, mehr Seelenkunde als Menschenkunde (73). His enthusiasm for the poets' "descriptive" psychology was consistent with his emphasis on concrete psychological observation and case studies.

Although Carus was neither the first nor the only thinker to be interested in Homeric psychology, his arguments were definitely original. Christoph Meiners, for example, mentioned Homer in his "Brief History of the Opinions of Primitive Peoples on the Nature of the Human Soul." His approach was anthropological and comparative, but he was interested in ideas on the soul less for what they revealed about past forms of psychic life than as the interpretive key to such apparently incomprehensible features of ancient and modern primitive peoples as their funerary rites, celebrations, fear of deformed creatures, beliefs concerning dreams and the transmigration of souls, or trances and magicians' predictions.[118] Dietrich Tiedemann, in his Spirit of Speculative Philosophy, examined the conceptual contents of Homeric poetry but lamented the impoverished state of primitive Greek philosophy, which, he claimed, offered only scanty notions about the soul, the world, and the divine. "They did not reflect on the faculties of the soul," he castigated, "and barely noticed the most important things. Homer knew only the affects, the passions, the appetites, and memory."[119] At best, the archaic Greeks developed "the most childlike ideas" about particularly striking phenomena; like all primitive peoples, for example, they took dreams to be real sensations.[120]

In the eighteenth century, Homeric psychology was the object of academic theses and publications.[121] Carus, who was familiar with authors like Meiners and Tiedemann, nevertheless confined his list of works on the topic to very recent publications (1792 to 1801). The list included some Explanatory Notes on Homer, a Griechische Sprachlehre, where he found useful remarks on the original meaning of idiomatic expressions relating

---

118. Christoph Meiners, "Kurze Geschichte der Meynungen roher Völker über die Natur der menschlichen Seelen," Göttingisches Historisches Magazin, 4. Stück, 1788, 742–758.

119. Tiedemann, Geist der spekulativen Philosophie, vol. 1, p. 4.

120. Ibid.

121. For example, Friedrich Wilhelm Sturz's Prolusio prima de vestigiis doctrinae de animi humani immortalitate in Homeri carminibus (Gera: in officina Rothiana, 1795); this was followed by two others, dated 1796 and 1797. Sturz, who taught rhetoric at the Gera gymnasium, followed Porphyry and observed that man in Homer possessed two souls, a sensitive and corporeal, and a rational and immaterial.

to the soul, two "Greek mythologies," which he criticized harshly, and the *Homeric Psychology* of a certain Karl Wilhelm Halbkart. Halbkart has remained so little known that one cannot say he "founded" an anthropological and philological way of conceiving Homeric psychology. He was nonethless one of the first to adopt this approach systematically. The contrast between Carus's verdict on the book and the judgment of Thomas Jahn, the only modern author (discussed below) who seems to have examined Halbkart's work closely, throws light on the parameters of the late eighteenth-century psycho-anthropological field.

At the time he published *Homeric Psychology*, Karl Wilhelm Halbkart (1765–1830), a writer, poet, and translator of Xenophon's *Anabasis*, was deputy director (*Conrector*) of a Silesian gymnasium. A disputation on the work was held at the philological seminar in Halle. This was not surprising, since the seminar had been established by Friedrich August Wolf, under whom, like many other German teachers at the time, Halbkart had studied. (He pointed out that he had been unable to consult Wolf's *Prolegomena ad Homerum* before the *Psychologia homerica* went to press.)[122] The very idea of reconstructing Homeric psychology on the basis of a study of language corresponded to one of the principles of Wolf's *Altertumswissenschaft*, namely that ancient languages should be studied as "monuments," not only for literary and linguistic pleasure, but also because language, far from being only the envelope and outer garb of ideas, was intimately linked with them.[123]

In *Homeric Psychology*, the divisions of *psychologia* and its traditional subject matter were juxtaposed with the conceptual and methodological contents of philology, ethnology, and the "history of humanity." Halbkart examined the system of terms related to the soul, the ideas on the soul as united with and separate from the body, the state and location of souls after death, and, last, Homer's opinion on the souls of nonhuman beings such as animals and nymphs. His reading relied on the analysis of words and concepts. He showed, for instance, that in Homer the soul was often named by synecdoche—as befitted a "poetic age"—after the name of the organ which appeared to be its seat: *thumos*, *noos*, *psukhē*, *menos*, and others. This prefigured his remarks on the pronoun *autos*: while modern philosophers

---

122. Karl Wilhelm Halbkart, *Psychologia homerica, seu de homerica circa animam vel cognitione vel opinione commentatio* (Züllichau: sumtibus Friderici Frommanni, 1796), "Praefatio"; henceforth referred to in the text by chapter and paragraph number.

123. "Nicht als modische Gewänder und Hüllen von Ideen, sondern als mit den Ideen innig verwachsene Kunstgebilde, sind uns diese Sprachen selbst eine Art von Denkmälern . . ."; Wolf, *Darstellung*, p. 98.

translated *autos* as *ego*, and identified it with the mind, its Homeric meaning was different, because Homer did not separate mind and body (2.3). This
would explain, in Halbkart's view, the subtlety of material detail in Homer's descriptions, which he considered so true to life that the reader actually
experiences the narrated actions—for instance, the clash between Agamemnon and Achilles, or the encounter between Ulysses and Telemachus (2.4,
2.19). From the Homeric focus on actions, Halbkart inferred that verbs had
been invented before nouns (2.6).[124] As described in Homer, however, human actions were not freely chosen but subject to the will of the gods (2.9).

Halbkart's interest in the imagination did not simply result from its
place in the traditional schema of the mental faculties but reflected the significance he attributed to this faculty in the history of humanity. Since archaic mental life drew particularly on the imagination and memory, it was
easier and more natural for primitive man to invent external causes than to
look for internal ones. Hence the constant intervention of the gods and the
belief people had in an external origin of dreamed apparitions (2.10, 2.11).
Drawing on Meiners, Halbkart maintained that these beliefs were shared
by all primitive peoples; he referred to the native Greenlanders' belief that
the rational soul could migrate while the vital soul remained in the body
(his source was a 1765 *History of Greenland* by David Cranz, a mission-
ary and historian of the Moravian Brethren). In Homer, however, the soul
(*psukhē, animus*) left the body at the moment of death. According to Halbkart, Homer could not have entertained ideas of immortality and eternity
such as those familiar to an enlightened age of mathematics and theoretical
philosophy. Rather (2.21), like Cranz's Greenlanders who count with their
fingers and toes until twenty and say "countless" thereafter, Homer could
imagine only a very long duration (that is how Halbkart interpreted the immortality attributed to the gold-and-silver dogs Hephaestos crafted to watch
over Alkinoos's house; *Od.* VII.91).

Whereas Homer attributed to death such power that not even the gods
could prevent it from striking those they loved (*Od.* III.236), he did not
elaborate the notion of fate as the Greek philosophers would later do (3.6).
Homeric ideas on the state of the soul after death could be found among
Canadian and Patagonian Indians (5.3). The dead mentioned in *Odyssey* XI
were shades with bodily form; they had not only the outward appearance of
their living counterparts but also their nature, character, and even behavior

---

124. Romilly also noted not only Homer's preference for verbs, but also translators' tendency
to replace them with nouns that interiorize actions and psychologize Homeric descriptions (*Patience*, p. 27n6, and p. 32n11).

(for example, the giant Orion continued to hunt the wild beasts he had killed while alive); the peoples of Southern Asia, or Captain James Cook's Tahitians, were reputed to hold the same beliefs (5.6). When Halbkart came across contradictions in the first *nekya* (the evocation of the dead in *Od.* XI)—where a parched Tantalus, Tityos with his liver ripped apart by vultures, or a sweat-drenched Sisyphus with flexed muscles did not look much like disembodied shadows—he was heartened by Meiners, and reported that, "still today," some nations attribute to souls not only a physical location and appearance but even limbs of flesh and blood (5.8).

All in all, Halbkart arrived at conclusions that seem, for the most part, accepted still today: Homer's language is more descriptive than analytic and depicts observable actions rather than internal reflection, as illustrated by Homer's preference for verbs; the characters' sense of self is external and bodily; Homer focuses on behavior, and instead of positing a unitary and substantial soul, uses various terms associated with bodily organs; the absence of free will and the subjection to the gods and fate are among the fundamental traits of the Homeric subject; last—and here Halbkart seems outdated to us—Homeric psychology and the language that incarnates it stem from the predominance of sensation and imagination in the Greek world of Homer's time, similar in this to the mental universe of the American Indian.

Halbkart's work is almost entirely absent from the vast literature on Homeric psychology published since the end of the eighteenth century and does not even figure in university theses bearing the same title.[125] It received short shrift from Carus, a point I shall return to. Then, in an 1877 comparative history of the ideas about man's state after death, the Jena theologian Edmund Spiess found in Halbkart an explanation for the ancient hero's fear of death (he dreaded the miserable life of a shade to which the soul was condemned in Hades).[126] It was not until over a century later that Halbkart's work received a more detailed analysis, in Thomas Jahn's thesis on the lexical and semantic field of soul and spirit in Homer.[127] Jahn singled out Halb-

125. I have examined two, but there are doubtless others. Emile Louis Hamel defended a *Thesis philosophica de psychologia homerica* at the Sorbonne (Paris: August Delalain, 1832) in which I could detect traces of Halbkart (who is nevertheless not mentioned). Emil Gotschlich defended his at Breslau (*Psychologia homerica sive Historia notionum psychologicarum apud Homerum: Dissertatio inauguralis philologica . . .* [Breslau: A. Neumann, 1864]). Abbott mentioned Halbkart in the chapter "Destiny of the soul" of his "Literature of the doctrine of a future life" (no. 1530).

126. Edmund Spiess, *Entwicklungsgeschichte der Vorstellungen vom Zustande nach dem Tode auf Grund vergleichender Religionsforschung* (1877; Graz: Akademische Druck- u. Verlagsanstalt, 1975), p. 283.

127. Thomas Jahn, *Zum Wortfeld "Seele-Geist" in der Sprache Homers* (Munich: Beck, 1987), pp. 125–127.

kart's observation that *psukhē* referred to life and an "appearance of body" (*corporis simulacrum*). He also emphasized the way Halbkart differentiated the parts of the soul, focused on *psukhē*, *noos*, and *thumos*, and considered that terms such as *phren*, *menos*, *etor*, or *kardia* represent organs or seats of the soul which could sometimes be substituted for the latter by synecdoche. According to Jahn, Halbkart identified the most relevant semantic material in Homer, launched the idea of a "discovery of the mind" a century and a half before Snell, and reached conclusions that, for the most part, remained valid. Such conclusions are indeed remarkably consistent with those of later authors, from Erwin Rhode to Jacqueline de Romilly. The one major disagreement concerns Halbkart's thesis that the Homeric subject was incapable of free and autonomous mental action, which Jahn sees as resulting from a naive assimilation between Homeric and primitive man.

Yet Halbkart arrived at valid semantic results through an analysis which presupposed just such an assimilation, and which led to a conclusion considered false. His philology was part of a science of antiquity with close links to a "history of humanity" that aimed to gain insight into remote periods through comparison with modern "primitives." How these were regarded therefore affected how Homer was interpreted. This is reflected in Carus's reservations about Halbkart, as well as in his own approach to Homeric psychology (128–129). He criticized Halbkart for focusing more on rational than on empirical psychology and for concerning himself with the soul's state after death to the detriment of its properties in the living human being. Moreover, in Carus's opinion, Halbkart did not distinguish sufficiently between pre- and post-Homeric times, or between popular beliefs and poetry or myth; moreover, he should have spent more time on the terminology of the soul and human nature, and less on locating the underworld. Had he done so, Carus concluded, Halbkart would have achieved a more comprehensive survey and a more natural classification of Homeric psychological ideas. Clearly, then, Carus did not question Halbkart's anthropological premises or his comparative method but rather his very conception of psychology. Halbkart was not concerned with psychology as a scientific discipline, yet he called Homer's discourse on the soul *psychologia*. Carus, on the other hand, sought to restrict this term to the empirical psychology he elaborated in the framework of an *Erfahrungsseelenlehre* and a cultural and mental history of humanity.

The Greeks were absolutely central in this context. They had discovered the spirit of observation and the psychological sense. They were "the first to create a pure science of man and a systematic psychology. They are the bridge between Asia and Europe, and they explain the psychological culture of modern Europe (*die neueuropäische psychologische Cultur*), which still today depends on the Greeks" (95). Their significance had political and climatic causes. The Greek nation was composed of autonomous peoples grouped into independent states. Like the German nation, Carus remarked, Greece had achieved universality because it possessed a single language. Besides, unlike the Egyptians or the Asians, the Greeks did not live in a hot and fertile climate that would have weakened them (95). Carus thought that the first *Bildungsmomente* in the prehistory of psychology involved strengthening the soul's autonomy through physical exercise and ordeals. This paved the way for the emergence of heroism, in which the soul divorced itself from the body and its sufferings. When foreigners later settled in various regions of ancient Greece and introduced new customs, they did not destroy or restrict the Greeks' freedom, nor did they curb their intellectual activity.

Still later, Carus continued, the domination of despots and priests did not stifle the Greeks' imagination (*Phantasie*). The *aoidos* was the real source of their humanity. Rhapsodists tempered raw feeling and embellished language; they were the "humanizers" (*Vermenschlicher*) of images, replacing the "grotesque," monstrous and animal forms of Eastern gods with the "most beautiful and majestic human forms" (96). They also created the ideals found in mythology and philosophy and made them poetic (*poetisierend*). As they traveled from island to island, they came into contact with other customs and languages; their senses were thereby refined, and their imagination enlivened. Their capacity for observation was further sharpened by the heterogeneity of the Greek nation, where almost all levels of culture (*Stufen der Bildung*) coexisted. The Greeks' taste for accounts of their ancestors' deeds nurtured a "historical sense" that became constitutive of their "sense of self" (*Selbstgefühl*, 98).

Greek psychology (as both doctrine and mentality) was accessible preeminently through language. Word and idea were so closely linked that the evolution of the meaning of a psychological term revealed the "periods of psychological development" (98). Yet, Carus urged, we must abandon "our scientific words" in order to restore the real original usage—it would be *unhistorisch* to proceed otherwise (98). Carus first reviewed key terms such as *phusis, psukhē, pneuma, noos, thumos, phren, kardia, etor,* and *menos*. He explained that, since the Homeric poems were the first materials available in the history of psychology, his inquiry was based largely on their vocabulary, which incorporated pre-Homeric words. Before Homer, psychological terms presupposed at least the movement of breathing; spirit (*Geist*) could be felt in the chest, as a bodily phenomenon (124). It was only when the need to observe men made itself felt that, according to Carus, Homer could emerge.

This need was generated by political circumstances. At the time of the Pelagians—those "Greek savages"—men were driven by an unbridled *Natur-Trieb*, particularly by the natural instinct of self-preservation. Murder and mistreatment were the rule, even against the weak, slaves, and women. That began to change as Phoenician influences spread from Eastern Attica, for instance in the protection given to women by marriage. Even though human sacrifice under religious supervision was practiced right up to the heroic age, feelings of sympathy, compassion, affection, and intimacy developed. The passions were tamed and the imagination awakened, leading to enthusiasm (*Be-geisterung*) and the first expressions of poetry, "outpourings of a passionate heart" (126). A primitive dance combined with pantomime and accompanied by "childlike music" acquired sacred status. As shown by the appearance of Orpheus, religion began to transcend animal instinct. The process of humanization continued in Asia Minor, where the Ionians from Athens had settled. Since major differences existed among Greek populations, the need to observe men (*das Bedürfniß der Menschenbeobachtung*) made itself felt with increasing urgency, particularly for traveler-poets (128). This was the context in which Homer appeared.

Throughout his historical account, Carus remained less interested in Homeric psychology as a set of concepts and terms concerning the soul than in Homer's anthropology (*homerische Menschenkunde*) as a whole, including the bard's pathognomonic and psychological observations, and his views on animals, man, the gods and their relations to humans, and the differences between men.

Carus noted that, although immortal, the gods were otherwise similar to humans. They had finite physical bodies and a human-like sensibility. They lived like humans, were subject to time and space, to error and chance, to nourishment and sleep. They had human-like feelings, needs and inclinations. There were of course differences: gods possessed more agile bodies, were more carefree, and their intellective faculties (the understanding, the imagination, and memory) were more powerful. They saw and heard better than humans, and they knew more because in the course of their long lives they accumulated more experience. Moreover, they could deify the dead or the living.

As for humans, Carus continued, Homer described them as speaking beings with short lives, characterized mainly by dominant feelings and inclinations: they were either fearful, mistrustful, and jealous or indecisive, capricious, and inconsiderate. Differences stemmed from their physical traits but also depended on the gods, who endowed humans at birth or intervened in other ways in their lives. Carus picked out a series of remarks in Homer concerning maternal influence: Hector was but a mortal suckled by a woman, whereas Achilles was suckled by Hera herself (*Il.* XXIV.58); Zeus accused Ares of being as inflexible and headstrong as his mother Hera (*Il.* V.892); the Myrmidons asked the wrathful Achilles if he had suckled bile at his mother's breast (*Il.* XVI.203). Other differences resulted from the level of development, occupations, and ways of life specific to each nation; among the Greeks, differences resulted from occupations (warrior, hunter, pirate, doctor, priest, poet, etc.), sex, social status, and age.

Homeric comparisons between animals and human beings were based on analogy but, according to Carus, presupposed some affinity and even a real similarity between the two. This was not to disparage mankind because something godlike was still attributed to animals; for instance, Achilles's immortal horses, a gift from Zeus to Peleus, lamented the death of their rider Patroclus (*Il.* XVII.426). As regards Homer's pathognomonic observations, Carus deemed them colorful, but unconvincing. He nonetheless acknowledged that the poetic works of an age in which inner impulses were externalized could prove useful for studying "natural" gestures and facial expressions later repressed by "conventional culture" (148), and he found in the *Iliad* and the *Odyssey* several examples in connection with the passions, especially anger, courage, and fear.

The Homeric poems shed light on certain psychological phenomena. Carus thought that the faculties philosophy would later identify were already differentiated, though not as separate processes, only as impulses and reactions related in particular to desires. Homer described actions later thinkers would link to the understanding. For example, several verbs expressed experience or prudence and referred to related phenomena, sometimes in a very physical way. "The faculty of cognition [*Erkenntnißvermögen*] was insight [*Einsicht*], and here the intrinsic relation between the childlike psychology of the past and religion is displayed most clearly, since insight is what differentiates gods and men most sharply" (150). The concrete nature of Homeric psychology was also evident regarding memory. In the absence of writing, memory played an important role, but it was treated as an activity, not an abstract faculty.

The imagination was intensely active in Homer's epics and often deceived and surprised protagonists. It seemed above all to be a sensory or emotional phenomenon provoked by a god, as when when Apollo appeared to Diomedes, driving him back with words and preventing him from pursuing Aeneas (*Il.* V.431–444). Dreams took place entirely in the outside world.

Penelope's dream, when she fell asleep in anguish after the suitors had set
sail in pursuit of Telemachus, took the form of a female phantom sent by
Athena (*Od.* IV.795–838). Premonition was similarly externalized: fear and
pain heralded misfortune, the words of the dying were true, lightning and
the flight of birds before dawn prefigured happiness, and, in short, "nature
was a prophet" (152).

For Carus, Homer's anthropology was heterogeneous and open ended.
It was based on the idea that, though the gods were close to human beings,
humans should not aspire to become like them. Even the most successful
individuals appeared to be passive, like shepherds, or to be relatives or favor-
ites of the gods, like heroes, or divinely inspired, like poets and sages. The
gods dominated humans and caused their misfortunes. The "divine in man"
(154) was a function of the body, and this was true for the gods themselves.
Psychological phenomena were mostly externalized and embodied in physical movement. The only phenomena governed by unchanging laws were the
physical signs of fate and death, as well as certain needs and passions. In
Homer's time, Carus concluded, the human being in his true or pure sense
had not yet been discovered: "Den Menschen hatte man eigentlich in seinem reinem [*sic*] Sinne doch noch nicht gefunden" (153).

* * *

Given the importance Carus attributed to language in fashioning and ex-
pressing mental phenomena, it may come as a surprise that he paid so little attention to vocabulary in his treatment of Homeric psychology. This
makes sense, however, since his aim was to write a history of psychology
that would unite the history of a discipline with the history of the human
psyche in different climatic, geographic, social, and political conditions.
Such a joint history of psychology and mentalities was to contribute to a
Geschichte der Menschheit irreducible to a history of words.

Carus's narrative follows a clear course from mythical ideas about the
soul to the empirical psychology of his time, from the poetic tales and fables
of antiquity to Enlightenment anthropologies and conjectural histories. His
narrative of this progress was itself part of the progress it narrated. In the
process of humanization, the human race became aware of its own mental
life. Such self-consciousness was generated in part by historical knowledge
of forms of mental life (more or less governed by imagination or by reason,
centered more or less on behavior or interiority), and by ideas and theories
about them. The history of the psyche and the historiography of psychology
thus became inseparable. Psychology was positioned at the heart of a total

science of man which was both diachronic and synchronic, historical and comparative. Man's self-understanding could develop—Carus thought that in the age of Homer man had yet to be discovered—and this development involved the creation of psychology as an empirical discipline. That is why the "psychological" ideas and authors that later textbooks relegated to the "prehistory" of the discipline were an integral part of Carus's history.

While Carus's historiography does not seem to have been taken up subsequently, some later scholars attempted to introduce psychology into history. In France, Ignace Meyerson (1888–1983) elaborated a psychology based on the historical study of a wide range of human creations: tools and techniques, arts and sciences, religions, languages, and social institutions.[128] His "historical psychology" never took off within the university, but it was from the outset close to the ideas of Lucien Febvre (1878–1956), one of the founders of the *Annales* school, and to the questions which inspired the French "history of mentalities" and the "new history" (for example, the question of how to reconstruct the emotional or affective life of past epochs).[129] These historical endeavors led to an anthropology of the past focused on mental and material universes, representations, attitudes, values, experiences, feelings, and the cultural practices which embodied them. Jean-Pierre Vernant, a friend and disciple of Meyerson, insisted on the importance of Meyerson's work and related to it his own historical anthropology of ancient Greece.[130] The handful of comparisons one can make between these initiatives and Carus's work do not, however, imply any direct descendance. Moreover, the teleological nature of Carus's historiography is fundamentally incompatible with historical anthropology, which seeks to understand the mental life of

128. Ignace Meyerson, *Les fonctions psychologiques et les œuvres* (1948; Paris: Albin Michel, 1995), afterword by Riccardo di Donato; Meyerson, *Écrits 1920–1983: Pour une psychologie historique* (Paris: Presses Universitaires de France, 1987), especially pt. 2; Françoise Parot, ed., *Pour une psychologie historique: Écrits en hommage à Ignace Meyerson* (Paris: Presses Universitaires de France, 1996).

129. Lucien Febvre, "Sensibility and history: How to reconstitute the emotional life of the past" (1941), in *A new kind of history: From the writings of Lucien Febvre*, ed. Peter Burke, trans. Keith Folca (New York, 1973); also "History and psychology" (1938), in Burke, *A new kind of history*.

130. Among the many examples that could be cited, see J.-P. Vernant, "Sur les recherches de psychologie comparative historique" (1960) and "De la psychologie historique à une anthropologie de la Grèce ancienne" (1989), in Vernant, *Passé et présent: Contributions à une psychologie historique*, ed. R. di Donato (Rome: Edizioni di storia e letteratura, 1995), vol. 1; Vernant, "The society of the gods " (1966), in *Myth and society in ancient Greece*, trans. Janet Lloyd (1974; Brighton: Harvester Press, 1979); and Vernant, "Mortels et immortels: Le corps divin" (1986), in *L'individu, la mort, l'amour: Soi-même et l'autre en Grèce ancienne* (Paris: Gallimard, 1989).

the past for itself, without trying to situate it on a path inevitably leading to modernity.

The normative nature of Carus's historiography was evident in his view of historical stages, which would become so popular in the nineteenth century. We need only mention the "law of three stages" of Auguste Comte (1798–1857), formulated in the first lesson of the *Lectures on Positive Philosophy* (1830): human understanding goes through a theological or fictional stage during which it seeks the innermost nature of things, as well as first and final causes; a metaphysical stage, merely a modification of the previous one, in which supernatural agents are replaced by abstract forces; and a scientific or positive stage, in which, abandoning the quest for ultimate causes or the origin and destination of the universe, man seeks to discover the laws really governing phenomena (namely, unchanging relations of succession and resemblance).

Not even in Germany did Carus's work leave a mark. Instead, one can identify the persistence of a set of theories and methods ranging from philology, mythology, and Heyne's science of antiquity to Wilhelm Wundt's *Völkerpsychologie* via the ideas of Wilhelm von Humboldt (1767–1835) on the mental development of humanity and the way language encapsulates a vision of the world. Among later thinkers one should mention Johann Friedrich Herbart (1776–1841) on the necessary bond between empirical psychology and the history of humanity, and the works of Moritz Lazarus (1824–1903) and Hajim Steinthal (1823–1899), whose *Zeitschrift für Völkerpsychologie und Sprachwissenschaft*, founded in 1860, declared in its very title the importance of language as the medium of thought and as a window onto the psychic development of humankind.[131]

Over the years, Wundt attributed increasing importance to the "psychology of primitive peoples" and finally considered it as indispensable to scientific psychology as experimental psychology.[132] Starting in 1900, the ten volumes of his *Völkerpsychologie* compiled ethnographic material relating to language, myth, and, to a lesser extent, customs (*Sitten*). In Wundt's view, these were the three major fields on the basis of which the psychological

---

131. Gustav Jahoda, *Crossroads between culture and mind: Continuities and change in theories of human nature* (New York: Harvester-Wheatsheaf, 1992), chap. 9. The founding texts are reproduced along with a helpful introduction in Georg Eckardt, ed., *Völkerpsychologie: Versuch eine Neuentdeckung: Texte von Lazarus, Steinthal und Wundt* (Weinheim: Psychologie Verlags Union, 1997).

132. Jahoda, *Crossroads*, chap. 11.

development of the human race could be reconstructed.[133] His *Elements of Folk Psychology* of 1912 laid out four stages (*Stufen*) of the "psychological history of mankind's development": primitive man, the totemic age, the age of heroes and gods, and the evolution toward humanity (*Entwicklung zur Humanität*).[134]

For Wundt, the problem that gave rise to *Völkerpsychologie* emerged in the fields of mythology and philology, when it became clear that the subjects addressed by the *Geisteswissenschaften* were all rooted in a "social community" (*Volksgemeinschaft*).[135] *Völkerpsychologie* was therefore concerned with the creations of the mind insofar as they resulted from life in society. To the extent that they presupposed human interaction, these creations could not be explained by the properties of individual consciousness. Methodologically, just as introspection (*Selbstbeobachtung*) could shed no light on the complex operations of thought, so the study of individual consciousness "is wholly incapable of giving us a history of the development of human thought, for it is conditioned by an earlier history about which it cannot of itself give us any knowledge."[136] That was why child psychology, in Wundt's view, was insufficient for understanding psychogenesis; children growing up in the civilized framework of a *Kulturvolk* were subject to influences which could not be separated from what spontaneously emerged in their consciousness. Since *Völkerpsychologie* described the stages of mental development and depicted the "true psychogenesis" of mankind, it constituted developmental psychology (*Entwicklungspsychologie*) par excellence.[137] For Wundt, folk psychology (in the sense of *Völkerpsychologie*) was based on ethnographic material but could follow two paths: either consider different fields separately (language, myth, arts, customs), although in reality they were constantly interacting, or, as in his *Elements*, examine various fields together within each stage.[138]

---

133. Gustav Jahoda, "Une esquisse de la *Völkerpsychologie* de Wundt," in Michel Kait and Geneviève Vermès, eds., *La psychologie des peuples et ses dérivés* (Paris: Centre national de documentation pédagogique, 1999).

134. This last stage involved a development toward the dissolution of national and, above all, of confessional barriers, a stage in which, Wundt claimed, "we still find ourselves today." Wilhelm Wundt, *Elements of folk psychology: Outlines of a psychological history of the development of mankind* (1912), trans. Edward Leroy Schaub (London: George Allen & Unwin, 1916), p. 10.

135. Ibid., p. 2.

136. Ibid., p. 3.

137. Ibid., p. 4. The translation cited here uses the older term "genetic psychology."

138. Ibid., pp. 6–7.

The historicity of the psyche never became a standard assumption of academic psychology. That is why *Völkerpsychologie*, if we disregard its most teleological, normative, and Eurocentric aspects, could seem an avant-garde project to those who wanted to give psychology an essentially (cross-)cultural basis. *Völkerpsychologie* was perhaps a missed opportunity, especially as it emerged at a time when psychology was becoming institutionalized. A century earlier, the work of Friedrich August Carus embodied the network of knowledge areas within which empirical psychology emerged in Germany. We cannot strictly talk of interdisciplinarity here, since that would imply the existence of distinct disciplines. Rather, we can see in Carus's work the ideal of an anthropology that would unite the historical and the unchanging, as well as the individual and the collective, and that did in fact draw together the very diverse materials out of which the Enlightenment dreamed of building a "science of man."

Anthropology's Place in the Encyclopedias

The history of a field of knowledge not only addresses theoretical, methodological, and institutional developments but also examines how the concept of the field became established, and the representations of the field's position and relationships within the totality of knowledge. Schemas for the classification of the sciences are particularly important in this connection. They go beyond the questions proper to particular sciences and often belong in an encyclopedic project linked to the search for a language or a logic common to all forms of knowledge. Precisely because such classifications are shaped by stylized representations of existing situations and are guided by metascientific ideals, they contribute to situating as well as defining disciplines. As Francis Bacon noted, "The received divisions of the sciences are suitable only for the received totality of the sciences"; that is why "we find in the intellectual as in the terrestrial globe cultivated tracts and wilderness side by side."[1] Nomenclatures and classifications are precisely ways of colonizing and appropriating such wilderness.

In the case of psychology, nomenclature and classification were agents of change.[2] The way psychology and anthropology switched roles is significant in this respect. Roughly until the end of the seventeenth century,

---

1. Francis Bacon, *Novum organum* (1620), in *The Instauratio Magna*, pt. 2, *Novum organum and associated texts*, ed. Graham Rees with Maria Wakely (Oxford: Clarendon Press, 2004), p. 27.

2. Claude M. J. Braun and Jacinthe M. C. Baribeau, "The classification of psychology among the sciences from Francis Bacon to Boniface Kedrov," *Journal of mind and behavior*, 5, 1984, 245–260; F. Vidal, "La psychologie dans l'ordre des sciences," *Revue de synthèse*, 4th ser., no. 3–4, 1994, 327–353. On the history of the classification of the sciences, see this article, Yeo's introduction to his *Encyclopaedic visions*, and their bibliographies.

anthropology was a branch of psychology, because psychology was the generic science of living beings (plants, animals, and man). The fact that in
the eighteenth century psychology became a branch of anthropology was
the sign of a profound transformation. The soul was no longer defined, in
Aristotelian fashion, as the form of a potentially living natural body, or as
source of vegetative, sensitive, and intellectual functions, but had become
a rational substance (*mens*, or mind) joined to the body. The new relationship between "psychology" and "anthropology" became the backbone of
the reform of both disciplines.

Encyclopedias give us a privileged vantage point from which to observe
these changes.[3] First, different encyclopedias bring to light culturally distinct situations.[4] Second—and this will be our focus—encyclopedias are instrumental in establishing epistemic frontiers through their classificatory
tables, trees of knowledge, texts on the organization of the disciplines, definitions, and cross-references. The lexical variations in the designation of a
field or a problem, as well as the links and tensions between the different
techniques employed to organize the encyclopedic material, point to what
enabled or hindered the development of a science and also reflect hesita-
tions in how disciplines were conceived.[5] Encyclopedias, in their pursuit of
unifying knowledge, may give the impression of working against specialization, but they actually confirm and reinforce divisions and contribute
to the crystallization into disciplines of dispersed knowledge, which they
consolidate through an alphabetical ordering.[6] Like practices which belie
a theory, the specialized content of the articles, the links created by cross-
references, and the explicit attribution of subject matter to particular fields
have greater epistemic purchase than the opening discourses, charts, and
trees of knowledge supposed to explain and justify encyclopedic projects.[7]

3. The organization of academies and libraries can also provide extremely useful information. See, for example, Lorraine Daston, "Classifications of knowledge in the age of Louis XIV,"
in David Lee Rubin, ed., *Sun King: The ascendancy of French culture during the reign of Louis
XIV* (Washington: Folger Shakespeare Library , 1992).

4. For the situation in England, see Yeo, *Encyclopaedic visions.*

5. Three articles by Roselyne Rey remain exemplary for this type of analysis: "La pathologie
mentale dans *l'Encyclopédie*: Définitions et distribution nosologique," *Recherches sur Diderot
et "l'Encyclopédie,"* 7, 1989, 51–70; "Le cas des sciences de la vie," *Recherches sur Diderot et
"l'Encyclopédie,"* 12, 1992, 41–58; and "Naissance de la biologie et redistribution des savoirs,"
Revue de synthèse, 4th ser., no. 1–2, 1994, 167–197. See also James Llana, "Natural history and
the *Encyclopédie,*" *Journal of the history of biology,* 33, 2000, 1–25.

6. Richard Yeo, "Reading encyclopedias: Science and the organization of knowledge in British dictionaries of arts and sciences, 1730–1850," *Isis,* 82, 1991, 24–49.

7. I refer to the encyclopedias as follows: the headword of the article in small capitals, followed—for the Paris and Yverdon *Encyclopédies*—by the classificatory terms or "field indicators"

ENLIGHTENMENT ENCYCLOPEDIAS

English-language Enlightenment encyclopedias were basically Lockean. Locke was, for instance, the major reference for the *Cyclopaedia* (1728) brought out by Ephraim Chambers (1680–1740).[8] Chambers divided knowledge into "scientifical" and "artificial" (techniques and applied sciences), maintained that sensation is its sole source (1, xv), and declared his intention "to trace the Progress of the Mind thro' the Whole" (1, v) in the manner of Locke. For example, he explained that the best way to understand the word "force" was to refer to the sensation and the "simple idea" of force (1, xix–xx). The "scientifical" approach proceeded "from Ideas and Things, to Words; . . . . In this way, we come from Ignorance to Knowledge; from simple and common Ideas, to complex ones" (1, xxii). Chambers espoused the new logic of faculties, which had shed Scholastic jargon and was based on knowledge of mental operations (LOGIC, 1, 469). Like "our famous Mr. Locke, and most of our latest *English* Philosophers," he rejected innate ideas and subscribed to the experimental procedures of the Corpuscular Philosophers (UNDERSTANDING, 2, 323).

Chambers situated psychology within anthropology and defined it as "a Discourse concerning the Soul" (PSYCHOLOGY, 2, 906). But neither of the two disciplines figured in his classificatory diagram or in his exposition of the organization of knowledge, where he tended to assimilate psychology to pneumatology, or the metaphysical science of the mind. Chambers's diagram of arts and sciences (1, ii) included metaphysics and pneumatology, but no "psychology"; likewise in the article PHILOSOPHY, despite the fact that it mentions both pneumatology and anthropology. Pneumatology and psychology could be distinguished by their object of study—respectively, the mind as "thinking Being" (MIND, 2, 552) and the soul as "a Spirit inclosed in an organiz'd body" (SOUL, 2, 98–99). But this distinction was nowhere elaborated.

Chambers's Lockean discourse went hand in hand with a critique of Descartes: the essence of the soul was not thought, but the fact of being "a spiritual substance, proper to inform, or animate a human Body, and by its Union with this Body, to constitute a reasonable Animal or Man" (SOUL,

---

in italics and in parenthesis (spelled out in full where there is an abbreviation in the original). This is followed by volume and page number.

8. Ephraim Chambers, *Cyclopaedia: or an Universal Dictionary of Arts and Sciences*, 2 vols. (London: J. and J. Knapton, 1728). See Lael Ely Bradshaw, "Ephraim Chambers' Cyclopaedia," in Frank A. Kafker, ed., *Notable encyclopaedias of the seventeenth and eighteenth centuries: Nine predecessors of the "Encyclopédie"* (Oxford: Voltaire Foundation, 1981).

2, 98–99). The article THINKING declared that the identification of the soul with thought had been superseded by Locke, and the entry THOUGHT was largely devoted to the Cartesian philosophy which, thanks to Newton and Locke (Chambers said), had been banished from England. In the article IDEA, Locke is deemed to have demonstrated once and for all the origin of ideas in sensation. Several other psychological articles (for example, ASSOCIATION, ERROUR, JUDGMENT, KNOWLEDGE, REASON, SENSE, and TRUTH) reveal the same orientation. Some, such as IMAGINATION and SPIRIT, assigned an important role to the physiology of the mental faculties. Overall, the *Cyclopaedia* leaves no doubt that the soul must be studied in the framework of the mechanist, corpuscular, "experimental" and Lockean "new philosophy." It does not, however, consolidate the science of the soul united with the body into a discipline. The division of knowledge presented in the preface, in which the subject matter of each science is laid out, linked the senses to physics, the mind and its faculties to metaphysics, ideas to logic, and the passions, pleasure, pain, and conscience to morals.[9]

The Scottish printer and naturalist William Smellie published the first edition of his *Encyclopaedia Britannica* in 1768–1771 in Edinburgh.[10] Smellie took issue with Diderot and d'Alembert's *Encyclopédie* (1751–1765) for dismembering the sciences through alphabetical ordering, rather than treating them as integrated wholes. Thus, his *Encyclopaedia* had major articles on an entire discipline followed by shorter ones devoted to terms and concepts of the science in question. Logic, metaphysics, moral philosophy, and theology were each the object of major articles, and each expressed the ambition, characteristic of the Scottish Enlightenment, to establish a "science of man."

Metaphysics—the subdivision of philosophy which studied nature and the properties of "thinking beings" (3, 174)—was divided into ontology, cosmology, anthropology, psychology, pneumatology, and theodicy. Anthropology included not only anatomy and physiology but also the metaphysical study of man, his essence, essential qualities, and necessary attributes

---

9. The second eighteenth-century edition, to which the Presbyterian minister Abraham Rees added many "modern improvements," was important for the history of English encyclopedias. But, as far as I can tell, it contains no new elements regarding the psycho-anthropological fields. See *Cyclopaedia* . . . (London: printed for J. J. and C. Rivington . . . , 1786–1788); on this edition, see Stephen Werner, "Abraham Rees's eighteenth-century *Cyclopaedia*," in Frank A. Kafker, ed., *Notable encyclopedias of the late eighteenth century: Eleven successors of the "Encyclopédie"* (Oxford: Voltaire Foundation, 1994).

10. William Smellie, ed., *Encyclopaedia Britannica; or, a Dictionary of Arts and Sciences* (Edinburgh: printed for A. Bell and C. Macfarquhar, 1768–1771). See F. A. Kafker, "William Smellie's edition of the *Encyclopaedia Britannica*," in Kafker, *Notable encyclopedias: Successors*.

(3, 175). It thus led to psychology, or the science of the soul in general and of the human soul in particular (3, 175). As we have seen, Smellie believed that psychology, despite sophisticated and abstract investigations, had never arrived at rational and well-founded conclusions. Similarly, he character-ized pneumatology as insidious and chimerical, because nothing can be known about spirits. Theodicy stood on less shaky ground (3, 175), since the workings of "the Deity" in the universe are more discernible than the manifestations of spirits. Following his critique of metaphysics and, with it, of what was called "psychology," Smellie (3, 176–203) offered an alternative, based on books II ("Of Ideas") and IV ("Of Knowledge and Opinion") of Locke's *Essay Concerning Human Understanding*.

The study of the human soul, wrote Smellie, no more belongs to pneu-matology than analyzing it as "mind" belongs to psychology (2, 984–1003). In fact, logic constitutes the "science or history of the human mind," orga-nized as it is into chapters on perception, judgment, reasoning, and method and containing the entire "history" of sensation and mental operations.[11] The normative aspect of morals (3, 270–309), like that of logic, was hence-forth to be grounded in experience. Moral philosophy relied on observation and based its arguments on "plain uncontroverted experiments"; when in-quiring into the obligations or destination of the human being, the question was no longer how man might have been created, but how he really was. The fact that the article SOUL (3, 618–619) dealt only with the immaterial-ity and immortality of the soul shows to what extent the *Encyclopaedia Britannica* excluded from "psychology" the empirical investigation of the mental faculties.

Later editions altered some terminology slightly, but the overall view-point remained unchanged. The additions to the article Metaphysics in the third edition (1787–1797) stated that anthropology, psychology, and pneu-matology could not be separated, because the only "created mind" of which we can acquire knowledge worthy of the name of "science" is our own.[12]

---

11. Buickerood, "The natural history of the understanding: Locke and the rise of facultative logic in the eighteenth century," *History and philosophy of logic*, 6, 1985, p. 185 and nn. 125 and 126, has shown that the article LOGIC copies one of the most popular Lockean logics of the eighteenth century, *Elements of logic* (1748), by William Duncan.

12. "Anthroposophy [i.e., anthropology], Psychology, and Pneumatology, if they be not words expressive of disctinctions where there is not difference, seem to be at least very needlessly dis-joined from each other. . . . if they denote our knowledge of all minds except the supreme, they are words of the same import; for of no created minds except our own can we acquire such knowl-edge as deserves the name of science." William Smellie, ed., *Encyclopaedia Britannica; or, a Dictionary of Arts and Sciences*, 3rd ed. (Philadelphia, T. Dobson, 1798) (same as ed. published in Edinburgh, 1787–1797), 11, 482.

Pneumatology therefore became "the science of the intellectual phenomena consequent on the operations or affections of our thinking principle"
(15, 83). In this psychologized form, pneumatology could legitimately be
integrated into a metaphysics conceived, in Lockean fashion, as tracing
knowledge back to sensation.

In the German-speaking realm, the major encyclopedic reference of the
first half of the eighteenth century was the *Universal-Lexicon* of Johann
Henrich Zedler (1706–1751), published in Leipzig from 1731 to 1750.[13] After
volume 17, as a result of financial problems, Zedler, a publisher, transferred
editorial responsibility to the Wolffian philosopher Carl Günther Ludovici
(1707–1778). Ludovici was a professor at the University of Leipzig and librarian of the *Deutsche Gesellschaft*, a society whose goal was to promote
the German language and literature. The vast articles on Wolff and his
philosophy (58, 549–678, 883–1232) borrowed generously from Ludovici's
textbooks, and many other entries were based on Wolff. The article Seelen-
Lehre dealt only with Wolff's psychology and highlighted its originality.
The novelty of Wolff's ideas was mentioned in relation to what the *Psychologia empirica* said about the possibility of a mathematical knowledge of
the soul (Psychometria).

The entry Metaphysick explained the place of psychology in Wolff's
metaphysics as a whole. The treatment of philosophical and psychological
material was not, however, thoroughly Wolffian. The articles in the *Universal-
Lexicon* tended to present divergent opinions without taking sides. In this
they remained close to the eclecticism characteristic of the *historia philosophica* of the period. For instance, the soul was treated in two articles:
Seele examined the soul's existence, essence, origin, freedom, and immortality, as well as its relation to the body, and mentioned only one *Psychology*, that of Broughton; Anima explained the traditional division into the
vegetative, sensitive, and rational soul and did not contrast pneumatology
with Wolffian psychology (Pneumatick). Clearly Zedler's intention was
not to link subjects systematically in order to constitute a system of knowledge. The articles on the soul, for example, referred neither to psychology
nor to anthropology, even though "psychology" was defined as the science
of the soul, and "anthropology" said to investigate man's bodily and spiritual condition (cf. Anthropologia).[14]

    13. Zedler, *Grosses vollständiges Universal-Lexicon*. See Peter E. Carels and Dan Flory, "Jo-
hann Heinrich Zedler's *Universal Lexicon*," in Kafker, *Notable encyclopaedias: Predecessors*.
    14. There exists a second major eighteenth-century German encyclopedia, the *Deutsche En-
cyclopädie*, edited by Heinrich Martin Gottfried Köster, a professor of history at the University of
Giessen. I do not deal with it here principally because it stops at "Ky." Köster was openly critical

Last, mention should be made of the major eighteenth-century Italian encyclopedia, the *Nuovo dizionario*, published between 1746 and 1751 by the jurist and naturalist Gianfrancesco Pivati (1689–1764), an archivist at the University of Padua and censor of the Venetian Republic.[15] Pivati privileged natural history, the physical sciences, medicine, and what we would call ethnography. He was a fervent advocate of experimental philosophy and often referred to papers from European academies, filling his articles with accounts of *sperienze* and *osservazioni*. However, as a citizen of a republic in which the Inquisition had considerable power, he was very cautious on sensitive theological issues. His articles on magic, the blacks, and the Jews expressed only in oblique ways his criticism of the persecution of witches, of slavery, or of certain Christian positions. His treatment of astronomy and the Inquisition was extremely short, and the article on trade carefully avoided any polemic around the Catholic condemnation of moneylending. The preliminary "Notizie" (1, xlv) took care to explain that metaphysics, which was divided into ontology and pneumatology, dealt with things separated from matter by their nature (God, angels, the human soul) or by abstraction (general ideas). But the encyclopedia itself did not contain any articles on these disciplines; the absence of any direct discussion of psychology or psychological material was thus perhaps due to political caution.

The article Anima, for instance, was cleverly put together: it began with a nominal definition of the soul as the internal principle of animate bodies and went on to discuss classical doctrines, Cartesian philosophy, and

---

of contemporary attempts to classify the sciences. In the article Encyclopädie, he maintained that a general system like Diderot and d'Alembert's was arbitrary and impossible to carry out successfully. He consequently put cross-references but no field indicators in his articles. Overall, the *Deutsche Encyclopädie* was opposed to skepticism and atheism but remained eclectic philosophically and very up to date in the "arts and sciences." A close reading of the existing articles would indicate that this was also the case for the anthropological and psychological fields (which are unfortunately not analyzed in the very comprehensive study by Goetschel et al., cited below). In the article Anthropologie, for example, it is "moral" anthropology—in its "general" branch (humankind and its physical and cultural variations) as well as in its "particular" branch (individual psychology)—which is designated the foundation of moral philosophy. See H. M. G. Köster, ed., *Deutsche Encyclopädie oder Allgemeines Real-Wörterbuch aller Künste und Wissenschaften* (Frankfurt-am-Main: Varrentrapp und Wenner, 1778–1809). Also, the studies by Willi Goetschel, Catriona Macleod, and Emery Snyder, "The *Deutsche Encyclopädie*," in Kafker, *Notable encyclopedias: Successors;* and Goetschel, Macleod, and Snyder, "The *Deutsche Encyclopädie* and encyclopedism in eighteenth-century Germany," in Clorinda Donato and Robert M. Maniquis, eds., *The "Encyclopédie" and the age of revolution* (Boston: G. K. Hall, 1992).

15. Gianfrancesco Pivati, ed., *Nuovo dizionario scientifico e curioso, sacro-profano* (Venice: Benedetto Milocco, 1746–1751), 10 vols. See Silvano Garofalo, *L'enciclopedismo italiano: Gianfrancesco Pivati* (Ravenna: Longo, 1980); and Garofalo, "Gianfrancesco Pivati's *Nuovo dizionario*," in Kafker, *Notable encyclopaedias: Predecessors.*

debates on the seat of the soul, before ending with a list of condemned theological propositions. Uomo began by stating that God endowed man with an immortal soul and then went straight on to the *mirabilissima macchina* of the body. The "Notizie" explain that metaphysics deals with immaterial things and is divided into ontology and pneumatology. But there are no articles on these disciplines. Fisica occupied sixty two-columned folios; the geography, history, and anthropology of the Philippines took up nine pages; even the study of physiognomy ("metoposcopy") was given more space than philosophy. Filosofia did little more than refer to Fisica and to the "Notizie," which talk about celestial dynamics and terrestrial mechanics, and about how sterile philosophy had been before the observations of the *famose Letterati* Galileo and Torricelli (1, li).

Each encyclopedia expressed a different intellectual universe. In England and Scotland the term "psychology" did not imply the idea of a new science; the discipline called "psychology" was simply one of the areas to be reorganized, and existed only in the framework of the "experimental" critique of logic and metaphysics. The investigation into psychological material was therefore restricted to these two disciplines, from a wholly Lockean perspective. In Germany, on the other hand, the concept of psychology gained consistency through Christian Wolff's philosophy, and the term came to name the empirical science of the soul. This science was not, however, exclusively Wolffian, since the German encyclopedists remained eclectic and sought to present a number of different points of view, as well as integrate the history of ideas and of philosophical schools into philosophy itself. As for Pivati's *Nuovo dizionario*, the absence of any direct treatment of issues relating to the soul can be attributed to the political and cultural life of Venice.

In short, the English encyclopedists favored the development of a Lockean empirical science of the human mind, which would replace metaphysics and serve as a foundation for logic; *psychologia* belonged precisely to the mental universe they sought to leave behind. Zedler, by contrast, presented psychology positively but without theorizing it as a new field of knowledge. His *Universal-Lexicon* was encyclopedic in scope, but it did not seek to establish a general order of knowledge.

In what follows, I propose a comparative diachronic analysis of two systems of knowledge conceived of as such: Diderot and d'Alembert's *Encyclopédie*, published from 1751 onward, and the *Encyclopédie* called "d'Yverdon," published between 1770 and 1776.[16] This choice is dictated

16. EP and EY, respectively. I will use the terms "Diderot and d'Alembert's *Encyclopédie*," "the Paris *Encyclopédie*," or simply "the *Encyclopédie*"; and "the Yverdon *Encyclopédie*" or "the

not only by the preeminent place of the *Encyclopédie* in the Enlightenment and its historiography but also by the intrinsic link between the two works—the latter being a revision of the former—and the fact that the contexts of their production and circulation were so markedly different. Comparing the two should illuminate certain processes and conditions that shaped the conceptual field of anthropology, within which psychology was reframed in the second half of the eighteenth century.

THE SYNTAX OF THE *ENCYCLOPÉDIES*

Each encyclopedia opens with a "Système figuré," a diagram arranging and interconnecting the different sciences. In principle, the concepts treated in the articles belong to the disciplines specified in the "Système figuré." In principle only, because these pictorial representations, conceived at the start of each editorial enterprise, ended up bearing little relation to the organization of knowledge as it emerged from other elements of the text: most obviously the headwords of the articles, highlighted by the use of capitals, but also the cross-references to other articles and the names of fields, sciences or disciplines to which each subject was explicitly attached. These names, which were placed in parentheses and italics after the headword, acted as classificatory terms or "field indicators."[17] For example, "PSYCHOLOGIE (*Métaphysique*)" in the Paris *Encyclopédie*. Some field indicators are *expanded* to comprise a series of names of interlocking fields, in order to allow readers to work their way through the encyclopedic system. For example (from the *Encyclopédie* again): "AME (*Ordre Encyclopédique. Entendement. Raison. Philosophie ou Science des Esprits, de Dieu, des Anges, de l'Ame*)." The path indicated suggests that the study of the soul belongs in pneumatology. Yet it does not stop there, since pneumatology is placed inside the "science of man," which in turn is a branch of general metaphysics. A subject may also be attached to several disciplines through a *combined*

Swiss *Encyclopédie*." When no authors are mentioned, it is because they are unknown; names in square brackets derive from John Lough, "The problem of the unsigned articles in the *Encyclopédie*," *Studies on Voltaire and the eighteenth century*, 32, 1965, 327–390.

17. Alain Cernuschi uses the terms "désignants encyclopédiques"; see *Penser la musique dans "l'Encyclopédie": Étude sur les enjeux de la musicographie et sur ses liens avec l'encyclopédisme* (Paris: Champion, 2000), p. 36. I earlier used "attributions catégorielles" (F. Vidal, "Anthropologie et Psychologie dans les encyclopédies d'Yverdon et de Paris: Esquisse de comparaison," *Annales Benjamin Constant*, no. 18–19, 1996, 139–151). Here I refer to the subject treated in an article as "attributed" or "attached" to a science, or as "falling within" or "belonging to" the discipline or area designated by the "field indicator."

field indicator: "IMAGINATION, IMAGINER (*Logique, Métaphysique, Littéra-*
ture & Beaux-Arts)."

Other subjects, by contrast, are not attached to any discipline. The "Preliminary Discourse" to the *Encyclopédie* claims that if the name of a science does not appear next to the article heading, the article's content should
enable the relevant science to be identified.[18] In practice, this is not quite
the case. For instance, given the definitions of the imagination in the article
PSYCHOLOGIE, if IMAGINATION had no field indicators, the reader would be
justified in placing it within psychology. Moreover, whereas field indicators
are meant to correspond to disciplines mentioned in the "Système figuré,"
there is often a discrepancy between the two. That is why the encyclopedic
order that emerges from the field indicators differs from the much more
schematic one displayed in the "Système figuré."

The field indicators show how, in the Paris *Encyclopédie,* anthropological subjects are absorbed by metaphysics, around which are constellated
philosophy, morals, logic, psychology, physiology, and the natural sciences.
These categorizations do not correspond to the "Système figuré" but are
consistent with the way Diderot and d'Alembert describe metaphysics in
their articles and in the "Preliminary Discourse." In the Yverdon *Encyclopédie,* by contrast, anthropology appears in its own name, and is placed at
the very heart of the human sciences, while psychology takes over many of
the subjects that the French encyclopedists placed within metaphysics.

Comparing the articles themselves is also instructive. The Paris *Encyclopédie* attaches anthropology to "animal economy," explains that the term
designates a "treatise on man," and refers to anthropography, or the anatomical description of man. This was consistent with the uses of the term
in the sixteenth and seventeenth centuries. However, like psychology, anthropology is omitted from the "Système figuré," and no article, except for
FEMME, is attached to it. By contrast, anthropology is present in the "Système
figuré" of the Yverdon *Encyclopédie,* is defined as that "important branch of
philosophical science which acquaints us with man considered from every
aspect," and is linked to philosophy, natural history, physiology, metaphysics, and psychology. As regards the subject matter of psychology and anthropology, "soul," for example, figures as a concept only within philosophy in
the Paris *Encyclopédie;* in the Yverdon *Encyclopédie,* however, "soul" falls

18. EP, "Discours préliminaire," 1, p. xviii. We refer to the French original. For an English
translation of the "Preliminary Discourse," including the "Detailed Explanation of the System of
Human Knowledge," see Jean le Rond d'Alembert, *Preliminary Discourse to the Encyclopedia of
Diderot,* trans. Richard N. Schwab with the collaboration of Walter E. Rex (Chicago: University
of Chicago Press, 1995), available free of charge at http://quod.lib.umich.edu/d/did/.

within metaphysics, pneumatology, psychology, and anthropology and is thus explicitly privileged as a key concept in the science of man.

THE PARIS AND YVERDON *ENCYCLOPÉDIES*

The Paris *Encyclopédie*, edited by Denis Diderot and (until 1759) Jean d'Alembert, comprised seventeen volumes of text and eleven of plates. The first seven volumes were published annually between 1751 and 1757; the remaining ten in December 1765 (with the false place-name of Neuchâtel). The volumes of plates were published from 1761 to 1772, at the rate of one per year.[19] The initial project was to translate and enrich Chambers's *Cyclopaedia*. In 1745, a group of "associated booksellers" appointed abbé Jean-Paul Gua de Malves, a mathematician and member of the Academy of Sciences, as the chief editor. In 1747, he was replaced by two of his close collaborators, Denis Diderot (1713–1784) and Jean-le-Rond d'Alembert (1717–1783).[20] Diderot's *Philosophical Thoughts* had already been condemned by the Paris *parlement* for being "contrary to religion and morality." He was soon to publish *The Indiscreet Jewels* (1748) and his *Letter on the Blind* (1749). In 1751, the police described him as "a young man who thinks he is ever so clever and prides himself on his impiety; very dangerous."[21] D'Alembert (whose first name comes from the church square on which his aristocratic mother, Madame de Tencin, had abandoned him) had already gained a European-wide reputation for his work on mathematics and geometry. He had published treatises on dynamics, equilibrium, and the movement of

---

19. The four volumes of text of the *Supplément à l'Encyclopédie* were published in 1776 (followed by a volume of plates in 1777), edited by the naturalist Jean-Baptiste-René Robinet. The *Supplément* should be clearly distinguished from the *Encyclopédie* itself, since most of the authors were different and many of the articles were actually taken from the Yverdon *Encyclopédie*. For these reasons I will not deal here with it here. See Kathleen Hardesty, *The Supplément to the "Encyclopédie"* (La Haye: Martinus Nijhoff, 1977).

20. There are some excellent introductions to the *Encyclopédie*: Franco Venturi, *Le origine dell'enciclopedia*, 2nd ed. (Turin: Einaudi, 1963); John Lough, *The "Encyclopédie"* (London: Longman, 1971); and François Moureau, *Le roman vrai de "l'Encyclopédie"* (Paris: Gallimard, 1990); see also the classics by Jacques Proust, *Diderot et "l'Encyclopédie"* (1962; Paris: Albin Michel, 1995); and Robert Darnton, *The business of enlightenment: A publishing history of the "Encyclopédie,"* 1775–1800 (Cambridge, MA: Harvard University Press, 1979). On individual authors, see Frank A. Kafker and Serena L. Kafker, *The Encyclopedists as individuals: A biographical dictionary of the authors of the "Encyclopédie"* (Oxford: Voltaire Foundation, 1988); and John Lough, "The contributors to the *Encyclopédie*," in Richard N. Schwab and Walter Rex, *Inventory of Diderot's "Encyclopédie,"* VII (Oxford: Voltaire Foundation, 1984).

21. Cited in Moureau, *Le roman vrai*, p. 43.

fluids and become a member of the Paris Academy of Sciences in 1741 and of the Academy of Berlin in 1746.

The *Encyclopédie* was immediately attacked in the Jesuits' *Journal de Trévoux*. In January 1752, the Sorbonne condemned the thesis of abbé Jean-Martin de Prades (ca. 1720–1782), which it had previously accepted. The abbé, a friend of Diderot's and author of the *Encyclopédie* article CERTI-TUDE, was accused of endorsing sensualism and maintaining that revealed religion was simply a "more developed" natural religion and that Christ's miraculous healings were comparable to Asclepius's and differed from false miracles only in that they had been prophesized.[22] In February, the first two volumes of the *Encyclopédie* were banned and the abbé de Prades fled into exile in Holland. The verdict rendered by the King's Council stated that the editors "thought fit to insert in these two volumes several statements endeavoring to destroy royal authority, and to champion the grounds of error, of the corruption of morals, of irreligion and unbelief."[23] The *Encyclopédie* continued to be published under the protection of Chrétien Guillaume de Lamoignon de Malesherbes, who, as director of the "Librairie," was in charge of overseeing the press and the book trade. It continued publication even after royal permission was withdrawn in 1759, in the aftermath of the condemnation of Claude-Adrien Helvétius's *De l'esprit* (*Essays on the Mind and Its Several Faculties*). In the same year, Pope Clement XIII had the *Encyclopédie* put on the Index of Forbidden Books.

With the expulsion of the Jesuits from France in 1762, the fiercest critics of the *Encyclopédie* were silenced. But they seemed simply to hand over their role to their Jansenist enemies and to the *antiphilosophes* of the review *L'Année littéraire*. Before that, the *Encyclopédie* had already been the target of satire, pamphlets, critiques, and refutations, many of which were penned by clergymen. In 1758–1759, the Jansenist Abraham-Joseph de Chaumeix's *Legitimate Prejudices against the Encyclopédie* established an ideological continuity among Locke, Helvétius, and the encyclopedists, making Locke—whose sensualism seemed to challenge the immateriality of the soul—into the author of "impious principles" acceptable only if one renounced one's Christianity.[24] The impression that the *Encyclopédie* was serving the interests of some cunning plot against religion and the estab-

---

22. See John S. Spink, "Introduction," in Denis Diderot, *Suite de l'Apologie de l'abbé de Prades*, in Diderot, *Œuvres complètes* (Paris: Hermann, 1978), vol. 4.

23. Ibid., p. 129.

24. Sylviane Albertan-Coppola, "Les *Préjugés légitimes* de Chaumeix ou l'*Encyclopédie* sous la loupe d'un apologiste," *Recherches sur Diderot et "l'Encyclopédie,"* 20, 1996, 149–158; Albertan-Coppola, "De Locke à Helvétius en passant par l'*Encyclopédie* ou faut-il 'casser le XVIIIᵉ siècle'?,"

lished order was reinforced by Diderot's explanation of cross-references: what one wished to criticize should be "respectfully presented" under the headword, but one should then "turn the system on its head" by referring to articles which developed "contrary truths." Diderot concluded as follows: "If such references of confirmation and refutation are foreseen well in advance, and skillfully prepared, they will give an encyclopedia the character which a good dictionary ought to possess: that of changing the common way of thinking."[25] Pierre Bayle's *Historical and Critical Dictionary* of 1696 provided a magnificent model, but the *Encyclopédie*, in fact, seldom used this strategy—especially since a publicly declared ploy can hardly be expected to work effectively. Subversive cross-references of course exist, as when, while claiming to record pagan accusations against the first Christians, ANTHROPOPHAGES refers to EUCHARISTIE, COMMUNION, AUTEL. But these did not form a consistent system. Diderot's remarks, however, did encourage the opinion that there was a concerted strategy to undermine the foundations of church and state.[26]

The Yverdon *Encyclopédie*'s authors shared with the French intellectual elite a vision of the world which gave pride of place to reason and to the origin of knowledge in sensation. As Protestants, they criticized the Roman Church but were not directly involved in anticlerical polemic. They were fervent Christians who believed in the compatibility of reason and faith and sought to preserve Revelation by making a sharp distinction between philosophy and theology. They systematically removed the controversial and anti-Christian elements of the Paris *Encyclopédie*; they were encyclopedic, but not *encyclopédistes*.[27] These differences shaped the form and contents of the psycho-anthropological field.

in Ulla Kölving and Irène Passeron, eds., *Sciences, musiques, Lumières: Mélanges offerts à Anne-Marie Chouillet* (Ferney-Voltaire: Centre international d'études du XVIIIᵉ siècle, 2002).

25. EP, ENCYCLOPÉDIE (*Philosophie*), by Diderot, 5, 636–648, p. 642.

26. Hans-Wolfgang Schneiders, "Le prétendu système des renvois dans l'*Encyclopédie*," in Edgar Mass and Peter-Eckhard Knabe, eds., *"L'Encyclopédie" et Diderot* (Cologne: DME-Verlag, 1985).

27. See the excellent summary by K. Hardesty Doig, "The Yverdon *Encyclopédie*," in Kafker, *Notable encyclopedias: Successors*; and, on the authors, Clorinda Donato and K. Hardesty Doig, "Notice sur les auteurs des quarante-huit volumes de 'discours' de l'*Encyclopédie* d'Yverdon," *Recherches sur Diderot et "l'Encyclopédie,"* 11, 1991, 133–141. The revival of interest in the Swiss *Encyclopédie* began with C. Donato, "Inventory of the *Encyclopédie d'Yverdon*: A comparative study with Diderot's *Encyclopédie*" (doctoral thesis, University of California, Los Angeles, 1987). The inventory part of Donato's dissertation is available online (http://net.c18.net/ey/), and the *Encyclopédie* itself has been published on DVD-Rom for PC (Paris: Champion, 2003).

The Yverdon *Encyclopédie* was conceived and published by Fortunato Bartolomeo de Felice (1723–1789).[28] Born in Rome, ordained in 1746, appointed in 1753 professor of experimental physics at the University of Naples, he moved in "reforming" circles, translated Descartes, Maupertuis, and the "Preliminary Discourse" of the *Encyclopédie*, and actively disseminated Newton and Leibniz. But in 1756, he helped a countess flee the convent in which her husband had had her locked away, and after traveling through France and Switzerland, the fugitives returned to Italy, where they were arrested. The countess returned to her convent, but de Felice eventually managed to escape from his monastery in Tuscany. Having made it to Padua, he was protected by the doctor and anatomist Giovanni-Battista Morgagni until the politically influential naturalist Albrecht von Haller arranged for him to settle in Bern. De Felice moved there in 1757 and never set foot in Italy again. He converted to Protestantism in 1759, married, and was granted citizenship of Neuchâtel (he later also became a *bourgeois* of Yverdon, which was under the jurisdiction of Bern). In Bern, where de Felice received the support of enlightened patricians, he ran a literary café and founded two periodicals of book reviews: one, in Italian, devoted to scientific and cultural works from Great Britain, France, and Germany, and the other, in Latin, devoted to Swiss and Italian publications.

De Felice settled in Yverdon in 1762 to open a branch of the Bern Typographical Society. He not only took on the immense task of reworking the *Encyclopédie*, but he also founded and headed an educational institute, wrote popularizing works on education, religion, literary history, and natural law, and published authors such as Cesare Beccaria (*On Crimes and Punishments*), Charles Bonnet (*The Contemplation of Nature*), Jean-Jacques Burlamaqui (*The Principles of Natural and Politic Law*), Morgagni (*The Seats and Causes of Diseases*), François Quesnay (*Physiocracy*), and Pietro Verri (*Thoughts on Happiness*). He was thus typical of the Swiss Enlightenment, preferring compromise to political or religious polemic, involv-

28. For older but still useful works, see Eugène Maccabez, *F. B. de Félice, 1723–1789, et son Encyclopédie, Yverdon 1770–1780* (Basel: E. Birkhäuser, 1903); Jean-Pierre Perret, *Les imprimeries d'Yverdon au XVIIe et XVIIIe siècle* (Lausanne: Roth, 1945); and Charly Guyot, *Le rayonnement de "l'Encyclopédie" en Suisse française* (Neuchâtel: Attinger, 1955). See also C. Donato, "L'*Encyclopédie* d'Yverdon et l'*Encyclopédie* de Diderot et de d'Alembert: Éléments pour une comparaison," *Annales Benjamin Constant*, 14, 1993, 75–83; Donato, "Fortunato Bartolomeo de Felice e l'edizione d'Yverdon dell'*Encyclopédie*," *Studi settecenteschi*, 16, 1996, 373–396; Henri Cornaz, "Fortunato Bartolomeo de Felice and the *Encyclopédie d'Yverdon*" (with details on the production of the EY), in Donato and Maniquis, *The "Encyclopédie"*; and Christian de Felice, *"L'Encyclopédie" d'Yverdon: Une encyclopédie suisse au siècle des Lumières* (Yverdon: Fondation de Felice, 1999).

ing himself in new forms of enlightened sociability ("literary" cafés and "economic" societies), encouraging communication across linguistically distinct regions, and promoting the spread of ideas, works and knowledge on a European scale.[29]

In that context, religion was the "primary culturogenic factor," more so than the confessional fragmentation which helps explain the cultural diversity of French-speaking Enlightenment Switzerland.[30] The Yverdon *Encyclopédie* was no exception: one of its explicit goals was to recast the Paris *Encyclopédie* by removing its *philosophe* spirit, but also the traces of Catholic theology, and harmonizing it with a broadly liberal and rationalist Protestantism. De Felice attributed the imperfections and ideological tactics of the French *Encyclopédie* to religious and political circumstances:

> The religion we profess and the government under which we have the good fortune to live allow us to paint a picture of human knowledge freely, as it is, without consulting other laws than those dictated by our esteem for the truth, our love of the good, in a word, by the insights of an enlightened reason, without fear of oppression by a power that is as monstrous as it is frightful and whose ignorance, fanaticism, superstition, pride, avarice, and cruelty stifle the truth or halt its progress. (1, ix)

In de Felice's view, the *Encyclopédie*'s shortcomings "bear no relation to the erudition and skill of the authors" (1, ix). In rewriting it, he sought to remedy these by adding and removing articles, using the cross-references for strictly didactic and encyclopedic purposes, and avoiding irony and sarcasm.

The starting point for the comparison between the two encyclopedias is de Felice's own explanation. He claims to follow the Paris *Encyclopédie* "article by article," marking with an "R" rewritten articles and with an "N" the new ones (which therefore have no equivalent headword in the Paris work). Some articles of the *Encyclopédie* were omitted, and others reproduced exactly, but with additions signaled by an asterisk, and often signed. Some unmarked articles were in fact modified, and signatures were omitted when articles underwent "more or less substantial changes, additions, or cuts which the original authors would perhaps not wish to ratify and which we do not have the right to impute to them" (1, xiii). These changes were generally extensive. The huge article ENCYCLOPÉDIE is emblematic of

29. Taylor, "The Enlightenment in Switzerland."
30. Rosset, "La vie littéraire et intellectuelle en pays romand," p. 201.

this: it was neither rewritten nor new, but it omitted the passage quoted above, in which Diderot describes the subversive use of cross-references.[31] While the Yverdon *Encyclopédie* was stylistically more austere and politically more conservative, it could also be critical, and its anticlerical, anti-Catholic propaganda could even be virulent. But it was always explicit in its denunciations.

The other major objective of the Yverdon *Encyclopédie* was to extend the cultural horizons of the original. It was clearly successful in this. The Yverdon *Encyclopédie* was more cosmopolitan than its French counterpart and contained more information, not only in the fields of biography, history, and geography but also in the most varied "sciences and arts," from Europe and elsewhere.

THE "SYSTÈMES FIGURÉS"

The way knowledge is organized in the encyclopedias is epitomized in their "Systèmes figurés." These diagrams define the "encyclopedic order" of the disciplines whose subject matter will be treated in the articles. Together with the cross-references, which serve "principally to indicate the connections between subjects,"[32] they help reestablish some of the coherence broken by the alphabetical arrangement, which severs the conceptual and genealogical links between fields. The "Systèmes figurés" generate discourses on the encyclopedia's goals but end up barely corresponding to the conceptual realities of the text. D'Alembert in his "Preliminary Discourse" and Diderot in the article ENCYCLOPÉDIE used a geographic simile: the encyclopedic order is a map of the world; the subcategories and articles are individual maps; the cross-references, itineraries. Just as each map employed a particular projection, so the encyclopedists were aware of the arbitrary character of their system; Chambers, who also used the territory and cartography metaphors, stressed this point. The system therefore had more of a pragmatic than an epistemic function.[33] That is why, while the "Systèmes figurés" were impressive introductory monuments, they were utimately not suited to represent the structures of knowledge.

Diderot and d'Alembert's "Système figuré des connoissances humaines" was derived from Bacon. The three principal Baconian branches—history,

---

31. See Donato, "Inventory," 1, 21–32, for this omission and the issue of cross-referencing.
32. EP, "Discours préliminaire," 1, p. xviii.
33. On this point, see Jacques Proust, "Diderot et le système des connaissances humaines," *Studies on Voltaire and the eighteenth century*, 256, 1988, 117–127.

poetry, and philosophy—were said to be based on the three faculties of the understanding: memory, imagination, and reason. The choice of the knowing subject as an organizational principle for the totality of the knowable corresponded to the encyclopedists' desire to take the human being as a "common center":

> It is the presence of man that makes the existence of beings worthy of interest, and what nobler intention could we have in the history of those beings than to be governed by this principle? Why should we not introduce man into our work as he is positioned in the universe? . . . This is what has led us to seek the general division to which we have submitted our work in the principal faculties of man.[34]

Nevertheless, as the encyclopedists themselves emphasized, their system differed on several accounts from Bacon's. Bacon organized the faculties of the understanding into the sequence memory-imagination-reason. The encycloedists, by contrast, placed the imagination after reason, so as to follow "the metaphysical order of the operations of the mind."[35] Their system differed from Bacon's "particularly in the philosophical branch."[36] It enlarged the field of reason, made philosophy into the main trunk of the tree of knowledge, and, above all, did away with "divine learning" or revealed theology, which ran parallel, in Bacon, to the field of human knowledge.[37]

As regards the human being, Bacon, as we have seen, proposed a "general science of the nature and state of man," one part of which was devoted to his miseries and prerogatives, the other to the "alliance" between soul and body. This second part included certain psychological themes, but no "psychology." Religion dealt with the rational soul, while another science addressed the soul as corporeal substance. Logic and morals concerned the understanding, reason, imagination, memory, appetite, and will.[38] Bacon's thought illustrated the dispersion of psychological subject matter, as well

---

34. EP, ENCYCLOPÉDIE (Philosophie), by Diderot, 5, 635–648, p. 641.

35. EP, "Discours préliminaire," 1, p. xxv.

36. EP, "Observations sur la division des sciences du chancelier Bacon," 1, p. li.

37. Bacon's distribution of knowledge can be found in The proficience and advancement of learning divine and human (1605), bk. 2, trans. Michael Kiernan (Oxford: Oxford University Press, 2000). On the relation between Bacon's and the encyclopedists' classifications, see Robert Darnton, "Philosophers trim the tree of knowledge: The epistemological strategy of the Encyclopédie," in The great cat massacre, and other episodes in French cultural history (New York: Basic Books, 1984).

38. Francis Bacon, De dignitate et augmentis scientiarum (1623), bk. 4, chap. 1 on the science of man in general, chap. 3 on the science of the human soul.

as the confusion between an empirical psychology and the normative disciplines of logic and morals. The English empiricists fell into an analogous confusion when they maintained that a word only had meaning if it could be attached to a sensation.[39] Diderot and d'Alembert's *Encyclopédie* too was characterized by the scattering of psychological material, and the conflation of the descriptive and the normative.

The *Encyclopédie* includes two graphic representations of the order of knowledge: one is a tabular plan (in the first volume of text); the other, a tree (in the first volume of tables), entitled "Essai d'une distribution généalogique des Sciences et des Arts Principaux" (Outline of a geneaological distribution of the principal sciences and arts). The tree has the same structure as the "Système figuré." The understanding forms the base of a trunk from which the three great branches of memory, reason, and imagination grow. Cartouches nailed to the branches provide information drawn principally from the "Explication détaillée du Système des connoissances humaines" (Detailed explanation of the system of human knowledge). The resulting plate is enormous, almost 1 m × 60 cm when folded out: spectacular and utterly unreadable.

The *Encyclopédie* contains an article PSYCHOLOGIE (*Métaphysique*), and the article PHILOSOPHIE includes psychology as a branch of philosophy. We will discuss these below, and now only note that psychology is absent from the "Système figuré." Figure 7.1 reproduces part of the *Encyclopédie*'s "Système figuré" and shows two of the three branches of philosophy. Philosophy is divided into the science of nature, the science of God, and the science of man. The science of nature includes ontology, mathematics, physics, and chemistry. The science of man is divided into logic (which has the same structure as the *Port-Royal Logic*) and morals. According to the definition of psychology—that "part of philosophy which treats of the human soul, defines its essence, and explains its operations"[40]—several logical subjects are in fact "psychological." This is because the encyclopedists wanted to give logic an empirical and nonformal character, because they gave cognitive operations normative value. The "art of thinking correctly, or making proper use of our rational faculties," depends on knowledge of these operations; thus, "in order to think correctly, it is necessary to perceive well, judge well, reason well, and link one's ideas methodically; from this it follows that apprehension or perception, judgment, reasoning, and method are

---

39. Anthony Quinton, *Francis Bacon* (Oxford: Oxford University Press, 1980), pp. 49–50; chap. 6 on the classification of the sciences.

40. EP, PSYCHOLOGIE (*Métaphysique*), 13, p. 544.

SCIENCE DE DIEU.
- THÉOLOGIE NATURELLE. / THÉOLOGIE RÉVÉLÉE. } RELIGION, D'où par abus SUPERSTITIONS.
- SCIENCE DES ESPRITS BIEN ET MAL FAISANS. } DIVINATION. MAGIE NOIRE.

SCIENCE DE L'HOMME.

PNEUMATOLOGIE ou SCIENCE DE L'AME { RAISONNABLE, SENSITIVE.

LOGIQUE.

ART DE PENSER.
- APPREHEN-SION. } SCIENCE DES IDÉES.
- JUGEMENT.... SCIENCE DES PROPOSITIONS.
- RAISONNE-MENT. } INDUCTION.
- ET MÉTHODE... { DÉMONS-TRATION. { ANALYSE. SYNTHESE.

ART DE RETENIR.
- MÉMOIRE... { NATURELLE ARTIFI-CIELLE. { PRÉNOTION. EMBLESME.
- SUPPLÉMENT DE LA MÉMOIRE. { ECRITURE. IMPRIMERIE. { ALPHABETH. { CHIFFRES.. { ARTS D'ÉCRIRE, D'IMPRIMER, DE LIRE, DE DÉCHIFFRER.} ORTHOGRAPHE.

GESTE.... { PANTOMIME. DÉCLAMATION.

SIGNES... { CARACTERES { IDEAUX. HIEROGLYPHI-QUES. HERALDIQUES ou BLASON.

ART DE COMMUNI-QUER.
- SCIENCE DE L'INSTRU-MENT DU DISCOURS. } GRAMMAIRE { PROSODIE. CONSTRUC-TION. SYNTAXE. PHILOLOGIE. CRITIQUE.
- PEDAGOGI-QUE. { CHOIX DES ETUDES. MANIERE D'ENSEIGNER.
- SCIENCE DES QUALITÉS DU DISCOURS. { RHÉTORIQUE. MÉCHANIQUE DE LA POESIE ou VERSIFICATION.

MORALE.

GENERALE { SCIENCE DU BIEN ET DU MAL EN GENERAL. DES DEVOIRS EN GENERAL. DE LA VERTU. DE LA NÉCESSITÉ D'ESTRE VERTUEUX, &c.

PARTICU-LIERE.
- SCIENCE DES LOIX, ou JURISPRU-DENCE { NATURELLE. ŒCONOMIQUE. POLITIQUE. } COMMERCE INTERIEUR, EXTERIEUR, DE TERRE, DE MER.

Fig. 7.1. Paris *Encyclopédie*. The "science of God" and the "science of man" (which, along with the "science of nature," form the branches of philosophy), from the "Système figuré des connoissances humaines."

the four fundamental articles of this art. It is from our reflections on these four operations of the mind that *logic* is formed."[41] Once again, we have here a perfect echo of the Port-Royal *Logic*.

Method and discourse were ways of ordering thoughts and words.[42] Apprehension, judgment, memory, and reasoning were treated as faculties or

41. EP, LOGIQUE (*Philosophie*), 9, 637–641, p. 637.
42. EP, MÉTHODE (*Arts & Sciences*), by Louis de Jaucourt; DISCOURS (*Belles-Lettres*), by Edme-François Mallet.

operations of the soul. Apprehension was identified with perception, the "first operation of the understanding," and consisted of the impression made by the senses on the soul.[43] Judgment was a "capacity of the soul to judge the appropriateness or inappropriateness of ideas," and reasoning proceeded by comparing ideas and was "nothing but a sequence of judgments which depend upon each other."[44] The article Mémoire emphasized the importance of differentiating imagination, memory, and reminiscence: "The first awakens the perceptions themselves, the second recalls only their signs and circumstances, and the last enables us to recognize those we have already had."[45] The first of the articles on the imagination is largely about literary works and the fine arts, but its author, Voltaire, related its subject matter—"the power which each sensible creature experiences in himself to represent in his mind sensible things"—to logic and metaphysics.[46] All of these faculties belong to the understanding, which "is nothing other than our soul itself, insofar as it conceives or receives ideas."[47] The article devoted to the understanding referred to Evidence and Sensations, both classified in metaphysics.[48] Logic, or at least its "contingent" parts, brought together much of the subject matter which would have been attached to psychology had the encyclopedic system been perfectly consistent.[49]

The "Detailed Explanation of the System of Human Knowledge" gives a better idea of the science of man. This science is structured according to the human faculties of the understanding and the will but includes disciplines—logic and metaphysics—whose task is not to describe but to preside over these faculties.[50] The normative function of logic dictates that the intellect and the question of the origin of knowledge be treated, and as such, logic is largely coextensive with the science of the soul. But the position of this science within the "Système figuré" (see fig. 7.1) is am-

43. EP, Appréhension (Ordre encyclopédique. Entendement. Raison. Philosophie ou science. Science de l'homme. Art de penser. Appréhension), by Claude Yvon; EP, Perception (Métaphysique).

44. EP, Jugement (Métaphysique), by Jaucourt; EP, Raisonnement (Logique & Métaphysique).

45. EP, Mémoire (Métaphysique), 10, 326–328, p. 327. See also EP, Réminiscence (Métaphysique).

46. EP, Imagination, Imaginer (Logique, Métaphysique, Litterature & Beaux-Arts), by Voltaire, 8, 560–563, p. 560.

47. EP, Entendement (Logique), taken from a manuscript by Jean Henri Samuel Formey.

48. EP, Evidence (Métaphysique); EP, Sensations (Métaphysique).

49. The sciences "of created spirits and of bodies" are said to be "contingent" in the sense that they proceed by analogy; the knowledge they provide is probable, not necessary. EP, Induction (Logique & Grammaire), 8, 686–690, p. 687.

50. EP, 1, p. xlviii.

biguous. Is it an autonomous division of philosophy, positioned where it is because of constraints in the layout, or does it in fact belong to the science of God or the science of man? According to the "Detailed Explanation," it is not the science of God which is divided into theology, the science of spirits, and the science of the soul, but rather pneumatology, which is divided into the science of God (including natural and revealed theology), the doctrine of spirits (angels and demons), and the science of the sensitive and rational human soul.[51] The science of the soul thus belonged to pneumatology, and that was indeed its place in the philosophy curriculum. This organization of disciplines and knowledge differs from the Wolffian scheme set out in the articles PHILOSOPHIE and PSYCHOLOGIE. It nevertheless corresponds approximately to how field indicators attach psychological topics to metaphysics, logic, philosophy, and morals. It may seem surprising that the "Système figuré" does not reflect the "analytic" method and the reconfiguration of metaphysics advocated in the "Preliminary Discourse"; the latter, however, recognized that the encyclopedic order was only "a sort of itemization of the knowledge one can acquire" and, consequently, "very different from the genealogical order of the operations of the mind."[52]

The "Système figuré" in the Yverdon *Encyclopédie* opens with general comments on the difficulty of arriving at "a division of the sciences that is precise enough to omit none, distinct enough to avoid confusion, and systematic enough to assign each its proper place." It follows that the Baconian and Parisian diagrams have to be revised. Since every mental faculty contributes to produce all types of knowledge, the sciences cannot be classified according to their "instrument" but only according to their object—which is why Bacon's divisions have to be abandoned.[53] The Yverdon system is emblematic of how the work as a whole recasts its material:[54] the objects of the sciences are grouped into three principal classes, in which facts, "known through simple observation or experiment," constitute the general object of history; the "relations between these facts" define that of philosophy; and natural signs or artificial symbols, "employed either to express facts and

51. Ibid.

52. EP, I, p. xix.

53. EP, "Explication détaillée du Système figuré des connoissances humaines," I, liii-lx, p. liii, henceforth referred to in the text as ED, followed by the page number. The "Système" and its "Explication" are almost certainly by de Felice; at the very least, they reflect the positions of the articles signed by him.

54. Alain Cernuschi, "L'arbre encyclopédique des connaissances: Figures, opérations, métamorphoses," in Roland Schaer, ed., *Tous les savoirs du monde: Encyclopédies et bibliothèques, de Sumer au XXIe siècle* (Paris: Bibliothèque nationale de France, 1996), p. 381.

their relations, with a view to preserving, extending, and communicating knowledge of them, or to imitate them for what in them is beautiful and appealing to the senses and the mind," form the general object of the liberal arts, here termed "symbolic and imitative art."

In conformity with the "Detailed Explanation," the "Système figuré" structured the disciplines hierarchically, according to the degree of dependence on their objects. On the left (fig. 7.2), the "history of facts" is divided into natural and "moral" history. The former deals with the phenomena of material beings, both celestial and terrestrial, from the elements to the animals. The latter studies the actions of "moral" or "spiritual" beings, that is, man as a protagonist in the political and intellectual spheres. One of its branches includes "literary history," which concerns the sciences and the arts, and includes the history of philosophy (ED, lv). Natural history and "moral" history together provide the "materials" of philosophy. On the right, "symbolic and imitative" art deals with the natural or artificial signs by which facts and their relations, that is, knowledge derived from history and philosophy, are "expressed." Symbolic art "teaches the use of language" (grammar, rhetoric, critique, hermeneutics, and philology), while "imitative art" embraces the fine arts, "or different ways of imitating nature," as well as aesthetics, "or the theory of taste."

Philosophy, or the "science of relations," reigns supreme in the middle of the table. It concerns either the "speculative relations" involved in physical and metaphysical "truths" or "practical relations or rules" for the soul and the body. Speculative philosophy uses materials provided by history and furnishes the materials of practical philosophy. The latter "applies *speculative* philosophy to the needs of man and instructs him in the *rules* he must follow to perfect his being and better his condition" (ED, lvii). Practical philosophy comprises rules for seeking truth by means of the understanding (logic) and for guiding action toward the good (moral philosophy). Morals is divided into general morals, natural and positive law, and morals properly speaking (including natural and revealed morality, as well as doctrines bearing on the domestic, pedagogical, and political fields). The rules for the body concern its health (various branches of medicine), "its needs and sources of comfort" ("economy," industry, "tools, instruments [and] machines," trade), and its "vigor and beauty" (orthopedics, gymnastics, bodily care).

As for speculative philosophy, it includes physics and metaphysics. The former concerns the relations between material beings and is divided into "physics properly speaking" and mathematics. "Physics properly speaking" is either "experimental" (empirical), and therefore concerned with "discovering the laws of nature through simple observation or experiments," or

PHILOSOPHIE
SCIENCE DES RAPPORTS.

SCIENCE DES RAPPORTS SPÉCULATIFS OU DES VÉRITÉS.

SCIENCE DES RAPPORTS PRATIQUES OU DES RÈGLES.

ART SYMBOLIQUE & IMITATIF, ou Art des Signes & de l'Imitation.

HISTOIRE DES FAITS, qui se divise en

NATURELLE, ou connoissance des faits concernant les Êtres matériels; qui sont

UNIFORMES. ⎰ CÉLESTE.
⎱ TERRESTRES. ⎰ + Élémens. Météores. Eaux. Minéraux. Végétaux. Animaux.

NON-UNIFORMES, MONSTRUEUX.

ET MORALE, ou connoissance des faits concernant les Êtres spirituels; classifé

Sources, en ⎰ Monumens ou Annalités. Mémoires ou Annales. Histoires proprement dites.

Êtres, ou Sujets ⎰ Générale. Particulière. Individuelle ou Biographique.

Tems, en ⎰ Ancienne. Moyenne. Moderne.

Actions, en ⎰ Politique, qui se subdivise en Politique, qui comprend celle de la Philosophe. Littéraire qui comprend celle de la Philosophe. Ecclésiastique, qui se ... Religieuse. Mythologique. ... proprement dite.

Méthode, en ⎰ Pure. Mélangée. Universelle.

PHYSIQUE, ou Science des rapports des Êtres matériels, qui se divise en

GÉNÉRALE.

PARTICULIÈRE.

PHYSIQUE proprement dite, ou Science des rapports des qualités, causes & effet, qui sont entre les Êtres matériels tant uniformes que non-uniformes & monstrueux.

des Élémens, Chymie, Alchymie; des Météores ou Météorologie; du globe, ou Théorie de la Terre; de ses productions ou Physiologie: qui comprend la Météorologie, Physiologie, Anatomie simple, comparée, &c.

CÉLESTE.

TERRESTRE.

MATHÉMATIQUES, ou Science des rapports de quantité.

PURES.

ARITHMÉTIQUE. GÉOMÉTRIE, qui se divise en ÉLÉMENTAIRE, qui comprend la Géométrie proprement dite & la Trigonométrie dont la pratique donne les TRANSCENDANTALE. ALGÈBRE, qui se subdivise en Élémentaire, ou Analytique, en ...

Longimétrie. Altimétrie. Planimétrie. Cubicilie. Stéréométrie.

⎰ Différentielle. Intégrale.

MIXTES.

MÉCHANIQUE. Statique. Dynamique. Hydrostatique. Hydrodynamique. Hydraulique. Aérométrie. Aéronomie.

COSMOGRAPHIE. Uranographie. Hydrographie. Géographie. Gnomonique. Chronologie.

OPTIQUE, dite, Perspective, Dioptrique, Catoptrique.

Pyrotechnie. Acoustique.

OPTIQUE. Optique p. dite, Perspective, Dioptrique, Catoptrique.

ARCHITECTURE. Civile, Navale, Militaire; à laquelle se rapporte l'attaque & la Défense des places.

MÉTAPHYSIQUE, ou Science des rapports abstraits, & des rapports des Êtres spirituels.

ONTOLOGIE, Science des rapports abstraits des Êtres, ou l'on considère ⎰ Le possible. L'Être. L'Essence. Le nécessaire. Le contingent. L'unité. L'espace. La durée. L'infini.

PSEUMATOLOGIE & PSYCHOLOGIE, Science des rapports des Êtres spirituels, qui comprend l'Anthropologie.

Facultés. ⎰ Entendement. Volonté. Matériel. Penchans.

De sa liaison avec le corps & la nature immortelle. ⎰ Réguliers. Irréguliers.

THÉOLOGIE, ou Science de Dieu, Divisée quant

aux ⎰ sources, en ⎰ Naturelle & Révélée.

sujets, en ⎰ Dogmatique. Positive. Élenchtique. Pastorale. Homilétique. Ascétique. Paracléctique.

usages, en ⎰ Symbolique. Systématique. Pazarétique. Historique. Polémique. Exégétique. &c.

méthodes, en

Pour l'ÂME.

RATIONELLE ou Logique pour l'Entendement. ⎰ Penser. Juger. Raisonner. Discourir. Inventer. Conjecturer. Découvrir & rappeller plus facilement les idées.

Qui lui apprend à

MORALE pour la volonté, qui se divise

MORALE GÉNÉRALE.

des Particuliers, distingués en Universel ou Universelle & Social qui est ⎰ Courage, Paternel, Domestique.

NATUREL.

Politique, autrement appelée des Gens, qui se sousdivise en ⎰ Naturel & nécessaire, qui comprend le ⎰ Droit public universel.

DROIT.

⎰ Divin. Humain, qui est ou Ecclésiastique Politique ou Civil, qui comprend le Droit Civil proprement dit.

POSITIF.

⎰ Féodal, de Commerce, &c. Écrit, ou non-écrit, d'où l'on dérive le Droit civil univerfel.

MORALE, proprement dite.

⎰ Naturelle. Révélée.

⎰ Commun, Particulier; Pédagogique Domestique & Politique ou Morale des Souverains ⎰ la Législation. la Police, l'économie publique, les Alliances, Eréité; Munitions, &c. &c.

⎰ autrement la Science du Gouvernement, intérieur & extérieur, qui a pour objet

---

Pour le CORPS.

Qui préscrit des regles

Pour sa santé, ce qui donne la MÉDECINE, ou ⎰ Hygiene, Pathologie, Séméiotique Thérapeutique, ⎰ Diete, Pharmacie, Chirurgie.

RURALE, à laquelle se rapporte; ⎰ l'Agriculture, la Pêche, &c. la Vétérinaire.

DOMESTIQUE POLITIQUE.

Pour ses besoins & ce qui donne l'OECONOMIE.

L'INDUSTRIE des Arts Métiers, Fabriques, Manufactures. ⎰ les Bâtimens, Ameublemens Vêtemens d'Agrémens, de Luxe.

Et qui sous se ⎰ des Outils, Instrumens & Machines, auffi la Médecine, les Sciences & Arts libéraux.

Le COMMERCE.

⎰ Nécessaires employés pour

pour sa vigueur & sa beauté, ce qui donne ⎰ l'Ortopédie, la Gymnastique, la Cosmétique.

SIGNES.

NATURELS qui sont ⎰ les cris confus, les sons imitatifs les mouvemens ou le Geste, les images.

ARTIFICIELS ou SYMBOLES qui sont ⎰ Hiéroglyphiques, Héraldiques, à Numismatiques Numériques, ou ⎰ Alphabétiques, qui sont ⎰ Lettres, Syllabes, Mots.

D'où résulte le Langage: ⎰ de vive voix, par l'Écriture.

ART SYMBOLIQUE Qui apprend à employer le Langage.

⎰ Langue, Prononciation, Écriture, Orthographie, Paléographie, Stegnographie, Crytographie, Rhétorique, La Critique.

Ou simplement, ce qui comprend la connoiffance de la

Ou exactement, ce qui donne la

Où à se corriger l'emploi; ce qui donne Ou à se critiquer; ce qui comprend ⎰ l'Hermeneutique, la Philologie.

ART IMITATIF

1º. LES BEAUX ARTS, ou les différentes manières d'imiter la Nature.

1º. Par le Langage,

L'Éloquence qui emploie ⎰ La Déclamation, l'Harmonie, les Tropes, Lyrique, Épique, Dramatique, Paraphique, Pathétique.

La Poésie.

La Prosodie.

2º. Par le Sont, La Musique ⎰ Vocale, Instrumentale.

3º. Par le Décors, ⎰ la Peinture, tracé sur le papier coloré ou non coloré, ou exprimé en relief, ou exécuté sur le terrein, ce qui donne la Sculpture, l'Architecture, qu'an ... donne ... quant au dessin.

IIº. L'AISTHÉTIQUE, ou Théorie du goût.

Fig. 7.2. Yverdon *Encyclopédie.* "Système figuré des connoissances humaines."

"theoretical," and therefore aimed at explaining "the results of these laws" and tracing effects "back to their hidden causes, which cannot be perceived by our senses" (ED, lv).

Metaphysics (fig. 7.3), or the "science of abstract relations and of relations between spiritual beings," is divided into ontology, pneumatology and psychology, and theology. It is, as de Felice rhapsodically describes it in a long article of distinctly Wolffian flavor, "the most sublime, the most excellent, and the most necessary of all the sciences."[55] Ontology deals with abstract notions and universal truths applicable to all beings. It therefore serves "as a basis and point of anchorage for all the other branches of philosophy" (ED, lvii). Placed after ontology, once again by virtue of the order of "dependence" of the objects of knowledge, is a cluster of disciplines—pneumatology, psychology, and anthropology—to which the "Detailed Explanation" gives the generic term of "pneumatology":

> The history of man naturally led him to investigate *spiritual* beings, which are the object of *pneumatology.* These investigations are nothing but the generalized results of the observations man makes concerning his own *soul*, and which are the object of the science called *psychology. Pneumatology*, understood as applicable to every spirit capable of existing, provides a theory of the understanding and the will, and of the ordinary, extraordinary, habitual, and general inclinations of immaterial beings. When pneumatology investigates the spiritual being in its union with the body to form the single whole we call *man*, it is named *anthropology.* (ED, lvi)

Alongside these three sciences taken together (pneumatology, psychology, and anthropology), there is theology, and particularly natural theology as *teleology*, or the science that "seeks to reflect upon the ends of all creatures, their links and hierarchy, as well as their admirable conformity with the ultimate purpose which the Creator reserves for this universe" (ED, lvi).

The "Système figuré" of the Yverdon *Encyclopédie*, and particularly the position of metaphysics within it, roughly reproduces Wolffian hierarchies. Just as Wolff organized the sciences to form an ideal chain of disciplines where one was tributary of the other for the definition of its objects and the deduction of its principles, so the Yverdon encyclopedia stressed how the

---

55. EY, MÉTAPHYSIQUE (*Philosophie*), R by de Felice, 28, 489a–501a, p. 490a.

MÉTAPHYSIQUE
ou Science des rapports abſtraits, & des
rapports des Etres ſpirituels,

qui ſe diviſe

ONTOLOGIE,
Science *des rap-*
ports abſtraits
des Etres,

où l'on conſidere
- Le poſſible,
- L'Etre,
- L'Eſſence,
- Le néceſſaire,
- Le contingent,
- L'unité,
- L'eſpace,
- La durée,
- Le fini,
- L'infini.

PNEUMATOLOGIE
& PSYCHOLOGIE.
Science *des rapports*
des Etres ſpirituels,
qui comprend l'*An-*
thropologie.

Où l'on traite de l'ame
& de ſes
- Facultés.
  - *Entendement.*
  - *Volonté.*
  - *Activité.*
- Penchans.
  - *Réguliers.*
  - *Irréguliers.*
  - *Habituels.*
- De ſa liaiſon
  avec le corps
  & ſa nature im-
  mortelle.

en

THÉOLOGIE.
ou Science de Dieu,
Diviſée quant

aux

ſources, en
- Naturelle &
- Révélée.

ſujets, en
- Herméneutique.
- Dogmatique.

buts, en
- Poſitive.
- Elenchtique.

uſages,
pratiques, en
- Paſtorale.
- Homilétique.
- Paraclétique.
- Aſcétique.

méthodes, en
- Cathéchetique.
- Symbolique.
- Syſtématique.
- Patriſtique.
- Hiſtorique.
- Scientifique.
- Exégétique. &c.

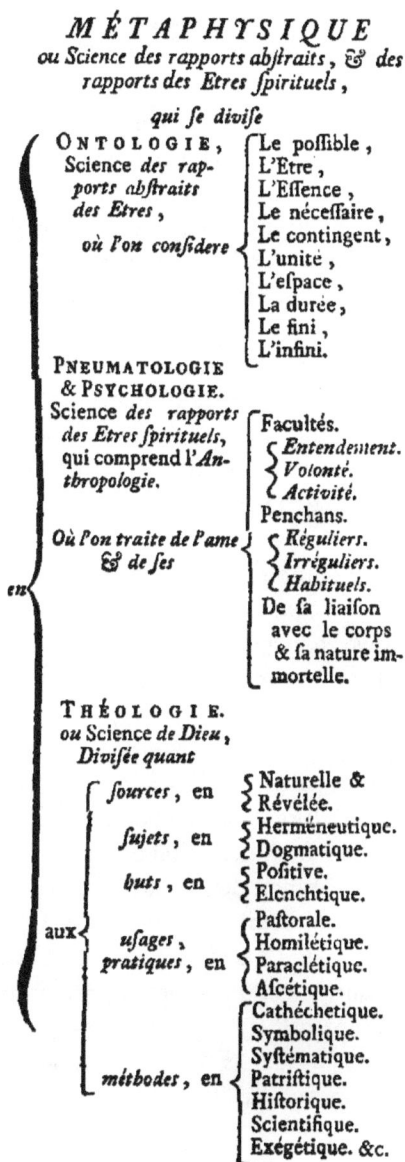

Fig. 7.3. Yverdon *Encyclopédie*. The different branches of metaphysics, from the "Système figuré des connoissances humaines."

objects of knowledge "depend" upon each other—history provides the material for philosophy, and philosophy, for "art." The Wolffian distinction between sciences of things and sciences of human action was reconfigured as the division between sciences of truths and sciences of rules (respectively, speculative and practical philosophy). For Wolff, the branch of philosophy dealing with things embraced ontology (on being in general), natural theology (on God), psychology (on the human soul), general cosmology (on the world in general), and physics (on bodies in particular), in that order. In the Yverdon *Encyclopédie*, speculative philosophy embraced physics ("physics properly speaking" and mathematics) and metaphysics.

Both the similarities and the divergences between the two schemes were particularly marked in the field of metaphysics. Whereas for Wolff metaphysics included ontology, pneumatology (itself embracing natural theology and psychology), and general cosmology, in the Swiss "Système figuré" it included ontology, pneumatology-psychology-anthropology as a unit, and theology. The separation between natural theology and psychology increased the latter's autonomy and linked it to anthropology. As we saw, according to the Yverdon "Detailed Explanation," pneumatology generalizes and extends psychology. Anthropology turns out to be the same science, but in its capacity to address the spiritual being in its union with the body to form the human being (ED, lvi).

The article ANTHROPOLOGIE in the Yverdon *Encyclopédie* shows clearly that the subject of anthropology is man as a composite of soul and body. This explains the juxtaposition of anthropology and psycho-pneumatology and the fact that the problem of the union of the two substances goes hand in hand with that of the immortality of the soul. The Swiss *Encyclopédie* emphasizes the "perfect conformity" on this point between Christian theology and "sound philosophy."[56] It is precisely the Yverdon *Encyclopédie*'s belief in such a conformity, and the desire to prove it, that distinguishes it from its Parisian counterpart.[57]

A comparison of the "Systèmes figurés" highlights this difference.[58] In Diderot and d'Alembert's version (fig. 7.1), the "science of God" is dan-

---

56. EY, AME (*Métaphysique, Pneumatologie, Psychologie, Antropologie*), R by Gabriel Mingard, 2, 338a–341a, p. 341a.

57. All critical works on the EY note this aspect of the work, as well as its coherent religious perspective; a detailed reading tends to confirm this: see, for example, C. Donato, "Rewriting heresy in the *Encyclopédie d'Yverdon*, 1770–1780," *Cromohs*, 7, 2002, 1–26, http://www.cromohs.unifi.it/7_2002/donato.html.

58. Alain Cernuschi, "La place du religieux dans le système des connaissances de l'*Encyclopédie* d'Yverdon," in Jean-Daniel Candaux, A. Cernuschi, C. Donato, and Jens Häseler,

gerously close to superstition, divination, and black magic, whereas this is not the case in de Felice's version (figs. 7.2 and 7.3). Moreover, in the Yverdon encyclopedia, philosophical discourses on God, spirits, the soul, and religion (pneumatology, psychology, and theology) are interrelated and clearly differentiated from the historical study of beliefs and institutions (religious, mythological, sacred, and ecclesiastical history). The objects of religious history (such as the separation of the churches, dogma, rites, forms of worship, orders, festivals, sects, and schisms) are "common to the false religions and the true one." Mythological history, by contrast, is "nothing but the particulars of pagan fables and at the same time of all the absurdities which were born from paganism, such as *superstition* in general and in particular, *judicial astrology, demonomania, magic, divination,* etc." (ED, lv). The separation of fields corresponds to the Protestant orientation of the Swiss *Encyclopédie,* while also reflecting the aim of instituting a psycho-anthropology that would be autonomous within metaphysics, yet fully integrated into a Christian framework.

In short, the differences between the "Systèmes figurés" are symptomatic of different goals, contexts, and positions. They also embody a paradox. The "modernization" of the nomenclature of the human sciences in the Swiss encyclopedia, in which the sciences of anthropology and psychology become central to the way in which knowledge about man is organized, went hand in hand not only with German cultural influence, but also with greater religious orthodoxy and political conservatism. Lexical innovation did not threaten a worldly order understood as linked to a transcendent one, but on the contrary reinforced the intellectual foundations of both. In the lexically more traditional Paris *Encyclopédie,* by contrast, the work's revolutionary thrust was all the more forceful for preserving older terms while inverting their sense.

ANTHROPOLOGY IN THE TEXT

The Paris *Encyclopédie* attached anthropology to "animal economy," defined it as a "treatise on man," and referred for more information to anthropography, or the anatomical description of man.[59] Animal economy

---

eds., *"L'Encyclopédie" d'Yverdon et sa résonance européenne: Contextes-contenus-continuités* (Geneva: Slatkine, 2005).

59. EP, ANTHROPOLOGIE ([*Œconomie animale*], ANTHROPOGRAPHIE [*Anatomie*]),I, p. 497. These two very short articles were written by the anatomist Pierre Tarin, one of the most prolific *Encyclopédie* contributors on anatomy and physiology. He wrote the very long article ANATOMIE, to which ANTHROPOGRAPHIE refers. For the text of ANTHROPOLOGIE, see appendix II.

was concerned with "the organization, the mechanisms, and all the functions and movements which maintain animals' life"; when these "cease entirely," death ensues.[60] The corresponding article gave an overview of these
functions in humans, citing the doctor Louis de Lacaze (1703–1765), and
principally the antimechanist theses of his *Idea of Physical and Moral Man*
(1755), which was to become crucial for the vitalist school of Montpellier.[61]
Anthropology was defined as the branch of animal economy which dealt
with humans, but it was used as a field indicator only for the main article
on women.[62] This was a long text by Paul-Joseph Barthez (1734–1806), a professor of medicine at Montpellier. The field indicator chosen for the entry
corresponds to the ideas Barthez would elaborate some twenty years later in
his *New Elements of the Science of Man* (1778). But the article itself lacks
coherence: Barthez addressed the difference between the masculine and
feminine reproductive organs and then discussed the supposed inferiority of
women, their education and capacities, and the image of woman conveyed
by various religious texts.

Anthropology, as defined in the *Encyclopédie*, would in fact best be illustrated by the first of the articles on man (HOMME), referred to in the very
first line of FEMME. The article is not attached to any particular branch of
knowledge, unlike the other articles with the same headword (which have
the field indicators "histoire naturelle," "anatomie," "materia medica,"
"morale," and "politique"). The absence of a classificatory term should not,
however, be taken as an omission; rather, it reflects the generic nature of
this first article, which refers to the others for the "different aspects" of the
human being and recognizes that the "different points of view which man
may adopt toward himself" could be multiplied to infinity.[63]

60. EP, ŒCONOMIE ANIMALE (*Médecine*), by Ménuret de Chambaud, 11, 360–366, p. 360.

61. Elizabeth Williams, *A cultural history of medical vitalism in Enlightenment Montpellier*
(Aldershot: Ashgate, 2003), pp. 151–154. Roselyne Rey mentions at several points Lacaze's importance for this school, as well as his relations with Ménuret, in her *Naissance et développement
du vitalisme en France de la deuxième moitié du 18e siècle à la fin du Premier Empire* (Oxford:
Voltaire Foundation, 2000).

62. EP, FEMME (*Anthropologie*), by Paul-Joseph Barthez, 6, pp. 468–471. The other articles
on women fall within natural law, morals, and jurisprudence and are followed by shorter entries
explaining expressions such as *femme authentiquée, coutumière,* and *divorcée,* as well as by longer ones such as FEMME MARIÉE and FEMME EN COUCHE. See the overview by Sara Ellen Procious
Malueg, "Women and the *Encyclopédie*," in Samia I. Spencer, ed., *French women and the age of
Enlightenment* (Bloomington: Indiana University Press, 1984). For a brief comparison of the two
encyclopedias on this topic, see Danielle Johnson-Cousin, "La 'construction' du féminin dans
l'*Encyclopédie* d'Yverdon et dans l'*Encyclopédie* de Paris," *Studies on Voltaire and the eighteenth century*, 304, 1992, 752–758.

63. EP, HOMME, 8, pp. 256–257, followed by HOMME (*Histoire naturelle*), both by Diderot.

Whereas HOMME suggested the possibility of a generic science of man, the "Science of man" in the *Encyclopédie*'s "Système figuré" divided the study of the human being clearly into a physical and a pneumatological part. This does not necessarily imply a contradiction between the "Système" and the articles, since tabular classifications of the sciences can only represent relations of hierarchy, belonging, or interdependence. However, despite the remarks in HOMME about the infinite number of possible "points of view" on man, and despite the spirit of the "Preliminary Discourse" in its entirety, the Paris *Encyclopédie* did not make anthropology into a central discipline in the organization of knowledge.

This is precisely what the Yverdon *Encyclopédie* did. As the tables below suggest, its superior coherence regarding psycho-anthropological nomenclature was, at least in part, due to the fact that most of the articles on psychology and anthropology were by the same author, Gabriel Mingard (1729–1786), a Calvinist minister from Lausanne, who used categories and cross-references consistently. Mingard was one of the main contributors to the Yverdon *Encyclopédie* on philosophy, anthropology, theology, and natural history.[64] He translated Pietro Verri's *Thoughts on Happiness*, was a founding member of the Lausanne Literary Society, and was a member of the Yverdon Economic Society. When addressing theological or philosophical questions, he usually appealed to reason rather than to institutional or textual authority. The articles unrelated to religion bear the signature "G.M."; others—those which, as de Felice put it, "could expose one to the absurd rantings of the bigots"—were signed "M.D.B." Charles Bonnet declared himself particularly satisfied with Mingard's articles on "rational philosophy." While the coherence of anthropology and psychology in the Yverdon *Encyclopédie* may be attributed to Mingard, it also corresponds to de Felice's wish to show more clearly than the Paris *Encyclopédie* the links between articles and between the different branches of each science.

The Yverdon *Encyclopédie* positioned anthropology at the very center of the sciences of the human being and moreover called for its further development.[65] In the article devoted to it, Mingard stressed its etymology: anthropology should not be identified with anatomy but with "that important branch of philosophical science which acquaints us with man considered in

---

64. See Étienne Hofmann, "Le pasteur Gabriel Mingard, collaborateur de l'*Encyclopédie* d'Yverdon: Matériaux pour l'étude de sa pensée," in Alain Clavien and Bertrand Müller, eds., *Le goût de l'histoire, des idées et des hommes: Mélanges J.-P. Aguet* (Lausanne: Éd. de l'Aire, 1996).

65. EY, ANTHROPOLOGIE (*Philosophie, Histoire Naturelle, Physiologie, Métaphysique, Psychologie*), N by Mingard, 3, pp. 22a–25b. Page number and column are given in the text. See the full text in appendix II.

all those aspects which may give us cause to reflect and become objects of our knowledge" (22a). The authors cited in the Paris *Encyclopédie* should have therefore called "anthropography" (rather than "anthropology") their works on the "animal economy" of man (23b). Mingard understood anthropology as "the science of human nature conceived only as regards those features that distinguish it from the science of brute beasts" (24a). These include mental faculties, moral and artistic capacity, language, and perfectibility. It is due to these features that the human being will forever remain "the Creator's masterpiece" (24a–25a).

Mingard developed his definition into a fully fledged program on the structure and tasks of the science of man "considered from every aspect":

> *Anthropology* would acquaint us with 1. The origin of man, 2. The different states through which he passes, 3. His qualities or affections, 4. His faculties or actions, in order to deduce from these 5. The knowledge of his nature, 6. Of his relations, 7. Of his destination, and 8. Of the rules he must observe in order to conform to the latter. *Anthropology* so described would be linked to all the sciences; it would borrow from them, or provide, scientific principles, and all its consequences would be applied for the benefit of man, that is, for his preservation, his perfection, and his happiness. (22a–b)

Mingard goes on:

> All the sciences serve to perfect *anthropology*, and the latter will not be perfect as long as the others have not reached their perfection, and they will not be useful unless we can apply them to the science of man as we have just described it. (22b)

Mingard attached anthropology to philosophy, natural history, physiology, metaphysics and psychology, which thereby became the primary or constitutive anthropological sciences. Although the field indicators of these sciences do not include anthropology, ANTHROPOLOGIE implies a two-way relation: anthropology belongs to all these sciences, and all these sciences together form anthropology (fig. 7.4).

The diagram in figure 7.4 summarizes these relationships. The two-headed arrows indicate mutual dependence and constitution; the remaining lines, other connections. Some disciplines are linked to anthropology through sciences which are constitutive of it, and to which these disciplines are explicitly attached. In such cases, the direction of the arrow indicates

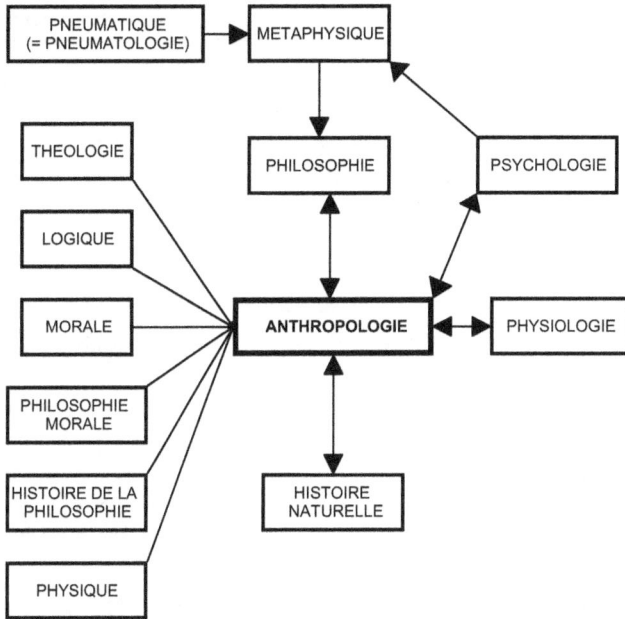

Fig. 7.4. Yverdon *Encyclopédie*. Relations between anthropology and other sciences.

that of the link: pneumatology, for instance, *belongs to* metaphysics. The lines without arrowheads connect anthropology to field indicators given in anthropological articles: PERFECTIBILITÉ, for example, is attached to metaphysics and to morals in addition to anthropology. (Table 7.1 lists the anthropological articles and fig. 7.5 shows how they are organized; they are discussed below.)

This is a highly schematic representation. If we took into account the disciplines to which the articles classed under psychology are attached, we would also have to link anthropology to "animal economy" and to medicine. A complete diagram would have to display not only all the primary anthropological disciplines but also the ones indicated through the cross-references—it would be entirely unmanageable. As it stands, the diagram depicts the most explicit relationships between anthropology and the other sciences and thus shows the position occupied by anthropology in the study of the human being.

Table 7.1 lists the anthropological articles in the Yverdon *Encyclopédie*. Three significant points emerge from the table. First, the preponderance of Gabriel Mingard, who emerges as responsible for the field's coherence.

TABLE 7.1. Articles in the Yverdon *Encyclopédie* that belong to anthropology

Headword	Other field indicators	Author (*)
ANTIPATHIE (R)	*Physiologie, Philosophie morale*	G.M.
APATHIE (R)	*Philosophie morale, Histoire de la philosophie*	G.M.
COMMERCE ÂME-CORPS (N)	*Psychologie*	G.M.
FEMME (R)		H.D.G.
HABITUDE (R)	*Morale*	G.M.
PERFECTIBILITÉ (N)	*Métaphysique, Morale*	G.M.
PHYSIQUE, MAL (N)	*Philosophie, Théologie, Physique, Philosophie morale*	G.M.
RÊVE (R)		G.M.
SYMPATHIE (R)	*Physiologie, Psychologie, Philosophie morale*	G.M.
TENTATION (R)	*Théologie, Morale*	M.D.B.

(*) G.M.: Gabriel Mingard, for the articles "unrelated to religion." M.D.B.: Gabriel Mingard, for the articles that "could expose one to the absurd rantings of the bigots." H.D.G.: Albrecht von Haller.

TABLE 7.2. Articles in the Yverdon *Encyclopédie* that belong to anthropology, and their position and authors in the Paris *Encyclopédie*

EY headword	EY other field indicators	EP field indicator	Author
ANTIPATHIE (R)	*Physiologie, Philosophie morale*	*Physiologie*	d'Alembert
APATHIE (R)	*Philosophie morale, Histoire de la philosophie*	[*Morale*]	Yvon
COMMERCE ÂME-CORPS (N)	*Psychologie*		
FEMME (R)		*Anthropologie*	Barthez
HABITUDE (R)	*Morale*	*Morale*	
PERFECTIBILITÉ (N)	*Métaphysique, Morale*		
PHYSIQUE, MAL (N)	*Philosophie, Théologie, Physique, Philosophie morale*	(1)	
RÊVE (R)		*Métaphysique*	
		Médecine	Malouin
SYMPATHIE (R)	*Physiologie, Psychologie, Philosophie morale*	*Physiologie*	Jaucourt
TENTATION (R)	*Théologie, Morale*	*Morale, Théologie*	

(1) EP contains an article MAL, LE (*Métaphysique*), by Jaucourt.

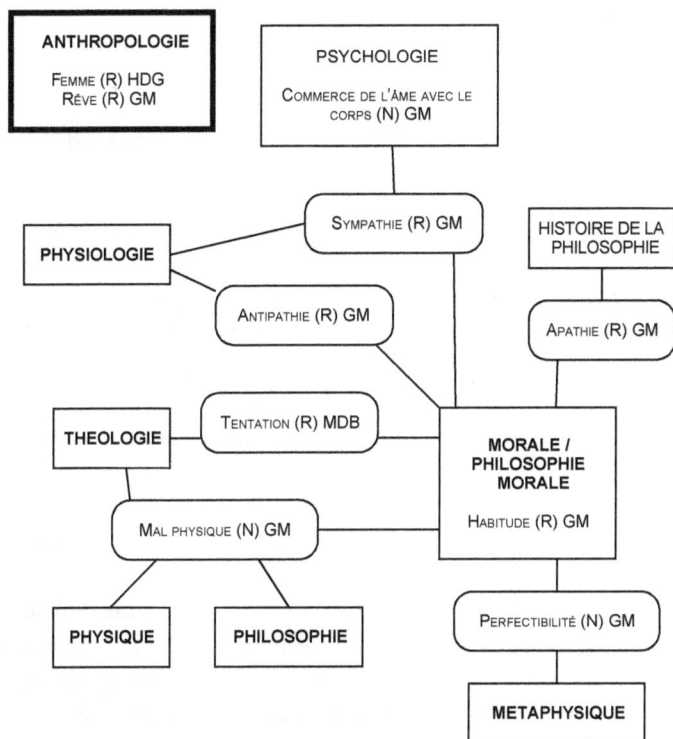

Fig. 7.5. Yverdon *Encyclopédie*. The anthropology articles and their links to other disciplines.

Second, this coherence itself, which results from the redistribution of the objects of study through the use of field indicators. Table 7.2 shows that the rewritten articles in the Yverdon *Encyclopédie* generally retain the Paris field indicators but add anthropology, thus integrating their topics into this discipline, and giving the discipline itself form and content by naming and itemizing its objects.

Last, we should note the extent of the overlap between anthropology and morals or moral philosophy—six articles out of nine. This is immediately visible in the diagram in figure 7.5, which schematizes the relationship by means of field indicators: *all* the articles are attached to anthropology; those which fall *only* within anthropology are placed in its box; the headwords in the boxes of other disciplines belong to that discipline *and* to anthropology; the articles attached to anthropology and to at least two other fields have

round-cornered boxes of their own and are linked to these other fields by
lines.

THE ANTHROPOLOGICAL TRANSFORMATION
OF MORALS

The new articles COMMERCE ÂME-CORPS, MAL PHYSIQUE, and PERFECTIBILITÉ
highlight the significance of the overlap between morals and anthropology.
I address the first below. The other two outline the place of man in a universe created by God, as well as his condition and possibilities within this
universe. The importance and nature of the issues addressed account for
the position of these articles at the intersection of several disciplines: MAL
PHYSIQUE at the juncture of philosophy, theology, anthropology, physics,
and moral philosophy, and PERFECTIBILITÉ at the juncture of metaphysics,
anthropology, and morals. This again demonstrates the Swiss *Encyclopédie*'s aim of giving the science of man a Christian framework and a moral
meaning.

In the article on physical ills, Mingard argued for a Leibnizian optimism,
a position widespread among Swiss Protestants and in the Wolffian circles
of German universities.[66] The illnesses which assail human beings, Mingard wrote, are often simply a consequence of excess; pain enables man to
become aware of his needs and seek to satisfy them; suffering motivates
him to use his talents and develop his capacities and ingenuity. In accordance with principles laid down in the articles OPTIMISME and MAL, Mingard stated that "everything has a destination; all the parts of the universe
are linked by relations," from which "the perfection of perfectible beings"
emerges (466b). Ills in general are part of this whole, with physical ills in
particular being but a "necessary consequence of the *physical* constitution
of the best of all worlds" (467a). In conclusion, there are "no such things as
physical ills that are truly ills" (467b). On the contrary, pain can be understood as one of the conditions for the realization of human perfectibility.

As for perfectibility ("a novel word in French," as Mingard noted), it
may involve developing new abilities or improving on already acquired capacities, but in either case it is a propensity shared by almost all beings in
the universe.[67] It is, however, an essentially human feature. Man's "fixed

66. EY, PHYSIQUE, MAL (*Philosophie, Théologie, Antropologie, Physique, Philosophie Morale*), N by Mingard, 33, pp. 466a–467b.

67. EY, PERFECTIBILITÉ (*Métaphysique, Antropologie, Morale*), N by Mingard, 33, 38a–41b,
pp. 39a–41b for the quotations in this paragraph. The term spread after Jean-Jacques Rousseau
used it in *Discourse on the origin of inequality* (1755). Insofar as perfectibility, man's "distinctive

and determined destination" requires him "to perfect himself in successive stages," from the fertilized seed to the acquisition of the "most sublime knowledge," in a movement which ceases only with the individual's decline and death. The very fact that man strives to surpass himself suggests that his destination, as intended by the Creator, "is to improve himself, and for this reason man has been made perfectible." Self-perfection is thus a "sacred obligation," and "perfectible man" must preserve his faculties, extend them, and use them to achieve his ultimate purpose. He should apply his faculties "to objects whose knowledge or enjoyment may more surely guide him toward his destination." Moreover, he must do so also for the sake of "bettering his fellow creatures," since the more perfect they become, the more they contribute to his own perfection and happiness. "In a word," Mingard concluded, "man must act in conformity with the end both of his entire person and of each of his faculties." This was a modest goal, since we should not strive to know the ultimate destination of things but content ourselves with never subjecting them to "ends contrary to their nature."[68]

Mingard clearly presupposed an order created by God, but he did not appeal to divine authority to justify his conclusions. The goal which the human being should pursue was not eschatological, but happiness and virtue on earth. The latter derive particularly from "attending to the soul"—which Mingard identified with "cultivating reason." As de Felice wrote, "man cannot hope for true happiness except by means of reason, and reason cannot lead him to this goal unless he attends to cultivating and perfecting his faculties."[69] Consequently, since suffering can advance human perfection and since perfectibility is "one of the essential particularities" of man, both are legitimate objects of anthropological reflection.

The rewritten articles confirm such a conception of anthropology, and show that Mingard's innovations are not purely lexical. For example, in the Paris *Encyclopédie*, apathy ("insensibility or privation of all passionate feeling or agitation of the mind") is a topic for the history of philosophy and religion; the article ends on a barb against Quietism, "apathy concealed behind a semblance of devoutness."[70] The Yverdon *Encyclopédie* also links

---

faculty," takes him out of his "original condition," Rousseau saw it as the root of human ills. Voltaire attacked this negative view in his *Essay on the customs and spirit of nations* (1756). He did not go so far as to say, as Condorcet later would, that perfectibility continues indefinitely, but he did not see it as a corrupting force, and argued that humans can improve themselves within limits defined by nature.

68. EY, TÉLÉOLOGIE, R by Mingard, 40, 310b–311a, p. 311a.

69. EY, AME, SOINS DE L' (*Droit Naturel*), N by de Felice, 8, pp. 341a–349a.

70. EP, APATHIE ([*Morale*]), by Yvon, 1, p. 522.

apathy to morals and the history of philosophy, but explores it further and brings out its relevance for understanding how man should live—which is indeed one of the aims of anthropology.[71]

The pair "antipathy-sympathy" undergoes a similar anthropological, and psychological reformulation. In Diderot and d'Alembert, antipathy was treated in a strictly "physical" framework, in which aversion between people was explored only so far as to say that its mechanism remains unknown.[72] De Felice, by contrast, carries out a more thorough analysis, and dwells on how antipathies are formed from ideas and habits acquired since child-hood.[73] Sympathy is the object of two articles in each encyclopedia. The first (which the Yverdon Encyclopédie takes over from the French work but to which it adds the field indicator Physique) briefly describes the capac-ity of certain bodies to be combined, stressing that one should not attri-bute this to a "metaphysical being or occult quality."[74] The second Yverdon article replaces the Paris SYMPATHIE (Physiologie) with a new article at-tached to anthropology, physiology, psychology and moral philosophy. In the Paris Encyclopédie, sympathy is first defined as that "profound meeting of minds" or "conformity of natural qualities, ideas, moods, and tempera-ments by which two well-matched souls seek, love, and become attached to each other, [and] become as one," but the article goes on to announce that it will not treat of this "felicitous bond," and focuses on the "anatomical physics" of the "mutual harmony which reigns between different parts of the human body by means of the nerves, which are marvelously arranged and distributed to this end."[75]

The Yverdon Encyclopédie, by contrast, analyzes the phenomenon itself. Like antipathy, sympathy can be explained by the structure of the sense organs, by habit, by the agreeable effect produced by the memory of an object, and, in the case of sympathy for living creatures, by the perception of "external signs of spiritual qualities." The key feature here is therefore the association of ideas involved in memory:

---

71. EY, APATHIE (Philosophie Morale, Anthropologie, Histoire de la Philosophie), R by Min-gard, 3, pp. 88a–91b.

72. EP, ANTIPATHIE (Physique), by d'Alembert, 1, pp. 510–511.

73. EY, ANTIPATHIE (Antropologie, Physiologie, Philosophie morale), R by Mingard, 3, pp. 52b–54b.

74. EP, SYMPATHIE, by d'Alembert, 15, pp. 735–736; EY, SYMPATHIE (Physique), 39, pp. 636a–636b.

75. EP, SYMPATHIE (Physiologie), by Jaucourt, 15, pp. 736–740.

If we lay eyes on a person we have never seen before, but who has certain features in common with someone we have loved, . . . we will feel goodwill and interest stirring in our soul, whose cause we probe in vain. . . . We can no longer give the name of *sympathy* to the vague liking, the indeterminate inclination of one sex for the other, no more than to the feeling that makes us desire refreshing drinks in the heat of summer.[76]

In this exploration of the psychological phenomenon of sympathy, physiological reductionism and moral considerations (also present in the articles on habit) are replaced by an associationist analysis.[77]

Mingard attached the article Rêve only to anthropology—perhaps by oversight, given the strongly psychological character of the subject. In the Paris *Encyclopédie*, Rêve (dream) is defined in a very short article as "a dream [*songe*] one has while sleeping." A better knowledge of the subject is needed "not only in medicine but also in metaphysics, due to the objections of the idealists"; in any case, "all the objects of dreams are clearly tricks of the imagination."[78] A second article, medical, explains how dreams serve in diagnosis and prognosis: stormy seas, for example, "foretell stomach upsets." The article reflects the most classical humoral theory,[79] and indeed, it was taken from the Dutchman Jodocus Lommius's *Medicinalium observationum*, which was published in 1560 but went through new editions and translations right up to the second half of the eighteenth century. Mingard, by contrast, discusses dreams in a psychological framework, as products of "memory, imagination, and the association of ideas."[80]

The articles on *rêve* are linked to those on *songe*. Under SONGE (*Métaphysique & Physiologie*), the Yverdon *Encyclopédie* reproduces the Paris article, which deals more with the bodily sensations and neurophysiology of dreams than with psychological mechanisms. In the *Encyclopédie*, other articles explore "venereal dreams" (in medicine), as well as dreams in biblical criticism, mythology, poetry, and modern history (in North Americans' "dream festivals"). The Yverdon *Encyclopédie* reproduced only the article

76. EY, SYMPATHIE (*Antropologie, Physiologie, Psychologie, Philosophie morale*), R by Mingard, 39, 636b–638b, p. 638b.
77. HABITUDE in EY is identical to the corresponding article in EP except for the (omitted) example of a nun who always fell ill as soon as she left the hospital where she worked.
78. EP, RÊVE (*Métaphysique*), 14, pp. 223.
79. EP, RÊVE (*Médecine*), by Ménuret de Chambaud, 14, pp. 223.
80. EY, RÊVE (*Anthropologie*), R by Mingard, 36, 678b–680a, p. 680a.

on biblical criticism, in which Oriental sages are said not to claim their dream interpretations are divinely inspired, and in which internal textual contradictions are evoked to cast doubt on the divine origin of Daniel's interpretation of Nebuchadnezzar's dream (Daniel 2).[81] Mingard set out to refute this. The scriptures, he explained, clearly show that God used dreams to reveal himself, either directly (as when Joseph was instructed to take Mary as his wife), or through an interpreter. But prophetic dreams, like miracles, are no longer necessary.[82] "The claim by some superstitious persons that God reveals himself to them in *dreams* [*songes*] is therefore unfounded."[83] In short, Mingard restricted *songe* to prophetic dreams, thus effectively isolating it and confining the question of dreams (*rêve*) to the psycho-anthropological field.

Overall, the Yverdon *Encyclopédie* sought to go beyond anatomy and physiology in order to grasp the specifically anthropological dimension of phenomena such as apathy, sympathy, habit, and dreams through their psychological mechanisms. Its reform of the nomenclature and classification of the sciences was thus reinforced by the contents of the articles.

These contents were not always obviously psychological or anthropological. For example, in both encyclopedias, temptation is a topic for theology and morals, but in Yverdon, the subject is treated in depth, without irony, and, above all, in anthropological terms. To my knowledge, the article TENTATION is the only one the Swiss encyclopedia attached simultaneously to both theology *and* anthropology or psychology. For Diderot and d'Alembert, being tempted meant "being led into or urged on to evil, due to the charms of the world, the lust of the flesh, or the evil designs of the devil." For Mingard, temptation became "all that by which we may be incited to do some evil action."[84] There was no need to invoke the devil to understand it. Why, Mingard wondered, "have recourse to such a little-understood agent, when we have within us and in the natural state of human affairs all we need to account for our misdeeds. . . . *Temptations* are due to our errors,

---

81. EP, Rêve (*Critique sacrée*), by Jaucourt, 15, pp. 357–358.

82. Like the Paris *Encyclopédie*, the Yverdon encyclopedia does not deal with this question. But it does subscribe to Charles Bonnet's theory that God preordained miracles as modifications of the laws of nature. It defines the essence of miracles as their independence from secondary causes and develops an anti-Spinozist and anti-Rousseauist defense of miracles through an interminable dialogue between a "wise theologian" and a "young disciple of the new philosophers." See EY, Miracle (*Théologie*), R, 28, pp. 756a–778a.

83. EY, Rêve (*Critique sacrée*), R by Mingard (signing as "M.D.B."), 39, 41b–44b, pp. 44a, 44b.

84. EP, Tentation (*Morale, Théologie*), 16, p. 140.

our habits, our inclinations, our passions, as well as to the circumstances in which we cannot help finding ourselves given that we are human beings who are on this earth and have appetites." In conclusion, Mingard advised that, in order to avoid temptation we must recall our duties and obligations, flee circumstances that may arouse passions, and remind ourselves through prayer "that we are ceaselessly under the gaze of the being who exhorts us to virtue, and who will as certainly punish those who violate virtue's laws as he will reward those who observe them."[85] God still keeps an eye on us, but temptation is a primarily anthropological issue. Thus, the Yverdon *Encyclopédie* participated in the moral and psycho-anthropological turn Protestantism was taking at the time, and this is doubtless why the article TENTATION bore the signature "M.D.B."

* * *

Mingard's observation that anthropology "acquaints us with man considered from every aspect" is not a tautology. The cross-references in his article Anthropologie sketch the entire anthropological field, beginning with the ultimate goal of the sciences of man, that is, self-knowledge (CONNAISSANCE DE SOI-MÊME), or "an accurate idea of man's origin, faculties, duties, rights and destination,"[86] followed by the means and principal objects of these sciences, senses and ideas, the soul and the body (SENS, IDÉES, ÂME, CORPS). The article then moves on to beauty, the fine arts, virtue, the moral sense, and morality, followed by language and speech (BEAUTÉ, BEAUX-ARTS, VERTU, SENS MORAL, MORALITÉ, LANGAGE, PAROLE). The enumeration highlights human perfectibility (PERFECTIBILITÉ) and the promise it holds out for perfection (PERFECTION) relating to morals, religion, society, happiness, and the life to come (MORALE, RELIGION, SOCIÉTÉ, BONHEUR, VIE À VENIR).[87]

Mingard additionally points to a future science—anthropology—as both an expression and a condition of perfectibility itself:

All the sciences man cultivates or can cultivate will always in some respect relate to *anthropology*, which is the most important of all the

85. EY, TENTATION (*Théologie, Anthropologie, Morale*), R by Mingard (signing as "M.D.B."), 40, pp. 423b–425a.

86. EY, CONNOISSANCE DE SOI-MÊME (*Morale*), N by de Felice, 10, 67b–70b, p. 67b.

87. EY, VIE À VENIR, N, 42, p. 288a, is made up of cross-references: "IMMORTALITÉ DE L'AME, PARADIS, ENFER, &c." The words in small capitals are the headwords of the articles to which ANTHROPOLOGIE refers.

sciences and the most worthy to claim his attention: Without this knowledge of ourselves, what perfection and what happiness may we hope to
attain?

Developing anthropology, he explains, is all the more necessary because
it is not yet the object of a "complete and systematic treatise." Its subject
matter is dispersed across the fields of the naturalist, the psychologist, and
the moralist. While no single work yet presents "the entire corpus of the
science of man," Mingard finds an "outline" of it in the "really philosophical" writings of Charles Bonnet. Since we treat of man in a fragmented way,
we forget he is neither a body nor a mind in isolation, but "a mixed being,
composed of an organized body and a rational soul, so intimately joined
together that these two substances form but a single individual whom we
know only in this composition, without any idea of the manner in which
the two substances could exist separately."[88]

Thus, the proper object of anthropology is the human as a *compositum*.
The fact that the only article to have anthropology and psychology together
as sole field indicators is COMMERCE DE L'ÂME AVEC LE CORPS, which deals
with the interaction of the soul and the body, is a sign of the status of psychology as the fundamental anthropological discipline. Anthropology was
the science of the human being as composed of two substances; psychology was the science of the soul joined to and interacting with the body. In
the eighteenth century psychology absorbed many of the issues hitherto
scattered in metaphysics, logic, and morals, by reformulating them in psychological terms and—in the case of the Swiss *Encyclopédie*—by situating
them in the perspective of Christian perfectibility.

88. EY, ANTHROPOLOGIE, pp. 22b–23b, for the quotations in this paragraph.

CHAPTER EIGHT

Human Perfectibility and the Primacy of Psychology

We have seen that, in the Yverdon *Encyclopédie*, all the sciences are linked to anthropology, and that anthropology, which draws on the results of all the sciences, produces knowledge through which human perfectibility may be realized. Perfectibility and, even more crucially, the interaction between the soul and the body are two of the God-given "laws" of human "organization." The reformulation of these "laws" in anthropological terms called for an empirical psychology, which in turn meant giving metaphysics, logic, and morals a psychological foundation.

PSYCHOLOGY IN THE PARIS *ENCYCLOPÉDIE*

In the Paris *Encyclopédie*, psychology is attached to metaphysics as the "part of philosophy which treats of the human soul, defines its essence, and explains its operations." The article briefly presents the Wolffian division between empirical and rational psychology and asserts that such a "dual method" produces "the most accurate proofs to which one can aspire." While "in ordinary [philosophy] classes, the doctrine of the soul is merely a part of pneumatology, or the doctrine of spirits, which is itself merely a part of metaphysics," Wolff transformed it into "a separate part of philosophy" which provides first principles for other sciences: for natural law, since the good or bad character of an action "can only be deduced from human nature and in particular from the properties of the soul"; for natural theology, since a "notion of divine attributes" can only be derived from knowledge of "the properties of our own soul, its imperfections and limitations"; for moral philosophy, since exhorting mankind to practice virtue and shun vice requires a knowledge of the "appetites of the soul"; and, last, for logic since, despite logic's primacy within philosophy, "it is still subordinate to psychology, insofar as it borrows principles from psychology without which it could not convey the difference between ideas or formulate the rules of reasoning, which are based on nature and the operations of the soul."[1]

Again following Wolff, the entry PHILOSOPHIE gave psychology as one of the principal branches of the discipline.[2] Other articles confirm this position. Diderot drew on Brucker's *Historia critica philosophiae* when discussing the "psychology" of various philosophical schools (ECLECTISME; PÉRIPATÉTICIENNE, PHILOSOPHIE; PLATONISME). The abbé Yvon, in his lengthy article AME, addressed four issues: the origin of the soul, its nature, its destiny, and "the beings in which it resides." Diderot added an account of the seat of the soul, but the treatment of the fourth issue was reserved for AME DES BETES (*Métaphysique*). Yvon presented different opinions and defended the soul's spirituality and immateriality. He also explained that its faculties are abstractions by means of which its functioning may be analyzed—they are merely the "various ways in which the single force which constitutes the essence of the soul may be applied." The major works on this subject, in Yvon's view, were Malebranche's *The Search after Truth*, Locke's *Essay*, and Wolff's "two Philosophies"—clearly an error for "Psychologies"—whose "psychological analysis" the abbé summarized in a few lines.[3]

The term "psychology" had been used occasionally in French from the sixteenth century, but it was rare. It is absent from Antoine Furetière's *Dictionnaire universel* (1690) and Thomas Corneille's *Dictionnaire des arts et des sciences* (1694). It appears in the *Dictionnaire de Trévoux* of 1732, but not in the third edition of Pierre Richelet's *Dictionnaire de la langue française* of the same year. It entered the *Dictionnaire de l'Académie française* in its fourth edition (1762), but years later it was still absent from abbé Jean-François Féraud's *Dictionnaire critique* (1787–1788). The article PSYCHOLOGIE in the *Encyclopédie* is one of the many signs of Wolff's presence

---

1. EP, PSYCHOLOGIE, 13, p. 543. See appendix II for the full text.

2. EP, PHILOSOPHIE, 12, pp. 511–515.

3. EP, AME (*Ordre Encyclopédique. Entendement. Raison. Philosophie ou Science des Esprits, de Dieu, des Anges, de l'Ame*), by Yvon, 1, 327–340, p. 338. CATALOGUE follows the old sense of psychology in defining it as a branch of "physiography," or the study of living beings. It assigns the study of spirit, thought, and the properties and operations of the rational soul to metaphysics as a branch of "spiritology." See CATALOGUE (*Littérature & Librairie*), 2, pp. 759–765, taken from a manuscript by abbé Gabriel Girard. Girard was the author of the famous *Synonymes François*, which often served as basis for the *Encyclopédie* articles bearing the field indicator *Grammaire*.

and confirms that while the encyclopedists disliked the Wolffian style and method, they did not cling to a monolithic ideology.

Christian Wolff was known in France through the *Institutions of Physics* of Madame du Châtelet (1706–1749). This anti-Newtonian textbook opened with a presentation of Wolff's metaphysics.[4] Summaries of Wolffian philosophy, including psychology, were published in the 1740s by the Huguenot pastors Jean des Champs (1707–1767) and Jean Henri Samuel Formey (1711–1797), along with an anonymous synopsis (praised in the Jesuits' *Journal de Trévoux* in April 1746) of the first part of Wolff's empirical psychology. Like des Champs, the anonymous author remarked that the term "psychology" must seem extraordinary to a French reader and consequently required explanation. Before becoming an Yverdon contributor, Formey, permanent secretary of the Berlin Academy of Sciences and a critic of the French *philosophes*, sold the manuscript of his Wolffian philosophical dictionary to the *Encyclopédie* publishers.[5]

The *Encyclopédie* thus played a significant role in the dissemination of Wolffian thought in France. Several important articles on logic, philosophy, metaphysics, and jurisprudence were taken from Formey, who sometimes— but without acknowledgment—copied them from Madame du Châtelet; still other articles translated Wolff directly.[6] But whereas some, such as PSYCHOLOGIE, faithfully summarized the German philosopher, others used him for subversive purposes. For example, in the article COSMOLOGIE, the famously anti-Wolffian d'Alembert referred to Wolff as a springboard to develop his own metaphysics, in the spirit of the "Preliminary Discourse."

---

4. Gabrielle-Emilie Le Tonnelier de Breteuil, Marquise du Châtelet, *Institutions physiques* (1741), ed. of 1742, in Wolff, *G.W.*, III.28. The marquise did not go into detail concerning how knowledge was ordered but highlighted the ontological importance of the principle of noncontradiction for the discovery of what is possible or impossible, and above all the principle of sufficient reason, "on which all contingent truths depend" (§ 8, p. 23). Voltaire, who lived with her at the time, deplored her applying "her mind to the weaving of such spider's webs"; François-Marie Arouet de Voltaire, letter to Maupertuis, 10 August 1741, in *The complete works of Voltaire*, ed. Theodore Bestermann (Geneva: Institut et Musée Voltaire, 1970), vol. 92, letter D2526, p. 95.

5. Marcu, "Un encyclopédiste oublié"; C. Donato, "Jean Henri Samuel Formey's contribution to the *Encyclopédie* d'Yverdon," in Fontius and Holzhey, *Schweizer im Berlin*. EP, SONGE, cited above, is a shortened version of Formey's "Essai sur les songes," *Histoire de l'Académie royale des sciences et des belles-lettres de Berlin*, 1746, 317–334.

6. Sonia Carboncini, "Lumières e Aufklärung: A proposito della presenza della filosofia di Christian Wolff nell'*Encyclopédie*," *Annali della Scuola Normale Superiore di Pisa (Classe di Lettere e Filosofia)*, 14, 1984,1297–1308; Carboncini, "L'*Encyclopédie* et Christian Wolff: A propos de quelques articles anonymes," *Les études philosophiques*, no. 4, 1987, 489–504.

Similarly, the *Encyclopédie*'s definition of metaphysics as the "science of the reasons for things" had a Wolffian flavor. Yet the article also stated that, insofar as metaphysics limited itself to "empty and abstract considerations on time, space, matter and spirit," it was a "contemptible" science.[7] In Diderot's view, Locke breathed new life into the classical axiom that "there is nothing in the intellect which was not previously in the senses"; the axiom implied that "every idea, when fully broken down, must be reducible to a sensible representation," and that "any expression which does not find outside our minds a sensible object to which it may be attached is void of meaning."[8] For Wolff, on the other hand, philosophy required a priori demonstrations and a syllogistic method. This was why the Wolffian style and approach were anathema to the encyclopedists.

Voltaire, who called Wolff a "German chatterbox," accused him of importing into Germany all the "horrors of Scholasticism" encumbered by a wealth of Leibnizian notions, such as sufficient reason, monads and indiscernibles, plus "all the scientific absurdities Leibniz invented out of sheer conceit, and which the Germans study because they are German."[9] Condillac accused Wolff of preferring synthesis to analysis and considered it "shameful" for Germany that such a man could create a school.[10] La Mettrie compared him with Locke and "true geniuses" such as Newton and Boerhaave, describing him as lost in a "labyrinth of errors" caused by oversystematizing.[11] The *Encyclopédie* itself complained that Wolff's method was "profoundly obscure and of abhorrent dryness, due to a barbaric and gothic pretense of demonstrative rigor and brevity," and accused it of having "extinguished good taste and ruined the best minds" in Germany.[12] Wolff and the Wolffians, however, regarded their method as the best way of achieving demonstrative certainty in philosophy.

A similar process of critical reformulation was at work as regards psychology. Despite its Wolffian article PSYCHOLOGIE, the *Encyclopédie* did

---

7. EP, MÉTAPHYSIQUE, 10, p. 440. And elsewhere: "As for metaphysics properly speaking, on which we believe we dwelled at too great a length in the first volumes, we will reduce it in future volumes to what is true and useful in it, that is, very little." EP, "Avertissement des éditeurs," 3, p. v.

8. EP, LOCKE, PHILOSOPHIE DE, by Diderot, 9, 625–627, p. 626.

9. Voltaire, letter to Maupertuis, 10 August 1741, in *Complete works*, vol. 92, p. 95.

10. Étienne Bonnot de Condillac, letter to Maupertuis, 12 August 1750, in Georges Le Roy, ed., *Œuvres philosophiques de Condillac* (Paris: Presses Universitaires de France, 1947–51), 2, p. 535.

11. Julien Offray de La Mettrie, *Abrégé des systèmes, pour faciliter l'intelligence du Traité de l'Ame* (first published in 1751), § IV, in *Œuvres philosophiques* (Paris: Fayard, 1987), vol. 1.

12. EP, RENVOI, 14, p. 123.

not call "psychology" the empirical study of the soul and, more important, it approached the soul in a resolutely non-Wolffian perspective.

Although we should not ask of the *Encyclopédie* more coherence than it actually has, it is significant that psychology is absent from the "Système figuré" and that no article uses *Psychologie* as field indicator. The encyclopedic tree places "pneumatology, or science of the soul" within the science of man. The article ESPRIT, to which PNEUMATIQUE (*Physique*) refers for pneumatology properly speaking, is attached to metaphysics. It refers to PENSÉE (since the mind is "a thinking and intelligent being"), DIEU, ANGE, and AME (for distinctions between an infinite and a finite spirit, separated from or united with a body), and to RAISONNEMENT and JUGEMENT (for the human mind as "thinking and rational substance"). All of these—with the exception of angels (exclusively theological) and the study of the soul (attached directly to pneumatology)—are the objects of metaphysics, some exclusively (mind, thought, judgment), others in conjunction with logic (reasoning) or with theology (God). More precisely, ESPRIT refers to "PENSÉE &c.," implying not only the article itself but all the cross-references it contains. Table 8.1 presents the resulting distribution of subjects and fields. Quite apart from the fact that they are not systematic and sometimes erroneous or indirect, the cross-references convey only a tiny part of the information available in the *Encyclopédie*.[13] But in the case which interests us here, other articles confirm the mainly metaphysical and logical classification of "psychological" material. ATTENTION, for example, belongs to logic; MÉMOIRE, to metaphysics; IMAGINATION, to both. These subjects are psychological at least to the extent that they are defined as "operations" or "powers" of the soul, and that, if we follow the *Encyclopédie*'s own definitions, they should receive the field indicator *Psychologie*. The author of PENSÉE states that his subject

> expresses all the operations of the soul. Hence I will call *thought* everything the soul experiences, whether through external impressions or the

13. Working with the online *Encyclopédie*, Gilles Blanchard and Mark Olsen have shown that some categories in the "Système figuré" are omitted from the cross-references. They also examine how closely related certain fields are to each other. As one might expect, certain groupings have no equivalent in the "Système figuré." This is the case, for example, of the links between natural history, on the one hand, and botany, chemistry, and mineralogy, on the other, fields which do not belong to the same main branch in the "Système." G. Blanchard and M. Olsen, "Le système des renvois de l'*Encyclopédie*: Une cartographie des structures de connaissances au XVIIIᵉ siècle," *Recherches sur Diderot et "l'Encyclopédie,"* 31–32, 2002, 45–70. On the cross-references more generally, see Schneiders, "Le prétendu système."

TABLE 8.1. Field indicators and cross-references for the articles ESPRIT and PENSÉE in the Paris *Encyclopédie*

Article	Field indicators				
	Métaphysique	*Logique*	*Théologie*	*[Pneumatologie]*	*Philosophie*
ESPRIT	X			[X]	
Articles referred to under ESPRIT:					
PENSÉE	X				
JUGEMENT	X				
RAISONNEMENT	X	X			
DIEU	X		X		
AME				X	
ANGE			X		
PENSÉE	X				
Articles referred to under PENSÉE:					
OPÉRATION		X			
PERCEPTION	X				
SENSATIONS	X				
CONSCIENCE	X				
IDÉE		X			X
NOTION		X			

use it makes of its own reflections. An *operation* is thought insofar as it can produce some change in the soul and by this means enlighten and guide it. A *perception* is the impression produced in us by the presence of objects. A *sensation* is that same impression insofar as it comes through the senses. *Consciousness* is becoming apprised of it [sensation]. An *idea* is becoming apprised of it [sensation] as an image. A *notion* is any idea which we ourselves have produced.[14]

All these concepts belong to logic or metaphysics. The choice of field indicator tends to depend on the extent to which mental functioning is stressed. If it is, the concept falls within metaphysics or within logic (insofar as logic

---

14. EP, PENSÉE (*Métaphysique*), 12, 308–309, pp. 308–309. "Operations" are called "acts of the mind," and they are apprehension or perception, judgment, reasoning, and method; EP, OPÉRATION (*Logique*), 11, p. 497.

is based on the study of the operations of the mind).[15] AME turns out to be the least strictly psychological article in this group, since it focuses on examining the soul's origin and arguing for its immateriality and immortality. This is borne out by the cross-references: IMMATÉRIALISME OU SPIRITUALITÉ (*Métaphysique*) and SPINOSISME;[16] for the seat of the soul, GLANDE PINÉALE, CERVEAU, and other anatomical articles; and TARENTULE ([*Histoire naturelle, Médecine*]), to illustrate the reciprocal effects of the soul and the body. In short, for Diderot and d'Alembert, the soul itself remains an object of pneumatology, while the study of its faculties or operations belongs to metaphysics, or to logic when the emphasis is on the role of these operations in thinking correctly.

PSYCHOLOGY IN THE YVERDON *ENCYCLOPÉDIE*

The Yverdon *Encyclopédie*, by contrast, redefined the same material so that it *could* be attached to psychology. For a start, psychology itself was developed in considerable detail.[17] According to the Yverdon classification of the sciences, psychology was a branch of pneumatology, which itself belonged in metaphysics. In Diderot and d'Alembert, the soul was likewise an object of pneumatology, which was also a branch of metaphysics. Moreover, both encyclopedias characterized Wolff as the founder of psychology understood as an autonomous discipline within philosophy. Only the Yverdon *Encyclopédie*, however, reorganized the disciplines in accordance with the Wolffian scheme. At the same time—and this is another similarity—de Felice criticized Wolff's "synthetic" excesses and the Scholastic architectonics of his system. He lambasted the "sophists, besotted with their systems," and while valuing Wolff, he considered that Locke had laid the "solid foundations" for the principles which could lead to "a more precise knowledge of man" and "the anatomy of his faculties."[18] Under PSYCHOLOGIE, Mingard mentioned Wolff alongside Bonnet and Condillac, but in his articles on psychology, the latter two were by far the most authoritative sources. Thus, the Yverdon *Encyclopédie* treated psychology as an autonomous field and systematically transposed into it subjects previously classed within logic or metaphysics.

15. EP, LOGIQUE (*Philosophie*), 9, pp. 637–641.

16. No such headword exists, but there is an article SPINOSA, PHILOSOPHIE DE (*Histoire de la philosophie*).

17. EY, PSYCHOLOGIE (*Métaphysique*), R by Mingard, 35, pp. 511b–513a. All quotations in the following three paragraphs are taken from this article. See appendix II for the full text.

18. EY, MÉTAPHYSIQUE, pp. 494a and 496a.

Such rearrangement was consistent with Mingard's position on the importance of psychology for the other sciences. He explained that the other components of pneumatology would be "absolutely unknown" were it not for psychology. We acquire "primitive ideas" either through the senses, by which we know bodies, or through reflection "on what is going on inside us," by which we may know our soul, and speculate on "the essence of spiritual substances [*esprits*]." It is by comparing ideas acquired via the senses and via reflection "that we are able to infer that what thinks, wills, and feels is not matter." Thanks to that demonstration, psychology becomes the foundation of all the philosophical sciences, and especially those that deal with man.

By comparing voluntary and involuntary actions, Mingard writes, we are able to differentiate between those resulting from the "reflective will of an intelligent being" and those due to chance. That is how we discover traces of an intelligent cause, whose "effects" prove it is far "above us." Psychology therefore "serves to guide us" toward theology. It also directs us toward morals, since it is "by studying the effects which the actions of other beings have on us" that we become able to judge if our own actions are acceptable or not. By "paying attention" to what happens inside us and to how we arrive at conviction, belief, opinion, or doubt, "we can discover all the propositions and rules of logic which, without psychology, would yet be undiscovered." Even politics, as the "art of leading men," cannot do without psychology, since psychology explains the determinants of human action.

"In a word," Mingard concludes, "what science or art deserving of our attention does not have *psychology* as its foundation, its source, and its guide?" Without knowledge of the soul,

> we can pass judgment on nothing, decide nothing, determine nothing, choose nothing, reject nothing, prefer nothing, and do nothing with certainty and without error. Psychology is consequently the first and most useful of all the sciences, the source, the principle, and the foundation of them all, as well as the guide which leads to each.

Psychology's positions rests largely on its absolutely unique method: unlike all the other sciences, it derives its principles and facts from the very objects of its investigations. It is through the operations of the soul that psychology is constituted as a "rational and systematic science," and these operations are also its object of study. Which is why, Mingard explains, Wolff made the distinction between rational and empirical psychology. As already mentioned, alongside Wolff's two *Psychologies*, Mingard recommends Charles

Bonnet's *Essay on Psychology* and his *Analytical Essay on the Faculties of the Soul*, as well as Condillac's *Treatise on the Sensations*—a choice which shows a clear preference for analysis over synthesis.

Such a methodological choice sheds light on the Swiss encyclopedists' attitude toward Wolff. On the one hand, the Yverdon "Système figuré" largely follows Wolff's classification of the sciences, and there is no shortage of praise for his philosophy. Wolff's *Ontology*, says de Felice, is a masterpiece that should be made available to all. On the other hand, he considers Wolff's writings to be "too long, at least by half,"[19] and does not even mention Wolff in the article MÉTHODE. He accuses "geometers"—who "should understand the advantages of analysis better than the other philosophers"—of being too fond of synthesis, attributes Locke's superiority to the fact that he was not one of them, and claims that analysis is the path to truth and the best method to present one's findings.[20] Similarly, Mingard recognizes Wolff's importance for psychology but attacks the foundations of the Wolffian system, criticizing its "unlimited use" of the principle of sufficient reason at the expense of other, as-yet-undiscovered, principles.[21]

Psychological concepts in both encyclopedias were organized according to the methodological and philosophical choices outlined above. In the French *Encyclopédie*, these concepts were integrated into logic and metaphysics and given a psychological dimension; in the Swiss one, the same notions tended to be made explicitly psychological, and to be treated according to the epistemic primacy of psychology and the new arrangement of knowledge this entailed. A certain idea of psychology thus informed the transformation of logic and metaphysics, and this transformation, with its redefinitions and reconceptualizations, in turn shaped psychology as an autonomous field of knowledge. Here we can find the "Mingard effect" at work, which we have already noted in the case of anthropology. Appendix III lists the articles in the Yverdon *Encyclopédie* which belong to psychology, and their place in the Paris original.[22] They are almost all by Mingard. However, this does not mean that the outcome is absolutely coherent: despite the best editorial intentions and a concerted effort to redistribute fields and concepts, many "psychological" subjects do not appear under psychology.

---

19. EY, WOLFF, N by de Felice, 42, 710b–712a, p. 711b.
20. EY, MÉTHODE (*Logique*), R by de Felice, 18, 538a–543b, p. 543b.
21. EY, CONJECTURE (*Philosophie, Logique*), R by Mingard, 11, 13a–16a, p. 16a.
22. I exclude EY, CHYME (*Psychologie*), R, whose field indicator is clearly a mistake for *Physiologie*; cf. EP, CHYME (*Anatomie, Physiologie*).

As in the case of anthropology, it is additionally impossible to form a complete corpus on the sole basis of field indicators. For example, among Mingard's own texts, Pensée is attached to metaphysics and physiology, and Rêve—a psychological article if there is one—only to anthropology.[23] De Felice's partly psychological articles retain the Paris field indicators: *Logic* for Raison; *Logic* (as in EP) and *Morals* for Entendement; *Metaphysics* (as in EP) and *Physiology* for Mémoire (as in EP); *Logic* for Association des idées, which has no field indicator in Diderot and d'Alembert.

Another example of the different choice of field indicators is provided by "physical influence" as an explanation of the relations between the soul and the body. Diderot and d'Alembert treat it chiefly in Influence (*Métaphysique*). In de Felice, the corresponding article also belongs to metaphysics, but the choice of field indicator corresponds to Mingard's distinction between the *metaphysical* question of the "union" of the soul and the body and the *anthropological* and *psychological* question of the "interaction" (*commerce*) between these two substances.

THE FIELDS CLAIMED FOR PSYCHOLOGY

As was the case with anthropology, the appropriation of disciplines for psychology was not brought about solely by naming or classifying but also by reworking the contents of the relevant sciences. We will illustrate this from the fields that were most extensively redefined in psychological terms, namely, metaphysics, logic, and morals. A comprehensive study of each of these disciplines would have to take into account a much greater number of articles, but the corpus defined here through the field indicators demonstrates the general drift of the changes made in the Yverdon *Encyclopédie*.

This corpus confirms, first of all, that the psychological transformation of concepts in the Yverdon *Encyclopédie* was never purely lexical. When the Paris *Encyclopédie* attaches Apprendre to *Grammaire*, for example, it simply defines the word and differentiates it from similar terms. The Yverdon counterpart, on the other hand, reproduces the original article but also adds a new one placing the subject inside psychology, and explaining that "to learn" designates the fact of acquiring knowledge and preserving it in one's memory, as well as, in the arts, the tendency of this knowledge to produce certain physical effects.[24] Diderot wrote that to learn is "to apply

---

23. EY, Pensée (*Métaphysique, Physiologie*), R by Mingard, 32, pp. 734a–737b; see chap. 7 above for Rêve.

24. EY, Apprendre (*Psychologie*), N by Mingard, 3, pp. 216a–216b.

oneself to becoming learned [*savant*]." Mingard's redefinition stresses the operations needed to realize what the verb simply indicates; a new article, APPLICATION (*Psychologie*), gives details of the mental acts necessary for learning.[25]

In the Yverdon *Encyclopédie*, the psychologizing process involves moving a topic into the sphere of psychology by the use of a field indicator, and analyzing it from the perspective of the "acts" or "states" of the soul associated with it, rather than from that of a taxonomy of mental faculties. The different categories are clearly spelled out: "The senses, imagination, attention, memory, pure understanding, etc., are faculties. Perception, judgment, reasoning, and discourse are operations. The idea I have, the abstraction I make, the attention I give to an object in the present moment are *acts*."[26] These acts are then placed in the teleological perspective of the anthropology of perfectibility.

From a methodological point of view, the Yverdon articles which are recast in psychological terms exemplify the procedures described in the article ANALYSE. In the wake of Condillac and other psycho-logicians, Mingard defines analysis as the "operation of the mind," which consists in breaking down a composite idea through abstraction;[27] this is one of the logical methods for "discovering the truth."[28] He does not dispute innate ideas or insist on the opposition between analysis and synthesis, as Yvon does in the *Encyclopédie*, but explores the concept's extension. Since, as Mingard explains, the analytic method is not completely uniform—instead of starting by breaking down the whole it may be necessary first to analyze a single part, or even something external to the object—he tries to formulate rules that would be always valid. The psychological articles, like ANALYSE itself, apply these rules. They examine the concept in question carefully, differentiate it from related ideas, avoid using undefined concepts, and systematically spell out and illustrate different meanings and applications. Above all, since analysis is "the most natural method for discovering the truth," it embodies procedures that correspond to what psychology itself demonstrates when it follows the analytic method: "Our first knowledge is of individual facts which we discover through the senses or reflection."[29]

---

25. EY, APPLICATION (*Psychologie*), N by Mingard, 3, pp. 205a–206a. There are no cross-references between APPLICATION and APPRENDRE.

26. EY, ACTE (*Grammaire, Métaphysique*), N, 1, p. 353.

27. EY, ANALYSE (*Métaphysique*), R by Mingard, 2, p. 491b.

28. EY, ANALYSE (*Logique*), R by Mingard, 2, 491b–495b, p. 492a.

29. Ibid., p. 495b.

For Mingard, as for Charles Bonnet, psychology concerns the soul's *activities* rather than its *faculties*. The latter, "considered abstractly," are the capacity to act, to produce an effect, and to "set oneself in motion spontaneously."[30] The error of those who divide them into animal, sensitive, and intellectual is, in Mingard's opinion, that the first two categories encompass only passive capacities, since only the soul is capable of initiating an action and giving motion to the body.[31] Action, he explained, "is a change of state whose cause is contained in the being that brings about this change." That is why the "logicians" designate the soul's operations by the term *action*.[32] As we shall see, such a focus on actions provided an opportunity to foreground human perfectibility and the obligations accompanying it.

METAPHYSICS

The psychological reformulation of metaphysics affected both nomenclature and subject matter. The lexical dimension was vital, and Mingard was extremely attentive to terminology. For example, he devoted three separate articles to the notion of "judge" (*arbitre*). Two belonged to moral philosophy and concerned decision makers, including God as the "supreme judge." The third belonged to psychology and was written, as Mingard said, from the perspective of someone wishing to "make known the various powers of the human soul," that is to say, the psychologist.[33] Table 8.2 shows the Yverdon articles attached to psychology and metaphysics, as well as the articles linked to metaphysics in the Paris *Encyclopédie and* at least to psychology in the Yverdon version. All are either rewritten or new yet have a corresponding entry in the Paris original. I will examine now a few of these articles, leaving the others for the ensuing sections on logic and morals. Table 8.2 shows that FIN alone has the field indicator *Métaphysique* accompanied *only* by *Psychologie*. Diderot's very short article defines *fin*, or end, as "the ultimate reason we have for acting, or which we regard as such," adding that this reason always is "individual happiness."[34] Insofar as an end is related to motives, it is a moral issue. In the Yverdon *Encyclopédie*, Mingard gives the subject greater metaphysical and psycho-

30. EY, FACULTÉ (*Métaphysique*), R by Mingard, 18, 241a–244b, p. 242a.

31. Ibid., p. 243b.

32. EY, ACTION (*Physique & Métaphysique*), N by Mingard, 1, 367a–371b, pp. 367a–b and 371b.

33. EY, ARBITRE (*Psychologie*), N by Mingard, 3, 295b–301b, p. 296b.

34. EP, FIN (*Morale*), by Diderot, 6, p. 810.

TABLE 8.2. The overlap between psychology and metaphysics

EY headword	Field indicators in addition to *Psychologie*	EP field indicators
AME (R)	*Métaphysique, Pneumatologie, Anthropologie*	[*Pneumatologie*]
APPARENCE (N)	*Métaphysique, Philosophie morale, Logique*	*Grammaire*
BEAUTÉ (R)	*Métaphysique, Morale*	
CONNOISSANCE (R)	*Métaphysique, Logique*	*Métaphysique*
CONSCIENCE (R)	*Logique*	*Métaphysique*
COUTUME (R)	*Physique, Philosophie morale*	*Morale*
DOUTE (N)		*Logique, Métaphysique*
FIN (N)	*Métaphysique*	*Morale*
PERCEPTION (R)		*Métaphysique*
SPONTANÉ, SPONTANÉE		*Métaphysique*

logical depth. Whereas Diderot simply addresses the reasons for action, Mingard argues that the notion of end presupposes a free and intelligent agent capable of anticipating and willing an effect, and aware of "how his actions relate to the beings on which he wishes to produce the effect he has envisaged."[35] Since these effects are contingent, it follows that the concept of end is incompatible both with the idea of chance and with that of necessity, and that, consequently, it is absurd to attribute ends to nature. Moreover, he explains, since the universe represents only one of many possible designs, it must be the effect of a "powerful, intelligent and free agent."[36] A free being is able to act spontaneously, that is, to will and carry out actions he could choose not to carry out.[37] Mingard's way of conceiving this metaphysical issue highlights the significance of the subject for his Christian teleological anthropology, which he explicitly sets against the materialism, determinism, and atheism of d'Holbach's *The System of Nature* (1770).

35. EP, FIN (*Métaphysique, Psychologie*), N by Mingard, 19, 273b–280b, p. 273b.
36. Ibid., p. 280b.
37. EY, SPONTANÉ, SPONTANÉE (*Philosophie, Métaphysique, Psychologie, Morale*), R by Mingard, 39, 242b–244b, pp. 243b–244a. EP, SPONTANÉE (*Grammaire*) restricts itself to the medical sense ("Spontaneous stools, which occur without the assistance of enemas or suppositories"). Mingard deems this usage improper because the concept should apply only to beings endowed with will and active in their own right.

Other issues are nonetheless situated entirely outside metaphysics. The psychologizing of the notion of perception, for example, starts with the heading itself: whereas the Paris *Encyclopédie* attaches it to metaphysics and borrows exclusively from Condillac, the Yverdon *Encyclopédie* attaches it solely to psychology. The brief article which replaces the Paris text is typical of Mingard in that it emphasizes "the state of the soul which feels some sort of modification."[38] But the subject is elaborated in two new articles on apperception, which are likewise classified only under psychology. This sort of redistribution illustrates Mingard's debt to Wolff, particularly as regards the importance of the soul's activity as a proper subject of psychology. The first article defines apperception as "the act by which the soul distinguishes itself from all the other objects of its perceptions," and so regards itself as the "subject" which has or can have perceptions.[39] Apperception is thus a reflexive act which constitutes personality and makes me say "it is *I* who am at present thinking, feeling, willing, acting."[40]

The second article, though less focused on conceptual clarification, insists upon the difference between *apercevoir* and *s'apercevoir*, that is, between the soul perceiving an external object and the soul being aware of what happens within itself. Mingard sides with Locke and Condillac against the followers of Descartes and Leibniz, for whom the soul can have unconscious perceptions. For him, the soul "is aware of [*s'apperçoit de*] everything that goes on within it." The perceptions of which the soul is presumably not conscious are those "to which it has not sufficiently turned its attention to remember them."[41]

Mingard thus not only argues his case on logical grounds (explaining, for example, that it would be contradictory to claim the soul has perceptions of which it is not aware) but also on specifically psychological ones. Rather than debating whether the soul is in principle really aware of all its perceptions, the issue becomes whether or not it has perceptions which it forgets so rapidly that "it has no knowledge of having had them."[42] Mingard cites examples but considers that they still do not prove "that we were not aware of our perceptions at the moment we really had them."[43]

38. EY, Perception (*Psychologie*), R by Mingard, 33, pp. 5b–6a.
39. EY, Apperception (*Psychologie*), N by Mingard, 3, 194b–195b, p. 194b.
40. Ibid., p. 195a.
41. EY, Apercevoir (*Psychologie*), N by Mingard, 3, 195b–197a, p. 196a.
42. EY, Conscience (*Philosophie*), R by Mingard, 11, 83b–85a, p. 84a.
43. Ibid., p. 84b. EP, Conscience (*Philosophie, Logique, Métaphysique*), by Jaucourt. The article summarizes in a paragraph the debate, mentioned above, between Leibnizians and Lock-

The appropriation of metaphysics for psychology therefore meant recasting certain questions hitherto classed within metaphysics through an empirical analysis of the mental functioning involved in them. Whereas this procedure was only partially adopted in Diderot and d'Alembert's *Encyclopédie*, and moreover retained the name of "metaphysics," it was systematic in the Yverdon work and was given the name of "psychology."

LOGIC

Logic, as both encyclopedias define it, is "the art of thinking correctly or of making proper use of our rational faculties by defining, subdividing, and reasoning."[44] De Felice rewrote the French article, with some omissions and substitutions—including a vicious passage on the "nit-picking of the old school"—but he did not discard the praise of Descartes, Malebranche, Condillac, and Locke, who was "the first to undertake an elucidation of the operations of the human mind, in perfect conformity with nature."[45] The appropriation of logic for psychology would basically be the continuation of this Lockean project. When the encyclopedists, both French and Swiss, proclaimed that the mind should henceforth be studied "not in order to discover its nature, but to gain knowledge of its operations,"[46] they were not embarking on an entirely new project. Their purpose, however, was to deal with the conflation of psychology and pneumatology so as to isolate the latter, and reinterpret the former in Lockean terms.

Table 8.3 lists the articles attached to psychology and logic in the Yverdon *Encyclopédie*, as well as the articles attached to logic in the Paris *Encyclopédie and* at least to psychology in the Yverdon text. The articles reformulated in psychological terms through the field indicator (or which are new but have corresponding headwords in the Paris *Encyclopédie*) are APPRÉHENSION, CONCEPTION, DOUTE, PASSION, PRÉJUGÉ, and OPINION.

---

eans. Both encyclopedias have articles on conscience, EP in natural law and morals, EY in moral philosophy.

44. EP, LOGIQUE (*Philosophie*), 9, 637–641, p. 638; EY, LOGIQUE (*Philosophie*), R by de Felice, 26, 507b–513a, p. 511a–b.

45. Ibid., p. 638 (EP), p. 511a–b (EY). Locke, of course, is superior to his predecessors because he investigates the operations of the mind in accordance with its nature, "without letting himself be drawn to opinions based on systems rather than on realities; in this, his philosophy is to the philosophies of Descartes and Malebranche as history is to novels"; ibid. The field indicator is *Philol.* in both encyclopedias; but this is clearly a typographical error. The switching of *l* and ∫, or long "s," in the italics of the time was fairly common; for example, one of the EY field indicators for APPARENCE is *Psychoſ.*, an abbreviation for *Psychologie*.

46. Ibid., p. 639 (EP), p. 512b (EY).

TABLE 8.3. The overlap between psychology and logic

EY headword	Field indicators in addition to *Psychologie*	EP field indicators
AFFIRMATION (N)	*Logique*	
APPARENCE (N)	*Métaphysique, Philosophie morale, Logique*	*Grammaire*
APPRÉHENSION (R)	*Logique*	*Art de penser* [= *Logique*]
COMBINAISON (N)	*Logique*	
CONCEPTION (R)		*Logique*
CONNAISSANCE (R)	*Métaphysique, Logique*	*Métaphysique*
DOUTE (N)		*Logique, Métaphysique*
OPINION (R)	*Logique*	*Logique*
PASSION (R)	*Morale*	(PASSIONS) *Philosophie, Logique, Morale*
PRÉJUGÉ (R)	*Logique, Morale*	*Logique*

AFFIRMATION, APPARENCE, and COMBINAISON are new articles and are placed at the intersection of logic and psychology. PASSION is a special case. Whereas the French stress the dangers of passion, including for thought—which explains why the field indicators *Philosophie* and *Morale* are given alongside *Logique*—the Swiss take a more positive view of the subject, focusing less on logic than on psychology. On the other hand, CONNOISSANCE, which in Diderot and d'Alembert belongs in metaphysics, is treated in de Felice as a topic not only for metaphysics but also for psychology and logic. The psychological inflection Mingard gives to CONNOISSANCE underscores its importance for logic. (PASSION and CONNOISSANCE are discussed below, along with other articles on morals and metaphysics.)

The raw materials of logic are ideas: "In order to think correctly, it is necessary to perceive well, judge well, reason well, and connect one's ideas with method."[47] The author of IDÉE (the same text in both encyclopedias) goes so far as to qualify his topic as "one of the most important in philosophy," since it could encompass the whole of logic.[48] It is also to logic that

47. EP, LOGIQUE, p. 637; EY, LOGIQUE, p. 508a.
48. EP, IDÉE (*Philosophie, Logique*), 8, 489–494, p. 489.

Mingard attaches his new articles on simple, composite, clear, distinct, and confused ideas.[49]

The psychological transformation of logic is accomplished by examining the mechanism of ideas, envisaged as actions of the mind. In the Paris *Encyclopédie*, Yvon defines apprehension as "the same thing as perception." The rest of the article contains a rather spiteful refutation of Malebranche's distinction between the perception of corporeal things by the imagination and that of spiritual things by the intellect. For Yvon, there is no essential difference between these two operations; thought is a single faculty, and it is spiritual.[50] Mingard, on the other hand, avoids the controversy and starts by clarifying terms: apprehension should be distinguished from understanding, conception, idea, and perception. More important, unlike perception understood as a passive impression, apprehension "always requires an active soul"; it is the "act by which the soul develops ideas of things."[51] Apprehending is therefore not perceiving or thinking but rather the act by which the soul forms the object of its thoughts.

Another—but opposite—example of the psychological transformation of logic is provided by Mingard's treatment of "affirmation." He describes it as a "state" (and not a "free action") in which the soul sees and "feels that it sees" that one idea "is contained" within another (*goodness* within *God*, for example, or *moral disorder* within *lie*). The soul thus perceives "that such a thing is."[52] Affirmation may depend on the will, as when, if my ideas are unclear, I decide to examine them. Although as a state of the soul, affirmation is known only to myself, I can formulate it in a proposition. Affirmative arguments are based on an "inner feeling,"[53] but the content of this feeling must first be proved and the dependence between ideas demonstrated. Thus, "the idea of not punishing the innocent is contained in the idea of justice; the idea of justice is contained in the idea of God; hence, the idea of God contains the idea of a Being who does not punish the innocent."[54]

The phenomenon of mental "conception" provides another example. In the *Encyclopédie*, the chevalier de Jaucourt treats it as identical to

---

49. EY, Composées, idées; Claire, idée; Distinct, te; Confuse, idée—all by Mingard and attached exclusively to logic.

50. EP, Appréhension (*Ordre encyclopédique. Entendement. Raison. Philosophie ou science. Science de l'homme. Art de penser. Appréhension*), by Yvon, 1. p. 555.

51. EY, Appréhension (*Psychologie, Logique*), R by Mingard, 3, 215b–216b, pp. 215b, 216a.

52. EY, Affirmation (*Logique, Psychologie*), N by Mingard, 1, 527b–529a, pp. 527b, 528b.

53. Ibid., p. 528b.

54. EY, Affirmatif, Raisonnement, N by Mingard, 1, 526b–527a, p. 527a.

"comprehension" and defines it as the operation by which the understanding "links" ideas and "grasps their different branches, relations, and sequence." He then goes no further than general remarks, such as "Since these sensations and perceptions are often not well organized . . . the mind does not grasp the relations between things in their true perspective," or "The soul is sometimes drawn from conception to conception by links between ideas which correspond to its own present interests."[55]

Even at that level of generality, Jaucourt's remarks are relevant to logic. His article, however, contains no analysis comparable to Mingard's in CONCEPTION. Mingard places the topic in psychology and defines conception as "the act or the capacity to perform the act through which the understanding represents to itself the object of a composite idea in a manner distinct enough to apprehend its external and internal relations, its principles, causes, and consequences."[56] He then enumerates, in analytic order, the faculties involved: those of entertaining simple ideas; of representing to oneself composite ideas clearly; of breaking them down through abstraction; of identifying their mutual relations; of recomposing them into a whole; of discovering the reasons for the existence of this whole; and of apprehending the effects and consequences which result from it, the goal it strives for, and how it attains it. That is the way to *conceive* of a demonstration, the mechanics of a pump, or a system of astronomy, botany, or natural history.[57] Such a psychological analysis has implications for the conduct of the understanding, since the ability to conceive well depends not only on the "perfection of the senses" but above all on the "frequent, regular, and repeated exercise" of attention, on meditation, method, and the "habit of distinct ideas, which is the fruit of practicing abstraction."[58] The article is characteristic of Mingard in that it articulates psychology with reflections on the moral finality of the individual and humanity (perfection and happiness) and reflects on how to realize those ends (here, through exercizing the cognitive faculties).

Ultimately, the anthropology of perfectibility is at stake in logic. As we have seen, perfectibility, as a capacity the Creator has bestowed upon humanity, entails the individual obligation to perfect oneself. Thinking correctly is a dimension of this obligation, and that is why the Yverdon

55. EP, CONCEPTION (*Logique*), by Jaucourt, 3, p. 803.
56. EY, CONCEPTION (*Psychologie*), R by Mingard, 10, 669b–671a, p. 670a.
57. Ibid., p. 670a–b.
58. Ibid., p. 670b–671a.

Encyclopédie thoroughly reviews issues of prejudice, opinion, doubt, and appearance, that is, various forms of knowledge and certainty.

As in Locke and Condillac, an exploration of the origin of knowledge and its operations is at the very heart of logic. The Yverdon *Encyclopédie* kept the article CONNOISSANCE from the *Encyclopédie* but added an introduction by Mingard. In the *Encyclopédie*, knowledge is defined, in Lockean fashion, as "the perception of the association and congruence, or of the contradiction and lack of congruence, between two of our ideas"; this of course occurs in our mind and by means of "acts" of the mind.[59] In Mingard's view, the Lockean definition, like the Wolffian ("the act of the soul that procures the notion or the idea of a thing"), only identifies "sorts" or "degrees" of knowledge. Knowledge has to be further defined as "the state of our mind insofar as it has a distinct idea of the existence, nature, and relations of a thing," or as the mind's representation "of the existence, qualities, faculties, state, relations, and destination of a thing."[60] Such redefinition emphasized the mind's empirical functioning—which is why the question of knowledge is psychological—as well as the importance of understanding this functioning for the art of thinking correctly—which is why it is logical. Mingard stressed that the project of investigating how and what humans know, and with what degree of certainty, implied the rejection of Pyrrhonism,[61] and that empirical psychology could even invalidate skepticism.

The way "prejudice" is analyzed exemplifies the two approaches. The Paris *Encyclopédie* attaches the subject only to logic and defines prejudice as "the false judgment that the soul forms of the nature of things after insufficient exercise of the intellectual faculties; this unfortunate fruit of ignorance taints the mind and blinds and fetters it." The article remains on the level of casuistry and reduces the reasons for prejudice to certain human inclinations. We have a "bias toward affirmative arguments," which makes us favor the first opinion we make and encourages superstition, or else the understanding tends to rush to extremes, attributing universality to a particular principle (if stars revolve in circles, none can describe an ellipse). While individual prejudices are determined by temperament, habit and age, public prejudices ("which may be deemed the apotheosis of error") result from adherence to old customs or new fashions. Factions or schools give rise to yet other prejudices (for example, the supreme authority formerly

59. EP, CONNOISSANCE (*Métaphysique*), 3, 889–898, p. 889.

60. EY, CONNOISSANCE (*Métaphysique, Psychologie, Logique*), R by Mingard, 10, 49b–67b, p. 49b. Mingard's contribution is a signed introduction, pp. 49b–50b.

61. Ibid., p. 50b.

vested in Aristotle), as do the passions and the senses.[62] Prejudice here is simply an obstacle to gaining knowledge of the truth; it is a logical problem whose different types can be identified, but its mechanisms are never really examined. It is only when man has rid himself of prejudices, says the *Encyclopédie*, that he can approach nature "with an unclouded mind and a pure heart."

The Yverdon *Encyclopédie* keeps the original text but adds a new article in which prejudice is not only a logical but also a psychological and moral topic. The link between different cognitive operations and their logical finality is made visible in the following definition: prejudice is not simply a "false judgment" resulting from "insufficient exercise of the intellectual faculties" but a "judgment passed on the truth or falsity of a proposition, before we are sufficiently acquainted with and have sufficiently examined the available proofs."[63] As such, a prejudice may be a true judgment; its moral and political consequences stem from the fact that we have not examined its basis. Thus has the clergy often sought to instill "the most erroneous prejudices" into the people. Nevertheless, Mingard claims that even false prejudices can be useful. He rehearses an argument widespread among the enlightened elite,[64] namely, that most people are guided by prejudice "because they are incapable of judging for themselves a great many speculative and practical truths, while needing to decide on them"; consequently, the "imperfect examination" they would conduct would be more harmful than assenting to prejudice. As a consequence, while one should not follow the example of the Roman Catholic "professors of falsehood," one should also fight against those who, having never examined anything for themselves, yield to unbelief, debasing as prejudices the most widely held religious beliefs and moral principles. It is better, says Mingard, to keep an ill-fitting bit on a mettlesome horse than to remove it, for the horse will then run completely out of control. This is why he endorsed the idea of hell, which bridles "criminal desires." Mingard thus departs from the *philosophe* spirit of the Paris *Encyclopédie* and does so by appealing to cognitive effects, psychological mechanisms, and moral consequences. He concludes that some prejudices are worthy of respect and recommends maintaining a prejudice

---

62. EP, PRÉJUGÉ (*Logique*), by Jaucourt, 13, pp. 284–285.

63. EY, PRÉJUGÉ (*Psychologie, Logique, Morale*), R by Mingard (signing as "M.D.B."), *Supplément*, 5, pp. 378b–393b.

64. Harvey Chisik, *The limits of reform in the Enlightenment: Attitudes towards the education of the lower classes in eighteenth-century France* (Princeton: Princeton University Press, 1981).

"if it causes less disorder by its existence than we would produce by destroying it."

In Diderot and d'Alembert, "opinion" is "the assent which the mind gives to propositions which do not at first sight appear to be true."[65] The article moves from the distinction between science (in the sense of certain knowledge) and opinion to that between science and faith. Since opinions on a subject involve doubt, they are incompatible with the science of the same subject, which banishes uncertainty; only in that "school of illusions" which is Scholastic philosophy has anyone ever claimed the opposite. The case is different for the relations between science and faith. When no "natural reason" can confirm a Christian truth, the latter must be rooted entirely in Revelation and therefore be partly obscure. But this obscurity does not imply doubt, since both science and faith "command unswerving and certain assent equally." In short, whereas in philosophy assent results from certainty and involves an "act of reason" motivated by the persuasiveness of the proofs, in religion assent is an act of faith dictated by "the will subjected to the authority of Revelation."[66] In its treatment of "opinion," the French article rejects this kind of authority through an appeal to logic.

The article rewritten by Mingard for the Yverdon *Encyclopédie* takes a different approach. It also concerns the opposition between opinion and science but treats it empirically. Similar definitions are used in the two articles, but Mingard's focus is the "state" of the soul in relation to an inadequately proven proposition or a "simply probable" truth. Just as one person's opinion may be another person's science, depending on how knowledgeable they are, so the opinions of one and the same individual may evolve.[67] But there are two categories of persons who always claim to be right: the skeptics and atheists, who defend opinions which gratify their passions while maintaining that they are truths; and the common people along with "the untold number of individuals who do not use their reason," for whom everything is certain.[68]

"Opinion" would therefore be an issue for logic and psychology, since if one becomes aware of its mechanisms, one can understand how it may be called into question.[69] What psychology has to say about opinion is not

65. EP, OPINION (*Logique*), 11, 506–507, p. 506.

66. Ibid., p. 507.

67. EY, OPINION (*Psychologie, Logique*), R by Mingard, 31, 254b–258b, p. 255b.

68. Ibid., p. 257a.

69. Mingard also contributed a long article, CERTITUDE (*Métaphysique, Logique, Morale*), N, 8, 432a–447a. Although certainty is defined as the state of the mind "when it is convinced that the judgment it passes is true and it is in no doubt about the truth of the proposition which expresses it" (432b), the article is much less psychological than the others considered here. It opens with

without consequences for an enlightened education. Mingard notes that most ideas and teachings in theology and philosophy are opinions. Driven by self-interest or pride, theologians and philosophers cluttered up learning with insoluble and irrelevant questions; they fueled disputes, controversies, and persecution and made it more difficult to "study the most necessary sciences." This was why, in Mingard's view, it was important never to present opinions to the public, but only "useful truths," and not to communicate to the "common people" opinions which should better remain among scholars.[70] In short, one should not raise doubts liable to undermine the most firmly established truths.

In the Paris *Encyclopédie*, the abbé Edme-François Mallet attaches the problem of doubt to logic and metaphysics. For the logical aspect, he explains the difference between "actual" doubt, where the mind remains undecided, and "methodical" or Cartesian doubt, where it voluntarily suspends its assent while it gathers irrefutable proofs of truths it already accepts.[71] Most of the article is devoted to the metaphysical aspect and sets methodical doubt against skeptical doubt, attacking skepticism and its doubting "shrouded in darkness." Such is atheists' doubt, for example; they refuse to believe "out of anger and brutishness, blindness and malice, and even out of whimsy, and because they want to doubt."[72] Whereas Mallet has no kind words for Pyrrhonism, Diderot sings the praises of Montaigne and Bayle and shows much sympathy for those whom Mallet calls a "sect of liars."[73] But neither Mallet's article on doubt nor Diderot's on Pyrrhonism investigates doubt as a psychological process.

---

an explanation of the five sources of certain knowledge and the sorts of certainty corresponding to them: the "inner feeling of what is going on inside us" (physical certainty), "seeing immediately the congruence between ideas" (metaphysical certainty), "the persuasiveness of arguments" (logical certainty), "the analogy between facts" (analogical certainty), and evidence (intellectual certainty). The issue of miracles is discussed principally in relation to the last two categories. This article, CERTITUDE, which Bonnet praised, was to have replaced the one from the *Encyclopédie* by the abbé de Prades, which de Felice kept, he claimed, only to satisfy "the entreaties of some friends" (431).

70. EY, OPINION, pp. 258a, b. Various articles on religion stress this point. For example, EY, ORIGINEL, R, 31, pp. 458a–461b, which cites disagreements on original sin and accuses the theologians of "not so much basing their system on the scriptures as of adjusting the scriptures to their system."

71. EP, DOUTE (*Logique, Métaphysique*), by Mallet, 5, 87–90, p. 87. Mallet, a professor of theology at the University of Paris, was a fervent royalist and quite conservative in religious matters. He contributed to the *Encyclopédie* more than two thousand articles.

72. Ibid., p. 88.

73. Ibid. See PYRRHONIENNE OU SCEPTIQUE, PHILOSOPHIE (*Histoire de la Philosophie*), [by Diderot], 13, 608–614.

In the Yverdon *Encyclopédie*, by contrast, Mingard devotes two articles to the topic, one psychological and one logical, printed in that sequence in order to show the relation of dependence between the two disciplines. The first treats doubt as a voluntary or involuntary state of the soul. In the case of involuntary doubt due to lack of knowledge, the understanding cannot decide about a proposition's truth value. Mingard's examples give his articles an empirical density that is generally absent from the corresponding entries in Diderot and d'Alembert. If, he writes, I have never heard of Amazons, I am in a state of ignorance of which I am not even aware. If I read their history in Herodotus, I may accept it as true, but if I were told that some authors believe it and others reject it as untrue, I would probably be left in doubt.[74] The examples highlight a fundamental difference between ignorance and doubt: a person is aware of doubting. They also show that doubt is not a judgment of the soul on an object it desires to know but a judgment on its own state and its incapacity to know truth with certainty.[75] Since this incapacity is not absolute, there are degrees of doubt, between which the mind hesitates, depending on the reasons being considered. Doubt is not itself this "wavering state" but the condition one arrives at after examining the relevant arguments.[76] Now, since doubt is disagreeable for the soul, we tend to avoid it. Sometimes, however, we choose to suspend our judgment, fearing that what we believed to be true may not be so. Such doubt is caused by the "desire to know the true and the fear of being mistaken." It can therefore be called "voluntary," but it is not entirely so, because it arises spontaneously whenever one realizes one might be mistaken.

As an antidote to "freethinkers" whose beliefs are unfounded and who doubt out of laziness, Mingard recommends the "wise doubt of the philosopher." But his article is not polemical and focuses on psychological analysis. The ensuing article, attached to logic, concerns the usefulness of doubt for pursuing truth. Mingard considers that it is "more blameworthy to remain in doubt due to lack of study than to remain in ignorance."[77] Whereas such doubt "dishonors" the person who does not attempt to educate himself, voluntary doubt "is necessary and praiseworthy." Philosophers recommend it "as a useful weapon against error, and as the most effective defense against prejudices and their dire consequences" (467b). However, if we are to preserve our "reason and innocence," voluntary doubt must be inspired by the

---

74. EY, DOUTE (*Psychologie*), N by Mingard, 14, 462a–466b, p. 462 ab.
75. Ibid., p. 463a.
76. Ibid., p. 464a.
77. EY, DOUTE (*Logique*), R by Mingard, 14, 466b–472a, p. 467ab.

love of truth, as in Descartes. Otherwise it becomes a prejudice, as with the numerous "doctors of erudition" who only seek to satisfy some personal passion (471b).

The examples cited in the psychological article on doubt—the existence of Amazons and of the female Pope Joan—are irrelevant to how the understanding proceeds in practice. In the logical article, however, the examples not only illustrate a line of reasoning but additionally apply it for didactic and moral purposes. First, Mingard explains, from the fact that a problem is insoluble for us we must not assume that it is insoluble for the human mind in general. Locke, for example, an "excellent philosopher," was unable to decide whether or not matter could think. Others, with "keener minds than his in this respect" (468a), dispelled the doubt. Second, notions that are eminently open to doubt—such as monads, preestablished harmony, or innate ideas—become "an object of mechanical belief through habit." Mingard concludes that Leibniz and Descartes themselves did not accept such ideas as truths "but only as the offspring of their imagination, which they defended only for their own amusement." This reassuring intepretation, generous toward those whom Mingard also labels "philosophers enthralled to a system," leads him to observe that "the assertive tone with which a person defends an opinion is not a certain sign that the latter lies beyond all doubt" (468b, 469a).

Furthermore, doubts may persist on certain issues without this destroying the certainty we have concerning their object. Would I ever call into doubt my moral freedom, which I feel intimately, "because I do not understand how, since I am free, an omniscient being can predict the determinations of my will?" Or the reciprocal influence of soul and body because I do not know of any system that accounts for it with certainty? Should I doubt Revelation because it expounds things that I am unable to explain? "These considerations," Mingard concludes, "must necessarily be applied in all cases in which necessary doubt is cast not on a subject's essence but only on secondary circumstances—on how, not what, the thing is" (469a–b).[78] Last, doubt should not be confused with negation, since the human mind is made in such a way that it can admit the truth of facts it cannot explain. "We call into doubt the doctrine which explains the facts to us; . . . and yet we are most certain of the facts that are its object." In short, "Doubting the

---

78. Mingard develops this last point eloquently in EY, COMMENT (*Métaphysique*), N, 10, pp. 465a–469a.

fact is doubting the manner [in which the object exists], but doubting the manner is not doubting the fact" (469b, 470a).

Mingard wishes here to underpin a defense of Christianity with a psychology of doubt and to draw out what this implies for the art of thinking correctly. The article on Pyrrhonism in the Yverdon *Encyclopédie* pursues this double, psychological and logico-critical, approach. It mentions some empirical sources of doubt (the senses, the diversity of laws, imagination, and customs), approves certain uses of skepticism as an antidote "to prejudices and bias," but decries the modern Pyrrhonists—above all Bayle—for the "baneful skill" with which they insidiously undermine the foundations of truths held to be certain.[79] The skeptic's art is based on one of the main sources of doubt, namely, appearance (in the dual sense of what is perceived and what appears to be true). Bayle is the prime antihero of the Yverdon *Encyclopédie*: no one is "less philosophical" than he, "or else philosophy is the mother of all arbitrariness and whimsy."[80]

Although APPARENCE in the Yverdon *Encyclopédie* seems to be a new article, it is in fact a thorough reworking of the Paris entry. Diderot contributed on the subject a very short article on "grammar," followed by a definition and comments, one of which attributed the source of error to the confusion between the true and "what only appears so."[81] This last issue was addressed elsewhere, for instance, in the article ERREUR (*Philosophie*) and EVIDENCE (*Métaphysique*). It surfaced again, in the discussion of errors caused by perception, in de Felice's article ERREUR (*Logique, Morale*) for the Swiss encyclopedia.

Mingard's goal was to explore how the senses are a source of error, and to this end he treated "appearance" as a phenomenon in which metaphysics, philosophy, morals, psychology, and logic converge. He started from the distinction between what goes on inside us when we feel sensations and what, in the external object, causes those impressions. The very fact that we receive an impression proves to Mingard that objects exist: "I would not see the *appearance* of a rose if there did not exist outside of me a being fashioned and modified in such a way as to reflect back the light with the modifications characterizing a rose."[82] Deforming mirrors or the stick

---

79. EY, PYRRHONISME (*Philosophie*), R, 35, 613b–619a, pp. 614b, 618b. The article could be by Mingard.

80. EY, BAYLE, PIERRE (*Histoire littéraire*), N, 5, 109b–111b, p. 111a.

81. EP, APPARENCE (*Grammaire*), by Diderot, 1, 543.

82. EY, APPARENCE (*Métaphysique, Philosophie, Morale, Psychologie, Logique*), N by Mingard, 3, 172a–176a, p. 173a.

which appears bent in water show that physical appearances are deceptive only when a new cause changes the way an object acts on the sensory organs. "Physical appearance," therefore, "does not deceive us naturally."[83]

We generally acquire new ideas by combining simple or composite ideas.[84] This mechanism explains the appearance of intellectual objects. This appearance is "only the first notion we form for ourselves of the congruence or contradiction between two confused ideas we have joined together to form a new concept."[85] Since we have not analyzed these ideas, we do not realize that from other angles they may be united or separated, mutually exclusive or inclusive; we say then that our judgment about them pertains *only to appearances*. Mingard goes on to illustrate the moral significance of this analysis. One can imagine that God could not "destine man for all eternity to woes without end" without man's having the right to complain. However, if we analyze the ideas contained in this judgment (*right, authority, absolute, man, God, eternity*), we see that the right to dispose of a thing and the quality of the Creator of this thing are mutually compatible (174a–b).

According to Mingard, physical and intellectual appearances generally stem from properties that really exist in the object. However, they are sometimes the product of circumstances which we take to be signs of the existence of objects different from those we perceive. This is the case—and one which illustrates Mingard's reformist inclinations—of a man accused of murder on the basis of coincidences (he carries a sword, happens to be passing by the corpse at the moment when it is discovered, goes pale when he realizes he is suspected, etc.), who is "barbarically" put to the question and has a confession wrenched out of him which proves his guilt. This deplorable procedure has a psychological basis: "We are led by analogy and by the association of ideas to view certain *appearances* as signs of things they do not represent at all, but which they sometimes accompany" (175a).

In short, the psychological analysis of "appearance," while leading to philosophical and metaphysical conclusions, basically demonstrates that psychology should become the foundation of logic, that it is an instrument of moral judgment, and that it guarantees equitable and rational action. The article APPARENCE, moreover, explains its subject matter while putting into practice its recommendations. It is a significant example of the connections

---

83. Ibid., p. 173b.
84. EY, COMBINAISON (*Logique, Psychologie*), N by Mingard, 10, pp. 398b–400a.
85. EY, APPARENCE, p. 174a.

Mingard seeks to create between disciplines, and the epistemic and moral value he attributes to psychology.[86]

"Perfection," as Mingard describes it, is "a being's capacity to correspond fully to its purpose." Beings improve themselves by acquiring "new useful powers" or by extending those they already possess.[87] Since a being's perfection resides in the perfection of its faculties, humans must exercise theirs and steer them toward the good "through knowledge of the true."[88] This implies that logic is much more than a tool for solving problems and for reasoning correctly. It is, rather, one of the instruments humans have to perfect themselves morally. There was of course nothing new about combining correct reasoning, an educated will, and good conduct, but claiming that psychology was the foundation of logic certainly strengthened the new claims of the discipline.

MORALS

Like logic and metaphysics, morals was reformulated in psychological terms by identifying, describing, and analyzing "acts" of the soul. Here, Mingard tends to stress the importance of knowledge in the determination of the moral acts of a free being who is nevertheless bound by God's will. Free will, for example, is therefore not "a blind spring, but a potential realized in actions which are always determined by more or less clear knowledge and more or less clear judgments." And since a "depraved person" cannot claim to have done wrong despite himself, nor a righteous one to have done good against his will, we find side by side "free will and man's frailty, the need for grace and the realm of Providence."[89]

---

86. The pastor Alexandre-César Chavannes, who was more conservative than Mingard in theological matters and more dogmatic in moral ones, contributed the articles APPARENCE DU MAL (Morale), N, 3, 176a–176b, and APPARENCE (Morale), N, 3, 176b–178a. The former stresses that Christians must not only abstain from evil but also from what appears to be evil; the latter highlights the need to be careful in judging others, and not to assess one's own merit on the basis of one's reputation or the verdict of the public.

87. EY, PERFECTION (Métaphysique), R by Mingard, 33, 41b–43a, pp. 41b and 43a. In line with Wolff's definition, Mingard emphasizes the teleological dimension of perfectibility. He elaborates on perfectibility separately, as we saw above, and turns it into a central anthropological concept. In the Paris Encyclopédie, perfection is "the harmony which reigns within the variety of several different things, which all work toward the same goal." However, it is not treated in terms of perfectibility or of man's ultimate purpose; EP, PERFECTION (Métaphysique), 12, pp. 351–352.

88. "Our faculties will therefore become perfected through the exercise of these powers directed toward the good through knowledge of the true." EY, FACULTÉ, p. 244b.

89. EY, ARBITRE, pp. 300b–301a.

TABLE 8.4. The overlap between psychology and morals (moral philosophy)

EY headword	Field indicators in addition to *Psychologie*	EP field indicators
APPARENCE (N)	*Métaphysique, Philosophie morale, Logique*	*Grammaire*
APPÉTIT (R)	*Physiologie*	*Morale*
BEAUTÉ (R)	*Métaphysique, Morale*	
CŒUR (N)	*Morale*	
CONCUPISCENCE (R)	*Morale*	*[Théologie]*
COUTUME (R)	*Physique, Philosophie morale*	*Morale*
FIN (N)	*Métaphysique*	*Morale*
PASSION (R)	*Morale*	(PASSIONS) *Philosophie, Logique, Morale*
PLAISIR (R)	*Philosophie morale*	*Morale*
PLAISIRS	*Philosophie morale*	
PUDEUR (R)	*Morale*	*Morale*
SPONTANÉ, SPONTANÉE (R)	*Philosophie, Métaphysique, Morale*	(SPONTANÉE) *Grammaire, Médecine*
SYMPATHIE (R)	*Anthropologie, Physiologie, Philosophie morale*	

Table 8.4 shows the articles in the Yverdon *Encyclopédie* attached to psychology and morals, as well as the articles attached to morals in the Paris *Encyclopédie and* at least to psychology in the Swiss encyclopedia. All the articles are either rewritten or new, but all have corresponding headwords in Diderot and d'Alembert.

We examine below how the psychologizing of morals in the Yverdon *Encyclopédie* presupposes a certain idea of perfection and aims to show how humankind's perfectibility should be realized.

Let us turn first to the issue of "custom." On an individual level, *common practice* is defined as that which a person does ordinarily, *habit* as the ability (acquired through repetition) to do something with ease, and *custom* as a "state" of the soul and body when they have become accustomed to a perception or a physical impression and are no longer struck by it as they previously were.[90] While the study of custom is a matter for "physics" and psychology, its consequences are particularly relevant to moral philosophy.

90. EY, COUTUME, USAGE, HABITUDE (*Grammaire*), R by Mingard, 12, pp. 191b–192b.

In its physical effects, custom reduces our capacity to experience certain sensations and thus reveals our considerable, but not limitless, ability to adapt. Since custom not only inures us to pain but also numbs our sensitivity to pleasure, we should accustom ourselves to "painful impressions" so that they do not become an obstacle to our happiness, but not to "agreeable sensations of pleasure," lest we lose our capacity to enjoy them.[91]

As regards its psychological effects, custom determines prejudices, tastes, and moral behavior. If we consider an absurd proposition true because we have been told so since childhood, this is not because its truth has been demonstrated to us but because we are used to neglecting contrary proofs. The same applies to taste and above all to moral behavior: "*Custom renders useless the precautions nature has taken to make us virtuous.*" This is why custom has a decisive influence "on the state of man, his progress toward perfection, and his moral character."[92] The topic of custom thus sheds light on the obstacles to the realization of perfectibility.

The topic of beauty, on the other hand, clarifies the notion of perfection. Diderot's short article BEAUTÉ addresses issues central to his philosophy and aesthetics. Whereas, the philosopher says, we usually consider beautiful whatever elicits "the perception of appealing relations," it is in fact "everything that elicits in us the perception of relations" that is beautiful.[93] Diderot's aesthetic is further developed in BEAU (*Métaphysique*). Mingard retains the heading to elaborate a different view, indebted to Wolff's notion of beauty as "the perceivable quality of perfection" or as the perfection of things, insofar as it can give us pleasure.[94] For Diderot, beauty is based on the "perception of relations"; Mingard attacks this as too vague and insufficiently demonstrated and defines beauty as "a relation of conformity between the parts and the perfection of the whole."[95] In Wolffian terms, "the beautiful is the perfect being, insofar as we observe its perfection."[96] We do not distinguish the beautiful from the ugly by instinct but by means of a knowledge, often unclear, of "what is appropriate in each case for the

---

91. EY, COUTUME (*Physique, Psychologie, Philosophie morale*), R by Mingard, 12, 192b–197a, p. 194a. Cf. EP, COUTUME (*Morale*), by Formey, 4, pp. 410–411—a moralist's approach that does not address psycho-physiological mechanisms at all.

92. Ibid., p. 196b.

93. EP, BEAUTÉ, by Diderot, 2, pp. 182.

94. "Pulchritudo consistit in perfectione rei, quatenus ea vi illius ad voluptatem in nobis producendam apta"; "quod sit rei aptitudo producendi in nobis voluptatem, vel, quod sit observabilitas perfectionis: etenim in hac observabilitate aptitudo ista consistit." Wolff, *Psychologia empirica*, §§ 544 and 545.

95. EY, BEAU, BELLE (*Métaphysique*), R by Mingard, 5, 117a–136a, p. 127b.

96. Ibid., p. 136a.

perfection of each thing, and what serves it best for the fulfillment of its purpose."[97] Mingard's BEAUTÉ develops this aesthetic theory, bringing it out of the purely metaphysical domain into psychology and morals.

Beauty, Mingard explains, is not a sensation caused by a physical quality or a material impression, but the "reflective feeling" we experience whenever we discover a being's perfection: "Wherever perfection exists, if the being is such that its perfection may be discovered by the senses or reflection, *the clear and visible features of this perfection will always be beauty.*"[98] This perfection resides in the being's "capacity to fulfill its purpose" (145a). Symmetry, for example, is a sign of the perfection of bodies, human or architectural, that stand upright. A building can nonetheless "be perfect but without *beauty*" (145b). Mingard's favorite example was Gothic buildings, which he loathed; their wings seemed to him "hugely bulky," beside which the central body appeared "too feeble a link to hold together what, if we abide by good taste, which is none other than the law of nature, should be no more than props" (146a–b).[99]

Mingard's teleological argument embodies a metaphysics of beauty suited to an anthropology of perfectibility. We know, for example, that the finalities of men and women are not the same. Experience then teaches us that the distinctive features of each (the man's visible muscles and joints, the woman's "rounded" limbs and the smoothness of her movements) are not "constitutive characteristics, but signs which express perfection." Once we have recognized these as such, they "become in our eyes really *beautiful*" (147b). Psychology and morals are here inseparable: beauty "is pleasing to the mind because it offers the mind distinct ideas which it illuminates and which are easy to grasp; it speaks to the heart because it signals a physical and moral capacity to contribute to our happiness" (149b). Beauty was created to indicate perfection to beings capable of perceiving it: "The Creator, who wanted to make us perfect, leads us to perfection through the powerful charm of *beauty*; it is through the perfection it heralds that beauty rightfully pleases us. . . . *Beauty* is created in order that we should delight in it, but only as the harbinger of perfection" (150a). In the article GOÛT, similar ideas are given a strong psychological inflection.[100]

97. Ibid., p. 127a–b.
98. EY, BEAUTÉ (*Métaphysique, Psychologie, Morale*), R by Mingard, 5, 143b–150b, p. 147a.
99. Cf. EY, BEAUX-ARTS (*Métaphysique*), N by Mingard, 5, 151a–154a, pp. 153b–154a.
100. EY, GOÛT (*Grammaire, Beaux Arts, Littérature, Psychologie, Philosophie*), R by Mingard, 22, pp. 31b–47b. Mingard substitutes the original article for another in which he again formulates his aesthetic ideas and examines the psychological conditions of refined taste: more

As we have seen, custom may block man's progress toward perfection, of which beauty is the external sign. From a Christian perspective, however, the roots of good and evil are to be found only within the human being. Mingard treats this principle more psychologically than theologically. Whereas theologians saw concupiscence as the source of original sin and evil actions, Mingard identifies two human goals—preservation of the individual and procreation—that are necessary yet require acts "in themselves disagreeable." Human beings experience needs that make them want to carry out these acts; the pleasure produced by the satisfaction of these needs compensates for "the unpleasantness of the acts required to fulfill man's purpose."[101] Concupiscence is therefore "a natural and necessary appetite" and is "depraved" and blameworthy only when it leads to breaking social and moral rules.

By emphasizing the psychological basis of concupiscence, Mingard takes the concept out of the purview of theology and places it in a teleological (but not eschatological) framework at the juncture of several other notions, themselves reworked in psychological terms. For example, he explores the psychological dimension of the sense of modesty, seeing it as a phenomenon prescribed by nature for the good of humanity, which "calls for chaste love and stable marriage." He imagines that the prohibition on eating the fruit of the tree of knowledge was perhaps a prohibition on Adam and Eve's "acting as husband and wife before having received the express permission of their Creator" and concludes that their sin "was that of two lovers who, thoughtlessly and precipitately, yielded to their passion."[102] Concupiscence therefore derives from needs that humans must satisfy to fulfill a divinely prescribed purpose.[103]

---

attentive observation, a finer sensibility, senses subtle enough to detect the slightest nuance, a vivid imagination, and an accurate memory. Cf. EP, Goût (*Grammaire, Littérature, Philosophie*), 7, pp. 761–770, composed of texts by Voltaire, Montesquieu, and d'Alembert (Mingard criticizes the latter at the end of his article).

101. EY, Concupiscence (*Psychologie, Morale*), R by Mingard, 10, 720a–721a, p. 720b. Mallet's approach in the *Encyclopédie* is quite different. He situates concupiscence in an exclusively "theological" perspective in which "immoderate desire, or coveting sensual things, [is] inherent in man since his Fall"; EP, Concupiscence ([*Théologie*]), by Mallet, 3, p. 832.

102. EY, Pudeur (*Psychologie, Morale*), R by Mingard, *Supplément*, 5, 452b–471a, pp. 467a and 454b. This article replaces EY, Pudeur (*Morale*), 35, pp. 532a–533a, which is identical to Jaucourt's in the *Encyclopédie*.

103. EY, Appétit (*Physiologie, Psychologie*), N by Mingard, 3, pp. 197b–202b. See also EY, Appéter (*Psychologie, Œconomie animale*), R by Mingard, 3, pp. 197a–197b. In his article for the *Encyclopédie*, Yvon merely replaces, without psychological elaboration, the Scholastic distinction between irascible and concupiscible appetites with the Wolffian hierarchy of sensitive and rational appetite (will). See EP, Appétit (*Morale*), by Yvon, 1, p. 549.

Both man and animals have sensitive appetites. In addition, humans have a "rational" appetite whose finality is the soul's perfection. In man, all appetites stem from the soul, which alone feels, wants, and desires.[104] If the soul were separated from the body, it would have only rational appetites. Since the Creator made man as a composite of soul and body, he wanted these two substances to "assist each other," and for the body, as "an instrument necessary to the soul's perfection," to be maintained "in the best possible condition."[105] Following from such considerations, Mingard argues that since the only goal we pursue in satisfying our appetites is to meet our needs and thereby fulfill our higher purpose, any desire that takes us beyond need must be suppressed.[106]

God, Mingard explains, gave man pleasure "as the agreeable and powerful source of his activity," and its correct use is an aspect of controlling the passions.[107] But whereas the French encyclopedia treats the passions from the viewpoint of the physician and moralist, and principally as dangers, the Swiss one reframes them in a psycho-teleological perspective.[108] Passion is, for Mingard, an "act of the soul that judges" the effect of an object on its happiness.[109]

The soul can of course be mistaken and attribute value to an object that has none: that is when dangerous passions arise, "the source of all our aberrations, our crimes, and our misfortunes."[110] In themselves, however, passions are as necessary as all the other faculties "through which the Creator allows man to fulfill his purpose."[111] As a consequence, neither concupiscence nor the passions or pleasure are intrinsically linked to sin, which Mingard simply defines as "any action which makes beings deviate from

104. Following in Wolff's footsteps, Mingard distinguishes between a lower and a higher appetitive faculty.

105. EY, Appétit, p. 200b.

106. EY, Appétit (*Morale*), R by Mingard, 3, pp. 202a–203a.

107. EY, Plaisir (*Psychologie, Philosophie morale*), R by Mingard, 33, 742b–745b, p. 742b. Mingard adopts here (pp. 745a–b) the psychology of pleasure developed by Johann Georg Sulzer in his articles on the origin of agreeable and disagreeable feelings, published in the *History* of the Berlin Academy for 1751 and 1752. EP, Plaisir (*Morale*), also notes the usefulness of pleasure. The article Plaisirs (*Psychologie, Philosophie morale*) in EY (*Supplément*, 5, pp. 291b–301a) is taken, without acknowledgment, from Merian's article on the duration and intensity of pleasure and pain (1766), mentioned above in chap. 4.

108. EP, Passions (*Philosophie, Logique, Morale*), 12, pp. 142–146; EP, Passions (*Médecine, Hygiène, Pathologie, Thérapeutique*), 12, pp. 149–150.

109. EY, Passion (*Psychologie, Morale*), R by Mingard, 32, 431a–435b, p. 431b.

110. Ibid., p. 434b.

111. Ibid., p. 432b.

their true destination."[112] The way they function obeys the physiological and psychological mechanisms of need and desire, which are necessary to a species endowed with free will and the capacity to choose.[113]

THE PSYCHO-ANTHROPOLOGY OF PERFECTIBILITY

We have seen that the psychological transformation of logic, morals, and metaphysics in the Yverdon *Encyclopédie* is achieved by means of the systematic use of the field indicator *Psychologie* and an analysis of the soul's "acts." The psychological articles avoid polemic and tend to put into practice the analytic procedures they theorize. Psychological faculties here emerge as the primary object and principal means of human perfectibility:

> Since the perfection of a being consists in the perfection of his *faculties*, that is, in their number and development, and since a being's happiness is a function of this perfection, it follows that our dependence on the first cause and the concern for our happiness oblige us to work, to the best of our capacities, at perfecting the *faculties* with which we have been endowed. But these *faculties* are perfected only through exercise, and this exercise improves them only when it is in harmony with our destination; from this ensues the obligation both to use our *faculties* and to use them as the author from whom we received them intended.[114]

The Christian religion of course remained the "august and wise mistress" of the Swiss encyclopedists.[115] However, Christian dogma counted for them considerably less than the anthropological teleology of perfectibility. God assigned the universe a destination we cannot entirely fathom, but with which we can comply by following our nature and the nature of things. Man's anthropological, rather than transcendent or eschatological, finality is a direct consequence of his perfectibility, which prescribes ends that are in harmony with human nature. As a result, when we attempt to perfect our

---

112. EY, Péché, R by Mingard (M.D.B.), 32, 581a–583b, p. 581b. In its narrowest theological sense, "sin" refers to the actions carried out against the known will of God and as such is the opposite of "good works." Mingard's article ends with a denunciation of the "Roman" distinction between mortal and venial sins, which he considers "imprudent" and detrimental to morality.

113. Cf. EY, Choix (*Psychologie*), N by Mingard, 9, pp. 497a–500b. Desire is described here as "the constant, universal, and sole motivation for all voluntary and spontaneous actions" (p. 497b). These actions are the only ones to involve choice, which is an "act of the soul."

114. EY, Faculté, p. 244a.

115. EY, Pyrrhonisme, p. 613.

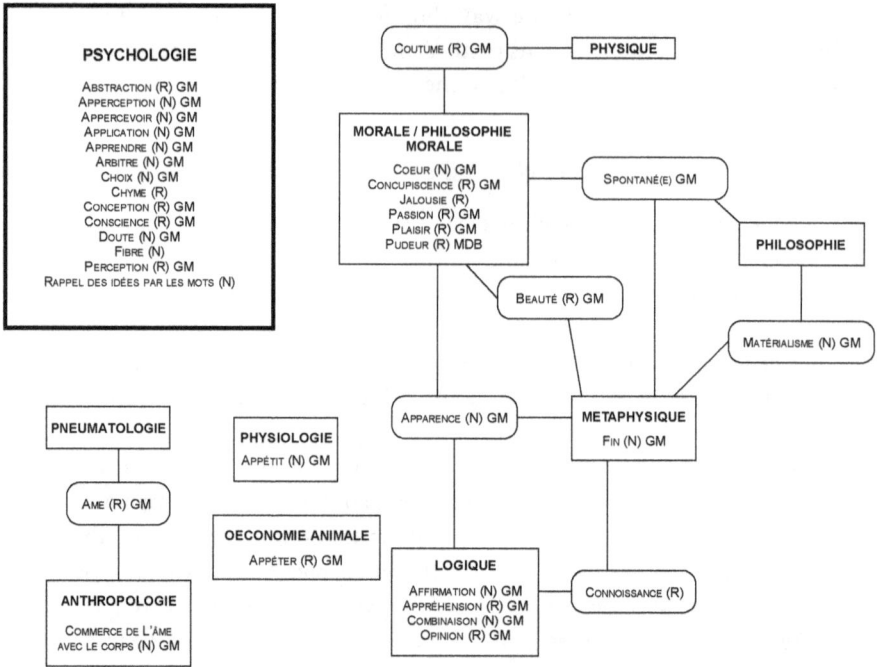

Fig. 8.1. Yverdon *Encyclopédie*. The articles on psychology and their links to other disciplines.

faculties, we do so in conformity with intrinsically human cognitive capacities and limitations. And since the soul's activity involves the faculties and is also the proper object of psychology, the latter is intrinsically linked to logic and morals, insofar as they are disciplines devoted to correct thought and action. Psychology can then use these two sciences to establish and strengthen its links with philosophy, metaphysics, and physics.

Figure 8.1 schematizes these relations by means of field indicators: *all* the articles featured are attached to psychology; those which belong *only* to psychology are placed in its box; the headwords in the boxes of other disciplines belong to that discipline *and* to psychology; the articles attached to psychology and to at least two other fields have round-cornered boxes of their own and are linked to these fields by lines. The diagram displays the extent of the overlap between psychology, on the one hand, and morals and logic, on the other. It also shows that the only notion that belongs to psychology and anthropology alone is "interaction between the soul and the body" (*commerce de l'âme avec le corps*). Since man is defined as a composite of these two substances, their interaction is a fundamental anthropological fact. This fact gives psychology its position as the first science within the science of man. Moreover, since the soul is knowable only through its activity as long as it is united with the body, an understanding of body-soul interaction is necessary for the realization of perfectibility, whose concept too belongs in anthropology.

The Union and Interaction of the Soul and the Body

The interaction of soul and body is at the very heart of the psycho-anthropological system of the Yverdon *Encyclopédie*. It refers to the relation of "mutual dependence" between the state and movements of each substance. Knowledge of this fact, says Mingard, "is an essential part of the natural history of mankind, and one of the fundamental principles of anthropology."[116] As a concept, it is sharply distinguished from the "union" of the two substances, which belongs exclusively to metaphysics.[117] Insofar as the interaction requires the union, and vice versa, the two notions are indeed inseparable, but they concern two different modes of knowledge that are not to be conflated.

In contrast to the Swiss encyclopedia, the *Encyclopédie* tends to elide the two questions, in a strategy that amounts to a radical critique of the very notion of a union between soul and body. The entry AME emphasizes that "while we are alive," the soul's functions depend on the body's organization and condition. This "mutual dependence of the body and of what thinks in man" is precisely the union of the two substances. Since Revelation and "sound philosophy" show that God freely willed it thus, the fact lies beyond doubt. Nevertheless, we have "no immediate idea of the dependence, union, or relation between these two things, *body* and *thought*."[118] While such a union is a revealed truth whose mode of operation remains "absolutely unknown," the fact of the union itself can be apprehended empirically. On the contrary, for Mingard, the union of the two substances is a purely metaphysical truth, and we can know only its "circumstances" or the "phenomena that manifest it"; in other words, we assume the union, but are able to know only the interaction of soul and body.[119] The soul "manifests itself as not perceiving anything except by means of the body, and this

116. EY, COMMERCE DE L'ÂME AVEC LE CORPS (*Anthropologie, Psychologie*), N by Mingard, 10, 505b–510a, p. 505b.

117. EY, UNION DE L'ÂME ET DU CORPS (*Métaphysique*), N by Mingard, 42, pp. 485b–487b.

118. EP, AME, I, p. 341.

119. EY, UNION DE L'ÂME ET DU CORPS, p. 486a.

is what our inner sentiment, as well as the current and shared daily experience of all of humanity tell us, and we can discover nothing else about its relation [to the body] than the existence of these facts."[120]

Empirical psychology therefore does not exhaust the question of the soul. That is why Mingard's article AME is also attached to metaphysics, pneumatology, psychology, and anthropology and sketches existing disagreements about the essence of the soul, its origin, relation to the body, and ultimate destiny, as well as about the soul of animals. The article takes the form of a general presentation containing numerous references to articles in which, as already mentioned, the reader should be able to find "not only what sound philosophy teaches us on each of these subjects but also its perfect conformity with what Christian theology imparts."[121] The Yverdon *Encyclopédie* could thus defend the immateriality and immortality of the soul while also positioning it at the very heart of anthropology—understood as the natural history of man—and warding off attacks against the principle of the soul's union with the body.

Diderot and d'Alembert's *Encyclopédie* takes a very different course. Some of the articles refer to the union of the soul with the body only in order to profess their ignorance of it or to dismiss certain topics. The author of COULEUR, for example, refuses to say anything about color as a sensation of the soul, because that "depends on the laws of the union of the soul with the body, about which we know nothing."[122] Why is it that the brain has no sensitivity and the nerves are the only sensitive parts in the body? "Since that depends on the laws of the union of the soul with the body, we can give no reason for it."[123] We acquire ideas by means of the senses: "How can objects which simply produce a movement in the nerves imprint *ideas* on the soul? In order to answer this question, we would have to . . . fathom the inexplicable mystery which is the wondrous union of these two substances."[124]

At other points, reference to the union or interaction of soul and body serves merely to explain or express the link between the physical and the mental. It is cited as an irrefutable principle, yet at times with the suggestion that it explains nothing at all and is mentioned simply to give prominence to a purely mechanistic approach. This is the case, for example, of

---

120. EY, APPERCEVOIR, p. 197a.
121. EY, PSYCHOLOGIE, p. 341b.
122. EP, COULEUR (*Physique*), by d'Alembert, 4, 327–333, p. 327.
123. EP, CERVEAU (*Anatomie*), by Tarin, 2, 862–864, p. 864.
124. EP, IDÉE (*Philosophie, Logique*), 8, 489–494, p. 489.

the "laws" of the soul-body union when they explain the function of pain, which is to warn the soul of harmful effects to the body.[125] Or else, when it is said that in order to understand the effects of a painting, "we must represent to ourselves this very close union of the soul with the body."[126] Thought always supposes certain sensory impressions, "independently of the habitual or actual disposition of the brain, and in accordance with the laws of the union of the soul with the body."[127] It is "the movement of the brain's fibers (together with the operations of the soul and consequently the laws of its union with the body) [which] determine internal sensations, ideas, imagination, judgment, and memory."[128] Likewise, such functions as voluntary movements are called *animal* "because they are said to derive from the interaction between the soul and the body; they cannot be performed (in man) without a joint operation of these two agents."[129] As regards vision, "we know from the law of the union of the soul with the body that certain perceptions of the soul necessarily result from certain movements of the body."[130]

The union of the two substances implies that the soul and the body affect each other, but this is the case for all organs. Exceptions would not "be in keeping with the wisdom and power of the Creator." As a consequence, the union must have altered the properties of the soul.[131] The French encyclopedists infer from this that the concepts of union and interaction between body and soul are irrelevant for "animal economy." The article MORT declares "the separation of the soul from the body is a mystery which is perhaps even more incomprehensible than their union. It is a theological dogma ratified by religion, and it consequently is beyond question, but it in no way accords with the insights of reason, nor is it based on any observation drawn from medicine."[132] In short, as the physician and chemist Paul-Jacques Malouin states in PHYSIOLOGIE, the "mutual interaction between the soul and the body" is

not only the most inconceivable idea there is, but additionally the most useless to the physician. The heat produced in the body would be

125. EP, DOULEUR ([*Médecine*]), 5, 83–87, p. 83.
126. EP, EXPRESSION (*Peinture*), by Claude-Henri Watelet, 6, 319, p. 319
127. EP, EDUCATION ([*Métaphysique*]), by César Chesneau du Marsais, 5, 397–403, p. 402.
128. EP, ŒCONOMIE ANIMALE (*Médecine*), 11, 360–366, p. 361.
129. Ibid., p. 362.
130. EP, VISION (*Optique*), 17, 343–347, p. 346.
131. EP, FACULTÉ VITALE, by Jean-Henri-Nicolas Bouillet, 6, 365–371, p. 367. See also p. 368.
132. EP, MORT (*Médecine*), by Paul-Jacques Malouin, 10, 718–727, p. 718.

conceivable even if man were but one substance. . . . Movement can be explained neither through the influence of the body nor the properties of the soul; there is nothing in the idea of the soul that can be found in the idea of movement. That is why heat and movement cannot be accounted for by the soul, and if, in order to explain voluntary movement, you say that it consists in the fact that the soul wills movement, you make it no clearer.[133]

Since recovery "occurs in the human body through the action of other bodies," the soul has no part in it, and "all the systems concerning its interaction with the body are useless." Malouin concludes that "whoever has healed the body should not worry about the soul; it always reliably recovers its functions as soon as the body recovers its own and removes the obstacles which seemed to prevent it from acting."[134]

Predictably, we find no such arguments in the Yverdon *Encyclopédie*, in which no article, to my knowledge, declares the uselessness of psychology for medicine. In PHYSIOLOGIE, the devout Albrecht von Haller focuses on a history of the discipline and claims to mention only scholars who carried out dissections or experiments.[135] The reworked article MORT reproduces the passage from the *Encyclopédie*, quoted above, about the incomprehensibility and uselessness for medicine of the notion a soul-body union; it nevertheless adds a long article to reassure the righteous Christian facing death.[136]

We have seen that in the Yverdon *Encyclopédie* the interaction between the soul and the body is "one of the fundamental principles of anthropology," and that "knowledge of this fact is an essential part of the natural history of mankind."[137] Their union is also a fact only materialists question. Most other people experience an "inner feeling" which convinces them that the impulses aroused in the body affect the soul and make it feel its own existence, and that, reciprocally, the soul acts on the body.[138] Yet, Mingard notes, when certain philosophers have to explain the modalities of this union, they immediately dismiss the possibility of any contact between the two sub-

---

133. EP, PHYSIOLOGIE ([*Médecine*]), 12, 537–538, p. 538.
134. Ibid.
135. EY, PHYSIOLOGIE, R by Haller, 33, pp. 418b–456b.
136. EY, MORT (*Histoire naturelle de l'homme, Morale*), 29, pp. 368b–369a. De Felice retains only the first paragraph from EP, rewrites the rest, and adds a subsidiary article, "Fear of death" (CRAINTE DE LA MORT, 369a–378a).
137. EY, COMMERCE DE L'ÂME AVEC LE CORPS, p. 505b.
138. EY, UNION DE L'ÂME ET DU CORPS, p. 486a.

stances and develop theories such as those of preestablished harmony and occasional causes. Since the soul is an "essentially thinking" substance and the brain is the organ of thought, the advocates of a reciprocal "physical influence" look for the neurological mechanisms of the union. Nevertheless, because this union "is a contingent and not a necessary fact" (given that the soul is not joined to every sort of matter), the question of the moment at which the soul unites with the body is also posed. Mingard considers these issues metaphysical, and relevant for a problem that is "an inexplicable mystery for philosophers and will probably remain so for a long time yet."[139]

Thus, anthropology and psychology assume the union of soul and body but address only its effects. These embody the interaction of the two substances, which, contrary to their union, can be empirically investigated. Since the soul's nature is unknowable, considerations about it do not help us to understand the interaction of the body and the soul, which must be inferred from our ideas of sensible objects. As de Felice explains when reviewing the theory of physical influence,

> Let us then frankly confess that synthetic or a priori investigation of the cause of the interaction between soul and body is impossible. It is by analysis that we can discover, if not its true cause, then at least the false causes or those [like physical influence] which do not conform to certain properties generally recognized as necessary by all philosophers.[140]

Mingard agrees with de Felice on this point. More important, he elaborates the issue from a resolutely psycho-anthropological standpoint. The interaction between the two substances can be deduced with certainty, he argues, not from some dogmatic definition but from experience itself. The very notion of "interaction" presupposes that the human being is composed of two different substances or principles. And it is "naturally, through the facts [we] witness and experience," that we differentiate the soul from the body and attribute to the former "intellectual and metaphysical acts" and to the latter physical and material movements.[141] At the same time, we feel that these two principles operate together, "and that their intimate union

---

139. Ibid., p. 487a.
140. EY, INFLUENCE PHYSIQUE (*Métaphysique*), N by de Felice, 24, 559a–560b, p. 559b. De Felice argues that one cannot examine the reciprocal action of the soul and the body as though this were the interaction of two bodies, because "since the substances are different, the nature of the actions must be different too"; p. 560a. A similar argument is put forward in EP, INFLUENCE (*Métaphysique*), 8, pp. 728–729.
141. EY, COMMERCE DE L'ÂME AVEC LE CORPS, pp. 505b and 506a.

results in one single individual." When we pay attention to what happens inside us, we perceive a "real correspondence" between the body and the soul, which we feel to be a "reciprocal influence."[142]

Mingard excludes from COMMERCE DE L'ÂME AVEC LE CORPS the debates on the definition of incorporeal substances, the capacity of matter to think and feel, and the reciprocal action of the soul and the body, referring to the article on "union" for possible explanations. The interaction should rather be approached from "the point of view of historians [i.e., natural philosophers], in order to describe its manifestations," like a traveler who relates what he sees and describes "the state of things, events, and facts without wishing to explain their hidden causes."[143] Causes are in this case beyond human grasp. We do not understand how sensory impressions convey ideas to the soul any better than the reciprocal action of the soul and the body: "I want to see, and I open my eyes without knowing how; I see, without knowing the mechanism which causes this." I know that the voluntary movement of my arm comes from the brain but not "how I act on my brain."[144] Everyone experiences the inner feeling which evinces the reality of this interaction. Mingard supports this statement with a reference to Wolff's remark in *Empirical Psychology* that ignorance of "how" things take place does not give us the right to deny facts "which it would seem our Creator wanted us to believe."[145]

Just as we accept the union of the two substances because we feel they interact, even though we do not know how they do it, so the psychologist admits the existence of a seat of the soul or organ of thought in the brain, even though the idea remains as inscrutable as a "dark labyrinth."[146] Since the psychologist cannot uncover "the secret of the mechanics" of the brain, he simply accepts a supposition deducible from incontrovertible facts.[147] This does not prevent him from hypothesizing that the soul acts on "sensitive fibers" that produce sensation when they move, or from applying this theory to particular psychological phenomena, such as the recall of ideas

142. Ibid., p. 506a.
143. Ibid., p. 506b.
144. Ibid., p. 509a–b.
145. Ibid., pp. 509b et 510a.
146. EY, ORGANE DE LA PENSÉE (*Physiologie, Psychologie*), N, 31, 413a–414a, p. 413a. The article reproduces Charles Bonnet's ideas and terminology; it is not signed, but the author could be Albrecht von Haller.
147. Ibid., pp. 413b, 414a.

through words.[148] Since, for the Yverdon *encyclopédistes*, recognizing the physical mechanisms of the interaction between soul and body does not lead to materialism, materialism is treated as a philosophical, metaphysical, *and* psychological topic.

The editors of the Paris *Encyclopédie* dispute the accusation of materialism leveled against them, and certain entries liable to this charge explicitly deny it.[149] On the one hand, the article IMMATÉRIALISME, which treats the issue in greatest depth, refers to AME for proof of the existence of immaterial substances. On the other hand, it stresses the materialism of ancient philosophy, discovers a materialist philosophy among early Christian theologians and medieval Scholastics, explains that believing the substance of God to be everywhere or that the soul is entirely in the brain amounts to considering them material, and emphasizes that we know neither what soul and matter are nor how they can be united.[150] By contrast, the Yverdon *Encyclopédie* criticizes materialism at length. It defends Locke but notes that it was rather the Leibnizians, the Wolffians, and Charles Bonnet who demonstrated the soul's immateriality. The soul's immateriality and immortality are mutually independent properties. Nonetheless, since the physical causes that destroy the body leave the soul intact, if the nature of the soul really is the same as God's, then "we can hope to strive eternally for the perfection of which he is the model, and for the beatitude he enjoys and of which he is the source."[151]

* * *

The Yverdon *Encyclopédie* reorganized knowledge about man by giving an anthropological and psychological basis to many areas and issues that Diderot and d'Alembert attached to metaphysics, logic, and morals. Such psychologizing was generally achieved in two ways. Disciplines and their subject matter were transposed into psychology by systematic use of the

---

148. EY, FIBRE (*Psychologie*), N, 19, pp. 37b–39b; EY, RAPPEL DES IDÉES PAR LES MOTS (*Psychologie*), N, 36, 126a–128b. These two articles, like ORGANE DE LA PENSÉE, reproduce the ideas and terminology of Charles Bonnet. They could be by Haller, the author of FIBRE (*Œconomie animale, Médecine*).

149. See, for example, the "Editors's Foreword" in EP, vol. 3.

150. EP, IMMATÉRIALISME ou SPIRITUALITÉ (*Métaphysique*), 8, pp. 571–574. MATIÈRE (*Métaphysique, Physique*) also refers to AME in its defense against the charge of materialism. The very brief article MATÉRIALISTES (*Théologie*) simply refers to MONDE, MATIÈRE, and the equally short SPINOSISTE (SPINOSA, PHILOSOPHIE DE, on the other hand, is a long critical essay).

151. EY, MATÉRIALISME (*Philosophie, Métaphysique, Psychologie*), N by Mingard, 27, 718a–736b, p. 736b.

field indicator *Psychologie*, and the subject matter itself was reformulated and analyzed empirically in terms of the "acts" of the soul, by examining the soul's activities, rather than producing a moral discourse or classifications of the mental faculties. Moreover, the Yverdon *Encyclopédie* linked knowledge of the soul to knowledge of the ultimate destination of the individual and humanity. Humans are obliged to perfect themselves because the Creator endowed them with perfectibility. By revealing how thought, appetites, and affects function, psychology assists man in fulfilling his higher purpose.

The Yverdon *Encyclopédie* thus expressed, in its own way, the same desire as Kant when, in the same period, he called on universities to admit empirical psychology as a university discipline. The ways in which the Swiss encyclopedia reorganized fields, and gave metaphysics, logic, and morals a psychological foundation, depended on a specific confessional, cultural, and political context. Nevertheless, the Yverdon *Encyclopédie* was highly cosmopolitan and illustrated a typically Swiss assimilation of French and German cultures. In this respect it embodied a widely shared vision. While it did not exhaust this vision, it certainly gave it prominence in a text of significant scope. The reformulation of knowledge in psychological terms may be what is meant when we speak of the Enlightenment as the "century of psychology," yet, more than the French *esprit philosophique*, it was a Swiss encyclopedia's Christian anthropology of perfectibility that ended up framing the psychology of the century.

CHAPTER NINE

Psychology, the Body, and Personal Identity

"Any philosophical system in which the human body does not play a
fundamental role is inept and ill adapted."
—Paul Valéry, "Soma et CEM," *Cahiers*

Eighteenth-century psychology was made possible, and was indeed called
for, when the soul-mind replaced the soul-form as part of the desintegra-
tion of Aristotelian frameworks of thought and the emergence of a mechanistic philosophy of matter and the world. This reconfiguration took root as
notions such as *fact, experience, observation,* and *experimentation* came to
embody new epistemic virtues. It involved the process of psychologization
that gave its name to the "century of psychology" and contributed to transforming the late-Renaissance *psychologia* and *scientia de anima* into the
empirical psychology of the eighteenth century. Enlightenment psychologists felt they were involved in an unprecedented project. They did not simply demarcate themselves from the ancients but went further, inventing a
bibliographic tradition and a history that articulated the relations between
the psychologies of the soul-form and of the soul-mind. They placed psychology within a general science of man and assimilated into anthropology
and psychology problems that had belonged to logic, metaphysics, and morals, thus demoting these disciplines in the order of the sciences. Moreover,
by the end of the eighteenth century, metascientific and historical thought
had been so well integrated into psychology that the historiography of psychology appeared inseparable from the progress of humanity toward the self-
consciousness often hailed as a hallmark of the Enlightenment and modern
subjectivity. The establishment of psychology as an empirical discipline of
the soul-mind was considered essential to the development of the human
psyche in conformity with mankind's inherent perfectibility.

Enlightenment psychology took two main forms. One turned its back on the language of the faculties in favor of case studies. It was not systematic and was principally made up of personal accounts and observations of phenomena such as madness, dreams, visions, split personality, child development, and moral, pedagogical, or aesthetic experience. This approach, exemplified by Karl Philipp Moritz's experiential psychology (*Erfahrungsseelenkunde*), fostered the constitution of a bourgeois identity and experience of the self. The other tendency, dominant among scholars who explicitly identified with psychology as an academic discipline, was, by contrast, systematic and in large part programmatic.

In principle these two forms were interlinked. Experiential psychology brought to light facts about the soul and its relation to the body that could contribute to self-knowledge, and thereby to the enlightened realization of human perfectibility. Systematic psychology pursued exactly the same goal, which was why it could claim to be the most useful of all the sciences and the cornerstone of the system of knowledge. Additionally, systematic psychologists considered that collecting and presenting individual empirical observations, as experiential psychology did, was of fundamental methodological importance. Yet a chasm separated the two psychologies.

The philosopher Jacob Friedrich Abel provides a good example of this separation. In 1786 he published a work he described as the first complete textbook of empirical psychology.[1] In addition to the methodological considerations mentioned above in chapter 4, Abel's *Introduction to Psychology* addressed the nature of the soul as simple substance, the senses, attention and imagination, different thought processes, including sensation, memory, consciousness, speech, and writing, belief, unbelief and will, and the operations of comparing, deducing, demonstrating, and explaining. The work was made up exclusively of definitions, principles, and "laws," with no concrete examples or empirical data.

These could be found elsewhere, in a *Collection and Explanation of Curious Phenomena Drawn from Human Life*, published in three volumes from 1784 to 1790. Abel maintained that this compilation showed the correct way of applying the theory of the soul to the practical knowledge of man (*praktische Menschenkenntnis*). It contained examples, explanations, and proofs of the rules set out in the *Introduction to Psychology*. Psychology's goal, Abel explained, is its application to the phenomena of human life,

---

1. Abel, *Einleitung in die Seelenlehre* (1786); see chap. 4.

and there is no science of man to which psychology cannot be applied.[2] Its ultimate purpose, he continued, is to educate people to wisdom and virtue and, through these, to happiness ("Bildung zu Weisheit und Tugend, und dadurch Glückseligkeit").[3] The empirical cases were extremely diverse: a relapse following a recollection, the state of the soul in catalepsy, the magic of childhood years, the "aesthetic" impression made by day, night, dusk, and the moon, the feeling one has on official holidays, two cases of split personality, prejudices, the mental, moral, and emotional history (*Geistes- und Herzens-Geschichte*) of an atheist, the wiles of the imagination—and more.

In the second volume of the *Collection*, which opened with the case of a male and a female thief, Abel summarized the role of life stories (*Lebensbeschreibungen*): they show "the growth and development of the faculties of the human soul and, through these, of man's happiness, if not in this world, then at least in the next."[4] This was why, he maintained, the case-history writer (*Geschichtsschreiber*) must seek to move his readers' hearts. The first volume of the *Collection* contained remarks on psychological method, and the need to begin by collecting and comparing observations before formulating general laws.[5] The general laws were taken up in the *Introduction to Psychology* and justified the accumulation of facts. Although Abel was true to his methodological principles, the presentation of facts on the one hand and systematic theorization on the other remained physically and conceptually separate; the former contained no theory, and the latter no empirical data. Moreover, whereas a substantial proportion of the empirical material for the science of the soul was derived from "anthropological" observations and individual experience, which took the form of factual existential narratives, the psychology which became established as a university discipline continued to be a physiology of the internal sense, or a discourse on beings whose existence as psychological subjects depended on self-consciousness.

2. "Ich behaupte so gar, daß keine Gattung der menschlichen Wissenschaften Kenntnisse ist, auf die die Psychologie nicht mittelbar oder unmittelbar angewendet werden kann"; [Jacob Friedrich Abel], *Sammlung und Erklärung merkwürdiger Erscheinungen aus dem menschlichen Leben* [pt.1] (Frankfurt, 1784), "Vorrede," pp. i–ii.

3. Ibid., p. xiii.

4. J. F. Abel, *Sammlung und Erklärung merkwürdiger Erscheinungen aus dem menschlichen Leben*, Zweiter Theil (Stuttgart: in der Erhardischen Buchhandlung, 1787), "Vorrede," p. [2].

5. Abel, *Sammlung* [pt. 1], pp. vi–vii.

Such a situation has consequences for the study of past psychological practices. Enlightenment psychologists often referred to individual cognitive processes, particularly the sense of self, attention, and observation, but there are no traces of how they actually worked. They did not leave notebooks, nor did they or others provide accounts of their practices. They did not describe in detail procedures to be applied, or propose norms for the presentation of results. In comparison with the paradigmatic brevity of the "fact" championed by the Scientific Revolution, the "fact" of Enlightenment experiential psychology was long, complex, and idiosyncratic.[6]

In eighteenth-century systematic psychology, on the other hand, facts usually took the form of generalizations and principles concerning the functioning of the mind, without reference to their empirical basis. Jean Trembley did of course call for Condillac's imaginary statue to be replaced by real children, but when it came to getting the reader on one's side, Condillac's strategy prevailed:

> I wish the reader to notice particularly that it is most important for him to put himself in imagination exactly in the place of the statue we are going to observe. He must enter into its life, begin where it begins, have but one single sense when it has only one, acquire only the ideas it acquires, contract only the habits it contracts: in a word he must fancy himself to become just what the statue is. The statue will judge things as we do only when it has all our senses and all our experience, and we can only judge as it judges when we suppose ourselves deprived of all that is wanting in it. I believe that the readers who put themselves exactly in its place will have no difficulty understanding this work [*Treatise on the Sensations*]; those who do not will meet with enormous difficulties.[7]

In order to believe the statue functions like us, we must first become like it. This practice of "putting oneself in the place of" ultimately means "putting oneself in the place of *oneself*." The psychologist, in his self-examination, as he directs his soul's attention onto the soul itself, is confronted with the limits of his nature and "organization," not only on the epistemological and cognitive levels but also on the anatomical and physiological ones. I cited in chapter 1 the first lines of Charles Bonnet's *Essay on Psychology*: "We know the soul only through its faculties; we know

6. L. Daston, "Perchè i fatti sono brevi?," *Quaderni storici*, 108, 2001, 745–770.

7. Condillac, "Advice of some importance to the reader," *Treatise on the sensations*, trans. Geraldine Carr (London: Favil Press, 1930), p. xxxxvii.

these faculties only through their effects. These effects manifest themselves through the intermediary of the body." When the Enlightenment psychologist examined what was going on in his own soul, he was aware that the crucial bodily organ both for the observer and for the phenomena observed was the one the soul used for its own activity—that is, the brain. In other words, as psychological theory itself maintained, the brain was that part of the body which psychologists used to carry out their investigations. Although psychologists were careful not to equate the brain with the thinking subject, it was nevertheless, in their view, the organ through which the soul functioned and apprehended itself, and therefore both the tool and object of psychological investigation.

THE SOUL, THE BODY, AND THE "COMPLETENESS OF THE NERVE"

Eighteenth-century psychology aspired to transform metaphysics into an "experimental physics of the soul." Whether in highlighting the activity of the soul and the evidence provided by the sense of self, or by stressing the way in which concepts derive from sensation, it expressed and reinforced the interiorized and objectified subject of Lockean and Cartesian philosophy. This subject was certainly not reducible to its capacity to will and to know—how could it be, since the capacity to feel was vital to it? But sensation took on meaning only if it could serve to elaborate thought. Gaining knowledge of the universe no longer implied discovering a preestablished order within oneself but constructing the universe as a representation that the soul developed in and through the brain on the basis of sensation. The psychological subject's very being came to be founded on a distance from itself, such that it could witness the activity of its own soul.

Self-consciousness, therefore, was the only really constitutive property of the self that took itself as an object; in Charles Taylor's words, the corresponding "I" was "punctual" or "neutral."[8] Characterized only by reflexivity, it was radically self-sufficient; it consisted in its capacity to separate itself from itself and perform an act of self-objectivation that transformed it into an entity which could be analyzed as an external object. As Taylor demonstrates, the notion of the punctual subject contains a paradox: since it is possible to adopt the point of view of a third person only from a first person position, radical objectivity "is only intelligible and accessible through

8. Charles Taylor, *Sources of the self: The making of the modern identity* (Cambridge, MA: Harvard University Press, 1989), p. 49.

radical subjectivity."[9] As the natural history of the soul united with the body, psychology both presupposed and consolidated this subject.

In other words, the subject of Enlightenment psychology was eminently a thinking subject, but it stemmed from a conception of the human being which was not reducible to the *cogito*. The different versions of anthropology which circulated in the eighteenth century—physical, psychological, medical, and pedagogical—shared the notion of the human being as composed of a soul and a body. Enlightenment psychology gave up trying to know substance in itself and tended to understand the soul as that which subsists behind mental states and phenomena. The question of the "union" of the body with the soul belonged to metaphysics and theology, while only their reciprocal "interaction" concerned psychology. This interaction was both a presupposition and an object of research: a presupposition because, except for materialists, it was considered a fundamental and incontrovertible fact, and an object of research because investigations into the structure, operations, and functions of each substance were deemed necessary for understanding their "commerce."

The eighteenth-century debate on the somatic seat of the soul (that is, the place where the two substances would interact) focused on locating it in the brain.[10] But apart from this debate, which became increasingly marginal over the century due to empirical and theoretical obstacles, medicine offered two approaches to the interaction between the soul and the body. One stressed the importance of explicitly taking this interaction into account; the other had to do with the very concept of "soul."

The first approach is well represented, but not exhausted, by the views of figures such as Georg Ernst Stahl or Robert Whytt (1714–1766), who, in different ways, considered the soul the basis of life and the principle of living creatures. For example, in a book that was a milestone in medical historiography, Kurt Sprengel (1766–1833), a professor of medicine at Halle, elaborated a "medico-psychological system" according to which there existed closer bonds between empirical psychology and the study of the human body than between the latter and mechanics or chemistry.[11] One could therefore discuss the soul in medicine without being an "animist,"

9. Ibid., p. 176.

10. Michael Hanger, *Homo cerebralis: Der Wandel vom Seelenorgan zum Gehirn* (Berlin: Berlin Verlag, 1997), chaps. 1–2.

11. Kurt Sprengel, *Versuch einer pragmatischen Geschichte der Arzneykunde*, vol. 5, pt. 1 ("Geschichte der theoretischen Arzneykunde im achtzehnten Jahrhundert"), 3rd ed. (Halle: Gebauer, 1828).

and without even comparing the soul to gravitation or attraction, through a combination of psychology and mechanics. David Hartley, for example, explained psychological functions through the vibrations of the nerves, while conceiving the soul itself as an immaterial substance. In such a perspective, by virtue of the nature of their objects, empirical psychology and the study of the human body were inseparable.

This can be further illustrated by the work of Samuel Auguste Tissot (1728–1797), the famous doctor from Lausanne. His investigation into the illnesses of "men of letters," as well as those "produced by masturbation," is fully representative of the psycho-medical beliefs of Enlightenment physicians concerning the union of soul and body, and the possibility of observing its effects. In *Onanism*, first published in 1760, Tissot asked why, in the case of masturbation, the faculties of the soul deteriorated at the same time as those of the body. The answer, he said, was related to the "insoluble question" of the mutual influence of the two substances—an issue, he declared, concerning which physicians could do no more than simply observe phenomena.[12] Tissot's was not a purely mechanistic system, since it assumed the interaction between the soul and body to be reciprocal, and it was not metaphysical, since, Tissot claimed, it was strictly based on observation.

The second type of medical approach to the soul attempted to identify the mechanisms of its union with the body, and above all to use them for etiological and diagnostic purposes. Tissot, again, considered that the soul and the body worked together to produce illness. He drew on a "psychosomatic" discourse, simultaneously based on a highly traditional humoral medicine and a more modern "neurology." The intersection between the humoral and neurological perspectives defined some of the limits of psychology at the time. For example, sexual excesses were said to be a cause of mania because coitus increased the quantity of blood in the brain, which in turn distended and weakened the nerves, thus making them more impressionable. In all the psychology and medicine of the period, including Tissot's, the nerve was a fundamental anatomical structure whose cultural importance was by no means restricted to the sciences of the body and the mind.[13] In the mid-eighteenth century, both ontologically and methodologically,

12. Samuel Auguste Tissot, *L'onanisme: Dissertation sur les maladies produites par la masturbation*, preface by Théodore Tarczylo (1760), ed. of 1768 (Paris: Le Sycomore, 1980).

13. George S. Rousseau, "Cultural history in a new key: Towards a semiotics of the nerve," in Joan H. Pittock and Andrew Wear, eds., *Interpretation and cultural history* (London: Macmillan, 1991).

the nervous system emerged as the "common matrix" of the sciences of the body and the mind.[14]

Whether represented as elastic fibers, vibrating chords, or fluid-conducting tubes, the nerves, and the brain where they were supposed to converged, were considered to be the link between the body and the soul, sensation and the understanding. Hence their omnipresence in the psycho-medical literature of the time. In *On the Health of Men of Letters* (1768), Tissot explained that stomach disorders affected the nerves, and it was the nerves which generated a vicious circle in which the mind harmed the body and the body harmed the mind. Depending on one's lifestyle, the nerves could turn too soft or too rigid, oscillate too rapidly or become incapable of performing their functions, the humors could become corrupted, and the brain take up fluids that were too mobile or not mobile enough. According to Tissot, thinking too intensely had the same effect as "a ligature applied to all the nerves."[15] Arguments based on fluids and on mechanics were combined as the case demanded. For instance, thinking kept the nerves "in a state of sustained action" for too long, dispersing the animal spirits necessary for their functioning and preventing the brain from producing more; too little "precious fluid" was left, and the remainder became adulterated, leading to a host of disorders. Without knowing it, Antonin Artaud hearkened back to Enlightenment psychology when he declared, in "Situation of the Flesh": "Above all else, there is the completeness of the nerve. A completeness that holds the whole of consciousness, and the hidden pathways of the spirit in the flesh."[16]

In short, eighteenth-century psychology often referred to the sciences that viewed the functioning of the body in terms of the hydraulics and mechanics of a nerve machine. It thereby became physiological or neurological. Yet even when it saw the "self" as residing in the brain, such a psychology was not necessarily materialist. In fact, while starting out from nerves and fibers, it could sometimes end up with the Christian theology of resurrection. Though rare, such a connection sheds light on the epistemic and moral problems that were ultimately at stake in Enlightenment empirical psychology.

14. Karl M. Figlio, "Theories of perception and the physiology of mind in the late eighteenth century," *History of science*, 12, 1975, 177–212.

15. Samuel Auguste Tissot, *De la santé des gens de lettres* (1768; Geneva: Slatkine, 1981), p. 36.

16. "Il y a par-dessus tout la complétude du nerf. Complétude qui tient toute la conscience, et les chemins occultes de l'esprit dans la chair." Antonin Artaud, "Position de la chair" (1925), in *Œuvres complètes*, vol. 1 (Paris: Gallimard, 1976).

PSYCHO-THEOLOGY AND "MODERN IDENTITY"

The brain of the *king of Prussia, who reigns gloriously,* would yield a codex of all the sciences.
—Guillaume Godart, *The Physics of the Human Soul* (1755)

Eighteenth-century empirical psychology did not cut itself off from religious debates on the soul, its immortality, and the life to come. Even if Enlightenment psychologists were not given to speculating on the soul as a substance separated from the body, they were nonetheless convinced that the empirical study of the soul united with the body was compatible with Christianity and could even be of service to it. Nonetheless, psychology was as distinct from metaphysical and theological defenses of the immortality and immateriality of the soul as it was from published or clandestine materialist literature. It aspired to be a natural science that relied on observation, experimentation, and "analysis," and to remain focused on the phenomenal.

In this respect, Charles Bonnet was an emblematic figure. An heir to Locke, he accepted that substances were unknowable and that only the manifestations of the soul, observable through the external or internal senses, in the self or the other, could be the object of empirical study. Psychology, as the natural science of the soul, gave up trying to solve the problem of the union of the soul with the body and dealt instead with their "interaction" at great length. This is why the apparent echoes of "Cartesian dualism" in eighteenth-century psychological treatises should be read less as statements of an irreducible duality than as reminders of an underlying unity.

Nevertheless, since psychology was part of "physics" and was increasingly close to neurology, it tended to receive materialist interpretations. Psychologists rebutted them by recalling the methodology informing their work, the origin of knowledge in sensation, and their own demonstrations of the immateriality of the soul.[17] Bonnet went even further, claiming to find in psychology a proof of the doctrine of the resurrection of the body:

---

17. Jean Trembley summarized their arguments eloquently: "Due to the fact that our knowledge is originally derived from the senses, that we can know the soul only by means of the body, that consequently philosophers have frequently been obliged to conceive psychology from the physical point of view and to reflect upon the brain, about which they had some concrete ideas, rather than upon the soul, of which they knew nothing other than that it is an active being, and last, due to the fact that they acknowledged the dependence of our faculties on the body, even though these are brought into play by the soul, due to the fact that it is the body which preserves impressions and not the soul, and that the same means by which the body is influenced also

If some of my readers were to find that I had made the soul too dependent on the body, I would ask them to consider that man is by nature
a *mixed* being, a being composed necessarily of two substances, one
spiritual and the other bodily. I would point out that this principle is so
precisely that of REVELATION that the doctrine of the resurrection of the
body is its immediate consequence. And this dogma, which has been so
clearly revealed to us, should not repel the Deist philosopher, far from
it, but should, on the contrary, appear to him as persuasive of the truth
of RELIGION, since it accords so perfectly with what we know with the
greatest certainty concerning the nature of our being.[18]

It is no coincidence that Bonnet, who from the eighteenth century onward
was one of the two authors (along with Hartley) celebrated for attempting to
link psychology to the anatomo-physiology of the nervous system, should
speculate on the physical and psychological conditions of possibility of the
resurrection of the flesh.[19]

The meaning of resurrection was for Bonnet profoundly personal.[20] The
portrait by the Danish painter Jens Juel, engraved as the frontispiece of Bonnet's *Complete Works*, shows him meditating on "the restitution and perfecting of living beings" (fig. 9.1).[21] On 18 May 1777 (the year the painting

---

influences the faculties of the soul—for these reasons these philosophers were considered to be
materialists in disguise. . . . In vain did they point out that they had proved the simplicity of the
soul in several ways, that they constantly attributed to it the origin of all feeling, of all thought
and of all action, that they considered the body to be only the instrument of the soul, and that an
instrument necessarily presupposes another being which manipulates it—all arguments fell on
deaf ears; people refused to see the truth and chose to see only the physical terms employed in
the psychological theory of man, and on the basis of a cursory survey of those words they decided
that the philosophers were materialists, and all counterarguments were branded as sophistry and
deception." Trembley, "Réponse à la question," p. 307.

18. Bonnet, *Essai de psychologie*, p. 1.

19. Cabanis (*Rapports*, p. 77), for instance, wrote that Bonnet "applied his anatomical knowledge directly to psychology on several occasions; and, although the results were not always successful, he at least managed to convey how the knowledge relating to the structure of organs was
connected with the knowledge relating to the nobler operations they carry out." Garat ("Analyse
de l'entendement," p. 164) was less indulgent. He deemed Bonnet's *Analytical Essay* to be "a
great work" but complained that beginning with the spirituality of man and ending with his
resurrection amounted to shrouding the beginning and the end "in obscurity."

20. On Bonnet's thought as a whole, including some remarks on the role of resurrection
within it, see Max Grober, "The natural history of heaven and the historical proofs of Christianity: *La Palingénésie philosophique* of Charles Bonnet," *Studies on Voltaire and the Eighteenth
Century*, 302, 1993, 233–255; Roselyne Rey, "La partie, le tout et l'individu: Science et philosophie dans l'œuvre de Charles Bonnet"; and F. Vidal, "La psychologie de Charles Bonnet comme
'miniature' de sa métaphysique," both in Buscaglia et al., *Charles Bonnet*.

21. Bonnet, *Œuvres*, vol. 1, pp. ix–x, note.

Fig. 9.1. Charles Bonnet. Frontispiece, engraving from an oil painting of 1777 by Jens Juel, in Charles Bonnet, *Œuvres d'histoire naturelle et de philosophie* (Neuchâtel, 1779–1783), vol. 1. "M. Juel," writes Bonnet, "has painted me while I was sunk in deep thought concerning the restitution & perfecting of living Beings" (*Œuvres*, vol. 1, pp. ix–x). The Bible is open to the First Epistle of Saint Paul to the Corinthians, at the point where one can read: "that which thou sowest is not quickened, except it die—O death, where is thy sting? O grave, where is thy victory?" (1 Corinthians 15:36 and 15:55). The inscription "Futuri spes virtutem alit" proclaims that hope in the future encourages virtue.

was completed), Bonnet's great friend Albrecht von Haller, on the eve of his death, bitterly lamented that what would shortly perish was not a life, but the contents of a brain: "Alas my brain, which shortly will be nothing but a clod of earth! I can hardly tolerate the idea that so many ideas accumulated over a long life should be lost as though they were but a child's reveries."[22]

Bonnet evoked his friend's distress with both hope and melancholy, the two states of feeling expressed by his portrait: "When one thinks what Leibniz, Newton, and Haller were, one cannot but wonder whether it is at all probable that death has deprived these great men for ever of the precious fruit of so many sleepless nights, so much thought and so much experience."[23] For Bonnet and others, the question of resurrection, quite apart from its personal meanings, was connected to a psycho-theology inseparable from the natural history of the soul and the issue of the body's role in defining who we are.

The doctrine of the resurrection of the flesh had aroused opposition from pagan philosophers from the very beginnings of Christianity. It seemed to render the new religion singularly unspiritual and raised seemingly insoluble problems: Why does one need a body to be oneself after death? What would the structure and material substance of the resurrected body be? Christian responses stressed the identity between the earthly and the resurrected body. The Scientific Revolution of the seventeenth century would alter the framework of the discussion: the concept of the material body was transformed by a new philosophy of matter, while a new philosophy of mind tended to reduce personal identity to a purely psychological continuity. These developments brought into yet sharper focus the unresolved question of the relation between the body and personal identity, which is still with us today. Across a spectrum of issues ranging from the unity of consciousness to multiple personalities, from millenarian asceticism to the narcissistic quest for well-being, or from indefinitely prolonged life support to medically assisted euthanasia, we are still confronted with the antinomy between the body as a property of the self and the body as constitutive of the very essence of the person.

But we should ask precisely what body is under scrutiny here. As personal identity became psychological, so death became a cognitive event. For psycho-physiologists like Haller and Bonnet, dying meant losing one's

22. Raymond Savioz, ed., *Mémoires autobiographiques de Charles Bonnet de Genève* (Paris: Vrin, 1948), p. 108.

23. C. Bonnet, *Essais sur la vie à venir*, ed. François de la Rive-Rilliet (Geneva and Paris, 1828), p. 7.

thoughts, even if, as Christians, they believed this loss would last only until the resurrection, when they would become themselves again, soul and body. If thoughts, however, are in the brain, and if the brain is the seat of memory and consciousness, then personal identity is situated within this organ. Such was Godart's view when he imagined himself dissecting celebrities and discovering Noah's Ark in Linnaeus's brain, an earthly paradise in the brain of the great French botanist Jussieu, and "a codex of all the sciences" in Frederick the Great's.[24] If this were the case, then could not the resurrection of the body be reduced to the resurrection of the brain? I address this eschatological question in conclusion because it is closely connected with the post-Aristotelian transformations of the sciences, including psychology, and because it represents at once a radical break with Christian tradition and a spectacular sign of its continuity. It is a question that reveals both the fundamental issues at stake in the sciences of the soul and the tensions characterizing the formation of "modern identity."[25]

THE BODY IN RESURRECTION

Faith in the resurrection of the body is inseparable from the mystery of the Incarnation, and an essential article of Christian dogma. As Saint Paul told the Corinthians: "But if there be no resurrection of the dead, then is Christ not risen: and if Christ be not risen, then is our preaching vain, and your faith is also vain" (1 Corinthians 15:13–14). From the first centuries of Christianity right up to the most recent Roman Catholic Catechism (1992) and beyond, this doctrine has always been deemed a mystery of faith. Since only Revelation can teach us something about it, it is useless, as Malebranche put it, "to ask a thousand physical and metaphysical questions about it."[26] Yet such questions were asked, for centuries, concerning the conditions of resurrection, and particularly the continuity between the earthly and the resurrected body.

From the outset, the Christian doctrine of resurrection met with a cool reception. Saint Paul was replying to the Sadducees when he declared that "that which thou sowest is not quickened, except it die: and that which

24. Godart, *La physique de l'âme*, p. 209.

25. These ideas are developed in F. Vidal, "Brains, bodies, selves, and science: Anthropologies of identity and the resurrection of the body," *Critical Inquiry*, 28, 2002, 930–974.

26. As expressed by the character of Théodore in the third of the "Dialogues on Death." Nicolas Malebranche, *Entretiens sur la métaphysique, sur la religion et sur la mort*, ed. of 1711, in *Œuvres*, ed. Geneviève Rodis-Lewis (Paris: Gallimard, 1992), vol. 2, p. 1038.

thou sowest, thou sowest not that body that shall be, but bare grain, it may chance of wheat, or of some other grain: but God giveth it a body as it hath pleased him, and to every seed his own body" (1 Corinthians 15:36–38). While the metaphor of the seed implied the necessity of death, it also posed the question of identity, in the form of a continuity between seed and plant. In resurrection, however, the body was to be transformed: "It is sown in corruption, it is raised in incorruption: It is sown in dishonour: it is raised in glory; it is sown in weakness; it is raised in power: it is sown a natural body; it is raised a spiritual body. There is a natural body, and there is a spiritual body" (ibid., 42–44). The resurrected body will be incorruptible, glorious, powerful, and spiritual. At the same time, it will be numerically identical to the earthly one, like the resurrected body of Christ, who asked his disciples, "Behold my hands and my feet, that it is I myself: handle me, and see; for a spirit hath not flesh and bones, as ye see me have" (Luke 24:39).[27] Throughout its history, Christianity would have to grapple with the oxymoron "spiritual body."

Paul's First Epistle to the Corinthians speaks of the resurrection of *the dead*. Anti-Christian polemic would take this in radical directions. In the second century, the Greek Platonist Celsus found it absurd to hanker after a putrefied body, and the wish to spring forth from the earth complete with one's flesh appeared to him a wish "which might be cherished by worms"—a disgusting doctrine, unworthy of God.[28] The rejoinders of Irenaeus of Lyons in *Against Heresies* and of Tertullian in *On the Resurrection of the Flesh* emphasized the resurrection of the *flesh* and highlighted the originality of the Christian position.[29] Likewise with the symbols of the faith: the Athanasian Creed (which dates back to the fourth or fifth century CE) stated that all men will be resurrected *with* their own bodies, and the fourth-century *Fides Damasi*, that we will be resurrected *in* our own flesh.

In addition to the difficulty of imagining how the dead body could be fabricated afresh, there was the question of how the original substance of the earthly body would be used. Since Christ had stated that "there shall not a hair of your head perish" (Luke 21:18), the question arose of whether all the nail and hair trimmings were going to be integrated back into the

27. "Numerical" identity implies identity of matter, which is where it differs from qualitative identity: when one bathes twice in the same river at the same place, one does not bathe twice in waters that have remained numerically the same.

28. Origen, *Contra Celsum*, V.14. Celsus's anti-Christian polemic, entitled *Alēthēs logos* [True discourse or True doctrine], is known only through Origen's citations.

29. Caroline Walker Bynum, *The resurrection of the body in Western Christianity*, 200–1336 (New York: Columbia University Press, 1995).

resurrected body. Theologians cast around for replies. Augustine, for instance, maintained that a body's earthly matter would not be reunited with the original parts of the body it had constituted. It would, rather, be like a metal statue reduced to its basic mass, and refashioned using the same quantity of matter. This way, even nail and hair trimmings would not be lost, "for they shall be changed into the same flesh, their substance being so altered as to preserve the proportion of the various parts of the body" (*City of God*, XXII, xix, 1). However absurd such questions may have appeared to some commentators, they did touch on an essential anthropological issue.

The last book of Augustine's *City of God* set out the issues Christians should address in replying to pagan objections. Why should we cast doubt on the possibility of transferring earthly bodies to an eternal abode? It is no more difficult for God to do this, wrote Augustine, than to unite the soul with the body. And why should we not believe in this, since we already believe that Christ is resurrected and has ascended to heaven? Some Platonist *ratiocinatores* may claim that an earthly body cannot remain on high because of its natural weight. Yet, Augustine asked, why should we deny that God can make the heavens into the dwelling place of immortal human bodies, when he has created birds? Is a craftsman not able to fashion a piece of metal such that it floats? As for aborted fetuses, it is impossible to know if they will be resurrected. What is certain, though, is that if they are counted as dead, they will be resurrected. Likewise, a fortiori, for children who die in infancy. But what body will they have? By a divine miracle, they will have the body they would have had had they been able to develop in accordance with the potential contained within their seed. In this way, each will receive a body that is really its own—except that it will be without defect and will be the age of Christ before his death. The sexes will be preserved, since gender is not a defect; but instead of arousing desire, the genital organs will only inspire praise of God's wisdom and bounty.

The problem acknowledged as the thorniest was that of the food chain: what will be the body of an individual eaten by a cannibal, or by an animal subsequently eaten by humans? Athenagoras, a second-century Greek apologist, explained in his *Treatise on the Resurrection of the Dead* that human flesh is incapable of assimilating human flesh. Two centuries later, Augustine disputed this theory, arguing that the flesh that has been consumed is exhaled into the air, whence God recalls it in order to restitute it to its original owner. For these thinkers, the numerical identity of the earthly and spiritual bodies remained fundamental.

The problems posed by this identity, however, were far from solved. In the thirteenth century, Thomas Aquinas integrated into Christianity

Aristotle's view of the union of soul-form and body-matter. Although Aquinas maintained that the soul and the body realize their essence only when united, he also thought that the dead body had no natural inclination to resurrect, and that its matter rejoined the soul by virtue of God's will. The resurrected body was numerically identical to the earthly body, since it was composed of the same matter—but the matter was the "same" only if animated by the same soul. Hence, the resurrected and the earthly person were identical only if they had the same soul.[30] This apparent spiritualization of identity did not, however, lead Aquinas to dismiss the body. On the contrary, when he examined the most controversial issues—cannibalism, and so forth—and the nature and composition of the glorious body, he not only reaffirmed the truly corporeal character of the resurrected body, but also the absolute requirement of a body for a human being's personal identity: "anima . . . non est totus homo, et anima mea non est ego."[31] Centuries later, in 1830, Ludwig Feuerbach inveighed against the ignoramuses and hypocrites who denied the sensory and physical nature of beatitude according to Christian doctrine.[32]

For Christian anthropology, in short, the self is not the soul, and the soul belongs to the definition of a person only insofar as it is joined to a body.[33] Christianity maintains that the human being is made up of a corruptible body and an immortal soul but does not accept that a person can exist other than composed of the two substances. The human being is not simply someone who has a body but someone whose existence is corporeal.[34]

---

30. Thomas Aquinas, *Summa theologiae*, suppl., questio 78, art. 3, and questio 79, art. 1-2.

31. Thomas Aquinas, *Super primam epistolam ad Corinthios lectura*, cap. XV, lectio II, n. 924, in *Super epistolas S. Pauli lectura*, ed. Raphael Cai (Turin: Marietti, 1953), vol. 1, p. 411.

32. Ludwig Feuerbach, *Thoughts on death and immortality* (1830), in *Thoughts on death and immortality: From the papers of a thinker, along with an appendix of theological-satirical epigrams*, trans. James A. Massey (Berkeley: University of California Press, 1980).

33. This is why Raymond Martin and John Barresi's *Naturalizing the soul: Self and personal identity in the eighteenth century* (London: Routledge, 2000) seems to me to be based on a misunderstanding. The authors contend that, in European philosophy, the self was assimilated to the soul qua immaterial substance right up to the end of the seventeenth century. This leads them to reduce pre-Lockean psychology to a dogmatic discourse on the essence of the soul. We have seen that this is by no means the case: psychology was recast in the course of the eighteenth century not due to the replacement of an immaterial and immortal substance by a "material mind" in the framework of a general process of secularization but due to the transition from the soul-form to the soul-mind.

34. Antoine Vergote, "Le corps: pensée contemporaine et catégories bibliques," *Revue théologique de Louvain*, 10, 1979, pp. 159-175; see also Caroline Walker Bynum, "Why all the fuss about the body? A medievalist's perspective," *Critical Inquiry*, 22, 1995, 1-33; James F.

Certain Christian discourses and practices can hardly be read as anything other than expressions of hatred of the flesh. Nevertheless, while Christianity never glorified the body for its own sake to the detriment of the soul, it always condemned the denigration of the flesh. Asceticism was a way of preparing the body to receive the Holy Spirit. As Peter Brown explains in his masterful work on sexual abstinence in the first centuries of the Christian era, what accounted for human wretchedness, even for writers in the Platonist tradition, was less the body in which the soul would be enclosed as in a prison than the human being's Fall, soul and body together.[35] In this light, the primary goal of sexual abstinence was not to repress desire but to master the fear of death by interrupting the succession of generations. So while abstinence has often been interpreted in terms of repression, its goal was to experience the body as a "temple of the Holy Ghost" (1 Corinthians 6:19) and to prepare it to be like that of the resurrected Christ.

THE LOSS OF THE BODY

The solutions arrived at by the church fathers and medieval theologians did not dispel the disquiet caused by the pagan objections to the resurrection doctrine. From the sixteenth century onward, it was "physics," the general science of nature, which took center stage in resurrection debates. Naturalist arguments, which had already been used to counter the objection of the food chain, became more widespread. Alchemists noted that when certain plants were reduced to ashes, they worked as seeds of new plants. This became the basis for palingenesis, the "new birth" of bodies that had been dissolved or reduced to dust. Palingenesis seemed to demonstrate nature's role in accomplishing the mystery of the resurrection. The naturalist explanations inspired by Paracelsus were supplemented by soteriological elements drawn from alchemical symbolism. Thus, in emblem books, mercury was both a metaphor and a model for the resurrection, since various operations enabled it to regain its original wholeness after it had been pulverized.

The terms of the debate underwent a radical transformation during the seventeenth century. First of all, the images of life blossoming out of death became untenable, with the discovery that there is no spontaneous genera-

---

Keenan, "Christian perspectives on the human body," *Theological studies*, 55, 1994, 330–346; and Gedaliahu G. Stroumsa, "*Caro salutis cardo*: Shaping the person in early Christian thought," *History of religions*, 30, 1990, 25–50.

35. Peter Brown, *The body and society: Men, women, and sexual renunciation in early Christianity* (New York: Columbia University Press, 1988).

tion *ex putri* and that only a living being can give birth to another living being. Embryology would henceforth take over to explain how the resurrected body was formed. Second, the mechanical and corpuscular view of matter that came to predominate postulated that natural phenomena could be explained by the action of uniform particles and their "accidents," such as rest, motion, figure, and position. For the English philosopher and chemist Robert Boyle (1627–1691), alchemy and mechanical philosophy lent support to the universal transmutability of substances. Indeed, if the ultimate material components of all things are the same, then any one thing can, under certain conditions, be transformed into any other. When applied to the resurrection, these principles foregrounded qualitative identity, rather than the numerical identity of bodily matter.[36]

Boyle used analogies from chemistry to suggest that a resurrected body could be deemed numerically identical to its corresponding earthly body even if it was not made up of exactly the same matter. It was essential that the mechanical particulars of the constituent particles should produce identical effects. In Boyle's reading, Ezekiel's vision (Ezekiel 37), in which the prophet sees bones become covered with flesh and transformed into living beings again, testified to the fact that a complete body could be formed out of just a portion of the original matter, which God would complete to make it into the same person who had died. As a result, the "sameness" (*idem*) of the earthly and resurrected body ceased to be the criterion by which to judge the identity (*ipse*) of the resurrected person.

A similar view applied in the domain of personal identity. In a chapter added to the second edition (1694) of his *Essay Concerning Human Understanding*, Locke, who was a friend of Boyle's, distinguished the *man* from the *person*. He defined a person's identity in terms of the continuity between consciousness and memory.[37] Identity, he said, resides in the consciousness

---

36. Robert Boyle, "Some physico-theological considerations about the possibility of the Resurrection" (1675), in *The Works of the Honourable Robert Boyle*, 2nd ed. (London: printed for J. and F. Rivington, 1772), vol. 4.

37. All the citations from Locke in this paragraph come from his *Essay*, bk. 2, chap. 27. See Udo Thiel, "Personal identity," in Garber and Ayers, *The Cambridge history*; Étienne Balibar, *John Locke: Identité et différence: L'invention de la conscience* (Paris: Seuil, 1998); Edwin McCann, "Locke's philosophy of body," in Chappell, *The Cambridge companion to Locke*; Nicholas Jolley, *Leibniz and Locke: A study of the "New Essays on Human Understanding"* (New York: Oxford University Press, 1984), chap. 7; and Raymond Martin, "Locke's psychology of personal identity," *Journal of the history of philosophy*, 38, 2000, 41–61. For debates relevant to the French context, see Perkins, *The concept of the self*, chap. 3; and Catherine Glyn Davies, *Conscience as consciousness: The idea of self-awareness in French philosophical writing from Descartes to Diderot* (Oxford: Voltaire Foundation, 1990).

that always accompanies thought and sensory experience; "as far as this consciousness can be extended backwards to any past action or thought, so far reaches the identity of that person." At the resurrection, there is consequently no need to have the same body in order to be the same person. If a person's consciousness remained attached to his little finger and the latter were to be separated from the rest of his body, then "it is evident the little finger would be the *person*, the *same person*; and the self then would have nothing to do with the rest of the body." Locke went on to explain that a prince's soul in a cobbler's body would be the same *person* as the prince, but a different *man*.[38] And, as far as resurrection is concerned, it is the person who counts. In his commentary on Saint Paul's Epistles, Locke therefore maintained that the resurrected body would be different from the earthly one. In the *Essay*, he states that the sentences passed at the Last Judgment can only be justified "by the consciousness all persons shall have, that they themselves, in what bodies soever they appear, or what substances soever that consciousness adheres to, are the same that committed those actions, and deserve that punishment for them." As Paul Ricœur has noted, Locke's vision pointed to "a conceptual reversal in which ipseity [*ipséité*, selfhood] tacitly replaced sameness [*mêmeté*]."[39]

The responses to Locke were of two sorts. Many critics were not prepared to reduce the anthropological problem of identity to the legal concept of the *person*, and those who accepted Locke's terms inflected the debate toward a psycho-physiology of personal identity. Some critics continued to call for the numerical identity of the earthly and the resurrected bodies; one wrote that "we cannot be the same Men unless we have the *same* Bodies. 'Tis a great Mistake to imagine that the *Identity* or *Sameness* of a *Man* consists wholly in the *sameness* of the *Soul*. If *Euphorbus*, and *Homer*, and *Ennius*, had had *one and the same* Soul, yet they would not have been *one and the same*, but Three distinct Men."[40] Locke's position was in fact

---

38. Christian Wolff defined the "person" in terms of the consciousness we have of being the same we were, that is, insofar as we preserve a "memory of self": "Da nun man eine Person nennet ein Ding, das sich bewust ist, es sey eben dasjenige, was vorher in diesem oder jenem Zustande gewesen" (*Vernünfftige Gedancken*, § 924); "*Persona dicitur ens, quod memoriam sui conservat*" (*Psychologia rationalis*, § 766). See Olivier-Pierre Rudolph, "Mémoire, réflexion et conscience chez Christian Wolff," *Revue philosophique*, 193, 2003, 351–360. For theories of consciousness and the self in Germany, see Falk Wunderlich, *Kant und die Bewußtseinstheorien des 18. Jahrhunderts* (Berlin: Walter de Gruyter, 2005), Part I.

39. Paul Ricœur, *Soi-même comme un autre* (Paris: Seuil, 1990), p. 151.

40. Humphrey Hody, *The resurrection of the (same) body asserted, from the traditions of the heathens, the ancient Jews, and the primitive church* . . . (London: printed for Awnsham and John Churchill, 1694), p. 218.

somewhat subtler. In common with most Enlightenment psychologists, he considered that the soul and the self were not identical. In amnesia, for example, the soul remained intact while the loss of memory led to the disintegration of the self.[41] Since, in any case, the self could not dispense with a material basis, scholars attempted to determine what part of the body must live on in order for a person to be resurrected as the same person he or she was when alive.

THE SEED AND THE BRAIN

Samuel Clarke (1675–1729), an English theologian and philosopher, illustrates the inflection toward psycho-physiology that took place at the beginning of the eighteenth century. Like Locke and Boyle, he was convinced that Christian dogma was consistent with reason. In his Boyle lectures of 1704–1705, Clarke argued against the idea "that, to constitute the same Body [at resurrection], there must be an exact Restitution of all and only the same Parts."[42] The body, he explained, develops out of *stamina originalia*, primordial filaments that make up its essential components and remain untouched by bodily changes. Clarke was tacitly following Marcello Malpighi (1628–1694). In his observations on the formation of the chick in the egg, Malpighi gave the name *stamen* to the first visible traces of the embryo and the beginnings of the organism as a whole. Let us suppose, he said, that the rest of the bodily substance comes from outside the body through nutrition and is continually replaced, whereas the originary filaments remain intact during one's lifetime, and even after death. It would then make no difference whether the remaining matter in the resurrected body was numerically identical to the matter of the earthly body or not, since the *stamina* constitute the body's physical essence. Clarke favored such a solution, others elaborated it further, and it came eventually to be considered self-evident.

The embryological explanation of the identity of the two bodies did not, however, settle the issue of personal identity. Clarke linked the two questions when he suggested that the mortal body is but the slough of some hidden principle, and that this principle, "possibly the present Seat of the *Soul*," shall at the resurrection "discover itself in its proper Form."[43] If the

---

41. Baertschi, *Les rapports*, pt. 2, chap. 4, § 2 in particuliar.

42. Samuel Clarke, *A Discourse concerning the Being and Attributes of God, the Obligations of Natural Religion, and the Truth and Certainty of the Christian Revelation*, 10th ed. (London, 1768), p 206.

43. Ibid., p. 207.

seat of the soul, which is necessarily that of memory and consciousness, coincided with the "seed" of the body (whether *stamina originalia* or something else), then such a seminal principle would explain how resurrected beings turn out to be physically and psychologically identical to the living persons they once were. The anonymous author of *Essays on Providence and on the Physical Possibility of Resurrection* (1719) believed the seminal theory would solve the problem of resurrection: if human beings really do come from an undying seed, then this seed could easily adopt a human form again and would actually be "the true *self.*"[44]

Charles Bonnet presents these embryologico-neurological speculations in their purest form. He started out from the notion that since man comprises a soul and a body, he can survive in his present state only as a composite being. Now, since the personality depends on memory, and memory on the brain, if man is to preserve his personality after death, "his soul must remain united with a body that death cannot destroy." To this end, God "may have enclosed in the earthly body this incorruptible body, in miniature, as the immediate seat of the soul and the instrument of all its operations."[45] The seat of the soul would thus be a "little ethereal machine," an "indestructible brain" which would function as "the seed of the future body."[46] In short, our present brain perhaps contains another brain "destined to develop in the other life" and to restore our personal identity when we are resurrected.[47] Given the specific properties of the glorious body, Bonnet imagined that the "miniature human body" formed in the seat of the soul (which was to become the spiritual body) was "very different from the one we know."[48]

The Basel mathematician Leonhard Euler (1707–1783) found Bonnet's "little machine" too fragile. He pointed out that since after death a covering of flesh would no longer protect it, it could well get damaged. And since that would destroy the personality, it would follow, absurdly, that "the soul can be preserved only by a perpetual miracle." Bonnet's reply was a clear expression of his desire to reconcile a physico-theological approach with empirical psychology:

44. Anonymous, *Essais sur la Providence et sur la possibilité physique de la Résurrection* (1719), "translated from the English of Dr. B," 2nd ed. (Amsterdam: chez E. J. Ledet, 1731), p. 226 and 232.

45. Savioz, *Mémoires autobiographiques*, p. 238.

46. Bonnet, *Essai analytique*, in *Œuvres*, vol. 4.1, p. 139, and ibid., n. 5.

47. Bonnet, *Contemplation de la nature* (1764), in *Œuvres*, vol. 6, p. 352.

48. Ibid., pp. 359–360. Bonnet develops the same arguments in his *Palingénésie philosophique, ou idées sur l'état passé et sur l'état futur des êtres vivans* (1769).

I am by no means grounding the immortality of the soul in the little organic machine, but, having stated that the soul is not the whole of man, and that it is the whole of man which is to be preserved, we must suppose that his soul remains united with a little organic machine. . . . The Gospel has revealed to us not the immortality of the soul but the immortality of man. What sense would the resurrection have, if the soul were the whole of man?[49]

When Bonnet mapped the embryology of the resurrected body onto a developmental psycho-physiology of personal identity, he imagined the seed of the glorious body as a brain. Voltaire's irony was not far off the mark when he commented that Bonnet "seems convinced that our bodies will be resurrected without stomachs and without the front and rear parts, but with *intellectual fibers*, and the finest of minds."[50]

In summary, Boyle's position was that the resurrected body required only part of the original bodily matter, Locke reduced personal identity to psychology, Clarke imagined that the "originary fibers" of the glorious body were also the seat of the soul, and Bonnet went further still in minimizing the importance of the flesh and replacing the resurrection of the body with the resurrection of a psychological identity by giving the seed of the glorious body the structure of a brain.

THE EMERGENCE OF THE CEREBRAL SUBJECT

From the seventeenth century on, the issue of personal psychological identity superseded that of the numerical identity of bodies. The series of amputations to which the body was subjected went hand in hand with an increasingly precise designation of the unit of flesh that should remain if a body were to be the body of a particular person. In the seventeenth century, the portion of matter in question was initially not specified, but it soon came to be defined as a seed. Since the problem of identity could not be solved by embryology, the seminal particle which was to produce the resurrected body took on the function of the seat of the soul, thus becoming capable of producing a whole psychosomatic unit. The question then arose of why one should bother with so much body anyway. Since only the

49. Savioz, *Mémoires autobiographiques*, p. 301.

50. Voltaire, *Dieu et les hommes, œuvre théologique mais raisonnable* . . . (1769), éd. Roland Mortier, in *Les œuvres complètes de Voltaire* (Oxford, Voltaire Foundation, 1994), vol. 69, p. 454.

psychological personality will be necessary at the Last Judgment, and since the brain is its organic basis, why should it not be brains rather than bodies that are resurrected?[51] And furthermore, if the *whole* brain is not necessary to constitute a person, then maybe only a fraction of it will be necessary to enjoy the beatific vision. Whatever fraction this is, it will have to contain the information that defines the self: "I" become my brain's data content. In this vein, a physicist suggested in 1994 that being resurrected as the same person will amount to having one's brain contents simulated by the super-powerful computers of a future age; once "resurrected," we will inhabit a Beyond going by the technical name of *cyberspace*.[52]

Eighteenth-century psychology did not go quite so far in the disincarna-tion of personal identity, but it certainly laid the foundations for a central anthropological figure of the twentieth and twenty-first centuries, namely, the cerebral subject.[53] Locke's view of "sameness" and of the relation be-tween psychological and corporeal identity implied a sharp decline in the role played by the body in comparison with the primordial importance of the corporeal existence of the self in the Christian tradition. This disem-bodiment was not, however, a full dematerialization. In Locke's thought experiments, memory and consciousness, the prerequisites of identity, re-mained attached to the little finger; eighteenth-century psychology trans-ferred them to the brain, identified as the seat of the soul. Unlike Christian anthropology, in which a soul separated from its body could not form a human being, here a human required only a brain in order to be a person. Advances in the neurosciences have since consolidated the conflation of personhood with "brainhood."[54]

In the nineteenth century, phrenology located psychological qualities in different "organs" of the brain surface, which in turn determined the cranial "bumps." Later anatomical and functional discoveries strengthened the belief in the brain as the seat of the self and lent support to research into the brains of geniuses, criminals, and the mentally ill, whose unusual, posi-tive or negative, mental characteristics were understood to reside in their

51. Detlef Bernhard Linke, "Gehirn, Seele und Auferstehung," *Evangelische Theologie*, 50, 1990, 128–135.

52. Frank J. Tipler, *The physics of immortality: Modern cosmology, God, and the resurrec-tion of the dead* (New York: Doubleday, 1994), chap. 9.

53. F. Vidal, "Le sujet cérébral: Une esquisse historique et conceptuelle," *Psychiatrie, sci-ences humaines et neurosciences*, 3, no. 11, 2005, 37–48.

54. F. Vidal, "Brainhood, anthropology of modernity," *History of the human sciences*, 22, 2009, 5–36.

gray matter.[55] Since then, digital representations produced by brain-imaging have acquired the status of unmediated and transparent facts, and of precise and realistic portraits of the kinds of people we are.[56]

A decisive point in this history, however, came before the development of such technologies. In the second half of the twentieth century certain Anglo-American analytic philosophers revived Lockean-style fictions, using, in particular, different figures of the brain transplant; to my knowledge, Sidney Shoemaker was the first to do so, in his book *Self-Knowledge and Self-Identity* (1963). Here the brain figures so powerfully as the somatic limit of the self—as the part of the body without which "I" am not "myself"—that for some time "brain surgical fictions" seemed indispensable for discussing personal identity.[57] Not all philosophers supported such a position,[58] but the very fact that they contested the ontology of the cerebral subject betrays its significance in contemporary culture.

Certain works of literary and cinematographic fiction prefigured the crucial role the brain would later play in philosophical thought experiments. This role was closely linked to the value and status of the brain in scientific research, and in contemporary culture since the 1960s (and especially since the 1990s). An International Brain Research Organization was created in 1960. Then, in the late 1970s, research into the way in which the mind "arises" out of brain functioning received the name of "cognitive neuroscience," which rapidly became an autonomous field of research. The years 1990 to 2000 were dubbed "Decade of the Brain," and the twenty-first century has already been proclaimed by some the century of the brain. These are just particularly visible expressions of the growing significance of the brain for our knowledge and representation of the human.

The brain has been defined as the ultimate limit of identity and even as consubstantial with the self. Moreover, it has come to represent one of the principal frontiers of science and the central challenge in our quest for

---

55. See, for example, Edwin Clarke and L. S. Jacyna, *Nineteenth-century origins of neuroscientific concepts* (Berkeley: University of California Press, 1987); Michael Hagner, *Geniale Gehirne: Zur Geschichte der Elitenhirnforschung* (Berlin: Wallstein, 2004); and Marc Renneville, *Le langage des crânes: Une histoire de la phrénologie* (Paris: Les Empêcheurs de tourner en rond, 2000).

56. Joseph Dumit, "Is it me or my brain? Depression and neuroscientific facts," *Journal of medical humanities*, 24, nos. 1/2, Summer 2003, 35–47; Dumit, *Picturing personhood: Brain scans and biomedical identity* (Princeton: Princeton University Press, 2004).

57. Stéphane Ferret, *Le philosophe et son scalpel: Le problème de l'identité personnelle* (Paris: Editions de Minuit, 1993), p. 11.

58. See, in particular, Kathleen V. Wilkes, *Real people: Personal identity without thought experiments* (1988; Oxford: Clarendon Press, 1999).

knowledge of what constitutes the human being qua human. Whether characterized as a "mind-brain" or in the terms of physicalist reductionism, the brain has acquired a definite ontological primacy, well beyond its function as the mind's cause, material foundation, or condition of possibility. One cannot change one's brain without changing who one is. Were some evil genius to decide to protect Rudolf Hess by giving him not only the teeth and fingerprints of Ramirez, but also Ramirez's brain, he would have gone too far; "he would not only have altered the physical appearance of one and the same person, but he would have replaced him quite simply with another."[59] The brain, however, invalidates the body as a criterion of identity only if one sees the body as a brainless entity. The cerebral criterion of personal identity is in fact based on a redefinition of the body as "what, materially speaking, is the basis of the person." As a consequence, the brain links bodily and psychological criteria, and the logical formula of the cerebral subject becomes "Person P is identical to person P^* if, and only if, P and P^* have one and the same functional brain."[60] Thus, a person made up of X's brain transplanted into Y's body *is* X; to have the same brain is to have the same body. If our self is a self-brain, then the brain is the "fundamental bastion of personal identity."[61]

Yet the brain is the bastion of personal identity because it provides the material foundations and conditions of possibility of the functions supposed to define the person—namely, memory and consciousness. Taking the brain as the measure of identity involves the body, but this measure is ultimately psychological. The same was true at the inception of the cerebral subject in eighteenth-century empirical psychology. The fact that at the time the psychological definition of identity was linked to the belief in an immaterial substance does not fundamentally alter its nature as a theory of man. Such a theory about the natural and necessary conditions for being a human person defines the moment of our death and—increasingly—the course of our lives.[62] In the eighteenth century, the brain's ontological primacy derived from the fact of its being the seat of the soul-mind; if, at the dawn of the twenty-first century, the word "soul" sometimes crops up in the neurosciences and neurophilosophy, it is undoubtedly because these disciplines

59. Ferret, *Le philosophe*, p. 16.
60. Ibid., p. 30.
61. Ibid., p. 91.
62. Thomas Schlich and Claudia Wiesemann, eds., *Hirntod: Zur Kulturgeschichte der Todesfeststellung* (Frankfurt: Suhrkamp, 2001); Robert Blank, *Brain policy: How the new neuroscience will change our lives and our politics* (Washington, DC: Georgetown University Press, 1999).

are still grappling with the problem of human identity in terms that derive directly from those first elaborated within Enlightenment empirical psychology. In 1760 Charles Bonnet wrote that "if a Huron's soul could have inherited Montesquieu's brain, Montesquieu would still be creating."[63] With minor adjustments, this radical early expression of the anthropology of the cerebral subject could be written today and would make as much or as little sense as it did then.

---

63. "Si l'Ame d'un Huron eut pu hériter du Cerveau de MONTESQUIEU, MONTESQUIEU créeroit encore." Bonnet, *Essai analytique*, § 771.

The Two Editions of Goclenius's *Psychologia*

The two editions of Rudolph Goclenius, *ΨΥΧΟΛΟΓΙΑ: hoc est, de hominis perfectione, animo, et in primis ortu hujus, commentationes ac disputationes quorundam Theologorum & Philosophorum nostrae aetatis . . .* (Marburg: ex officina typographica Pauli Egenolphi).

Author	Title	(1)	(2)
Hermannus Vultejus	*De perfectione hominis Philosophica*	1, 2	R
Franciscus Junius	*An Animus hominis propagetur a parentibus*	1, 2	R
Iohannes Iacobus Grynaeus	*Eiusdem quaestionis argumentosa, eaque modesta explicatio*	1, 2	R
Iohannes Iacobus Colerus	*Quaestio theologica et philosophica, num anima sit ex traduce, an vero a Deo quotidie inspiretur, ex veterum & recentium scriptis, quam diligentissime collecta*	1, 2	R
Iulius Caesar Scaliger	*Sententia de eodem argumento*	2	R
Hieronymus Savonarola	*De eodem*	2	R
Hyeronimus Zanchius	*De origine Animorum*	2	R
Casparus Peucerus	*De essentia, natura et ortu animi hominis*	1, 2	A
Egidius Hunnius	*An etiamnum per inspirationem hominibus infundantur illorum animae*	1, 2	A
Petrus Monavius Lascovius	*An animae rationales, sicut corpora, per seminalem traducem propagentur, an vero quotidie a Deo creatae, corporibus nascentum infundantur?*	1, 2	(3)

Continued

Author	Title	(1)	(2)
Rudolphus Hospinianus	*Oratio . . . quo affirmatur: Animam esse totam in toto, & in qualibet ejus parte totam*	1, 2	(4)
Timotheus Brightus	*De Traduce*	2	A
Ioannes Ludovicus Havenreuterus	*Sitne animus nobis ingeneratus a Deo, necne*	1, 2	A
Rudolph Goclenius	*De ortu animi*	1, 2	R
Nicolaus Taurellus	[Extract from *Philosophiae triumphus* (1573)]	2	A (5)

(1) 1 = 1st edition, 1590; 2 = 2nd edition, 1597.
(2) Position on traducianism: R (rejects), A (accepts).
(3) Broaches traducianism without taking sides.
(4) Does not discuss the origin and transmission of the *animus*, but defends the position that the soul exists undivided in all parts of the body.
(5) A purely physiological argument. Comment by Goclenius: "Vult igitur doctissimus hic vir animam existere e materia."

ANTHROPOLOGIE and PSYCHOLOGIE in the Paris and Yverdon *Encyclopédies*

ANTHROPOLOGY

DIDEROT AND D'ALEMBERT'S *ENCYCLOPÉDIE*, VOL. I, P. 497

ANTHROPOLOGIE, dans l'œconomie animale; c'est un traité de l'homme. Ce mot vient du Grec ανθρωπος, homme, λογος, traité. Teichmeyer nous a donné un traité de l'œconomie animale, qu'il a intitulé *Anthropologia*, in-4º. imprimé à Genes en 1739. Drake nous a aussi laissé une *Anthropologie* en Anglois, in-8º. 3 vol. imprimée à Londres en 1707 & 1727. Voyez ANTRO-POGRAPHIE. (L [= Pierre Tarin])

YVERDON *ENCYCLOPÉDIE*, VOL. 3, PP. 22A–25B

ANTHROPOLOGIE, (N), s.f. (*Philos[ophie]*, *Hist[oire] Nat[urelle]*, *Physiol[ogie]*, *Métaph[ysique]*, *Psychologie*). Ce mot vient du grec ανθρωπος, homme, λογος, discours, traité. Littéralement il signifie, Traité de l'homme. Ce mot devroit être employé pour désigner, conformément à son éthymologie, cette branche importante de la science philosophique, qui nous fait connoître l'homme considéré sous toutes les faces, qui peuvent offrir des idées à notre esprit, et devenir l'objet de nos connoissances. Ainsi l'anthropologie nous apprendroit à connoître Iº. l'origine de l'homme, 2º. les divers états par lesquels il passe, 3º. ses qualités ou affections, 4º. ses facultés ou actions, pour en déduire, 5º. la connoissance de sa nature, 6º. de ses relations, 7º. de sa destination, et 8º. des regles auxquelles il doit se conformer pour y répondre convenablement. L'anthropologie tiendroit ainsi à toutes les sciences; elle en emprunteroit, ou en fourniroit les principes, et en rapporteroit à l'homme toutes les conséquences pour son utilité, c'est-à-dire, pour sa conservation, sa perfection et son bonheur. Les sciences en effet, les arts et les métiers, ne sont que les connoissances spéculatives et pratiques que l'homme acquiert; il n'acquiert de connoissances qu'autant que leurs objets ont avec lui quelques rapports qui les mettent à sa portée, et qui les lui rendent intéressans par leur influence réelle ou supposée sur son état, ensuite de ce qu'il est, et de ce qu'il peut ou doit devenir. Lors même que l'homme semble s'occuper de recherches qui ne paroissent avoir aucune influence sut son état, il y est toujours porté par sa nature, par sa constitution, par le sentiment de son état, de ses besoins, de ses facultés, de ses relations, par le desir et l'espérance de trouver dans les objets qu'il étudie, des moyens d'atteindre quel que nouveau degré de bonheur. Toutes les sciences que l'homme cultive ou pourra cultiver, tiendront toujours par quelque endroit à l'anthropologie, qui est la plus importante des sciences, la plus digne d'occuper l'homme: sans cette connoissance de nous-mêmes, quelle perfection, quelle félicité pourrons-nous atteindre? Toutes les sciences servent à perfectionner l'anthropologie, et celle-ci ne sera parfaite, qu'autant que les autres seront parvenues à leur perfection, et elles ne seront utiles qu'autant que nous les rapporterons à la science de l'homme telle que nous venons de la décrire. v. Connoissance de soi-même.

Un traité complet et systématique d'anthropologie est encore un ouvrage à faire: peut-être les matériaux en sont tout trouvés, et sont épars dans les divers traités qui existent; mais ils ne sont pas rassemblés encore, ni disposés dans l'ordre et sous les rapports qui pourroient offrir dans leur réunion, le corps entier de la science de l'homme. Là l'homme n'est envisagé par le naturaliste que comme un corps organisé, purement physique et matériel. Ici le Psychologue le considére comme un pur esprit. Tel moraliste lui prescrit ses devoirs, comme si l'homme ne soutenoit qu'un certain nombre de relations, tandis qu'il en soutient un grand nombre d'autres qui modifient sensiblement sa capacité, et qui bornent ou étendent l'exercice de ses forces beaucoup en delà ou en deçà des loix qu'on lui prescrit. Tel autre détermine la destination de l'homme, non sur ce qu'indique sa nature envisagée sur toutes ses faces, mais d'après quelque système imaginé par un esprit qui n'a pas vu tout l'ensemble de l'homme, toutes ses relations, toutes ses dépendances. Il en est peu qui se soient souvenus, en parlant de l'homme, qu'il n'étoit pas un certain corps organisé de telle maniere, ni une certaine substance intelligente, qui a par elle-même telle capacité; mais que l'homme étoit l'ouvrage d'une sagesse infinie, qui n'a jamais rien fait que pour le plus grand bien, et qui a fait de l'homme, un être mixte, composé d'un corps organisé et d'une ame raisonnable, unis ensemble si intimement que ces deux substances ne forment qu'un seul individu, que nous ne connoissons que sous cette composition, sans avoir aucune idée de la maniere dont pourroient exister séparément ces deux substances. Nous ne saurions en effet imaginer quelles idées, quelles sensations, quelles connoissances nous aurions sans le secours de nos sens, de nos organes, qui sont notre corps; ni quelles actions nous serions capables de faire sans ce principe immatériel d'intelligence et d'activité qui est notre ame. v. SENS, IDÉES. La maniere de cette union du corps et de l'ame, le méchanisme de leur influence mutuelle, réelle ou apparente, ne sauroit être comprise. Qui pourroit cependant nier cette union, et soutenir que nous ne sentons pas cette influence réciproque du corps sur le principe intelligent, sensible et actif, et de ce principe d'intelligence, de sensibilité et d'activité sur le corps? v. AME, CORPS, INFLUENCE PHYSIQUE, ACTION. S'il n'existe point encore de traité d'anthropologie complet, on en trouvera en quelque sorte une esquisse, une analyse, dans les ouvrages vraiment philosophiques de l'illustre M. Bonnet de Geneve. *Essai de Psychologie, Essai analytique sur les facultés de l'ame, Considérations sur les corps organisés, Palyngénésie philosophique, Contemplation de la nature, Recherches philosophiques sur les preuves du Christianisme.* C'est un esprit tel que celui de cet excellent observateur, qui sait voir et saisir si bien l'ensemble et le rapports des objets divers qu'il considére; tirer des faits connus les conséquences qui peuvent en découler; associer la métaphysique la plus profonde avec les observations du Physicien le plus exact, analyser avec tant d'ordre et de clarté les sujets qu'il étudie; qui pourroit nous tracer un systême bien lié d'anthropologie.

Quelques Auteurs ont pris ce mot dans un sens beaucoup plus resserré, et ne désignent par-là que la seule œconomie animale de l'homme, en sorte que cette science n'est qu'une branche de la physiologie, tels ont été Teychmeyer, qui a donné sous le titre d'*Anthropologia,* un volume in-4º. imprimé à Genes en 1739, qui n'est qu'un traité d'œconomie animale; et Drake, Auteur Anglois, qui a publié dans sa langue, sous ce titre, un ouvrage de même genre, en trois volumes in-8º. imprimés à Londres en 1707 et 1727. L'un et l'autre auroient dû plutôt, comme Riolan le fils, intituler ces traités *anthropographie,* ou *description de l'homme.* v. ANTHROPOGRAPHIE.

Il est enfin quelques auteurs qui entendent par anthropologie, la science de la nature humaine envisagée uniquement sous les seuls traits qui la distinguent de celle des animaux brutes. Ceux d'entre les philosophes qui ne reconnoissent point d'ame chez les bêtes, les regardant comme de purs automates, rapportent à l'anthropologie tout ce qui fait l'objet de la psychologie, ou de la science de l'ame. Ceux qui croient que les betes ont une ame qui ne differe de celle de l'homme qu'en degré de capacité et de perfection, ne prennent pour objet de cette science que les seuls traits qui élévent l'homme au dessus des brutes; ces derniers reconnoissant dans celles-ci des

perceptions, des sensations, une sorte de pensée, la connoissance de leur
état et de l'existence des êtres qui les environnent, des idées assez claires
pour distinguer les objets les uns des autres, pour en appercevoir les propriétés qui intéressent leur conservation, leur bien-être, la satisfaction de leurs
besoins, une volonté, une liberté, le pouvoir de faire des actions spontanées,
des affections, des passions; découvrant dans les bêtes des fibres nerveuses,
organes des sens, et dont l'extrêmité intérieure va comme chez l'homme
se rendre au cerveau; un corps calleux, dans ce cerveau, qui paroît y être
comme chez nous, le centre de la sensibilité, et le siege du principe actif qui
réagit sur ces fibres et fait agir le corps. v. CORPS CALLEUX, SENSORIUM. Ils
se sont persuadés que le principe de ce que nous voyons faire aux bêtes, de-
voit être de même nature que celui des opérations de même genre que nous
exécutons. Or s'il est vrai que les bêtes sentent, veulent, agissent, ont des
idées etc., ces diverses capacités que nous croyons voir en elles, ne peuvent
pas plus être, chez la brute que chez l'homme, des propriétés de la matiere:
elles supposent en elles comme en nous, un principe immatériel, une ame.
v. MATIÈRE PENSÉE, IDÉE, SENTIMENT, VOLONTÉ.

Dans cette supposition, l'anthropologie ne considérera dans l'homme
que les traits de supériorité, de perfection, qui élévent l'homme au dessus
des bêtes animées comme lui. Ils trouvent en conséquence que l'homme
differe de la brute, I^o. par la faculté de faire des abstractions; soit physiques,
en décomposant les idées individuelles composées; soit métaphysiques, en
généralisant les idées individuelles. v. ABSTRACTIONS, Idées ABSTRAITES.
2°. Par l'imagination, qui réunissant diverses idées que la nature ne nous
offroit jamais que séparées, en forme de nouveaux assemblages qui donnent naissance aux nouvelles inventions, fruit de l'industrie. v. IMAGINA-
TION, INVENTION, ANALYSE. 3°. Par la curiosité, ou le desir de connoître et
d'approfondir tout ce qui s'offre à nous, lors même qu'il ne paroît avoir
aucune influence sur notre bien-être; mais uniquement parce que nous sentons le besoin de connoître. v. CURIOSITÉ, CONNOITRE. 4°. Par le sens moral,
qui nous rend capables d'appercevoir et de sentir la beauté, dans les proportions, la symétrie, les rapports et l'harmonie, soit physiques, soit morales;
d'où naissent les beaux arts et la vertu. v. BEAUTÉ, BEAUX ARTS, VERTU, SENS
MORAL, MORALITÉ. 5°. Par la faculté de parler, de fixer par des mots ou des
signes quelconques, dans notre mémoire, toutes nos idées, soit simples, soit
composées, soit naturelles, soit abstraites, ce qui nous met en état de les
comparer et de raisonner. v. LANGAGE, PAROLE. 6°. Enfin par la perfectibilité
qui résulte de tous ces traits particuliers et propres à la nature humaine,
c'est-à-dire, par la capacité que ces prérogatives nous donnent de perfectionner chaque jour notre intelligence, notre volonté, notre sensibilité, notre

activité, en un mot, toutes nos qualités et nos facultés, sans que l'on puisse déterminer encore un point de perfection au delà duquel l'homme ne puisse pas atteindre à force de soins, de travail et de tems. v. PERFECTIBILITÉ, PERFECTION. Voy. aussi HOMME, AME, ANIMAL, BÊTE.

Quelque système que l'on embrasse à l'égard de ce qui fait la différence entre l'homme et la brute, celle-ci reste toujours dans un degré prodigieux d'infériorité au dessous de celui-là. L'homme n'est pas rabaissé, quand même la brute seroit douée de la qualité d'être sentant, voulant et agissant avec spontanéité. L'homme n'est pas relevé quand même on refusera une ame à la bête. Toujours l'homme sera le chef-d'œuvre du Créateur, parmi les êtres qui sont sur la terre. Lui seul dominera sur les animaux, et les fera servir à ses besoins, à ses plaisirs; lui seul sentira son excellence, s'élévera par la pensée vers son Dieu, et entrera en relation avec lui par la religion. Lui seul étudiera, et apprendra à connoître la nature, et à voir dans ses productions les traits visible de l'infinie perfection de l'Etre invisible qui a tout fait. Lui seul cultivera les sciences et les beaux arts, et saura, par les divers métiers, rendre sa vie plus agréable, en satisfaisant avec plus de facilité à ses besoins. Lui seul, après avoir connu les faits, en découvrira les rapports, en calculera les quantités, en dévinera les causes. Lui seul jugeant de la convenance de ses actions par des rapports métaphysiques, les approuvera ou les blâmera, indépendamment d'un intérêt physique. Lui seul sera capable de vertu, sentira la justice des loix, l'obligation de s'y soumettre. Lui seul formera des sociétés régulieres avec ses semblables, et donnera par là naissance à de nouvelles obligations qu'il saura rendre sacrée par ses réflexions. Lui seul enfin prévoira sa destination, perçant dans l'avenir y verra au delà de la mort, une perspective encourageante de bonheur, que la foi rendra efficace pour porter à faire le bien que sa conscience lui recommande, que sa raison approuve, et qui lui est prescrit par la loi de son Dieu. v. HOMME, BÊTE, MORALE, RELIGION, SOCIÉTÉ, BONHEUR, VIE À VENIR. (G[abriel] M[ingard])

PSYCHOLOGY

DIDEROT AND D'ALEMBERT'S *ENCYCLOPÉDIE*, VOL. 13, P. 543

PSYCHOLOGIE (a), s. f. (*Métaphysique*) partie de la Philosophie, qui traite de l'ame humaine, qui en définit l'essence, & qui rend raison de ses opérations. On peut la diviser en Psychologie empirique, ou expérimentale, & Psychologie raisonnée. La premiere tire de l'expérience les principes, par lesquels elle explique ce qui se passe dans l'ame, & la Psychologie raisonnée, tirant de ces principes d'expérience une définition de l'ame, déduit, ensuite de cette définition, les diverses facultés & opérations qui conviennent a l'ame. C'est la double méthode à posteriori & à priori, dont l'accord produit la démonstration la plus exacte que l'on puisse prétendre. La Psychologie fournit des principes à diverses autres parties de la Philosophie, au droit naturel (b), à la Théologie naturelle (c), à la Philosophie pratique (d), & à la Logique (e). Rien de plus propre que l'étude de la Psychologie, pour remplir des plaisirs les plus vifs, un esprit qui aime les connoissances solides & utiles. C'est le plus grand bonheur dont l'homme soit susceptible ici bas, consistant dans la connoissance de la vérité, en tant qu'elle est liée avec la pratique de la vertu, on ne sauroit y arriver sans une connoissance préalable à l'ame, qui est appellée à acquérir ces connoissances, & à pratiquer ces vertus.

Notes

(a) Psychologie, dans les cours ordinaires, la doctrine de l'ame n'est qu'une partie de la Pneumatologie ou doctrine des esprits, qui n'est elle même qu'une partie de la Métaphysique. Mais M. Wolff dans la disposition philosophique de son cours, a fait de la Psychologie une partie distincte de la Philosophie, à laquelle il a consacré deux volumes; l'un pour la Psychologie empyrique; l'autre pour la Psychologie raisonnée, & il a placé cette tractation immédiatement après sa Cosmologie, parce qu'il en découle des principes pour presque toutes les autres parties, comme les notes suivantes le justifient.

(b) Au droit naturel. On démontre dans le droit naturel, quelles sont les bonnes & les mauvaises actions. Or la raison de cette qualification des actions, ne peut se deduire que de la nature humaine, & en particulier des propriétés de l'ame. La connoissance de l'ame doit précéder l'étude du droit naturel.

(c) A la Théologie naturelle. Nous ne pouvons arriver à la notion des attributs divins, qu'en dégageant la notion des propriétés de notre ame, de ses imperfections & de ses limitations. Il faut donc commencer par acquérir dans la Psychologie, des idées distinctes de ce qui convient à notre ame, pour en abstraire les principes généraux, qui déterminent ce qui convient à tous les esprits, & par conséquent à Dieu.

(d) A la Philosophie pratique. L'Etique ou la Morale a pour objet principal d'engager les hommes à pratiquer les vertus, & à fuir les vices, c'est-à-dire, de déterminer en général les appétits de l'ame d'une maniere convenable. Qui ne voit donc que cette détermination des appétits demande qu'on se représente distinctement la substance dans laquelle ils résident?

(e) A la Logique. Quoique par des raisons particulieres, on ait conservé à la Logique le premier rang entre les parties de la Philosophie, elle ne laisse pas d'étre subordonnée à la Psychologie, entant qu'elle lui emprunte des principes sans lesquels elle ne pourroit faire sentir la différence des idées, ni établir les regles du raisonnement qui sont fondées sur la nature & les opérations de l'ame.

YVERDON *ENCYCLOPÉDIE*, VOL. 35, PP. 511B–513A

PSYCHOLOGIE, (R), s.f. (*Métaphysique*). Ce mot formé des mots grecs, ψυχε l'ame, & de λογος, discours, signifie cette partie de la philosophie qui nous enseigne tout ce qu'on peut savoir de l'ame humaine: elle est une branche de la pneumatologie, qui a pour objet les esprits en général; celle-ci les considere uniquement entant qu'esprits, & s'occupe à découvrir qu'elle est leur nature essentielle, leurs qualités, leurs facultés distinctives, ce qu'ils peuvent produire, & ce qu'ils peuvent souffrir d'effets. Sous ce genre d'esprits, on comprend Dieu, les anges, & les ames humaines. Quand nous venons à considérer Dieu seul, cette branche de pneumatologie rentre dans la théologie naturelle. Nous connoissons trop peu les anges pour que ce que nous en dirions de particulier, pût former une science digne réellement de tenir un rang dans la philosophie, & de faire une branche de la métaphysique: on l'a traitée cependant dans les siecles passés sous le nom de démonologie. Quand on traite de l'ame humaine seule, on donne au recueil des découvertes faites sur ce sujet, le nom particulier de psychologie ou science de l'ame.

Si l'on y fait bien attention, on trouvera que la psychologie est la premiere partie connue de la pneumatologie, celle sans laquelle les autres nous seroient absolument inconnues.

Deux seuls moyens nous sont fournis pour acquerir toutes nos idées primitives, les sens qui nous font connoître les objets corporels, & notre réflexion qui nous donne la connoissance de nous-mêmes, de notre intelligence, des facultés & des qualités de notre ame, de son état & de ses actions actuelles; c'est par-là seulement que nous acquerons l'idée de l'intelligence, de la volonté, du sentiment; c'est d'après ce que nous éprouvons en nous-mêmes, que nous venons à nous faire des notions de moralité, de vertu & de vice, de perfections & d'imperfections morales, d'activité, d'action, de liberté, de choix, de science, d'ignorance, &c. C'est en réflechissant sur ce qui se passe en nous, & sur ces diverses notions qui supposent nécessairement l'intelligence, que nous venons à pouvoir raisonner sur l'essence des esprits: c'est en comparant les idées que nos sens nous donnent de la matiere, en qui nous ne voyons nulle trace de pensée, avec celle que la conscience nous donne de ce qui se passe en nous, que nous parvenons à juger que ce qui pense, veut & sent, n'est pas matiere. v. MATIERE.

C'est d'après nos actions sur la matiere, sur les corps qui nous environnent, c'est d'après le caractere que nous imprimons à ces actions quand nous les faisons à dessein, comparées avec celles que nous faisons sans dessein, que nous venons à distinguer les actions dues à la volonté réfléchie d'une intelligence, d'avec celles qui sont faites par hasard; & que nous découvrons dans l'arrangement & les changemens successifs de ce monde, les traces d'une intelligence qui l'a arrangé & qui le dirige. Le caractere de ces effets, comparé avec celui des effets que nous produisons, nous fait juger de la supériorité de cette cause intelligente au-dessus de nous, & des attributs dont elle est ornée. Ainsi la psychologie sert de guide vers la théologie.

Nous voyons des hommes qui nous ressemblent, tout nous dit qu'ils sont comme nous, & ce n'est qu'en nous connoissant nous-mêmes, en étudiant les effets que produisent sur nous les actions des autres êtres, que nous sommes en état de juger quelles actions de notre part peuvent convenir ou ne convenir pas à l'égard de nos semblables. Ainsi la psychologie nous conduit à la morale. C'est en faisant attention à ce qui se passe à notre égard, quand nous acquerons des connoissances, quand nous appercevons le vrai avec évidence, que nous dissipons les ténèbres de l'ignorance, & les nuages de l'erreur; que nous produisons en nous la conviction, la croyance, l'opinion, le doute; c'est en tenant compte des moyens employés pour produire en nous ces effets, que nous parvenons à découvrir toutes les propositions & les regles de la logique, qui sans la psychologie seroit encore à découvrir. Je ne saurois déterminer ma conduite envers moi même, m'en faire un systême, découvrir ma destination, trouver la route du bonheur, si je n'étudie pas tout ce que je puis connoître de mon ame, de ses facultés & de ses qualités, de sa liaison avec le corps, & de sa dépendance des sens & des organes, de son état, de ses rélations, & de sa destination; c'est en étudiant ce qui fait impression sur moi, ce qui me détermine à agir ou à ne pas agir, que je découvre tous les principes de l'art de conduire les hommes, ou de la politique. En un mot, quelle est la science ou l'art qui mérite notre attention, qui n'ait pas la psychologie pour base, pour principe & pour guide? C'est à nous que nous rapportons tout, c'est d'après l'influence des choses sur nous que nous les approuvons ou les condamnons; c'est donc le rapport des choses avec nous qui nous les rend intéressantes; & sans la connoissance de la nature, des facultés, des qualités, de l'état, des rélations, & de la destination de l'ame humaine, nous ne pouvons rien prononcer, rien décider, rien déterminer, rien choisir, rien rejetter, rien préférer, rien faire surement & sans erreur. La psychologie est ainsi la premiere, la plus utile de toutes les sciences, la source, le principe, le fondement de toutes, & le guide qui conduit vers chacune d'elles. Le sage eut donc raison de donner pour premier des préceptes celui de nous connoître nous-mêmes.

Cette science ne pouvant pas puiser ses principes & les faits d'après lesquels elle raisonne dans une science qui puisse être supposée la précéder, ne doit & ne peut pas les chercher hors de l'objet même de ses recherches. Elle est donc obligée avant que de raisonner, de chercher des faits qui lui servent de principes, & ces faits sont ceux-là uniquement que lui fournit la propre expérience de l'ame; elle doit étudier ce qui se passe dans l'ame, recueillir toutes les observations, & en tenir compte; de-là naît une premiere partie de la psychologie que les Wolfiens nomment la psychologie empirique. Ces faits une fois connus, l'homme cherche par la comparaison qu'il en fait à en découvrir le principe, à connoître quelle est l'essence de l'ame, quelles sont les regles auxquelles ses opérations sont assujetties, de quelles nouvelles opérations elle est capable, quels nouveaux effets il peut en attendre; de-là naît une seconde partie qu'on nomme la psychologie raisonnée. Celle-là sert de base à celle-ci; comme la physique expérimentale est la base de la physique raisonnée & systêmatique. v. ANTROPOLOGIE, MÉTAPHYSIQUE, AME, & tous les articles dépendans de ce dernier. Les meilleurs ouvrages à étudier sur cette matiere sont Wolfii *Psychologia empirica* & *Psychologia rationalis*, Bonnet, *Essai de psychologie*, & *Essai analytique sur les facultés de l'ame*, Condillac, *Traité des sensations*. (G[abriel] M[ingard])

Articles in the Yverdon *Encyclopédie* Belonging to Psychology and Their Place in the Paris *Encyclopédie*

EY headword	Other field indicators	Author	EP field indicators	Author
ABSTRACTION (R)		G.M.		Dumarsais
AFFIRMATION (N)	*Logique*	G.M.		
AME (R)	*Métaphysique, Pneumatologie, Anthropologie*	G.M.	*Pneumatologie*	Yvon
ANTHROPOLOGIE	*Philosophie, Histoire Naturelle, Physiologie, Métaphysique*	G.M.	*Théologie*	Mallet
APPARENCE (N)	*Métaphysique, Philosophie morale, Logique*	G.M.	*Grammaire*	Diderot [1]
APPERCEPTION (N)		G.M.		
APPERCEVOIR (N)		G.M.		
APPÉTER (R)	*Oeconomie animale*	G.M.		
APPÉTIT (R)	*Physiologie*	G.M.	*Morale*	Yvon
APPLICATION (N)		G.M.	[2]	
APPRÉHENSION (R)	*Logique*	G.M.	*Logique*	Yvon
APPRENDRE (N)		G.M.	*Grammaire*	Diderot
ARBITRE (N)		G.M.	*Droit*	Toussaint
BEAUTÉ (R)	*Métaphysique*	G.M.		Diderot
CHOIX (N)		G.M.	*Grammaire*	Diderot
			[*Beaux-Arts*]	Landois
CŒUR (N)	*Morale*	G.M.		
COMBINAISON (N)	*Logique*	G.M.		

Continued

EY headword	Other field indicators	Author	EP field indicators	Author
COMMERCE ÂME-CORPS (N)	*Anthropologie*	G.M.		
CONCEPTION (R)		G.M.	*Logique*	Jaucourt
CONCUPISCENCE (R)	*Morale*	G.M.	*Théologie* [3]	Mallet
CONNOISSANCE (R)	*Métaphysique, Logique*	G.M.	*Métaphysique*	
CONSCIENCE (R)	*Logique*	de Felice	*Métaphysique*	Jaucourt
COUTUME (R)	*Physique, Philosophie morale*	G.M.	*Morale*	Formey
DOUTE (N)		G.M.	*Logique, Métaphysique*	
FIBRE (N)			*Anatomie*	Barthez
			Economie animale, Médecine	d'Aumont
FIN (N)	*Métaphysique*	G.M.	*Grammaire*	Diderot
			Morale	Diderot
MATÉRIALISME (N)	*Philosophie, Métaphysique*	G.M.		
OPINION (R)	*Logique*	G.M.	*Logique*	
ORGANE DE LA PENSÉE (N)	*Physiologie*			
PASSION (R)	*Morale*	G.M.	(PASSIONS) *Philosophie, Logique, Morale*	
PRÉJUGÉ (R) [4]	*Logique, Morale*	M.D.B.	*Logique*	Jaucourt
PUDEUR (R) [4]	*Morale*	M.D.B.	*Morale*	Jaucourt
RAPPEL DES IDÉES PAR LES MOTS (N)				
SPONTANÉ, SPONTANÉE (R)	*Philosophie, Métaphysique, Morale*		(SPONTANÉE) *Grammaire, Médecine*	
SYMPATHIE (R)	*Anthropologie, Physiologie, Philosophie morale*			

Note: G.M. = Gabriel Mingard for the articles "unrelated to religion." M.D.B. = Gabriel Mingard for the articles that "could expose one to the absurd rantings of the bigots."
[1] Definition of the word, followed by other articles.
[2] Several articles, principally about applying one science to another.
[3] "For Theologians, means . . ."
[4] In the *Supplément* to EY. The articles in EY reproduce those of EP.

BIBLIOGRAPHY

Abbot, Ezra, "Literature of the doctrine of a future life: or, A catalogue of works relating to the nature, origin, and destiny of the soul," in W. R. Alger, *A critical history of the doctrine of a future life* (Philadelphia: George W. Childs, 1864).

Abel, Jacob Friedrich, *Einleitung in die Seelenlehre* (1786; Hildesheim: Olms, 1985).

———, *Sammlung und Erklärung merkwürdiger Erscheinungen aus dem menschlichen Leben* (Frankfurt, 1784).

———, *Sammlung und Erklärung merkwürdiger Erscheinungen aus dem menschlichen Leben*, Zweiter Theil (Stuttgart: in der Erhardischen Buchhandlung, 1787).

———, *Sammlung und Erklärung merkwürdiger Erscheinungen aus dem menschlichen Leben*, Dritter Theil (Stuttgart: in der Erhard- und Löflundischen Buchhandlung, 1790).

Adelung, Johann Christoph, *Versuch einer Geschichte der Cultur des menschlichen Geschlechts* (Leipzig: Hertel, 1782).

Albertan-Coppola, Sylviane, "De Locke à Helvétius en passant par *l'Encyclopédie*; ou, Faut-il 'casser le XVIIIᵉ siècle,'" in Ulla Kölving and Irène Passeron, eds., *Sciences, musiques, Lumières: Mélanges offerts à Anne-Marie Chouillet* (Ferney-Voltaire: Centre international d'études du XVIIIᵉ siècle, 2002).

———, "Les préjugés légitimes de Chaumeix; ou, *L'Encyclopédie* sous la loupe d'un apologiste," *Recherches sur Diderot et "l'Encyclopédie,"* 20, 1996, 149–158.

Alembert, Jean Le Rond d', *Preliminary discourse to the Encyclopedia of Diderot*, trans. Richard N. Schwab and Walter E. Rex (Indianapolis: Bobbs-Merrill, 1963).

Allen, Don Cameron, *Mysteriously meant: The rediscovery of pagan symbolism and allegorical interpretation in the Renaissance* (Baltimore: Johns Hopkins Press, 1970).

Alsted, Johann Heinrich, *Encyclopaedia* (1630; Stuttgart: Frommann-Holzboog, 1989–1990).

Anderson, Walter, *The philosophy of ancient Greece investigated, in its origin and progress, to the aeras of its greatest celebrity, in the Ionian, Italic, and Athenian schools . . .* (Edinburgh: printed by Smellie, 1791).

Anonymous, *Essais sur la Providence et sur la possibilité physique de la Résurrection* (1719), "translated from the English of Dr. B," 2nd ed. (Amsterdam: chez E. J. Ledet, 1731).

————, *Psychologie ou Traité sur l'âme: contenant les connoissances, que nous en donne l'expérience* (1745), in Wolff, *Gesammelte Werke*, ser. III, vol. 46.

————, "Ueber die ersten psychologischen Versuche bei den Griechen, vorzüglich des Plato und Aristoteles," *Psychologisches Magazin*, 1, 1790, 298–388.

Aquinas, Thomas d' (Saint), *Super primam epistolam ad Corinthios lectura*, in *Super epistolas S. Pauli lectura*, ed. Raphael Cai (Turin: Marietti, 1953), vol. 1.

Ariew, Roger, and Marjorie Greene, "The Cartesian destiny of form and matter," *Early science and medicine*, 2, 1997, 300–325.

Aristotle, *On the soul (De anima)*, trans. J. A. Smith, in *The complete works of Aristotle: The revised Oxford translation*, ed. Jonathan Barnes (Princeton: Princeton University Press, 1984).

————, *Sense and sensibilia* (from *Parva naturalia*), trans. J. I. Beare, in *The complete works of Aristotle: The revised Oxford translation*, ed. Jonathan Barnes (Princeton: Princeton University Press, 1984).

Armando, Luigi Antonello, *L'invenzione della psicologia: Saggio sull'opera storiografica di E. G. Boring* (Rome: Nuove Edizioni Romane, 1988).

Arriaga, Roderico de, "Disputationes in tres libros Aristotelis De Anima," in *Cursus philosophicus* (1632; Paris: apud Jacobum Quesnel, 1639).

Ash, Mitchell G., "The self-presentation of a discipline: History of psychology in the United States between pedagogy and scholarship," in Loren Graham, Wolf Lepenies, and Peter Weingart, eds., *Functions and uses of disciplinary histories* (Dordrecht: Reidel, 1983).

Azouvi, François, "Genèse du corps propre chez Malebranche, Condillac, Lignac et Maine de Biran," *Archives de philosophie*, 45, 1982, 85–107.

————, ed., *L'institution de la raison: La révolution culturelle des Idéologues* (Paris: Vrin, 1992).

————, *Maine de Biran: La science de l'homme* (Paris: Vrin, 1995).

Bacon, Francis, *The Instauratio Magna*, pt. 2, *Novum organum and associated texts*, ed. Graham Rees with Maria Wakely (Oxford: Clarendon Press, 2004).

————, *The proficience and advancement of learning divine and human* (1605), trans. Michael Kiernan (Oxford: Oxford University Press, 2000).

Badaloni, Nicola, *Antonio Conti: Un abate libero pensatore tra Newton e Voltaire* (Milan: Feltrinelli, 1968).

Baertschi, Bernard, "La conception de la conscience développée par Mérian," in Fontius and Holzhey, *Schweizer im Berlin*.

————, *Les rapports de l'âme et du corps: Descartes, Diderot et Maine de Biran* (Paris: Vrin, 1992).

Baldwin, James Mark, "Autobiography," in Carl Murchison, ed., *A history of psychology in autobiography* (Worcester, MA: Clark University Press, 1930), vol. 1.

————, *History of psychology: A sketch and an interpretation* (New York: G. P. Putnam's Sons, 1913).

Balibar, Étienne, *John Locke: Identité et différence: L'invention de la conscience* (Paris: Seuil, 1998).

Barbieri, Ludovico, *Trattato di psicologia nel quale si ragiona della natura dell'anime umane, e degli altri spiriti, della loro eccellenza sopra i corpi, della intelligenza, della volontà, della immortalità ecc.* (Venice: Valvasenese, 1756).

Bardili, Christoph Gottfried, *Epochen der vorzüglichen Philosophischen Begriffe, nebst den nöthigen Beylagen*, Erster Theil, *Epochen der Ideen von einem Geist, von Gott und der menschlichen Seele: System und Aechtheit der beiden Pythagoreer, Ocellus und Timäus* (1788; Brussels: Culture et civilisation, 1970).

Barnouw, Jeffrey, "The philosophical achievement and historical significance of Johann Nicolas Tetens," *Studies in eighteenth-century culture*, 9, 1979, 301–335.

————, "Psychologie empirique et épistémologie dans les *Philosophische Versuche* de Tetens," *Archives de philosophie*, 46, 1983, 271–289.

Baumeister, Friedrich Christian, *Philosophia definitiva hoc est definitiones philosophicae ex systemate Lib. bar. a Wolf . . .* (1735), ed. of 1775, in Wolff, *Gesammelte Werke*, ser. III, vol. 7.

Baumgarten, Alexander Gottlieb, *Metaphysica* (1739), ed. of 1779 (Hildesheim: Olms, 1982).

————, *Aesthetica*, Latin-German, ed. and trans. Dagmar Mirbach (1750; Hamburg: Felix Meiner, 2007).

Beattie, James, *Elements of moral science* (Edinburgh: printed for T. Cadell, 1790).

Beccaria, Cesare, *Dei delitti e delle pene* (1764), ed. Franco Venturi (1965; Turin: Einaudi, 1994).

————, *On crimes and punishments*, trans. Graeme R. Newman and Pietro Marongiu (New Brunswick, NJ: Transaction Publishers, 2009).

————, *Ricerche intorno alla natura dello stile* (1770), in *Opere*, ed. Sergio Romegnoli (Florence: Sansoni, 1958).

Beck, Johann Tobias, *Umriss der biblischen Seelenlehre: Ein Versuch* (1843; Stuttgart: J. F. Steinkopf, 1871).

Beck, Lewis White, *Early German philosophy: Kant and his predecessors* (1969; Bristol: Thoemmes Press, 1996).

Becker, Carl L., *The heavenly city of the eighteenth-century philosophers* (1932; New Haven: Yale University Press, 1955).

Bellucci, Dino, *Science de la nature et Réformation: La physique au service de la Réforme dans l'enseignement de Philippe Mélanchthon* (Rome: Edizioni Vivere In, 1998).

Benoni Debrun, François-Joseph, *Cours de psychologie: Traité de psychographie* (1801), in Serge Nicolas, ed., *Un cours de psychologie durant la Révolution française de 1789: Le traité de psychographie de Benoni Debrun* (Paris: L'Harmattan, 2003).

Bérard, Victor, *Les navigations d'Ulysse*, vol. 1, *Ithaque et la Grèce des Achéens* (Paris: Armand Colin, 1927).

Berengo, Mario, ed., *Giornali veneziani del Settecento* (Milan: Feltrinelli, 1962).

Bergamo, Mino, *L'anatomia dell'anima: Da François de Sales a Fénelon* (Bologna: Il Mulino, 1991).

Berlin, Isaiah, *The Age of Enlightenment* (1956; New York: New American Library, 1984).

Berthoud, Gérald, "Une 'science générale de l'homme': L'œuvre d'Alexandre-César Chavannes," *Annales Benjamin Constant*, 13, 1992, 29–41.

Bilfinger, Georg Bernhard, *Dilucidationes philosophicae de Deo, anima humana, mundo, et generalibus rerum affectionis* (1725), in Wolff, *Gesammelte Werke*, ser. III, vol. 18.

Blackwell, Constance, "Epicurus and Boyle, Le Clerc and Locke: 'Ideas' and their redefinition in Jacob Brucker's *Historia philosophica doctrinae de ideis*," in Marta Fattori, ed., *Il vocabolario della République des Lettres: Terminologia filosofica e storia della filosofia: Problemi di metodo* (Florence: Olschki, 1997).

———, "Jacob Brucker's theory of knowledge and the history of natural philosophy," in Schmidt-Biggemann and Stammen, *Jacob Brucker*.

———, "Thales Philosophus: The beginning of philosophy as a discipline," in Kelley, *History and the disciplines*.

Blair, Hugh, *A critical dissertation on the poems of Ossian, the son of Fingal*, 2nd ed. (1765; New York: Garland, 1970).

Blanchard, Gilles, and Mark Olsen, "Le système des renvois de *l'Encyclopédie*: Une cartographie des structures de connaissances au XVIIIᵉ siècle," *Recherches sur Diderot et "l'Encyclopédie,"* 31–32, 2002, 45–70.

Blank, Robert, *Brain policy: How the new neuroscience will change our lives and our politics* (Washington, DC: Georgetown University Press, 1999).

Bloom, Paul, "Body and soul: A partnership re-evaluated," *International herald tribune*, 21 September 2004, 9.

Blumenthal, Arthur L., "A reappraisal of Wilhelm Wundt," *American psychologist*, 30, 1975, 1081–1088.

———, "Wilhelm Wundt: Problems of interpretation," in Wolfgang G. Bringmann and Ryan D. Tweney, eds., *Wundt studies* (Toronto, C. J. Hogrefe, 1980).

Bödeker, Hans Erich, "Konzept und Klassifikation der Wissenschaften bei Johann Georg Sulzer (1720–1779)," in Fontius and Holzhey, *Schweizer im Berlin*.

Bödeker, Hans Erich, et al., eds., *Aufklärung und Geschichte: Studien zur deutschen Geschichtswissenschaft im 18. Jahrhundert* (Göttingen: Vandenhoeck & Ruprecht, 1992).

Böhme, Jacob, *ΨΥΧΟΛΟΓΙΑ vera I[acobi] B[oehmi] T[eutonici] XL: Quaestionibus explicata et rerum publicarum vero regimini: ac earum MAIESTATICO IURI applicata . . .* , trans. Johannes Angelius Werdenhagen (Amsterdam: Apud Iohann. Ianßonium, 1632).

———, *PSYCHOLOGIA VERA, oder Vierzig Fragen von der Seelen, ihrem Urstande, Essenz, Wesen, Natur und Eigenschaft was sie von Ewigkeit in Ewigkeit sey . . .* , in J. Böhme, *Sämtliche Schriften*, ed. August Faust (Stuttgart: Fr. Frommans Verlag, 1942), vol. 3.

Böhme, Joachim, *Die Seele und das Ich im homerischen Epos* (Leipzig: B. G. Teubner, 1929).

Bonnet, Charles, "Analyse abrégée de l'*Essai analytique*" (1779), in Bonnet, *Œuvres*, tome 7 (= vol. 9).

———, *Considérations sur les corps organisés* (1776), in Bonnet, *Œuvres*, tome 3 (= vol. 3).

———, *Contemplation de la nature* (1764), in Bonnet, *Œuvres*, tomes 4.1 and 4.2 (= vols. 4 and 5).

———, *Essai analytique sur les facultés de l'âme* (1760), in Bonnet, *Œuvres*, tome 6 (= vol. 8).

————, *Essai de psychologie* (1754), in Bonnet, *Œuvres*, tome 8 (= vol. 10).

————, *Essais sur la vie à venir*, ed. François de la Rive-Rilliet (Geneva, 1828).

————, *Œuvres d'histoire naturelle et de philosophie* (Neuchâtel: Samuel Fauche, 1779–1783), *in quarto*.

Bonno, Gabriel, *Les relations intellectuelles de Locke avec la France* (Berkeley: University of California Press, 1955).

Boring, Edwin G., *A history of experimental psychology* (1929; New York: Appleton-Century-Crofts, 1957).

————, "A note on the origin of the word 'psychology,'" *Journal of the history of the behavioral sciences*, 2, 1966, 167.

Bourdieu, Pierre, "Le champ scientifique et les conditions sociales du progrès de la raison," *Sociologie et sociétés*, 7, 1975, 91–117.

Boyle, Robert, "Some physico-theological considerations about the possibility of the Resurrection" (1675), in *The Works of the Honourable Robert Boyle*, 2nd ed. (London: printed for J. and F. Rivington, 1772).

Bradshaw, Lael Ely, "Ephraim Chambers' *Cyclopaedia*," in Kafker, *Notable encyclopedias: Nine predecessors*.

Braun, Claude M. J., and Jacinthe M. C. Baribeau, "The classification of psychology among the sciences from Francis Bacon to Boniface Kedrov," *Journal of mind and behavior*, 5, 1984, 245–260.

Braun, Lucien, *Histoire de l'histoire de la philosophie* (Paris: Ophrys, 1973).

Braunschweiger, David, *Die Lehre von der Aufmerksamkeit in der Psychologie des 18. Jahrhunderts* (Leipzig: Hermann Haacke, 1899).

Bremmer, Jan N., *The early Greek concept of the soul* (1983; Princeton: Princeton University Press, 1993).

Bretschneider, Carl Gottlieb, ed., *Philippi Melanthonis Opera quae supersunt omnia*, Corpus reformatorum, vol. 13 (1846; New York: Johnson Reprint Corp., 1963).

Brett, George Sydney, *A history of psychology* (1912–1921; Bristol: Thoemmes Press, 1998).

Brock, Adrian, "Something old, something new: The 'reappraisal' of Wilhelm Wundt in textbooks," *Theory and psychology*, 3, 1993, 235–242.

Brockliss, Laurence W. B., "Curricula," in Ridder-Symoens, *A history of the university*.

————, *French higher education in the seventeenth and eighteenth century: A cultural history* (Oxford: Clarendon Press, 1987).

Broughton, John, *Psychologia: or, An Account of the Nature of the Rational Soul . . .* (1703; Bristol: Thoemmes Press, 1990).

Brown, Peter, *The body and society: Men, women, and sexual renunciation in earlier Christianity* (New York: Columbia University Press, 1988).

Brozek, Joseph, "*Psychologia* of Marcus Marulus (1450–1524): Evidence in printed works and estimated date of origin," *Episteme*, 7, 1973, 125–131.

Brucker, Johann Jacob, *Historia critica philosophiae a mundi incunabulis usque ad nostram aetatem deducta*, 4 vols. and appendix (1742–47; Hildesheim: Olms, 1975).

————, *Historia philosophica doctrinae de ideis qua tum veterum imprimis graecorum tum recentiorum philosophorum placita enarrantur* (Augsburg: apud Dav. Raym. Mertz, et I. Iac. Mayer, 1723).

————, *Kurtze Fragen aus der Philosophischen Historie* . . . (Ulm: bey Daniel Bartholomäi und Sohn, 1731–35).

Bruhlmeier, Daniel, Helmut Holzhey, and Vilem Mudroch, eds., *Schottische Aufklärung: A hotbed of genius* (Berlin: Akademie Verlag, 1996).

Bryson, Gladys, *Man and society: The Scottish inquiry of the eighteenth century* (Princeton: Princeton University Press, 1945).

Buddeus, Johann Franz, *Introductio ad historiam philosophiam ebraeorum* (1702; Halle: Impensis Orphanotrophei, 1720).

Budé, Guillaume, *De l'institution du prince* . . . (Paris: Imprimé à L'Arrivour . . . par maistre Nicole, 1547).

Buffière, Félix, *Les mythes d'Homère et la pensée grecque* (Paris: Belles Lettres, 1956).

Buickerood, James G., "The natural history of the understanding: Locke and the rise of facultative logic in the eighteenth century," *History and philosophy of logic*, 6, 1985, 157–190.

Burdorf, Dieter, and Wolfgang Schweickard, with the collaboration of Annette Gerstenberg, eds., *Die schöne Verwirrung der Phantasie: Antike Mythologie und Kunst um 1800* (Tübingen: Francke Verlag, 1998).

Burke, Peter, ed., *A new kind of history: From the writings of Lucien Febvre*, trans. Keith Folca (London: Routledge and Kegan Paul, 1973).

Buscaglia, Marino, René Sigrist, Jacques Trembley, and Jean Wüest, eds., *Charles Bonnet savant et philosophe (1720–1793)* (Geneva: Editions Passé Présent, 1994).

Bynum, Caroline Walker, *The resurrection of the body in Western Christianity, 200–1336* (New York: Columbia University Press, 1995).

————, "Why all the fuss about the body? A medievalist's perspective," *Critical inquiry*, 22, 1995, 1–33.

Cabanis, Pierre-Jean-Georges, "Lettre sur un passage de la *Décade philosophique* et en général sur la perfectibilité de l'esprit humain" (1809), in Cabanis, *Œuvres philosophiques*, vol. 2.

————, *Œuvres philosophiques*, ed. Claude Lehec and Jean Cazeneuve (Paris: Presses Universitaires de France, 1956).

————, *Rapports du physique et du moral de l'homme* (1800), in Cabanis, *Œuvres philosophiques*.

Caianiello, Silvia, "La lecture de Vico dans l'historicisme allemand," *L'art de comprendre*, 7, 1998, 139–167.

Campo, Mariano, *Cristiano Wolff e il razionalismo precritico* (1939), in Wolff, *Gesammelte Werke*, ser. III, vol. 9.

Canguilhem, Georges, "Qu'est-ce que la psychologie?" (1956), in *Études d'histoire et de philosophie des sciences* (Paris: Vrin, 1994).

Carboncini, Sonia, "*L'Encyclopédie* et Christian Wolff: A propos de quelques articles anonymes," *Les études philosophiques*, no. 4, 1987, 489–504.

————, "Lumières e Aufklärung: A proposito della presenza della filosofia di Christian Wolff nell'*Encyclopédie*," *Annali della Scuola Normale Superiore di Pisa (Classe di Lettere e Filosofia)*, 14, 1984, 1297–1308.

Carels, Peter E., and Dan Flory, "Johann Heinrich Zedler's *Universal-Lexicon*," in Kafker, *Notable encyclopedias: Nine predecessors*.

Carhart, Michael, *The science of culture in Enlightenment Germany* (Cambridge, MA: Harvard University Press, 2008).

Carpintero, Helio, *Historia de la psicología en España* (Madrid: EUDEMA, 1994).

Carpov, Jacob, *Psychologia sacratissima: Hoc est de ANIMA CHRISTI hominis in se spectata commentatio theologico-philosophica* (Frankfurt: Io. Adam. Melchior, 1738).

Carrard, Benjamin, *Essai . . . sur cette Question: Qu'est-ce qui est requis dans l'Art d'Observer; & jusques-où cet Art contribute-t-il à perfectionner l'Entendement?* (Amsterdam: chez Marc-Michel Rey, 1777).

Carrive, Paulette, "L'idée d''histoire naturelle de l'humanité' chez les philosophes écossais du XVIIIᵉ siècle," in Olivier Bloch, Bernard Balan, and P. Carrive, eds., *Entre forme et histoire: La formation de la notion de développement à l'âge classique* (Paris: Méridiens Klincksieck, 1988).

Carter, Richard B., *Descartes' medical philosophy: The organic solution to the mind-body problem* (Baltimore: Johns Hopkins University Press, 1983).

Carus, Friedrich August, *Geschichte der Psychologie*, introduction by Rolf Jeschonnek (1808; Berlin: Springer Verlag, 1990).

———, *Ideen zur Geschichte der Menschheit*, in Carus, *Nachgelassene Werke*, vol. 6.

———, *Nachgelassene Werke*, ed. Ferdinand Hand (Leipzig: bei Iohann Ambrosius Bart und Paul Gotthelf Kummer, 1808–1810).

———, *Psychologie der Hebräer*, in Carus, *Nachgelassene Werke*, vol. 5.

Casmann, Otto, *Psychologia anthropologica; sive animae humanae doctrina . . .* (Hanau: apud Guilielmum Antonium, impensis Petri Fischeri Fr., 1594).

Cassirer, Ernst, *The philosophy of the Enlightenment*, trans. Fritz C. A . Koelln and James P. Pettegrove (Boston: Beacon Press, 1961).

Cernuschi, Alain, "L'arbre encyclopédique des connaissances: Figures, opérations, métamorphoses," in Roland Schaer, ed., *Tous les savoirs du monde: Encyclopédies et bibliothèques, de Sumer au XXIe siècle* (Paris: Bibliothèque nationale de France / Flammarion, 1996).

———, *Penser la musique dans "l'Encyclopédie": Étude sur les enjeux de la musicographie et sur ses liens avec l'encyclopédisme* (Paris: Champion, 2000).

———, "La place du religieux dans le système des connaissances de *l'Encyclopédie* d'Yverdon," in Jean-Daniel Candaux, A. Cernuschi, Clorinda Donato, and Jens Häseler, eds., *"L'Encyclopédie" d'Yverdon et sa résonance européenne: Contextes-contenus-continuités* (Geneva: Slatkine, 2005).

Cerullo, John C., "E. G. Boring: Reflections on a discipline builder," *American journal of psychology*, 101, 1988, 561–575.

Cesarotti, Melchiore, *L'Iliade di Omero*, pt. 1, *Ragionamento preliminare storico-critico*, in *Opere* (Pisa: dalla Tipografia della Società letteraria, 1802), vol. 6.

Chambers, Ephraim, *Cyclopaedia: or an Universal Dictionary of Arts and Science* (London: J. and J. Knapton, 1728).

Chappell, Vere, ed., *The Cambridge companion to Locke* (New York: Cambridge University Press, 1994).

Chartier, Roger, *The cultural origins of the French Revolution* (1990), trans. Lydia G. Cochrane (Durham: Duke University Press, 1991).

Chauvin, Stephanus, *Lexicon philosophicum* (1692), ed. of 1713 (Düsseldorf: Stern-Verlag Janssen, 1967).

Chavannes, Alexandre-César, *Anthropologie ou science générale de l'homme pour servir d'introduction à l'étude de la philosophie & des langues, & de guide dans le plan d'éducation intellectuelle* (Lausanne, 1788).

———, *Anthropologie ou science générale de l'homme pour servir d'introduction à l'étude de la philosophie et des langues, et de guide dans le plan d'éducation intellectuelle,* 13 manuscript vols., undated, Département des manuscrits, Bibliothèque cantonale et universitaire, Lausanne.

———, *Essai sur l'éducation intellectuelle, avec le projet d'une science nouvelle* (Lausanne, 1787).

Chiarini, Gioachino, "Ch. G. Heyne e gli inizi dello studio scientifico della mitologia," *Lares,* 55, 1989, 317–331.

Chisik, Harvey, *The limits of reform in the Enlightenment: Attitudes towards the education of the lower classes in eighteenth-century France* (Princeton: Princeton University Press, 1981).

Claparède, Édouard, *L'association des idées* (Paris: Octave Doin, 1903).

Clark, William, Jan Golinski, and Simon Schaffer, eds., *The sciences in enlightened Europe* (Chicago: University of Chicago Press, 1999).

Clarke, Edwin, and L. S. Jacyna, *Nineteenth-century origins of neuroscientific concepts* (Berkeley: University of California Press, 1987).

Clarke, Samuel, *A Discourse concerning the Being and Attributes of God, the Obligations of Natural Religion, and the Truth and Certainty of the Christian Revelation,* 10th ed. (London, 1768).

Clerc, Nicolas-Gabriel, *Abrégé des études de l'homme fait, en faveur de l'homme à former* (Paris: chez Maradan, 1789).

Condillac, Étienne Bonnot de, "Des progrès de l'art de raisonner" (in *Histoire moderne,* bk. XX, chap. 12, 1775; in *Cours d'études pour l'instruction du Prince de Parme,* 1768–1773), in Condillac, *Œuvres philosophiques.*

———, *Essai sur l'origine des connaissances humaines* (1746), in *Œuvres complètes de Condillac* (Paris: de l'imprimerie de Ch. Houel, 1798), vol. 1.

———, *Œuvres philosophiques de Condillac,* ed. Georges Le Roy (Paris: Presses Universitaires de France, 1947–1951).

———, *Treatise on the sensations,* trans. Geraldine Carr (London: Favil Press, 1930).

Condorcet, Jean-Antoine-Nicolas Caritat, marquis de, "Petits résumés sur l'histoire de l'éducation" (1774), in *Réflexions et notes sur l'éducation,* ed. Manuela Albertone (Naples: Bibliopolis, 1983).

Conti, Antonio, *Prose e poesie* (Venice: presso Giambattista Pasquali, 1739–1756).

———, *Scritti filosofici,* ed. Nicola Badaloni (Naples: Fulvio Rossi, 1972).

Corr, Charles A., "Christian Wolff's distinction between empirical and rational psychology," *Studia leibnitiana supplementa,* 14, 1975, 195–215.

Cottingham, John, *A Descartes dictionary* (Oxford: Blackwell, 1993).

Court de Gébelin, Antoine, *Monde primitif, analysé et comparé avec le monde moderne, considéré dans son génie allégorique et dans les allégories auxquelles conduisit ce génie* (Paris: chez l'Auteur . . . , 1777).

Courtine, Jean-François, "Le projet suarézien de la métaphysique," *Archives de philosophie*, 42, 1979, 235–274.

Coward, William, *The just scrutiny: or, a Serious enquiry into the modern notions of the soul, I, Consider'd as breath of life, or a power (not immaterial substance) united to body, according to the H. Scriptures, II, As a principle naturally mortal, but immortaliz'd by its union with the baptismal spirit, according to Platonisme lately Christianiz'd* . . . (London: Printed for John Chantry, 1705).

———, *Second thoughts concerning human soul, demonstrating the notion of human soul, as believ'd to be a spiritual immortal substance, united to human body, to be a plain heathenish invention, and not consonant to the principles of philosophy, reason, or religion; but the ground only of many absurd, and superstitious opinions, abominable to the reformed churches, and derogatory in general to true Christianity* (London: printed for R. Basset, 1702).

Crichton, Alexander, *An inquiry into the nature and origin of mental derangement* (London: printed for T. Cadell, 1798), vol. 1.

Cristofolini, Paolo, *Vico et l'histoire* (Paris: Presses Universitaires de France, 1995).

Cromaziano, Agatopisto, *Kritische Geschichte der Revolutionen der Philosophie in den drey letzten Jahrhunderten* (1766–1772), trans. Karl Heinrich Heidenreich, ed. of 1791 (Brussels: Culture et civilisation, 1968).

da Cruz Pontes, J. M., "Le problème de l'origine de l'âme de la patristique à la solution thomiste," *Revue de théologie ancienne et medieval*, 31, 1964, 175–229.

Dannenberg, Lutz, Sandra Pott, Jörg Schönert, and Friedrich Vollhardt, eds., *Säkularisierung in den Wissenschaften seit der Frühen Neuzeit*, vol. 2, *Zwischen christlicher Apologetik und methodologischem Atheismus: Wissenschaftsprozesse im Zeitraum von 1500 bis 1800* (Berlin: Walter de Gruyter, 2002).

Dannhauer, Johann Conrad, *Collegium psychologicum, in quo maxime controversae quaestiones, circa libros tres Aristotelis De Anima, proponuntur, ventilantur, explicantur* (Strasbourg: Typis Josiae Staedelii, 1660).

Danziger, Kurt, "The positivist repudiation of Wundt," *Journal of the history of the behavioral sciences*, 15, 1979, 205–230.

Darnton, Robert, *The business of enlightenment: A publishing history of the "Encyclopédie," 1775–1800* (Cambridge, MA: Harvard University Press, 1979).

———, "Philosophers trim the tree of knowledge: The epistemological strategy of the Encyclopédie," in *The great cat massacre, and other episodes in French cultural history* (New York: Basic Books, 1984).

Daston, Lorraine, "Attention and the values of nature in the Enlightenment," in L. Daston and F. Vidal, eds., *The moral authority of nature* (Chicago: University of Chicago Press, 2004).

———, "Baconian facts, academic civility, and the prehistory of objectivity," *Annals of scholarship*, 8, 1991, 337–363.

———, "Classifications of knowledge in the age of Louis XIV," in David Lee Rubin, ed., *Sun King: The ascendancy of French culture during the reign of Louis XIV* (Washington, DC: Folger Shakespeare Library, 1992).

———, *Eine kurze Geschichte der wissenschaftlichen Aufmerksamkeit* (Munich: Carl-Friedrich-von-Siemens-Stiftung, 2001).

——, "Perchè i fatti sono brevi?," *Quaderni storici*, 108, 2001, 745–770.

Davies, Catherine Glyn, *Conscience as consciousness: The idea of self-awareness in French philosophical writing from Descartes to Diderot* (Oxford: Voltaire Foundation, 1990).

De Angelis, Simone, "Zwischen *generatio* und *creatio*: Zum Problem der Genese der Seele um 1600—Rudolph Goclenius, Julius Caesar Scaliger, Fortunio Liceti," in Dannenberg et al., *Säkularisierung*.

Dear, Peter, *Revolutionizing the sciences: European knowledge and its ambitions, 1500–1700* (Princeton: Princeton University Press, 2001).

de Beer, E. S., ed., *The correspondence of John Locke* (Oxford: Clarendon Press, 1976–1989).

Delitzsch, Franz, *System der biblischen Psychologie* (1855; Leipzig: Dörffling und Franke, 1861).

Delon, Michel, ed., *Dictionnaire européen des Lumières* (Paris: Presses Universitaires de France, 1997).

Del Torre, Maria Assunta, *Le origine moderne della storiografia filosofica* (Florence: La nuova Italia, 1976).

Delval, Juan, and Juan Carlos Gómez, "Dietrich Tiedemann: La psicología del niño hace doscientos años," *Infancia y aprendizaje*, 41, 1988, 9–30.

Demarée, Gaston R., "Guillaume Lambert Godart: Médecin, philosophe et météorologiste: Un savant oublié du XVIIIᵉ siècle," *Ciel et terre*, 109, 1993, 47–51.

Descartes, René, *Meditations on first philosophy*, in *The philosophical writings of Descartes*, vol. 2, trans. John Cottingham (1988; Cambridge: Cambridge University Press, 1996).

——, *Méditations touchant la première philosophie dans lesquelles l'existence de Dieu et la distinction réelle entre l'âme et le corps de l'homme sont démontrées* (Latin ed. 1641; French ed. 1647), in Descartes, *Œuvres*, vol. 9.1.

——, *Œuvres, Correspondance*, ed. Charles Adam and Paul Tannery (1897–1913; Paris: Vrin, 1982–1991).

——, *Les passions de l'âme* (1649), in Descartes, *Œuvres*, vol. 11.

——, *Treatise of man*, trans. Thomas Steele Hall (Cambridge, MA: Harvard University Press, 1972).

Des Champs, Jean, *Cours abrégé de la philosophie wolffienne, en forme de lettres*, vol. 2, pt. 1, *Psychologie expérimentale*, pt. 2, *Psychologie raisonnée* (1747), in Wolff, *Gesammelte Werke*, ser. III, vol. 13.

Des Chene, Dennis, *Life's form: Late Aristotelian conceptions of the soul* (Ithaca: Cornell University Press, 2000).

——, *Spirits and clocks: Machine and organism in Descartes* (Ithaca: Cornell University Press, 2001).

Dessoir, Max, *Abriß einer Geschichte der Psychologie* (Heidelberg: Winter, 1911).

——, *Geschichte der neueren deutschen Psychologie*, 2nd ed. (1902; Amsterdam: E. J. Bonset, 1964).

Destutt de Tracy, Antoine Louis Claude, *Sur un système méthodique de Bibliographie* (1797), in *Mémoire sur la faculté de penser; De la métaphysique de Kant et autres textes* (Paris: Fayard, 1992).

Dewey, John, "The new psychology," *Andover review*, 2, 1884, 278–289.

Diamond, Solomon, "What Marulus meant by 'psychologia,'" *Storia e critica della psicologia*, 5, 1984, 407–412.

Diderot, Denis, *Suite de l'Apologie de l'abbé de Prades*, in *Œuvres complètes* (Paris: Hermann, 1978), vol. 4.

Diderot, Denis, and Jean Le Rond d'Alembert, eds., *Encyclopédie, ou Dictionnaire raisonné des sciences, des arts et des métiers* (Paris: Briasson . . . , 1751–1765).

Diele, Heidrun, " 'Man kann sich nun immer mehr mit ihr abgeben . . .': Tagebuch eines Vaters über seine 1794 geborene Tochter," *BIOS*, 13, 2000, 125–134.

Dilthey, Wilhelm, "Die Funktion der Anthropologie in der Kultur des 16. und 17. Jahrhunderts" (1904), in *Gesammelte Schriften*, vol. 2, *Weltanschauung und Analyse des Menschen seit Renaissance und Reformation* (Stuttgart: B. G. Teubner; Göttingen: Vandenhoeck & Ruprecht, 1991).

Dodds, Eric R., *The Greeks and the irrational* (Boston: Beacon Press, 1951).

Domínguez Molto, Adolfo, *El abate D. Juan Andrés Morell (Un erudito del siglo XVIII)* (Alicante: Instituto de estudios alicantinos, 1978).

Donato, Clorinda, "*L'Encyclopédie* d'Yverdon et *l'Encyclopédie* de Diderot et d'Alembert: Éléments pour une comparaison," *Annales Benjamin Constant*, 14, 1993, 75–83.

———, "Fortunato Bartolomeo de Felice et l'edizione d'Yverdon dell'*Encyclopédie*," *Studi settecenteschi*, 16, 1996, 373–396.

———, "Inventory of the *Encyclopédie d'Yverdon*: A comparative study with Diderot's *Encyclopédie*" (doctoral thesis, University of California, Los Angeles, 1987).

———, "Rewriting heresy in the *Encyclopédie d'Yverdon*, 1770–1780," *Cromohs*, 7, 2002, 1–26.

Donato, Clorinda, and Kathleen Hardesty Doig, "Notice sur les auteurs des quarante-huit volumes de 'discours' de *l'Encyclopédie* d'Yverdon," *Recherches sur Diderot et "l'Encyclopédie*," 11, 1991, 133–141.

Donato, Clorinda, and Robert M. Maniquis, eds., *The "Encyclopédie" and the age of revolution* (Boston: G. K. Hall, 1992).

du Châtelet, Gabrielle Emilie Le Tonnelier de Breteuil, marquise, *Institutions physiques* (1741), in Wolff, *Gesammelte Werke*, ser. III, vol. 28 (ed. of 1742).

Dumit, Joseph, "Is it me or my brain? Depression and neuroscientific facts," *Journal of medical humanities*, 24, 2003, 35–47.

———, *Picturing personhood: Brain scans and biomedical identity* (Princeton: Princeton University Press, 2004).

du Pleix, Scipion, *La Métaphysique, ou science surnaturelle* (1610), ed. of 1640 (Paris: Fayard, 1992).

———, *La Physique, ou science des choses naturelles* (1603), ed. of 1640 (Paris: Fayard, 1990).

Ebbinghaus, Hermann, *Abriß der Psychologie* (1908; Leipzig: Veit, 1910).

Eckardt, Georg, ed. *Völkerpsychologie: Versuch eine Neuentdeckung: Texte von Lazarus, Steinthal und Wundt* (Weinheim: Psychologie Verlags Union, 1997).

Eckardt, Georg, and Matthias John, "Anthropologische und psychologische Zeitschriften um 1800," in Eckardt et al., *Anthropologie*.

Eckardt, Georg, Matthias John, Temilo van Zantwijk, and Paul Ziche, eds., *Anthropologie und empirische Psychologie um 1800* (Cologne: Böhlau, 2001).

École, Jean, "De la nature de l'âme, de la déduction de ses facultés, de ses rapports avec le corps, ou la *Psychologia rationalis* de Christian Wolff," *Giornale di metafisica*, 24, 1969, 499–531 (also in École, *Introduction*).

———, "De la notion de philosophie expérimentale chez Wolff," *Les études philoso-phiques*, no. 4, 1979, 397–406.

———, "De quelques difficultés à propos des notions d'a posteriori et d'a priori chez Wolff," *Teoresi*, no. 1–2, 1976, 25–34.

———, "Des rapports de l'expérience et de la raison dans l'analyse de l'âme ou la *Psychologia empirica* de Christian Wolff," *Giornale di metafisica*, 21, 1966, 589–617 (also in École, *Introduction*).

———, *Index auctorum et locorum Scripturae Sacrae ad quos Wolffius in opere meta-physico et logico remittit* (1985), in Wolff, *Gesammelte Werke*, ser. III, vol. 10.

———, *Introduction à l'opus metaphysicum de Christian Wolff* (Paris: Vrin, 1985).

———, *La métaphysique de Christian Wolff* (1990), in Wolff, *Gesammelte Werke*, ser. III, vol. 12.

———, "Wolffius redivivus," *Revue de synthèse*, no. 116, 1984, 483–501.

Egger, Johann, *[Sefer Ha-Nefesh] Sive Psychologia rabbinica, quae agit de mentis huma-nae natura & praecipue ejus extremis, ex mente magistrorum Judaeorum accedit hinc inde sententia Mohammedis & Arabum* (Basel: Typis Joh. Ludov. Brandmülleri, 1719).

Eichhorn, Johann Gottfried, *Einleitung in das Alte Testament* (1780–1783; Leipzig: Weid-mannische Buchhandlung, 1803).

Elsky, Martin, "Reorganizing the encyclopaedia: Vives and Ramus on Aristotle and the Scholastics," in Glyn P. Norton, ed., *The Cambridge history of literary criticism*, vol. 3, *The Renaissance* (Cambridge: Cambridge University Press, 1999).

Erbse, Hartmut, "Nachlese zur homerischen Psychologie," *Hermes*, 118, 1990, 1–17.

Ernesti, Johann Heinrich Martin, *Enzyklopädisches Handbuch einer allgemeinen Ge-schichte der Philosophie und ihrer Literatur* (1807; Düsseldorf: Stern-Verlag Janssen, 1972).

Fabricius, Johann Andreas, *Abriß einer allgemeinen Historie der Gelehrsamkeit* (1752–1754; Hildesheim: Olms, 1978).

Farr, Robert M., "Wilhelm Wundt (1832–1920) and the origins of psychology as an experi-mental and social science," *British journal of social psychology*, 22, 1983, 289–301.

Febvre, Lucien, "History and psychology" (1938), in Burke, *A new kind of history*.

———, "Sensibility and history: How to reconstitute the emotional life of the past" (1941), in Burke, *A new kind of history*.

Feldman, Burton, and Robert D. Richardson, *The rise of modern mythology* (Blooming-ton: Indiana Press, 1972).

Felice, Christian de, *"L'Encyclopédie" d'Yverdon: Une encyclopédie suisse au siècle des Lumières* (Yverdon: Fondation De Felice, 1999).

Felice, Fortunato Bartolomeo de, ed., *Encyclopédie, ou Dictionnaire universel raisonné des connoissances humaines* (Yverdon: Société Typographique, 1770–75), *Supplé-ment* (1775–1776).

Fernández García, Mariano, *Lexicon scholasticum philosophico-theologicum* (1910; Hildesheim: Olms, 1988).

Ferret, Stéphane, *Le philosophe et son scalpel: Le problème de l'identité personnelle* (Paris: Editions de Minuit, 1993).

Ferri, Luigi, *La psychologie de l'association depuis Hobbes jusqu'à nos jours* (Paris: Germer Baillière, 1883).

Ferrone, Vincenzo, *Scienza, natura, religione: Mondo newtoniano e cultura italiana nel primo Settecento* (Naples: Jovene, 1982).

Feuerbach, Ludwig, *Thoughts on death and immortality* (1830), in *Thoughts on death and immortality: From the papers of a thinker, along with an appendix of theological-satirical epigrams*, trans. James A. Massey (Berkeley: University of California Press, 1980).

Feuerhahn, Wolf, "Entre métaphysique, mathématique, optique et physiologie: La psychométrie au XVIIIᵉ siècle," *Revue philosophique*, 193, 2003, 279–292.

———, "Die Wolffsche Psychometrie," in Oliver-Pierre Rudolph and Jean-François Goubet, eds., *Die Psychologie Christian Wolffs: Systematische und historische Untersuchungen* (Tübingen: Max Niemeyer, 2004).

Figlio, Karl M., "Theories of perception and the physiology of mind in the late eighteenth century," *History of science*, 12, 1975, 177–212.

Findlen, Paula, "The formation of a scientific community: Natural history in sixteenth-century Italy," in Anthony Grafton and Nancy Siraisi, eds., *Natural particulars: Nature and the disciplines in Renaissance Europe* (Cambridge, MA: MIT Press, 1999).

Finger, Otto, *Von der Materialität der Seele: Beitrag zur Geschichte des Materialismus und Atheismus in Deutschland der zweiten Hälfte des 18. Jahrhundert* (Berlin: Akademie-Verlag, 1961).

Finsler, Georg, *Homer in der Neuzeit von Dante bis Goethe: Italien, Frankreich, England, Deutschland* (Leipzig: Teubner, 1912).

Fischer, Klaus P., "John Locke in the German Enlightenment: An interpretation," *Journal of the history of ideas*, 35, 1975, 431–446.

Foerster, Donald M., *Homer in English criticism: The historical approach in the eighteenth century* (New Haven: Yale University Press, 1947).

Fontius, Martin, and Helmut Holzhey, eds., *Schweizer im Berlin des 18. Jahrhunderts* (Berlin: Akademie Verlag, 1996).

Formey, Jean Henri Samuel, *La Belle Wolfienne*, vol. 5, *Psychologie expérimentale* (1753), in Wolff, *Gesammelte Werke*, ser. III, vol. 16.

Fornaro, Sotera, "Homer in den deutschen und französischen Aufklärung," in Veit Elm, Günther Lothes, and Vanessa de Senarclens, eds., *Die Antike der Moderne: Von Umgang mit der Antike im Europa des 18. Jahrhunderts* (Hanover: Wehrhahn, 2009).

———, "Lo "studio degli antichi," 1793–1807," *Quaderni di storia*, 22, 1996, 109–124.

Foucault, Michel, *The order of things: An archaeology of the human sciences* (1966; London: Routledge, 1997).

Fox, Christopher, "Defining eighteenth-century psychology: Some problems and perspectives," in Christopher Fox, ed., *Psychology and literature in the eighteenth century* (New York: AMS Press, 1987).

Frank, Günter, "Philipp Melanchthons 'Liber de anima' und die Etablierung der frühneuzeitlichen Anthropologie," in Michael Beyer et Günther Wartenberg with the collaboration of Hans-Peter Hasse, eds., *Humanismus und Wittenberger Reformation* (Leipzig: Evangelische Verlagsanstalt, 1996).

Freedman, Joseph S., "Aristotle and the content of philosophical instruction at Central European schools and universities during the Reformation era (1500–1650)," in Freedman, *Philosophy and the arts.*

———, "Classifications of philosophy, the sciences, and the arts in sixteenth- and seventeenth-century Europe," in Freedman, *Philosophy and the arts.*

———, "Encyclopedic philosophical writings in Central Europe during the high and late Renaissance (ca. 1500–ca.1700)," in Freedman, *Philosophy and the arts.*

———, *European academic philosophy in the late sixteenth and early seventeenth centuries: The life, significance, and philosophy of Clemens Timpler (1563/4–1624)* (Hildesheim: Olms, 1988).

———, *Philosophy and the arts in Central Europe, 1500–1700* (Aldershot: Ashgate/Variorum series, 1999).

Freigius, Johann Thomas, *Quaestiones physicae: In quibus, Methodus doctrinam Physicam legitime docendi, describendique rudi Minerua descripta est, libris XXXV* (Basel: per Sebastianum Henricpetri, 1579).

Fülleborn, Georg Gustav, ed., "Abriss einer Geschichte und Literatur der Physiognomik," *Beyträge*, 8. Stück, 1797, 1–180.

———, *Beyträge zur Geschichte der Philosophie* (1791–1796; Brussels: Culture et civilisation, 1968).

———, "Kurze Geschichte der Philosophie," *Beyträge*, 3. Stück, 1793, 3–51.

———, "Thomas Campanella ueber die menschliche Erkenntniss," *Beyträge*, 6. Stück, 1795, 124–162.

———, "Über den einflus anderer Wissenschaften und äuserer Verhältnisse auf die Philosophie und dieser auf jene," *Beyträge*, 3. Stück, 1793, 52–69.

———, "Vermischte Bemerkungen zur Geschichte der Philosophie (I)," *Beyträge*, 7. Stück, 1796, 173–188.

———, "Versuch einer Uebersicht der neuesten Entdekungen in der Philosophie," *Beyträge*, 2. Stück, 1792, 102–135.

Garat, [Dominique-Joseph], "Analyse de l'entendement," in *Séances des Écoles Normales, recueillies par des sténographes, et revues par les professeurs*, new ed. (Paris: À l'Imprimerie du Cercle Social, 1800), vol. 1.

Garber, Daniel, "Soul and mind: Life and thought in the seventeenth century," in Garber and Ayers, *The Cambridge history of seventeenth-century philosophy.*

Garber, Daniel, and Michael Ayers, eds., *The Cambridge history of seventeenth-century philosophy* (Cambridge: Cambridge University Press, 1998).

Garber, Jörn, "Von der 'anthropologischen Geschichte des philosophierenden Geistes' zur *Geschichte der Menschheit* (Friedrich August Carus)," in Garber and Thoma, *Zwischen Empirisierung und Konstruktionsleistung.*

Garber, Jörn, and Heinz Thoma, eds., *Zwischen Empirisierung und Konstruktionsleistung: Anthropologie im 18. Jahrhundert* (Tübingen: Max Niemeyer, 2004).

Garin, Eugenio, "'ΕΝΔΕΛΕΧΕΙΑ' e 'ΕΝΤΕΛΕΧΕΙΑ' nelle discussioni umanistiche," *Atene e Roma*, 5, 1937, 177–187.

Garofalo, Silvano, *L'enciclopedismo italiano: Gianfrancesco Pivati* (Ravenna: Longo, 1980).

Gay, Peter, *The Enlightenment: An interpretation* (New York: W. W. Norton, 1969).

Genovesi, Antonio, *Delle scienze metafisiche per gli giovanetti* (1766; Naples: Angelo Coda, 1791).

———, *Disciplinarum metaphysicarum elementa mathematicum in morem adornata* (Venice: Remondini, 1779).

———, *Elementa metaphysicae mathematicum in morem adornata* (1st ed., 1743), 2nd ed. (Naples: typis Benedicti, et I. Gessari, 1751).

———, *La logica per gli giovanetti* (1766; Naples: Stamperia Abbaziana, 1790).

Gentile, Giovanni, *Storia della filosofia italiana dal Genovesi al Gallupi* (1929), in *Opere* (Florence: Sanzoni, 1935), vol. 18.

Gerhard, Johann Ernest, praes., autor Iohannes Steinhusius, *Ψυχολογια sive disquisitio de statu animae separatae* (Jena: literis Johannis Wertheri, 1663).

Gerhard, Johannes, praes., respond. Wolfgang Ernest Tüntzel, *Ψυχολογια generalis: H. e. Disquisitio de statu animarum post mortem* (Jena: Typis Ernesti Steinmanni, 1633).

[Gesner, Conrad], ed., *Ioannis Lodovici Vives valentini de Anima & vita Libri tres: Eiusdem argumenti Viti Amerbachii de Anima Libri III; Philippi Melanthonis Liber unus; His accedit nunc primum Conradi Gesner de Anima liber . . .* (Zurich: apud Jacobum Gesnerum, [1563]).

Gesner, Conrad, *De Anima liber, sententiosa brevitate, veluiti per tabulas et aphorismos ut plurimum conscriptus, philosophiae & medicinae studiosis accommodatus* (Zurich: apud Jacobum Gesnerum, [1563]).

Getto, Giovanni, *Storia delle storie letterarie* (Florence: Sansoni, 1969).

Gilbert, Neal W., *Renaissance conceptions of method* (New York: Columbia University Press, 1960).

Gil Colomer, Rafael, "El Tratado de 'anima' de Francisco Suárez," *Convivium*, 17–18, 1964, 128–141.

Gill, Christopher, *Personality in Greek epic, tragedy, and philosophy: The self in dialogue* (Oxford: Clarendon Press, 1996).

Gilson, Étienne, and Thomas D. Langan, *Modern philosophy: Descartes to Kant* (New York: Random House, 1964).

Giuntini, Chiara, *La chimica della mente: Associazione delle idee e scienza della natura umana da Locke a Spencer* (Florence: Le Lettere, 1995).

Goclenius, Rudolph, *Disputatio philosophica quadripartita: Prima Logica: De natura relatorum; Secunda Psychologica, de quaestione: cur accretio in adultis cesset; Tertia Ethica: An ignorantia excuset peccatum; Quarta Historica & Physica: an iris fuerit ante diluvium* (Marburg: typis Pauli Egenolphi, 1596).

———, *Lexicon philosophicum* (1613; Hildesheim: Olms, 1980).

———, *ΨΥΧΟΛΟΓΙΑ: hoc est, de hominis perfectione, animo, et in primis ortu hujus, commentationes ac disputationes quorundam Theologorum & Philosophorum nostrae aetatis . . .* (Marburg: ex officina typographica Pauli Egenolphi, 1590; 2nd ed., 1597).

Godart, Guillaume-Lambert, *La physique de l'âme humaine* (Berlin: Aux dépens de la Compagnie, 1755).

Goetschel, Willi, Catriona Macleod, and Emery Snyder, "The *Deutsche Encyclopädie* and encyclopedism in eighteenth-century Germany," in Donato and Maniquis, *The "Encyclopédie."*

———, "The *Deutsche Encyclopädie*," in Kafker, *Notable encyclopedias: Eleven successors.*

Goguet, Antoine-Yves, *The Origin Of Laws, Arts, and Sciences, And Their Progress Among The Most Ancient of Nations*, translated from the French (1758; Edinburgh: Printed for George Robinson . . . , 1775).

Goldstein, Jan, "Bringing the psyche into scientific focus," in Theodore M. Porter and Dorothy Ross, eds., *The Cambridge history of science*, vol. 7, *The modern social sciences* (New York: Cambridge University Press, 2003).

———, "Foucault among the sociologists: The 'disciplines' and the history of the professions," *History and theory*, 23, 1984, 170–192.

Golinski, Jan V., *Making natural knowledge: Constructivism and the history of science* (Cambridge: Cambridge University Press, 1998).

———, "Science in the Enlightenment," *History of science*, 24, 1986, 411–424.

Gotschlich, Emil, *Psychologia homerica sive Historia notionum psychologicarum apud Homerum: Dissertatio inauguralis philologica . . .* (Breslau: A. Neumann, 1864).

Grafton, Anthony, "Prolegomena to Friedrich August Wolf (1981)," in *Defenders of the text: The traditions of scholarship in an age of science, 1450–1800* (Cambridge, MA: Harvard University Press, 1991).

———, "The world of the polyhistors: Humanism and encyclopedism" (1985), in *Bring out the dead: The past as revelation* (Cambridge, MA: Harvard University Press, 2001).

Grandière, Marcel, *L'idéal pédagogique en France au dix-huitième siècle* (Oxford: Voltaire Foundation, 1998).

———, "Le sensualisme dans les traités d'éducation à la fin du XVIIIᵉ siècle (1760–1789): Quelques aspects," in Hubert Hannoun and Anne-Marie Drouin-Hans, eds., *Pour une philosophie de l'éducation* (Dijon: CNDP, 1994).

Gräße, Johann Georg Theodor, *Bibliotheca psychologica oder Verzeichniß der wichtigsten über das Wesen der Menschen- und Thierseelen und die Unsterblichkeitslehre handelnden Schriftsteller älterer und neurere Zeit . . .* (1845; Amsterdam: E. J. Bonset, 1968).

Gregory, Tullio, "Sul lessico filosofico latino del Seicento e del Settecento," *Lexicon philosophicum: Quaderni di terminologia filosofica e storia delle idee*, 5, 1991, 1–20.

Grell, Chantal, *Le dix-huitième siècle et l'antiquité en France, 1680–1789* (Oxford: Voltaire Foundation, 1995).

———, "Troie et la Troade de la Renaissance à Schliemann," *Journal des savants*, 1981, 47–76.

Grober, Max, "The natural history of heaven and the historical proofs of Christianity: *La Palingénésie philosophique* of Charles Bonnet," *Studies on Voltaire and the eighteenth century*, 302, 1993, 233–255.

Grün, Johann, *Liber de anima DN Philippi Melanthoni in diagrammata methodica digestus . . .* (Wittenberg: In Officina Typographica Simonis Gronenbergii, 1580).

Gusdorf, Georges, *L'avènement des sciences humaines au siècle des lumières* (Paris: Payot, 1973).

————, *La révolution galiléenne* (Paris: Payot, 1969).

Guy, Alain, "Ramón Campos, disciple de Condillac," in José Luis Abellán et al., *Pensée hispanique et philosophie française des Lumières* (Toulouse: Association des publications de l'Université de Toulouse-Le Mirail, 1980).

Guyot, Charly, *Le rayonnement de "l'Encyclopédie" en Suisse française* (Neuchâtel: Attinger, 1955).

Hagen, Gottlieb Friedrich, *Dissertatio mathematica de mensurandis viribus propriis atque alienis* (Giessen: Müller, 1733).

————, *Meditationes philosophicae de methodo mathematica . . .* (1734), in Wolff, *Gesammelte Werke*, ser. III, vol. 82.

————, *Programma de mesurandis viribus intellectibus* (Halle: Hendel, 1734).

Hagner, Michael, *Geniale Gehirne: Zur Geschichte der Elitenhirnforschung* (Berlin: Wallstein, 2004).

————, *Homo cerebralis: Der Wandel vom Seelenorgan zum Gehirn* (Berlin: Berlin Verlag, 1997).

————, "Toward a history of attention in culture and science," *Modern language notes,* 118, 2003, 670–687.

Halbkart, Karl Wilhelm, *Psychologia homerica, seu de homerica circa animam vel cognitione vel opinione commentatio* (Züllichau: sumtibus Friderici Frommanni, 1796).

Hamel, Emile Louis, *Thesis philosophica de psychologia homerica* (Paris: August Delalain, 1832).

Hamilton, William, "Contribution towards a history of the doctrine of mental suggestion or association," in W. Hamilton, ed., *The works of Thomas Reid* (Edinburgh: Maclachlan and Stewart, 1863), vol. 2.

Hankins, Thomas, *Science and the Enlightenment* (Cambridge: Cambridge University Press, 1985).

Hardesty, Kathleen, *The Supplément to the "Encyclopédie"* (La Haye: Martinus Nijhoff, 1977).

Hardesty Doig, Kathleen, "The Yverdon *Encyclopédie*," in Kafker, *Notable encyclopedias: Eleven successors.*

Harrison, E. L., "Notes on Homeric psychology," *Phoenix*, 14, 1960, 63–80.

Hartley, David, *Observations on man, his frame, his duty, and his expectations* (1749; London: printed for T. Tegg and Son, 1834).

Harvey, E. Ruth, *The inward wits: Psychological theory in the Middle Ages and the Renaissance* (London: Warburg Institute, 1975).

Häseler, Jens, "Johann Bernhard Merian—ein Schweizer Philosoph an der Berliner Akademie," in Fontius and Holzhey, *Schweizer im Berlin.*

Hatfield, Gary, "Attention in early scientific psychology," in Richard D. Wright, ed., *Visual attention* (New York: Oxford University Press, 1998).

————, "The cognitive faculties," in Garber and Ayers, *The Cambridge history of seventeenth-century philosophy.*

————, "Descartes' physiology and its relation to his psychology," in John Cottingham, ed., *The Cambridge companion to Descartes* (New York: Cambridge University Press, 1992).

————, "Empirical, rational, and transcendental psychology: Psychology as science and as philosophy," in Paul Guyer, ed., *The Cambridge companion to Kant* (Cambridge: Cambridge University Press, 1992).

———, "Kant and empirical psychology in the 18th century," *Psychological science*, 9, 1998, 423–428.

———, *The natural and the normative: Theories of spatial perception from Kant to Helmholtz* (Cambridge, MA: MIT Press, 1990).

———, "Remaking the science of mind: Psychology as a natural science," in Christopher Fox, Roy Porter, and Robert Wokler, eds., *Inventing human science: Eighteenth-century domains* (Berkeley: University of California Press, 1994).

Hawenreuter, Johann Ludwig, *ΨΥΧΟΛΟΓΙΑ: sive Philosophica de animo ΣΥΖΗΤΕΣΙΣ, ex libris tribus Aristotelis περω ψυχες, excerpta* . . . (Strasburg: Antonius Bertramus, 1591).

———, *Theses ex praecipuis philosophiae partibus* . . . (Strasburg: Antonius Bertramus, 1611).

Hazard, Paul, *La crise de la conscience européenne* (1935; Paris: Gallimard, 1961).

Head, Brian W., "The origins of 'la science sociale' in France, 1770–1800," *Australian journal of French studies*, 19, 1982, 115–132.

Hegel, Georg Wilhelm Friedrich, *Lectures on the history of philosophy*, trans. Elizabeth S. Haldane and Francis H. Simson (London: Kegan Paul, Trench, Trübner, 1896).

Helm, Jürgen, "Die 'spiritus' in der medizinischen Tradition und in Melanchthons 'Liber de anima,'" in Günther Frank and Stefan Rhein, eds., *Melanchthon und die Naturwissenschaften seiner Zeit* (Sigmaringen: Jan Thorbecke Verlag, 1998).

———, "Zwischen Aristotelismus, Protestantismus und zeitgenössischer Medizin: Philipp Melanchthons Lehrbuch *De anima* (1540/1552)," in Jürgen Leonhardt, ed., *Melanchthon und das Lehrbuch des 16. Jahrhunderts* (Rostock: Universität Rostock Philosophische Fakultät, 1997).

Hennings, Justus Christian, *Geschichte von den Seelen der Menschen und Thiere: Pragmatisch entworfen* (Halle: bey J. J. Gebauers Witwe und J. Jac. Gebauer, 1774).

Hepp, Noémie, *Homère en France au XVIIe siècle* (Paris: Klincksieck, 1968).

Heraclitus, *Allégories d'Homère*, ed. and trans. F. Buffière (Paris: Belles Lettres, 1962).

———, *Homeric problems*, ed. and trans. Donald A. Russell and David Konstan (Atlanta: Society of Biblical Literature, 2005).

Herder, Johann Gottfried, *Herders sämmtliche Werke*, ed. Bernhard Suphan (1887–1913; Hildesheim: Olms, 1962).

———, "Homer, ein Günstling der Zeit" (1795), in *Herders sämmtliche Werke*, vol. 18.

———, "Homer und Ossian" (1795), in *Herders sämmtliche Werke*, vol. 18.

———, *Ideen zur Philosophie der Geschichte der Menschheit* (1784–1791), in *Herders sämmtliche Werke*, vol. 14.

———, *Vom Geist der Ebräischen Poesie: Eine Einleitung für die Liebhaber derselben, und der ältesten Geschichte des menschlichen Geschlechts* (1782–1783), in *Herders sämmtliche Werke*, vols. 11–12.

Heumann, Christoph August, *Conspectus reipublicae literariae, sive Via ad historiam literariam iuventuti studiosae aperta* (1718; Hanover: apud haeredes Nic. Foerster et filii, 1763).

———, "Eintheilung der Historiae Philosophicae," *Acta philosophorum, sive gründliche Nachrichten aus der Historia philosophica*, 3. Stück, 1715, 462–472.

Heyne, Christian Gottlob, "De origine et caussis fabularum homericarum commentatio," *Novi commentarii Societatis Regiae scientiarum Gottingensis*, 8, 1777, 34–58.

———, "Inquiry into the causes of fables or the physics of ancient myths" (1764), in Feldman and Richardson, *The rise of modern mythology* (extracts trans. from "Quaestio de causis fabularum seu mythorum veterum physicis," in Heyne, *Opuscula*, vol. 1).

———, "An interpretation of the language of myths or symbols traced to their reasons and causes and thence to forms and rules" (1807), in Feldman and Richardson, *The rise of modern mythology* (excerpts trans. from "Sermonis mythici seu symbolici interpretatio ad causas et rationes ductasque inde regulas revocata," *Commentationes Societatis Regiae scientiarum Gottingensis*, 1807).

———, *Opuscula academica collecta et animadversionibus locupletata* (1785–1812; Hildesheim: Olms, 1997).

———, review of *An essay on the original genius of Homer*, by R. Wood, *Göttingische Anzeigen von gelehrten Sachen*, 32. Stück, 1770, 257–270.

———, "Vita antiquioris Graeciae ex ferorum et barbarorum populorum comparatione illustrata: Commentatio II" (1779), in Heyne, *Opuscula*, vol. 3 (trans. in Pandolfi, "Civiltà antiche").

———, "Vita antiquissimorum hominum, Graeciae maxime, ex ferorum et barbarorum populorum comparatione illustrata: Commentatio I" (1779), in Heyne, *Opuscula*, vol. 3 (trans. in Pandolfi, "Civiltà antiche").

Hilgard, Ernst R., David E. Leary, and Gregory R. McGuire, "The history of psychology: A survey and critical assessment," *Annual review of psychology*, 42, 1991, 79–107.

Hinske, Norbert, "Wolffs empirische Psychologie und Kants pragmatische Anthropologie: Zur Diskussion über die Anfänge der Anthropologie im 18. Jahrhundert," *Aufklärung*, 11, 1999, 97–107.

Hippius, Fabian, ΨΥΧΟΛΟΓΙΑ *Physica, sive de corpore animato, Libri quatuor, toti ex Aristotele desumti, morborum saltem doctrinis ex Medicis scriptis adiecta . . .* (Frankfurt: Typis Wolffgangi Richteri, sumptibus Ioannis Spiessij, 1600).

Hißmann, Michael, *Anleitung zur Kenntniss der auserlesenen Litteratur in allen Theilen der Philosophie* (Göttingen: im Verlage der Meyerschen Buchhandlung, 1778).

———, *Geschichte der Lehre der Association der Ideen, nebst einem Anhang vom Unterschied unter associierten und zusammentgesezten Begriffen, und den Ideenreyhen* (Göttingen: im Verlag Victorinus Voßpiegel und Sohn, 1777).

———, ed., *Magazin für die Philosophie und ihre Geschichte* (Göttingen: im Verlage der Meyerschen Buchhandlung, 1778–1783).

———, *Psychologische Versuche, ein Beytrag zur esoterischen Logik* (Frankfurt, 1777.

Hody, Humphrey, *The resurrection of the (same) body asserted, from the traditions of the heathens, the ancient Jews, and the primitive church . . .* (London: printed for Awnsham and John Churchill, 1694).

Hofmann, Étienne, "Le pasteur Gabriel Mingard, collaborateur de l'*Encyclopédie* d'Yverdon: Matériaux pour l'étude de sa pensée," in Alain Clavien and Bertrand Müller, eds., *Le goût de l'histoire, des idées et des hommes: Mélanges J.-P. Aguet* (Lausanne: Ed. de l'Aire, 1996).

Holzapfel, Wolfgang, and Georg Eckardt, "Philipp Melanchthon's psychological thinking under the influence of humanism, Reformation and empirical orientation," *Revista de historia de la psicología*, 20, 1999, 5–34.

Holzhey, Helmut, "Seele (§ 4. Neuzeit)," in Ritter et al., *Historisches Wörterbuch*, vol. 9.

Holzhey, Helmut, and Simone Zurbuchen, "Die Schweiz zwischen deutscher und französischer Aufklärung," in Werner Schneiders, ed., *Aufklärung als Mission: Akzeptanzprobleme und Kommunikationsdefizite* (Marburg: Hitzeroth, 1993).

Honour, Hugh, *Neo-classicism* (Middlesex: Penguin Books, 1979).

Horkheimer, Max, and Theodor Adorno, *Dialectic of enlightenment*, trans. John Cumming (1947; London: Verso, New Left Books, 1999).

Hotson, Howard, *Johann Heinrich Alsted, 1588–1638: Between Renaissance, Reformation, and universal reform* (New York: Oxford University Press, 2000).

Hume, David, *A Treatise of human nature: being an attempt to introduce the experimental method of reasoning into moral subjects* (1739–1740), ed., L. A. Selby-Bigge, rev. P. H. Nidditch (Oxford: Clarendon Press, 1978).

Hunter, Ian, *Rival Enlightenments: Civil and metaphysical philosophy in early modern Germany* (Cambridge: Cambridge University Press, 2001).

Hutchinson, Ross, *Locke in France, 1688–1734* (Oxford: Voltaire Foundation, 1991).

Ijsewijn, Josef, *Companion to Neo-Latin studies*; pt. 1, *History and diffusion of Neo-Latin literature*; pt. 2 (with Dirk Sacré), *Literary, linguistic, philological and editorial questions* (1990; Louvain: Leuven University Press, 1998).

Im Hof, Ulrich, *Aufklärung in der Schweiz* (Bern: Francke, 1970).

Iselin, Isaak, *Über die Geschichte der Menschheit* (1786; Hildesheim: Olms, 1976).

Jahn, Thomas, *Zum Wortfeld "Seele-Geist" in der Sprache Homers* (Munich: Beck, 1987).

Jahnke, Jürgen, "Neuere Arbeiten zur Psychologie im 18. Jahrhundert: Historiographische Probleme, Ergebnisse und Tendenzen," *Psychologie und Geschichte*, 2, 1990, 19–24.

———, "Psychologie im 18. Jahrhundert: Literaturbericht 1980 bis 1989," *Das achtzehnte Jahrhundert*, 14, 1990, 253–278.

Jahoda, Gustav, *Crossroads between culture and mind: Continuities and change in theories of human nature* (New York: Harvester-Wheatsheaf, 1992).

———, "Une esquisse de la Völkerpsychologie de Wundt," in Michel Kait and Geneviève Vermès, eds., *La psychologie des peuples et ses dérivés* (Paris: Centre national de documentation pédagogique, 1999).

Janssens-Knorsch, U. "Jean Deschamps, Wolff-Übersetzer und 'Aléthophile français' am Hofe Friedrichs des Großen," in Schneiders, *Christian Wolff*.

Jardine, Nicholas, *The scenes of inquiry: On the reality of questions in the sciences* (Oxford: Clarendon Press, 1991).

Jasenas, Michael, *A history of the bibliography of philosophy* (Hildesheim: Olms, 1973).

Jiménez García, Antonio, "Las traducciones de Condillac y el desarrollo del sensismo en España," in Antonio Heredia, ed., *Actas del VI seminario de historia de la filosofía española e iberoamericana* (Salamanca: Ediciones Universidad de Salamanca, 1990).

Johnson-Cousin, Danielle, "La 'construction' du féminin dans *l'Encyclopédie* d'Yverdon et dans *l'Encyclopédie* de Paris," *Studies on Voltaire and the eighteenth century*, 304, 1992, 752–758.

Jolley, Nicholas, *Leibniz and Locke: A study of the "New Essays on Human Understanding"* (New York: Oxford University Press, 1984).

Jones, Peter, ed., *Philosophy and science in the Scottish Enlightenment* (Edinburgh: John Donald, 1988).

———, ed., *The "science of man" in the Scottish Enlightenment: Hume, Reid and their contemporaries* (Edinburgh: Edinburgh University Press, 1989).

Kafker, Frank A., ed., *Notable encyclopedias of the late eighteenth century: Eleven successors of the "Encyclopédie"* (Oxford: Voltaire Foundation, 1994).

———, *Notable encyclopedias of the seventeenth and eighteenth centuries: Nine predecessors of the "Encyclopédie"* (Oxford: Voltaire Foundation, 1981).

———, "William Smellie's edition of the *Encyclopaedia Britannica*," in Kafker, *Notable encyclopedias: Eleven successors*.

———, and Serena L. Kafker, *The Encyclopedists as individuals: A biographical dictionary of the authors of the "Encyclopédie"* (Oxford: Voltaire Foundation, 1988).

Kahle, Ludwig Martin, *Bibliothecae philosophicae Struvianae emendatae, continuatae atque ultra dimidiam partem* (Göttingen: impensis Vandhoeket Cunonis, 1740).

Kant, Immanuel, *Anthropology from a pragmatic point of view*, trans. Robert B. Louden (Cambridge: Cambridge University Press, 2007).

———, *Critique de la raison pure* (1781/1787), trans. Alexandre J.-L. Delamarre and François Marty, in *Œuvres philosophiques*, ed. Ferdinand Alquié (Paris: Gallimard, 1980), vol. 1.

———, *Gesammelte Schriften* (Berlin: Walter de Gruyter, 1968).

———, *Leçons de métaphysique*, trans. Monique Castillo (Paris: Livre de poche, 1993).

———, *Lectures on metaphysics*, trans. Karl Ameriks and Steve Naragon (Cambridge: Cambridge University Press, 1977).

———, "Metaphysik L_1: Kosmologie, Psychologie, Theologie nach Pölitz," in *Vorlesungen über Metaphysik und Rationaltheologie*, in *Gesammelte Schriften*, vol. 28.1 (= *Vorlesungen*, vol. 5.1).

Kaufmann, Doris, *Aufklärung, bürgerliche Selbsterfahrung und die "Erfindung" der Psychiatrie in Deutschland 1770–1850* (Göttingen: Vandenhoeck & Ruprecht, 1995).

———, "Dreams and self-consciousness: Mapping the mind in the late eighteenth and nineteenth centuries," in Lorraine Daston, ed., *Biographies of scientific objects* (Chicago: University of Chicago Press, 2000).

Keenan, James F., "Christian perspectives on the human body," *Theological studies*, 55, 1994, 330–346.

Kelley, Donald R., ed., *History and the disciplines: The reclassification of knowledge in early modern Europe* (Rochester: University of Rochester Press, 1997).

———, "The problem of knowledge and the concept of discipline," in Kelley, *History and the disciplines*.

Kemp, Simon, *Medieval psychology* (New York: Greenwood Press, 1990).

Kersting, Christa, *Die Genese der Pädagogik im 18. Jahrhundert: Campes "Allgemeine Revision" im Kontext der neuzeitlichen Wissenschaft* (Weinheim: Deutscher Studien-Verlag, 1992).

Kessler, Eckhard, "The intellective soul," in Schmitt et al., *The Cambridge history of Renaissance philosophy*.

Klemm, Otto, *A history of psychology*, trans. Emil Carl Wilm and Rudolf Pintner (1911; New York: Charles Scribner's Sons, 1914).

Klibansky, Raymond, Erwin Panofsky, and Fritz Saxl, *Saturn and melancholy: Studies in the history of natural philosophy, religion and art* (1964; Nendeln, Lichtenstein: Thomas Nelson and Sons, 1979).

Knebel, Sven K., "Scientia de Anima: Die Seele in der Scholastik," in Gerd Jüttemann, Michael Sonntag, and Christoph Wulf, eds., *Die Seele: Ihre Geschichte im Abendland* (Weinheim: Psychologie-Verlag, 1991).

———, "Scotists vs. Thomists: What seventeenth-century Scholastic psychology was about," *Modern schoolman*, 74, 1997, 219–226.

Knellwolf, Christa, "The science of man, " in Martin Fitzpatric, Peter Jones, Christa Knellwolf, and Iain McCalman, eds., *The Enlightenment world* (London: Routledge, 2004).

Knight, Isabel F., *The geometric spirit: The abbé de Condillac and the French Enlightenment* (New Haven: Yale University Presss, 1968).

Knorr Cetina, Karin, *Epistemic cultures: How the sciences make knowledge* (Cambridge, MA: Harvard University Press, 1999).

Koch, Hans-Theodor, "Bartholomäus Schönborn (1530–1585): Melanchthons de anima als medizinisches Lehrbuch," in Heinz Scheible, ed., *Melanchthon in seinen Schülern* (Wiesbaden: Harrassowitz Verlag, 1997).

Körber, Christian Albrecht, *Versuch einer Ausmessung menschlicher Seelen und aller einfachen endlichen Dingen überhaupt, Wie solche der innern Beschaffenheit derselben, gemäß ins Werck zu richten ist, wenn man ihre Kräffte, Vermögen und Würckungen recht will kennen lernen* (Halle: in der Lüderwaldischen Buchhandlung, 1746).

Korfmann, Manfred et al., *Troia: Traum und Wirklichkeit* (Stuttgart: Konrad Theiss Verlag, 2001).

Kors, Alan Charles, ed., *Encyclopedia of the Enlightenment* (New York: Oxford University Press, 2003).

Kosenina, Alexander, *Ernst Platners Anthropologie und Philosophie: Der philosophische Arzt und seine Wirkung auf Johann Karl Wezel und Jean Paul* (Würzburg: Königshausen & Neumann, 1989).

Köster, Heinrich Martin Gottfried, ed., *Deutsche Encyclopädie oder Allgemeines Real-Wörterbuch aller Künste und Wissenschaften* (Frankfurt-am-Main: Varrentrapp und Wenner, 1778–1809).

Koyré, Alexandre, *La philosophie de Jacob Boehme* (1929; Paris: Vrin, 1971).

Kramer, S., "Ontologie," in Ritter et al., *Historisches Wörterbuch*.

Krauss, Werner, *Zur Anthropologie des 18. Jahrhunderts: Die Frühgeschichte der Menschheit im Blickpunkt der Aufklärung*, ed. Hans Kortum and Christa Gohrisch (1978; Frankfurt: Ullstein, 1987).

Krstic, Kruno, "Marko Marulic—the author of the term 'psychology,'" *Acta Instituti Psychologici Universitatis Zagrabiensis*, 36, 1964, 7–13.

Krüger, Johann Gottlob, *Naturlehre: Zweyter Theil, welcher die Physiologie oder Lehre von dem Leben und der Gesundheit der Menschen in sich fasset* (Halle: Carl Hermann Hemmerde, 1743).

————, *Naturlehre: Dritter Theil, welcher die Pathologie, oder Lehre von den Kranck-heiten in sich fasset* (1749; Halle: Carl Hermann Hemmerde, 1755).

————, *Versuch einer Experimental-Seelenlehre* (Halle: Carl Hermann Hemmerde, 1756).

Kuehn, Manfred, *Scottish common sense in Germany, 1768–1800: A contribution to the history of critical philosophy* (Kingston: McGill-Queen's University Press, 1987).

Kugel, James L., *The idea of biblical poetry: Parallelism and its history* (New Haven: Yale University Press, 1981).

Kühne-Bertram, Gudrun, "Aspekte der Geschichte und der Bedeutungen des Begriffs 'Pragmatisch' in den Philosophischen Wissenschaften des ausgehenden 18. und des 19. Jahrhunderts," *Archiv für Begriffsgeschichte*, 27, 1983, 158–186.

Kusukawa, Sachiko, "Between the *De anima* and dialectics: A prolegomenon to Philippo-Ramism," in Paul Richard Blum, Constance Blackwell, and Charles Lohr, eds., *Sapientiam amemus: Humanismus und Aristotelismus in der Renaissance: Festschrift für Eckhard Keßler zum 60. Geburtstag* (Munich: Wilhelm Fink, 1999).

————, "Lutheran uses of Aristotle: A comparison between Jacob Schegk and Philip Melanchthon," in C. Blackwell and S. Kusukawa, eds., *Philosophy in the sixteenth and seventeenth centuries: Conversations with Aristotle* (Aldershot: Ashgate, 1999).

————, *The transformation of natural philosophy: The case of Philipp Melanchthon* (Cambridge: Cambridge University Press, 1995).

Laehr, Heinrich, *Die Literatur der Psychiatrie, Neurologie und Psychologie von 1459–1799* (Berlin: Reimer, 1900).

Lafitau, Joseph-François, *Customs of the American Indians compared with the Customs of Primitive Times* (1724), ed. and trans. William N. Fenton and Elizabeth L. Moore (Toronto: Champlain Society, 1974).

Lamberton, Robert, *Homer the theologian: Neoplatonist allegorical reading and the growth of the epic tradition* (Berkeley: University of California Press, 1986).

Lamberton, Robert, and John J. Keaney, eds., *Homer's ancient readers: The hermeneutics of Greek epic's earliest exegetes* (Princeton: Princeton University Press, 1992).

La Mettrie, Julien Offray de, *Abrégé des systèmes, pour faciliter l'intelligence du Traité de l'Ame* (1751), in *Œuvres philosophiques* (Paris: Fayard, 1987), vol. 1.

Lampe, G. W. H., *A Patristic Greek lexicon* (Oxford: Clarendon Press, 1968).

Lapointe, François H., "Origin and evolution of the term 'psychology,'" *American psychologist*, 25, 1970, 640–646.

————, "Who originated the term 'psychology,'" *Journal of the history of the behavioral sciences*, 8, 1972, 328–335.

Laudan, Rachel, "Histories of the sciences and their uses: A review to 1913," *History of science*, 31, 1993, 1–21.

Laudin, Gérard, "Changements de paradigme dans l'historiographie allemande: Les origines de l'histoire de l'humanité dans les 'histoires universelles' des années 1760–1820," in Chantal Grell and Jean-Michel Dufays, eds., *Pratiques et concepts de l'histoire en Europe XVIe–XVIIIe siècle* (Paris: Presses de l'Université de Paris–Sorbonne, 1990).

Leary, David, "Immanuel Kant and the development of modern psychology," in Mitchell G. Ash and William R. Woodward, eds., *The problematic science: Psychology in the nineteenth century* (New York: Praeger, 1982).

———, "The philosophical development of the conception of psychology in Germany, 1780–1850," *Journal of the history of the behavioral sciences*, 14, 1978, 113–121.

Le Clerc, Jean, *Pneumatologia*, in *Opera philosophica* (1692; Amsterdam: apud Joan. Ludov. de Lorme, 1710), vol. 2.

Lecoq, Anne-Marie, ed., *La Querelle des Anciens et des Modernes, XVIIe–XVIIIe siècles* (Paris: Gallimard, 2001).

Le Goff, F., *De la philosophie de l'abbé de Lignac* (Paris: Hachette, 1863).

Leijenhorst, Cees, and Johannes M. M. H. Thijseen, "The tradition of Aristotelian natural philosophy: Two theses and seventeen answers," in C. Lüthy, C. Leijenhorst, and J. M. M. H. Thijseen, eds., *The dynamics of Aristotelian natural philosophy from antiquity to the seventeenth century* (Leiden: Brill, 2002).

Leinsle, Ulrich Gottfried, *Das Ding und die Methode: Methodische Konstitution und Gegenstand der frühen protestantischen Metaphysik*, 2 vols. (Augsburg: Maro Verlag, 1985).

———, *Reformversuche protestantischer Metaphysik im Zeitalter des Rationalismus* (Augsburg: Maro Verlag, 1988).

Lelarge de Lignac, Joseph Adrien, *Le témoignage du sens intime et de l'expérience, opposé à la foi profane et ridicule des fatalistes modernes* (Auxerre: François Fournier, 1760).

Lenoir, Timothy, "The discipline of nature and the nature of disciplines," in Ellen Messer-Davidow, David R. Shumway, and David J. Sylvan, eds., *Knowledges: Historical and critical studies in disciplinarity* (Charlottesville: University Press of Virginia, 1993).

———, *Instituting science: The cultural production of scientific disciplines* (Stanford: Stanford University Press, 1997).

Le Roy, Alphonse, "La philosophie au pays de Liège (XVIIe et XVIIIe siècles)," *Bulletin de l'Institut archéologique liégeois*, 1860, 1–157.

Letoublon, Françoise, and Catherine Volpilhac-Auger, with the collaboration of Daniel Sangsue, eds., *Homère en France après la Querelle (1715–1900)* (Paris: Champion, 1999).

Leventhal, Robert S., "The emergence of philological discourse in the German states, 1770–1810," *Isis*, 77, 1986, 243–260.

Liceti, Fortunio, ΨΥΧΟΛΟΓΙΑ ΑΝΘΡΟΠΙΝΗ, *sive de ortu animae humanae libri III: In quibus multa Arcana ac Secreta Naturae, tum de Semine, tum de Foetu, ut & assimilatione parentum & liberorum, panduntur ac revelatur* . . . (Frankfurt: Johann Saur, 1606).

Liébault, Jean, *Trois livres appartenans aux infirmitez et maladies des femmes* (Lyon: Jean Veyrat, 1598).

Lindeboom, G. A., *Descartes and medicine* (Amsterdam: Rodopi, 1979).

Linden, Maretta, *Untersuchungen zum Anthropologiebegriff des 18. Jahrhunderts* (Bern: Lang, 1976).

Lindenberger, Ulman, and Paul B. Baltes, "Lifespan psychology: In honor of Johann Nicolaus Tetens (1736–1807)," *Zeitschrift für Psychologie*, 207, 1999, 299–323.

Linke, Detlef Bernhard, "Gehirn, Seele und Auferstehung," *Evangelische Theologie*, 50, 1990, 128–135.

Lipenius, Martin, *Bibliotheca realis philosophica omnium materiarum, rerum, & titulo-rum, in universo totitus philosophiae ambitu occurentium* (1682; Hildesheim: Olms, 1967).

Littman, Richard A., "Psychology's histories: Some new ones and a bit about their prede-cessors: An essay review," *Journal of the history of the behavioral sciences*, 17, 1981, 516–532.

Llana, James, "Natural history and the *Encyclopédie*," *Journal of the history of biology*, 33, 2000, 1–25.

Locke, John, *Essay concerning human understanding* (1690; 2nd ed., 1694), ed. Peter H. Nidditch (New York: Oxford University Press, 1979).

Lohr, Charles H., "The sixteenth-century transformation of the Aristotelian division of the speculative sciences," in Donald R. Kelley and Richard H. Popkin, eds., *The shapes of knowledge from the Renaissance to the Enlightenment* (Dordrecht: Klu-wer, 1991).

Longo, Mario, *Historia philosophiae philosophica: Teorie e metodi della storia della filosofia tra Seicento e Settecento* (Verona: IPL, 1986).

Lord, Albert, *The Singer of Tales* (Cambridge, MA: Harvard University Press, 1960).

Lossius, Johann Christian, *Neues philosophisches allgemeines Real-Lexikon oder Wörter-buch der gesammten philosophischen Wissenschaften* (Erfurt: bei J. E. G. Rudolphi, 1805).

Lough, John, "The contributors to the *Encyclopédie*," in Richard N. Schwab and Walter Rex, eds., *Inventory of Diderot's "Encyclopédie," VII* (Oxford: Voltaire Foundation, 1984).

———, *The "Encyclopédie"* (London: Longman, 1971).

———, "The problem of the unsigned articles in the *Encyclopédie*," *Studies on Voltaire and the eighteenth century*, 32, 1965, 327–390.

Löwenbrück, Anna-Ruth, "Johann David Michaelis et les débuts de la critique biblique," in Yvon Belaval and Dominique Bourel, eds., *Le siècle des Lumières et la Bible* (Paris: Beauchesne, 1986).

Lüthy, Christoph, and William R. Newman, "'Matter' and 'form': By way of a preface," *Early science and medicine*, 2, 1997, 215–226.

Maccabez, Eugène, *F. B. de Félice, 1723–1789, et son encyclopédie, Yverdon 1770–1780* (Basel: E. Birkhäuser, 1903).

Macpherson, James, *The poems of Ossian and related works*, ed. Howard Gaskill, intro-duction by Fiona Stafford (Edinburgh: Edinburgh University Press, 1996).

Magirus, Johannes, *Anthropologia, hoc est: Commentarius eruditissimus in aureum Philippi Melanchthonis libellum de Anima; Completus & locupletus Opera Georgii Caufungeri* (Frankfurt: Wolfgang Richter, 1603).

Mahnke, Dietrich, "Beiträge zur Geistesgeschichte Niedersachsens," pt. 1, "Der Stader Rektor Casmann," *Stader Archiv*, Neue Folge, Heft 4, 1914, 142–190.

Maistre, Joseph de, *St. Petersburg dialogues* (1821), trans. Richard A. Lebrun (Montreal: McGill-Queen's University Press, 1993).

Malebranche, Nicolas, *Entretiens sur la métaphysique, sur la religion et sur la mort*, ed. of 1711, in *Œuvres*, ed. Geneviève Rodis-Lewis (Paris: Gallimard, 1992), vol. 2.

Malueg, Sara Ellen Procious, "Women and the *Encyclopédie*," in Samia I. Spencer, ed., *French women and the Age of Enlightenment* (Bloomington: Indiana University Press, 1984).

Marat, Jean Paul, *De l'homme, ou des principes et des loix de l'influence de l'âme sur le corps, et du corps sur l'âme* (Amsterdam: chez Marc-Michel Rey, 1775).

———, *Essay on the Human Soul* (London: printed for T. Becket, 1772),

———, *A philosophical essay on man: Being an attempt to investigate the principles and laws of the reciprocal influence of the soul on the body* (London: printed for J. Ridley and T. Payne, 1773).

———, *Über den Menschen oder die Prinzipien und Gesetze des Einflusses der Seele auf den Körper und des Körpers auf die Seele*, trans. Joachim Wilke, ed. G. M. Tripp (Weinheim: VCH, Acta Humaniora, 1992).

Marcu, Eva, "Un encyclopédiste oublié: Formey," *Revue d'histoire littéraire de la France*, 53, 1953, 298–305.

Marino, Luigi, *I maestri della Germania: Göttingen 1770–1820* (Turin: Einaudi, 1975).

———, *Praeceptores Germaniae: Göttingen 1770–1820*, trans. Brigitte Szabó-Bechstein (Göttingen: Vandenhoeck & Ruprecht, 1995).

Marquard, Odo, "Anthropologie," in Ritter et al., *Historisches Wörterbuch*, vol. 1.

Martin, Raymond, "Locke's psychology of personal identity," *Journal of the history of philosophy*, 38, 2000, 41–61.

Martin, Raymond, and John Barresi, *Naturalizing the soul: Self and personal identity in the eighteenth century* (London: Routledge, 2000).

Marx, Jacques, "L'art d'observer au XVIIIᵉ siècle: Jean Senebier and Charles Bonnet," *Janus*, 61, 1974, 201–220.

———, *Charles Bonnet contre les Lumières 1738–1850* (Oxford: Voltaire Foundation, 1976).

Masseau, Didier, *Les ennemis des philosophes: L'anti-philosophe au temps des Lumières* (Paris: Fayard, 2000).

Massimi, Marina, "Marcus Marulus, i suoi maestri e la 'Psychologia de ratione animae humanae,'" *Storia e critica della psicologia*, 4, 1983, 27–41.

Mauchart, Immanuel David, "Ideen zu einer Psychologie der Bibel," *Allgemeines Repertorium für empirische Psychologie und verwandte Wissenschaften*, 6, 1801, 3–41.

McCann, Edwin, "Locke's philosophy of body," in Chappell, *The Cambridge companion to Locke*.

McClelland, I. L., *Benito Jerónimo Feijóo* (New York: Twayne, 1969).

McCracken, Charles, "Knowledge of the soul," in Garber and Ayers, *The Cambridge history of seventeenth-century philosophy*.

Mecacci, Luciano, "Primi usi della parola 'psicologia' tra Settecento e Ottocento in Italia: Breve nota fino al 1830," in *Teoria e modelli*, 8, no. 3, 2004, 31–39.

Meiners, Christoph, *Grundriß der Geschichte der Menschheit* (1785), ed. of 1793 (Königstein: Scriptor, 1981).

———, "Kurze Geschichte der Meynungen roher Völker über die Natur der menschlichen Seelen," *Göttingisches historisches Magazin*, 4. Stück, 1788, 742–758.

Meißner, Heinrich Adam, *Philosophisches Lexicon . . . aus . . . Christian Wolffens sämtlichen teutschen Schrifften . . .* (1737; Düsseldorf: Stern-Verlag Janssen, 1970).

Melanchthon, Philipp, *Commentarius de anima* (Wittenberg, 1550).

———, *Liber de anima* (1552), in *Philippi Melanthonis Opera quae supersunt omnia*, Corpus reformatorum, vol. 13 (1846), ed. Carl Gottlieb Bretschneider (New York: Johnson Reprint Corp., 1963).

———, *Loci communes* (1521), ed. Horst Georg Pöhlmann (Gütersloh: Gütersloher Verlagshaus, 1997).

———, *A Melanchthon reader*, ed. Ralph Keen (New York: Peter Lang, 1988).

Mengal, Paul, "La constitution de la psychologie comme domaine de savoir aux XVIᵉ et XVIIᵉ siècles," *Revue d'histoire des sciences humaines*, 2, 2000, 5–27.

———, "La mythistoire de la psychologie," *Césure*, no. 2, 1992, 127–146.

———, *La naissance de la psychologie* (Paris: L' Harmattan, 2005).

———, "Naissances de la psychologie: La nature et l'esprit," *Revue de synthèse*, 4ᵉ série, no. 3–4, 1994, 355–373.

———, "Pour une histoire de la psychologie," *Revue de synthèse*, 4ᵉ série, no. 3–4, 1988, 485–497.

Mercer, Christia, "The vitality and importance of early modern Aristotelianism," in Tom Sorell, ed., *The rise of modern philosophy: The tension between the new and traditional philosophies from Machiavelli to Leibniz* (Oxford: Clarendon Press, 1993).

Mercier-Faivre, Anne-Marie, *Un supplément à "l'Encyclopédie": Le "Monde primitif" d'Antoine Court de Gébelin: Suivi d'une édition du "Génie allégorique et symbolique de l'antiquité," extrait du "Monde primitif," 1773* (Paris: Champion, 1999).

Merian, Johann Bernhard, "Sur la durée et sur l'intensité du plaisir et de la peine," *Histoire de l'Académie Royale des Sciences et des Belles-Lettres de Berlin*, 1766, 381–400.

Merker, Nicolao, "Cristiano Wolff e la metodologia del razionalismo," *Rivista critica di storia della filosofia*, 22, 1967, 271–293; 23, 1968, 21–38.

Métraux, Alexandre, "An essay on the early beginnings of psychometrics," in Georg Eckardt and Lothar Sprung, eds., *Advances in historiography of psychology* (Berlin: VEB Deutscher Verlag der Wissenschaften, 1983).

Meyer, Annette, "Das Projekt einer 'Natural history of man' in der schottischen Aufklärung," *Storia della storiografia*, 39, 2001, 93–102.

———, "The experience of human diversity and the search for unity: Concepts of mankind in the late Enlightenment," *Studi settecenteschi*, 21, 2001, 245–264.

Meyerson, Ignace, *Écrits 1920–1983: Pour une psychologie historique* (Paris: Presses Universitaires de France, 1987).

———, *Les fonctions psychologiques et les œuvres* (1948), afterword by Riccardo di Donato (Paris: Albin Michel, 1995).

Michael, Emily, "Renaissance theories of body, soul, and mind," in John P. Wright and Paul Potter, eds., *Psyche and Soma: Physicians and metaphysicians on the mind-body problem from antiquity to Enlightenment* (Oxford: Clarendon Press, 2000).

Micraelius, Johannes, *Lexicon philosophicum terminorum philosophis usitatorum* (1662; Düsseldorf: Stern-Verlag Janssen, 1966).

———, praes., Caspar Voigth, auctor & respondens, *Psychologia per theses succinctas de anima humana ejuisque potentiis et operationibus, quibus multae, eaque difficiles quaestiones circa illam doctrinam occurentes tanguntur . . .* (Stettin: Typis Georgii Gœtschii, [ca. 1650]).

Mijuskovic, Ben Lazare, *The Achilles of rationalist arguments: The simplicity, unity, and identity of thought and soul from the Cambridge Platonists to Kant: A study in the history of an argument* (The Hague; Martinus Nijhoff, 1974).

Mischel, Theodor, "Kant and the possibility of a science of psychology," *Monist* 51 (1967): 599–622.

Monboddo, James Burnett, Lord, *Antient metaphysics*, vol. 3, *Containing the history and philosophy of men: With a preface containing the history of antient philosophy, both in antient and later times . . .* (Edinburgh: Cadell, 1784).

Montesquieu, Charles de Secondat, baron de, "Essai sur les causes qui peuvent affecter les esprits et les caractères" (before 1748), in *Œuvres*, ed. Roger Caillois (Paris: Gallimard, 1951), vol. 2.

Moravia, Sergio, "The capture of the invisible: For a (pre)history of psychology in 18th-century France," *Journal of the history of the behavioral sciences*, 19, 1983, 370–378.

———, *Filosofia e scienze umane nell'età dei lumi* (Florence: Sansoni, 1982).

———, *Il pensiero degli Idéologues: Scienza e filosofia in Francia (1780–1815)* (Florence: La nuova Italia, 1974).

———, *La scienza dell'uomo nel Settecento* (Bari: Laterza, 1978).

Moritz, Karl Philipp, ed. *ΓΝΩΘΙ ΣΑΥΤΟΝ oder Magazin zur Erfahrungs-Seelenkunde als ein Lesebuch für Gelehrte und Ungelehrte* (1783–1793; Nördlingen: Franz Greno, 1986).

Moureau, François, *Le roman vrai de "l'Encyclopédie"* (Paris: Gallimard, 1990).

Müller, Martin, "Methoden psychologischer Forschung und Diagnostik im Deutschland des 18. Jahrhunderts," in Horst Gundlach, ed., *Arbeiten zur Psychologiegeschichte* (Göttingen: Hogrefe, 1994).

Muller, Richard A., *Dictionary of Latin and Greek theological terms: Drawn principally from Protestant Scholastic theology* (Grand Rapids, MI: Baker Books, 1985).

Müller-Brettel, Marianne, and Roger A. Dixon, "Johann Nicolas Tetens: A forgotten father of developmental psychology?," *International journal of behavioral development*, 13, no. 2, 1990, 215–230.

Muratori, Ludovico Antonio, *Della forza della fantasia umana* (1745), ed. Carlo Pogliano (Florence: Giunti, 1995).

Nast, Johann Jakob Heinrich, *Ueber Homers Sprache aus dem Gesichtspunkt ihrer Analogie mit der allgemeinen Kinder- und Volks-Sprache* (Stuttgart: bei Johann Benedikt Metzler, 1801).

Nicolas, Serge, *Histoire de la psychologie* (Paris: Dunod, 2001).

Noreña, Carlos G., *Juan Luis Vives and the emotions* (Carbondale: Southern Illinois University Press, 1989).

Nüsslein, Franz Anton, *Grundlinien der allgemeinen Psychologie zum Gebrauche der Vorlesungen* (Mainz: bey Florian Kupferberg, 1821).

Nutton, Vivian, "The anatomy of the soul in early Renaissance medicine," in G. R. Dunstan, ed., *The human embryo: Aristotle and the Arabic and European traditions* (Exeter: University of Exeter Press, 1990).

———, "Wittenberg anatomy," in Ole Peter Grell and Andrew Cunningham, eds., *Medicine and the Reformation* (London: Routledge, 1993).

O'Donnell, John M., "The crisis of experimentalism in the 1920s: E. G. Boring and his uses of history," *American psychologist*, 34, 1979, 289–295.

Olson, David R., *The world on paper: The conceptual and cognitive implications of writing and reading* (New York: Cambridge University Press, 1994).

Olson, Richard, *The emergence of the social sciences, 1642–1792* (New York: Twayne, 1993).

———, "The human sciences," in Porter, *The Cambridge history of science*, vol. 4.

O'Neal, John C., *The authority of experience: Sensationist theory in the French Enlightenment* (University Park: Pennsylvania State University Press, 1996).

O'Neil, W. M., "The Wundt myths," *Australian journal of psychology*, 36, 1984, 285–289.

Ong, Walter J., *Ramus, method, and the decay of dialogue: From the art of discourse to the art of reason* (Cambridge, MA: Harvard University Press, 1958).

Onians, Richard Broxton, *The origins of European thought about the body, the mind, the soul, the world, time and fate* (Cambridge: Cambridge University Press, 1951).

Outram, Dorinda, *The Enlightenment* (New York: Cambridge University Press, 1995).

Palmer, Robert R., *Catholics and unbelievers in 18th-century France* (1939; New York: Cooper Square Publishers, 1961).

Pancera, Carlo, "L'importanza dei testi scolastici di Francesco Soave (1743–1806)," in Luciana Bellatalla, ed., *Maestri, didattica e dirigenza nell'Italia dell'Ottocento* (Ferrara: Tecomproject, 2000).

Pandolfi, Claudia, "Civiltà antiche e selvaggi moderni: Due dissertazioni di Christian Gottlob Heyne," *I castelli di Yale*, 2, 1997, 253–287.

Park, Katharine, "The organic soul," in Schmitt et al., *The Cambridge history of Renaissance philosophy*.

Park, Katharine, and Eckhard Kessler, "The concept of psychology," in Schmitt et al., *The Cambridge history of Renaissance philosophy*.

Parot, Françoise, ed., *Pour une psychologie historique: Écrits en hommage à Ignace Meyerson* (Paris: Presses Universitaires de France, 1996).

Parry, Milman, "The traditional epithet in Homer" (1928), in *The making of Homeric verse: The collected papers of Milman Parry*, ed. and trans. Adam Parry (Oxford: Clarendon Press, 1971).

Perkins, Jean A., *The concept of the self in the French Enlightenment* (Geneva: Droz, 1969).

Perret, Jean-Pierre, *Les imprimeries d'Yverdon au XVIIe et XVIIIe siècle* (Lausanne: Roth, 1945).

Petersen, Peter, *Geschichte der aristotelischen Philosophie im protestantischen Deutschland* (Leipzig: Felix Meiner, 1921).

Phylopsyches, Alethius, *ΨΥΧΗΛΟΓΙΑ; or Second Thoughts on second thoughts: Being a discourse fully proving from Scripture, the Writings of the Learned Ethnicks, Fathers of the Church, Philosophy, and the Dictates of right Reason, the separate existence of the Soul . . .* (London: Printed and Sold by John Nutt, n.d. [1702 or 1703]).

Piquer, Andrés, *Lógica moderna, o Arte de hallar la verdad, y perficionar la razón* (Valencia: En la Oficina de Joseph García, 1747).

Pitassi, Maria-Cristina, "Jean Le Clerc bon tâcheron de la philosophie: L'enseignement philosophique à la fin du XVIIIᵉ siècle," *Lias*, 10, 1983, 105–122.

Pivati, Gianfrancesco, ed., *Nuovo dizionario scientifico e curioso, sacro-profano* (Venice: Benedetto Milocco, 1746–1751).

Platner, Ernst, *Anthropologie für Aerzte und Weltweise* (1772; Hildesheim: Olms, 1998).

———, *Neue Anthropologie für Aerzte und Weltweise: Mit besonderer Rücksicht auf Physiologie, Pathologie, Moralphilosophie und Aesthetik* (Leipzig: bey Siegfried Lebrecht Crusius, 1790).

———, *Philosophische Aphorismen nebst einigen Anleitungen zur philosophischen Geschichte*, 2nd ed. (1793–1800; Brussels: Culture et Civilisation, 1970).

Popp, Walter, *Kritische Bemerkungen zur Associationstheorie*, 1. Theil, *Kritische Entwickelung des Associationsproblems* (Leipzig: Johann Ambrosius Barth, 1913).

Porphyry, *On the cave of the nymphs*, trans. and with introduction by Robert Lamberton (Barrytown, NY: Station Hill Press, 1983).

Porter, Roy, ed., *The Cambridge history of science*, vol. 4, *Eighteenth-century science* (Cambridge: Cambridge University Press, 2003).

———, "Psychology," in John W. Yolton, ed., *The Blackwell companion to the Enlightenment* (Oxford: Blackwell, 1992).

Porter, Roy, and Mikulas Teich, eds., *The Enlightenment in national context* (Cambridge: Cambridge University Press, 1981).

Priestley, Joseph, *The history of the philosophical doctrine concerning the origin of the soul, and the nature of matter* (1777; New York: Garland, 1976).

Proust, Jacques, *Diderot et "l'Encyclopédie"* (1962; Paris: Albin Michel, 1995).

———, "Diderot et le système des connaissances humaines," *Studies on Voltaire and the eighteenth century*, 256, 1988, 117–127.

Pucci, Giuseppe, *Il passato prossimo: La scienza dell'antichità alle origine della cultura moderna* (Rome: La Nuova Italia Scientifica, 1993).

Puech, Michel, "Tetens et la crise de la métaphysique allemande en 1775 (*Über die allgemeine speculativische Philosophie*)," *Revue philosophique*, no. 1, 1992, 3–29.

Quinton, Anthony, *Francis Bacon* (Oxford: Oxford University Press, 1980).

Ramul, Konstantin, "The problem of measurement in the psychology of the eighteenth century," *American psychologist*, 15, 1960, 256–265.

Ratcliff, Marc, "Une métaphysique de la méthode chez Charles Bonnet," in Buscaglia et al., *Charles Bonnet*.

Raulet, Gérard, ed., *Aufklärung: Les Lumières allemandes* (Paris: Flammarion, 1995).

Redfield, James, "Le sentiment homérique du moi," *Le genre humain*, 12, 1985, 93–111.

Reed, Edward S., "Descartes' corporeal ideas hypothesis and the origin of scientific psychology," *Review of metaphysics*, 35, 1982, 731–752.

Rees, Abraham, *Cyclopaedia . . .* (Londres: printed for J. J. and C. Rivington, 1786–1788).

Régis, Pierre Sylvain, *Système de philosophie, contenant la logique, métaphysique, physique et morale* (Lyon: chez Anisson, Posuel et Rigaud, 1691).

Remaud, Olivier, *Les archives de l'humanité: Essai sur la philosophie de Vico* (Paris: Seuil, 2004).

Renneville, Marc, *Le langage des crânes: Une histoire de la phrénologie* (Paris: Les Empêcheurs de penser en rond, 2000).

Rey, Roselyne, "Le cas des sciences de la vie," *Recherches sur Diderot et "l'Encyclopédie,"* 12, 1992, 41–58.

————, "Naissance de la biologie et redistribution des savoirs," *Revue de synthèse,* 4ᵉ série, no. 1–2, 1994, 167–197.

————, *Naissance et développement du vitalisme en France de la deuxième moitié du 18e siècle à la fin du Premier Empire* (Oxford: Voltaire Foundation, 2000).

————, "La partie, le tout et l'individu: Science et philosophie dans l'œuvre de Charles Bonnet," in Buscaglia et al., *Charles Bonnet.*

————, "La pathologie mentale dans *l'Encyclopédie*: Définitions et distribution nosologique," *Recherches sur Diderot et "l'Encyclopédie,"* 7, 1989, 51–70.

Ribot, Théodule, *La psychologie allemande contemporaine: Ecole expérimentale* (Paris: Baillière, 1879).

————, *La psychologie anglaise contemporaine (école expérimentale),* 2nd ed. (Paris: Baillière, 1875).

Richards, Graham, "The absence of psychology in the eighteenth century: A linguistic perspective," *Studies in history and philosophy of science,* 23, 1992, 195–211.

————, *Mental machinery: The origins and consequences of psychological ideas,* pt. 1, *1600–1850* (Baltimore: Johns Hopkins University Press, 1992).

————, *Putting psychology in its place: An introduction from a critical historical perspective* (London: Routledge, 1996).

Richards, Robert J., "Christian Wolff's prolegomena to *Empirical and Rational Psychology*: Translation and commentary," *Proceedings of the American Philosophical Society,* 124, 1980, 227–239.

Ricœur, Paul, *Soi-même comme un autre* (Paris: Seuil, 1990).

Ricuperati, Giuseppe, "Le categorie di periodizzazione e il Settecento: Per una introduzione storiografica," *Studi settecenteschi,* 14, 1994, 9–106.

————, "Illuminismo e Settecento dal dopoguerra a oggi," in Guiseppe Ricuperati, ed., *La reinvenzione dei Lumi: Percorsi storiografici del Novecento* (Florence: Leo S. Olschki, 2000).

Ridder-Symoens, Hilde de, ed., *A history of the university in Europe,* vol. 2, *Universities in early modern Europe (1500–1800)* (Cambridge: Cambridge University Press, 1996).

Riedel, Wolfgang, "Erster Psychologismus: Umbau des Seelenbegriffs in der deutschen Spätaufklärung," in Garber and Thoma, *Zwischen Empirisierung und Konstruktionsleistung.*

Riedl, Clare C., "Suarez and the organization of learning," in Gerard Smith, ed., *Jesuit thinkers of the Renaissance* (Milwaukee: Marquette University Press, 1939).

Risse, Wilhelm, *Bibliographia philosophica vetus,* pars 5, *De anima* (Hildesheim: Olms, 1998).

Ritter, Joachim, Karlfried Gründer, and Gottfried Gabriel, eds., *Historisches Wörterbuch der Philosophie* (Darmstadt: Wissenschaftliche Buchgesellschaft, 1971).

Rodis-Lewis, Geneviève, *Descartes: Textes et débats* (Paris: Livre de poche, 1984).

Rodríguez Aranda, Luis, "La recepción e influjo de la filosofía de Locke en España," *Revista de filosofía,* 14, 1955, 359–381.

Rohde, Erwin, *Psyche: The Cult of Souls and the Belief in Immortality among the Greeks* (1894), trans. W. B. Hillis (London: Routledge and Kegan Paul, 1925).

Romilly, Jacqueline de, *"Patience, mon cœur": L'essor de la psychologie dans la littérature grecque classique* (Paris: Belles Lettres, 1984).

Roos, Magnus Friedrich, *Fundamenta psychologiae ex Sacra Scriptura sic collecta, ut dicta eius De Anima eiusque facultatibus agentia collecta, digesta atque explicata sint* (Tübingen: sumptibus Lud. Friedr. Fuessi, 1769).

Rosset, François, "La vie littéraire et intellectuelle en pays romand au XVIIIᵉ siècle," in Roger Francillon, ed., *Histoire littéraire de la Suisse romande*, vol. 1, *Du moyen âge à 1815* (Lausanne: Payot, 1996).

Rousseau, George S., "Cultural history in a new key: Towards a semiotics of the nerve," in Joan H. Pittock and Andrew Wear, eds., *Interpretation and cultural history* (London: Macmillan, 1991).

———, "Psychology," in George S. Rousseau and Roy Porter, eds., *The ferment of knowledge: Studies in the historiography of eighteenth-century science* (Cambridge: Cambridge University Press, 1980).

Rudolph, Olivier-Pierre, "Mémoire, réflexion et conscience chez Christian Wolff," *Revue philosophique*, 193, 2003, 351–360.

Rump, Johann, *Melanchthons Psychologie (seine Schrift de anima) in ihrer Abhängigkeit von Aristoteles und Galenos* (Kiel: G. Marquardsen, 1897).

Rupp-Eisenreich, Britta, "Des choses occultes en histoire des sciences humaines: Le destin de la 'science nouvelle' de Christoph Meiners," in B. Rupp-Eisenreich and Patrick Menget, eds., "L'anthropologie: Points d'histoire," special issue, *L'ethnographie*, 79, no. 90–91 1983, 131–183.

Russo, Joseph, and Bennett Simon, "Homeric psychology and the oral epic tradition," *Journal of the history of ideas*, 29, 1968, 483–498.

Sánchez-Blanco Parody, Francisco, *Europa y el pensamiento español del siglo XVIII* (Madrid: Alianza, 1991).

Sancipriano, Mario, "La pensée anthropologique de J. L. Vivès: L'entéléchie," in August Buck, ed., *Juan Luis Vives: Arbeitsgespräch Wolfenbüttel* (Hamburg: Dr. Ernst Hauswedell & Co., 1981).

Santinello, Giovanni, *Storia delle storie generali della filosofia*, vol. 2, *Dall'età cartesiana a Brucker*; vol. 3, *Il secondo illuminismo e l'età kantiana* (Brescia: La Scuola, 1981; Padua: Antenore, 1988).

Savigny, Christophe de, *Tableaux accomplis de tous les arts liberaux, contenans brieuement et clerement par singuliere methode de doctrine, vne generale et sommaire partition des dicts arts, amassez et reduicts en ordre pour le soulagement et profit de la ieunesse* (Paris: par Iean & François de Gourmont freres, 1587).

Savioz, Raymond, ed., *Mémoires autobiographiques de Charles Bonnet de Genève* (Paris: Vrin, 1948).

———, *La philosophie de Charles Bonnet de Genève* (Paris: Vrin, 1948).

Scheerer, Eckart, "Die Berliner Psychologie zur Zeit der Aufklärung (1764–1806)," in Lothar Sprung and Wolfgang Schönpflug, eds., *Zur Geschichte der Psychologie in Berlin* (Frankfurt: Peter Lang, 1992).

———, "Psychologie," in Ritter et al., *Historisches Wörterbuch*, vol. 7.

Schelling, Friedrich Wilhelm Joseph, "Ueber Mythen, historische Sagen und Philosopheme der ältesten Welt" (1793), in *Werke I*, vol. 1, ed. Wilhelm G. Jacobs et al. (Stuttgart: Frommann-Holzboog, 1976).

Scherzer, Johann Adam, *Vade mecum sive Manuale philosophicum* (1654), ed. of 1675 (Stuttgart: Fromann-Holzboog, 1996).

Schlich, Thomas, and Claudia Wiesemann, eds. *Hirntod: Zur Kulturgeschichte der Todesfeststellung* (Frankfurt: Suhrkamp, 2001).

Schliemann, Heinrich, *Ilios: The city and country of the Trojans: The results of researches and discoveries on the site of Troy and throughout the Troad in the years 1871, 72, 73, 78, 79: Including an autobiography of the author* (London: John Murray, 1880).

Schløsler, Jørn, *John Locke et les philosophes français: La critique des idées innées en France au dix-huitième siècle* (Oxford: Voltaire Foundation, 1997).

Schmaltz, Tad M., *Radical Cartesianism: The French reception of Descartes* (New York: Cambridge University Press, 2002).

Schmidt, James, ed., *What is enlightenment? Eighteenth-century answers and twentieth-century questions* (Berkeley: University of California Press, 1996).

Schmidt-Biggemann, Wilhelm, "Jacob Bruckers philosophiegeschichtliches Konzept," in Schmidt-Biggemann and Stammen, *Jacob Brucker*.

———, "New structures of knowledge," in Ridder-Symoens, *A history of the university*.

———, *Topica universalis: Eine Modellgeschichte humanistischer und barocker Wissenschaft* (Hamburg: Felix Meiner, 1983).

Schmidt-Biggemann, Wilhelm, and Theo Stammen, eds., *Jacob Brucker (1696–1770): Philosoph und Historiker der europäischen Aufklärung* (1696–1770; Berlin: Akademie Verlag, 1998).

Schmitt, Charles B., *Aristotle and the Renaissance* (Cambridge, MA: Harvard University Press, 1983).

———, "The rise of the philosophical textbook," in Schmitt et al., *The Cambridge history of Renaissance philosophy*.

Charles B. Schmitt, Quentin Skinner, Eckhard Kessler, and Jill Kraye, eds., *The Cambridge history of Renaissance philosophy* (New York: Cambridge University Press, 1990).

Schneiders, Hans-Wolfgang, "Le prétendu système des renvois dans l'Encyclopédie," in Edgar Mass and Peter-Eckhard Knabe, eds., *"L'Encyclopédie" et Diderot* (Cologne: dme-Verlag, 1985).

Schneiders, Werner, ed., *Christian Wolff, 1679–1754: Interpretationen zu seiner Philosophie und deren Wirkung* (Hamburg: F. Meiner Verlag, 1983).

———, *Lexikon der Aufklärung: Deutschland und Europa* (1995; Munich: C. H. Beck, 2001).

Schöndorf, Harald, "Der Leib und sein Verhältnis zur Seele bei Ernst Platner," *Theologie und Philosophie*, 60, 1985, 77–87.

Schrimpf, G., "Disciplina," in Ritter et al., *Historisches Wörterbuch*, vol. 2.

Schulerus, Johannes, *Philosophia novo methodo explicata: Cujus Pars prior continet Excercitationum philosophicarum Libros VI: Metaphysicam, Theologiam naturalem, Angelogra[p]hiam, Psychologiam, Logicam, Ethicam: In quibus, Assertis purioris Philosophiae principiis, Aristotelis Peripateticorumque errores passim refutantur . . .* (The Hague: Ex Typographia Adriani Vlacq, 1663).

Schüling, Hermann, *Bibliographie der psychologischen Literatur des 16. Jahrhunderts* (Hildesheim: Olms, 1967).

———, *Bibliographisches Handbuch zur Geschichte der Psychologie: Das 17. Jahrhundert* (Giessen: Universitätsbibliothek, 1964).

Schütz, Christian Gottfried, "Betrachtungen über die verschiednen Methoden der Psychologie; nebst einem kritischen Auszuge aus des Hrn. Abt von Condillac Traité des sensations," in Charles Bonnet, *Analytischer Versuch über die Seelenkräfte*, trans. C. G. Schütz (Bremen: Johann Henrich Cramer, 1770), vol. 1.

Schweling, Johann Eberhard, *Philosophiae tomus pneumaticus: Continens Psychologiam, Theologiam naturalem, neque non Angelographiam* (Bremen: typis Hermanni Braueri, 1695).

Scola, Giovanni, "Lettere ai seguaci del sistema sintetico" (1781), in Mario Berengo, ed., *Giornali veneziani del Settecento* (Milan: Feltrinelli, 1962).

Scripture, Edward Wheeler, *The new psychology* (New York: Charles Scribner's Sons, 1897).

Seiler, Georg Friedrich, *Animadversiones ad psychologiam sacram* (Erlangen: Typis Cammererianis, 1778–1787).

———, *Biblische Hermeneutik, oder Grundsätze und Regeln zur Erklärung der heiligen Schrift des Alten und Neuen Testaments* (Erlangen: Bibelanstalt, 1800).

Sell, Alan P. F., *John Locke and the eighteenth-century divines* (Cardiff: University of Wales Press, 1997).

Senebier, Jean, *Essai sur l'art d'observer et de faire des expériences*, 2nd ed. (Geneva: J. J. Paschoud, 1802).

Shapin, Steven, *The Scientific Revolution* (Chicago: University of Chicago Press, 1996).

Sigrist, René, *L'essor de la science moderne à Genève* (Lausanne: Presses polytechniques et universitaires romandes, 2004).

Simon, Bennett, and Herbert Weiner, "Models of the mind and mental illness in ancient Greece," pt. 1, "The Homeric model of mind," *Journal of the history of the behavioral sciences*, 2, 1966, 303–314.

Simone, Franco, "La notion d'encyclopédie: Élément caractéristique de la Renaissance française," in Peter Sharratt, ed., *French Renaissance studies, 1540–70: Humanism and the encyclopedia* (Edinburgh: Edinburgh University Press, 1976).

Simonsuuri, Kirsti, *Homer's original genius: Eighteenth-century notions of the early Greek epic (1688–1798)* (New York: Cambridge University Press, 1979).

Smellie, William, ed., *Encyclopaedia Britannica; or, a Dictionary of Arts and Sciences* (Edinburgh: printed for A. Bell and C. Macfarquhar, 1768–1771).

———, ed., *Encyclopaedia Britannica; or, a Dictionary of Arts and Sciences*, 3rd ed. (Philadelphia: T. Dobson, 1798) (same as ed. published in Edinburgh, 1787–1797).

Smith, Roger, "Does the history of psychology have a subject?," *History of the human sciences*, 1, 1988, 147–177.

———, *The Fontana history of the human sciences* (London: Fontana Press, 1997).

Snell, Bruno, *La découverte de l'esprit: La genèse de la pensée européenne chez les Grecs* (1946), trans. Marianne Charrière and Pascale Escaig (Combas: Ed. de l'Eclat, 1994).

Snellius, Rudolph, *In aureum Philippi Melanchthonis de anima, vel potius de hominis physiologia, libellum, commentationes utilissimae . . . : Accesserunt D. Rodolphi Goclenii . . . theses quaedam ac disputationes de psychologicis selectissimae* (Frankfurt: Peter Fischer, 1596).

———, *Snellio-ramaeum philosophiae syntagma . . .* (Frankfurt: Peter Fischer, 1596).

Soave, Francesco, *Istituzioni di metafisica*, in *Istituzioni di metafisica, parte prima (Psicologia)*, in *Istituzioni di logica, metafisica ed etica* (1804), rev. and expanded ed. (Venice: Nella Stamperia di Sebastiano Valle, 1813), vol. 3.

Sophocles, Evangelinus Apostolides, *Greek lexikon of the Roman and Byzantine periods (from B.C. 146 to A.D. 1100)* (Boston: Little, Brown and Co., 1870).

Souter, Alexander, *A glossary of later Latin to 600 A.D.* (Oxford: Clarendon Press, 1949).

Spiess, Edmund, *Entwicklungsgeschichte der Vorstellungen vom Zustande nach dem Tode auf Grund vergleichender Religionsforschung* (1877; Graz: Akademische Druck- u. Verlagsanstalt, 1975).

Sprengel, Kurt, *Histoire de la médecine, depuis son origine jusqu'au dix-neuvième siècle* (1792–1799), trans. A. J. L. Jourdan, rev. E. F. M. Bosquillon (Paris: chez Deterville et [chez] Th. Desoer, 1815–1820).

Staël, Germaine de, *De la littérature considérée dans ses rapports avec les institutions sociales* (1800), ed. Gérard Gengembre and Jean Goldzink (Paris: Flammarion, 1991).

Starobinski, Jean, "Fable and mythology in the seventeenth and eighteenth centuries" (1981), in *Blessings in disguise; or, The morality of evil*, trans. Arthur Goldhammer (Cambridge, MA: Harvard University Press, 1993).

———, "Panorama succinct des sciences psychologiques entre 1575 et 1625," *Gesnerus*, 37, 1980, 3–16.

Stewart, Dugald, *Dissertation, exhibiting a general view of the progress of metaphysical, ethical, and political philosophy, since the revival of letters in Europe* (1811/1821), in *The works* (Cambridge: Hilliard and Brown, 1829), vol. 6.

Stichweh, Rudolf, *Zur Entstehung des modernen Systems wissenschaftlicher Disziplinen: Physik in Deutschland 1740–1890* (Frankfurt: Suhrkamp, 1984).

Stiening, Gideon, "Psychologie," in Barbara Bauer, ed., *Melanchthon und die Marburger Professoren (1527–1627)* (Marburg: Universitätsbibliothek, 1999).

———, "Verweltlichung der Anthropologie im 17. Jahrhundert? Von Casmann und Magirus zu Descartes und Hobbes," in Dannenberg et al., *Säkularisierung.*

Stockhausen, Johann Christoph, *Critischer Entwurf einer auserlesenen Bibliothek für die Liebhaber der Philosophie und schönen Wissenschaften* (1751; Berlin: bey Haude und Spener, 1771).

Stocking, George W., "On the limits of 'presentism' and 'historicism' in the historiography of the behavioral sciences," *Journal of the history of the behavioral sciences*, 1, 1965, 211–217.

Stolle, Gottlieb, *Anleitung zur Historie der Gelehrheit, denen zum besten, so den Freyen-Künsten und der Philosophie obliegen* (1718; Jena: in Verlegung Johannes Meyers seel. Erben, 1736).

Strieder, Friedrich Wilhelm, et al., *Grundlage zu einer hessischen Gelehrten- und Schriftsteller-Geschichte seit der Reformation bis auf gegenwärtige Zeiten* (Cassel: Cramer, 1781–1868).

Stroumsa, Gedaliahu G., "Caro salutis cardo: Shaping the person in early Christian thought," *History of religions*, 30, 1990, 25–50.

Struve, Burkhard Gotthelf, *Bibliotheca philosophica in suas classes distributa* (1704; Iena: apud Ern. Claudium Baillar, 1707).

Sturm, Thomas, "Kant on empirical psychology: How not to investigate the human mind," in Eric Watkins, ed., *Kant and the sciences* (New York: Oxford University Press, 2001).

——, *Kant und die Wissenschaften vom Menschen* (Hamburg: Mentis Verlag, 2009).

Sturz, Friedrich Wilhelm, *Prolusio prima de vestigiis doctrinae de animi humani immortalitate in Homeri carminibus* (Gera: in officina Rothiana, 1795).

Suárez, Francisco, *Commentaria una cum quaestionibus in libros Aristotelis De anima* (1572), ed. and trans. Salvador Castellote et al. (Madrid: Sociedad de estudios y publicaciones, 1978–1981).

——, *De anima* (1621), in *Opera omnia* (Paris: L. Vivès, 1856), vol. 3.

——, *Disputaciones metafísicas* (1597), ed. and trans. Sergio Rábade Romeo, Salvador Caballero Sánchez, and Antonio Puigcerver Zanón (Madrid: Gredos, 1960–1966).

Sullivan, Shirley Darcus, *Psychological activity in Homer: A study of phren* (Ottawa: Carleton University Press, 1988).

——, "A multi-faceted term: Psyche in Homer, the Homeric Hymns, and Hesiod," *SFIC*, 6, 1988, 151–180.

Sulzer, Johann Georg, *Kurzer Begriff aller Wissenschaften und andern Theile der Gelehrsamkeit, worin nach seinem Inhalt, Nuzen und Vollkommenheit kürzlich beschrieben wird*, 2nd ed. (Leipzig: bey Johann Christian Langenheim, 1759).

Taylor, Charles, *Sources of the self: The making of the modern identity* (Cambridge, MA: Harvard University Press, 1989).

Taylor, Samuel S. B., "The Enlightenment in Switzerland," in Porter and Teich, *The Enlightenment*.

Temkin, Owsei, *Galenism: Rise and decline of a medical philosophy* (Ithaca: Cornell University Press, 1973).

Tenneman, Wilhelm Gottlieb, *Geschichte der Philosophie* (Leipzig: bei Johann Ambrosius Bart, 1798–1819).

Tetens, Johann Nicolas, *Saggi filosofici e scriti minori*, ed. Raffaele Ciafardone (L'Aquila: L. U. Japadre, 1983).

——, *Über die allgemeine speculativische Philosophie* (1775), in *Über die allgemeine speculativische Philosophie: Philosophische Versuche über die menschliche Natur und ihre Entwickelung*, ed. Wilhelm Uebele (Berlin: Verlag von Reuther & Reichard, 1913).

Thiel, Udo, "Personal identity," in Garber and Ayers, *The Cambridge history of seventeenth-century philosophy*.

Thomman, Marcel, "Wolff, Christian," in Denis Huisman, ed., *Dictionnaire des philosophes* (Paris: Presses Universitaires de France, 1984).

Thomson, Ann, "From *l'histoire naturelle de l'homme* to the natural history of mankind," *British journal for eighteenth-century studies*, 9, 1986, 73–80.

Thümmig, Ludwig Philipp, *Institutiones philosophiae wolfianae, in usos academicos adornatae* (1725–1726), in Wolff, *Gesammelte Werke*, ser. III, vol. 19.

Tiedemann, Dietrich, *Geist der spekulativen Philosophie* (1791–97; Brussels: Culture et civilisation, 1969).

Timpler, Clemens, *Metaphysicae systema methodicum* (1604; Hanau: apud Haeredes Guilielmi Antonii, 1612).

————, *Physicae seu philosophiae naturalis systema methodicum: Pars tertiam & postrema Physicae, complectens Empsychologiam; Hoc est, Doctrinam de corporibus naturalibus animatis* (Hanau: apud Haeredes Guilielmi Antonii, 1610).

Tipler, Frank J., *The physics of immortality: Modern cosmology, God, and the resurrection of the dead* (New York: Doubleday, 1994).

Tissot, Samuel Auguste, *L'onanisme: Dissertation sur les maladies produites par la masturbation*, preface by Théodore Tarczylo (1760), ed. of 1768 (Paris: Le Sycomore, 1980).

————, *De la santé des gens de lettres* (1768; Geneva: Slatkine, 1981).

Tomasselli, Sylvana, "Studying eighteenth century psychology," *History of science*, 29, 1991, 102–104.

Tonelli, Giorgio "Wolff, Christian," in Paul Edwards, ed., *The encyclopedia of philosophy* (New York: Macmillan, 1967), vol. 8.

Trembley, Jean, "Observations sur l'attraction & l'équilibre des Sphéroïdes," *Mémoires de l'Académie Royale des Sciences et Belles-Lettres de Berlin*, 1799–1800, 68–109.

————, "Réponse à la question, proposée par la Société de Harlem: Quelle est l'Utilité de la Science Psychologique dans l'éducation & la direction de l'Homme, & relativement au bonheur des Sociétés? Et quelle serait la meilleure maniere de perfectionner cette belle Science, & d'accroitre ses progrès?," *Verhandelingen, uitgegeeven door de Hollandsche Maatschappye der Weetenschappen, te Haarlem*, 20, no. 1, 1781, 1–310.

Tschirnhaus, Ehrenfried Walter von, *Médecine de l'esprit ou préceptes généraux de l'art de découvrir* (1687), trans. Jean-Paul Wurtz (Paris: Ophrys, 1980).

————, *Medicina mentis* (1687), ed. and trans. Johannes Haussleiter et al. (Leipzig: J. A. Barth, 1963).

————, *Medicina mentis* (1687), trans. Lucio Pepe and Manuela Sana (Naples: Guida, 1987).

Vande Kemp, Hendrika, "A note on the term 'psychology' in English titles: Predecessors of Rauch," *Journal of the history of the behavioral sciences*, 19, 1983, 185.

————, "Origin and evolution of the term *psychology*: Addenda," *American psychologist*, 35, 1980, 774.

Venturi, Franco, *Le origine dell'enciclopedia*, 2nd ed. (Turin: Einaudi, 1963).

Vergote, Antoine, "Le corps: Pensée contemporaine et catégories bibliques," *Revue théologique de Louvain*, 10, 1979, 159–175.

Vernant, Jean-Pierre, *L'individu, la mort, l'amour: Soi-même et l'autre en Grèce ancienne* (Paris: Gallimard, 1989).

————, *Myth and society in ancient Greece*, trans. Janet Lloyd (1974; Brighton: Harvester Press, 1979).

————, *Passé et présent: Contributions à une psychologie historique*, ed. Riccardo di Donato (Rome: Edizioni di storia e letteratura, 1995).

Verney, Luís António, *Verdadeiro método de estudar para ser util à Republica, e à Igreja: Proporcionado ao estilo, e necesidade de Portugal* (1746), ed. António Salgado Júnior (Lisbon: Livraria Sá da Costa, 1949–1952).

Verri, Pietro, *Discorso sull'indole del piacere e del dolore* (1773; expanded ed., 1781), ed. Silvia Contarini (Rome: Carocci, 2001).

Vico, Giambattista, *New science: Principles of the new science concerning the common nature of nations*, 3rd ed. (1744), trans. David Marsh, introduction by Anthony Grafton (London: Penguin Books, 2001).

Vidal, Fernando, "Anthropologie et psychologie dans les encyclopédies d'Yverdon et de Paris: Esquisse de comparaison," *Annales Benjamin Constant*, no. 18–19, 1996, 139–151.

———, "Brainhood, anthropology of modernity, "*History of the human sciences*, 22, 2009, 5–36.

———, "Brains, bodies, selves, and science: Anthropologies of identity and the resurrection of the body," *Critical inquiry*, 28, 2002, 930–974.

———, "Jean Starobinski: The history of psychiatry as the cultural history of consciousness," in Mark S. Micale and Roy Porter, eds., *Discovering the history of psychiatry* (New York: Oxford University Press, 1999).

———, "Psychologie empirique et méthodologie des sciences au siècle des Lumières: L'exemple de Jean Trembley," *Archives des sciences*, 57, no. 1, 2004, 17–39.

———, "La psychologie dans l'ordre des sciences," *Revue de synthèse*, 4ᵉ série, no. 3–4, 1994, 327–353.

———, "La psychologie de Charles Bonnet comme 'miniature' de sa métaphysique," in Buscaglia et al., *Charles Bonnet*.

———, "Psychology in the eighteenth-century: A view from encyclopaedias," *History of the human sciences*, 6, 1993, 89–119.

———, "La 'science de l'homme': Désirs d'unité et juxtapositions encyclopédiques," in Claude Blanckaert et al., eds., *L'histoire des sciences de l'homme: Trajectoire, enjeux et questions vives* (Paris: L'Harmattan, 1999).

———, "Soul," in Kors, *Encyclopedia*, vol. 4.

Vila, Anne C., *Enlightenment and pathology: Sensibility in the literature and medicine of eighteenth-century France* (Baltimore: Johns Hopkins University Press, 1998).

Vittadello, A. M., "Expérience et raison dans la psychologie de Christian Wolff," *Revue philosophique de Louvain*, 71, 1973, 488–510.

Vives, Juan Luis, *De anima et vita* (1538; Valencia: Ayuntamiento de Valencia, 1992).

———, *Ioannis Lodovici Vivis "De anima et vita,"* ed. and trans. Mario Sancipriano (Padua: Gregoriana, 1974).

———, *Juan Luis Vives, Valenciano, De anima et vita: El alma y la vida*, trans. Ismael Roca (Valencia: Ajuntament de València, 1992).

Volney, Constantin François de, *A view of the soil and climate of the United States of America* (1804), trans. C. B. Brown (New York: Hafner, 1968).

Voltaire, *Dieu et les hommes, œuvre théologique mais raisonnable . . .* (1769), in *Les œuvres complètes de Voltaire*, ed. Roland Mortier (Oxford: Voltaire Foundation, 1994), vol 69.

———, "On Mr. Locke," in *Philosophical letters* (1734), trans. Ernest Dilworth (Indianapolis: Bobbs-Merrill, 1961).

von Engelhardt, Dietrich, *Historisches Bewußtsein in der Naturwissenschaft von der Aufklärung bis zum Positivismus* (Freiburg: Karl Alber, 1979).

von Wille, Dagmar, *Lessico filosofico della Frühaufklärung: Christian Thomasius, Christian Wolff, Johann Georg Walch* (Rome: Ed. dell'Ateneo, 1991).

Walch, Johann Georg, *Philosophisches Lexikon* (1726), ed. Justus Christian Hennings (1775; Hildesheim: Olms, 1968).

Walker, Daniel P., "Medical spirits and God and the soul," in Marta Fattori et Massimo Bianchi, eds., *Spiritus* (Rome: Edizioni dell'Ateneo, 1984).

Walser, Heinrich, *Institutiones philosophicae*, Liber 3, *Psychologia* (Augsburg: M. Rieger, 1791).

———, "Syllabus scriptorum in materia psychologica celebrium," in Walser, *Institutiones philosophicae*.

Waquet, Françoise, ed., *Mapping the world of learning: The polyhistor of Daniel Georg Morhof* (Wiesbaden: Harrassowitz Verlag, 2000).

Warren, Howard C., *A history of association psychology* (London: Constable, 1921).

Waszek, Norbert, "Le cadre européen de l'historiographie allemande à l'époque des Lumières et la philosophie de l'histoire de Kant," in Myriam Bienenstock, ed., *La philosophie de l'histoire: Héritage des Lumières dans l'idéalisme allemand?* (Tours: Université François Rabelais, 2001).

Werner, Stephen, "Abraham Rees's eighteenth-century *Cyclopaedia*," in Kafker, *Notable encyclopedias: Eleven successors*.

Weyer, Jost, *Chemiegeschichtsschreibung von Wiegleb (1790) bis Partington (1970)* (Hildesheim: Gerstenberg, 1974).

Wilkes, Kathleen V., *Real people: Personal identity without thought experiments* (Oxford: Clarendon Press, 1988. 1999).

Will, Georg Andreas, *Oratio sollemnis de Aesthetica veterum* (Altdorf: sumtibus Laurentii Schupfelii, 1756).

Williams, Williams, *A cultural history of medical vitalism in Enlightenment Montpellier* (Aldershot: Ashgate, 2003).

Winckelmann, Johann Joachim, *Reflections on the Painting and Sculpture of the Greeks: with Instructions for the Connoisseur, and an Essay on Grace in Works of Art*, trans. Henry Fusseli (London: printed for the Translator, and sold by A. Millar, 1765).

Winter, Eduard, et al., eds., *Ehrenfried Walter von Tschirnhaus und die Frühaufklärung in Mittel- und Osteuropa* (Berlin: Akademie Verlag, 1961).

Wolf, Friedrich August, *Darstellung der Altertumwissenschaft nach Begriff, Umfang, Zweck und Wert* (1807; Berlin: Akademie-Verlag, 1985).

———, *Esposizione della scienza dell'antichità secondo concetto, estensione, scopo e valore* (1807); ed. Salvatore Cerasuolo (Naples: Bibliopolis, 1999).

———, "Giambattista Vico über den Homer," *Museum der Alterthums-Wissenschaft*, 1, 1807, 555–570.

———, *Prolegomena to Homer* (1795), trans., with introduction and notes by A. Grafton, Glenn W. Most, and James E. G. Zetzel (Princeton: Princeton University Press, 1985).

Wolff, Christian, *Gesammelte Werke*, ed. Jean École, ser. I, *Deutsche Schriften*; ser. II, *Lateinische Schriften*; ser. III, *Materialien und Dokumente* (Hildesheim: Olms, 1965–).

———, *Philosophia practica universalis, methodo scientifica pertractata, Pars prior . . .* (1738), in Wolff, *Gesammelte Werke*, ser. II, vol. 10.

———, *Philosophia rationalis sive Logica, methodo scientica pertractata et ad usum scientiarum atque vitae aptata: Praemittitur Discursus praeliminaris de Philosophia in genere* (1728), ed. of 1740, in Wolff, *Gesammelte Werke*, ser. II, vol. 1.

———, *Preliminary discourse on philosophy in general* (1728), trans. Richard J. Blackwell (New York: Bobbs-Merrill, 1963).

———, *Psychologia empirica, methodo scientifica pertractata, quae ea, quae de anima humana indubia experientiae fide constant, continentur et ad solidam universae philosophiae practicae ac theologiae naturalis tractationem via sternitur* (1732), ed. of 1738, in Wolff, *Gesammelte Werke*, ser. II, vol. 5.

———, *Psychologia rationalis methodo scientifica pertractata, qua ea, quae de anima humana indubia experientiae fide innotescunt, per essentiam et naturam animae explicantur, et ad intimiorem naturae ejusque autoris cognitionem profutura proponitur* (1734), ed. of 1740, in Wolff, *Gesammelte Werke*, ser. II, vol. 6.

———, *Vernünfftige Gedancken von Gott, der Welt und der Seelen des Menschen, auch allen Dingen überhaupt* (1720), in Wolff, *Gesammelte Werke*, ser. I, vol. 2.

Wollgast, Siegfried, *Philosophie in Deutschland zwischen Reformation und Aufklärung 1550–1650* (Berlin: Akademie-Verlag, 1988).

Wood, Paul B., *The Aberdeen Enlightenment: The arts curriculum in the 18th century* (Aberdeen: Aberdeen University Press, 1993).

———, "The natural history of man in the Scottish Enlightenment," *History of science*, 28, 1990, 89–123.

———, "Science, philosophy, and the mind," in Porter, *The Cambridge history of science*, vol. 4.

Wood, Robert, *An Essay On The Original Genius and Writings of Homer: With A Comparative State Of The Ancient And The Present State Of The Troade* (1769; Washington, DC: McGrath Publishing Co., 1973).

Woolhouse, R. W., *Locke* (Minneapolis: University of Minnesota Press, 1983).

Wörger, Franz, *Psychologia Salomonis: Post tot interpretum sinistras detorsiones à Varenio dextrè explicata, & adversos Spinosa aliorumque Atheorum inanes cavillationes adhuc uberius defensa* (Hamburg: apud Gothofredum Schultzen, 1686).

Wunderlich, Falk, *Kant und die Bewußtseinstheorien des 18. Jahrhunderts* (Berlin: Walter de Gruyter, 2005).

Wundt, Wilhelm, *Elements of folk psychology: Outlines of a psychological history of the development of mankind* (1912), trans. Edward Leroy Schaub (London: George Allen and Unwin, 1916).

———, *Völkerpsychologie: Eine Untersuchung der Entwicklungsgesetze von Sprache, Mythus und Sitte* (1905; Leipzig: Alfred Kröner Verlag, 1920).

Wundt, Max, *Die deutsche Schulmetaphysik des 17. Jahrhunderts* (Tübingen: J. C. B. Mohr, 1939).

Yeo, Richard, *Encyclopaedic visions: Scientific dictionaries and Enlightenment culture* (Cambridge: Cambridge University Press, 2001).

———, "Reading encyclopedias: Science and the organization of knowledge in British dictionaries of arts and sciences, 1730–1850," *Isis*, 82, 1991, 24–49.

Yolton, John W., *Locke and French materialism* (Oxford: Clarendon Press, 1991).

———, *Locke: An introduction* (Oxford: Blackwell, 1985).

———, *A Locke dictionary* (Oxford: Blackwell, 1993).

———, *Thinking matter: Materialism in eighteenth-century Britain* (Oxford: Blackwell, 1984).

Young, Robert M., "Association of ideas," in Philip P. Wiener, *Dictionary of the history of ideas* (New York: Scribner's, 1973–1974), vol. 1.

———, "Scholarship and the history of the behavioral sciences," *History of science*, 5, 1966, 1–51.

Zacchia, Paolo, *Die Beseelung des menschlichen Fötus: Buch IX, Kapitel 1 der "Questiones medico-legales*," ed. and trans. Beatrix Spitzer (Cologne: Böhlau, 2002).

———, *Quaestionum medico-legalium tomus posterior: quo continentur liber nonus et decimus . . .* (Lyon: Ioan. Ant. Huguetan, & Marci-Ant. Ravaud, 1661).

Zambelli, Paola, "Antonio Genovesi and eighteenth-century empiricism in Italy," *Journal of the history of philosophy*, 16, 1978, 195–208.

———, *La formazione filosofica di Antonio Genovesi* (Naples: Morano, 1972).

Zedler, Johann Heinrich [then Carl Günther Ludovici], ed., *Grosses vollständiges Universal-Lexicon aller Wissenschaften und Künste* (1732–1750; Graz: Akademische Druck- und Verlags-Anstalt, 1993).

Zelle, Carsten, "Experimentalseelenlehre und Ehrfahrungseelenkunde: Zur Unterscheidung von Erfahrung, Beobachtung und Experiment bei Johann Gottlob Krüger und Karl Philipp Moritz," in Zelle, *"Vernünftige Ärzte."*

———, "Experiment, experience and observation in eighteenth-century anthropology and psychology: The examples of Krüger's *Experimentalseelenlehre* and Moritz's *Erfahrungseelenkunde*," *Orbis litterarum*, 56, 2001, 93–105.

———, "Sinnlichkeit und Therapie: Zur Gleichsprünglichkeit von Ästhetik und Anthropologie um 1750," in Zelle, *"Vernünftige Ärzte."*

———, ed., *"Vernünftige Ärzte": Hallesche Psychomedizinischer und die Anfänge der Anthropologie in der deutschsprachigen Frühaufklärung* (Tübingen: Max Niemeyer, 2001).

Zurbuchen, Simone, "Iselin, Isaak," in Knud Haakonssen, ed., *The Cambridge history of eighteenth-century philosophy* (New York: Cambridge University Press, 2006), vol. 2.